Springer INdAM Series

Volume 10

Editor-in-Chief

V. Ancona

Series Editors

P. Cannarsa
C. Canuto
G. Coletti
P. Marcellini
G. Patrizio
T. Ruggeri
E. Strickland
A. Verra

More information about this series at
http://www.springer.com/series/10283

Angelo Favini • Genni Fragnelli •
Rosa Maria Mininni
Editors

New Prospects in Direct, Inverse and Control Problems for Evolution Equations

Editors
Angelo Favini
Dipartimento di Matematica
Università di Bologna
Bologna
Italy

Genni Fragnelli
Dipartimento di Matematica
Università di Bari
Bari
Italy

Rosa Maria Mininni
Dipartimento di Matematica
Università di Bari
Bari
Italy

ISSN 2281-518X   ISSN 2281-5198 (electronic)
Springer INdAM Series
ISBN 978-3-319-36464-3   ISBN 978-3-319-11406-4 (eBook)
DOI 10.1007/978-3-319-11406-4
Springer Cham Heidelberg New York Dordrecht London

Printed on acid-free paper

Springer is part of Springer Science+Business Media (www.springer.com)

During the preparation of this volume, Alfredo Lorenzi, one of the organizers of the meeting "Differential Equations, Inverse Problems and Control Theory", passed away. We would like to dedicate this work to him. His death left an immeasurable emptiness for those who knew him as a student, professor, colleague or friend, or as a father or husband. Alfredo, we shall miss you immensely and we shall always remember you with your smile.

Alfredo Lorenzi (1944-2013)

Preface

The International Conference *Differential Equations, Inverse Problems and Control Theory* took place at the Palazzone in Cortona (Italy), from June 16 to 21, 2013. The conference, organized by Angelo Favini and Alfredo Lorenzi, was held in collaboration with the Mathematics Department of the University of Bologna and with INdAM (Istituto Nazionale di Alta Matematica). It was attended by about 40 mathematicians from universities in a variety of countries, including France, Germany, Israel, Italy, Japan, Romania, and the USA.

As is well known, applied sciences consider situations in which one observes the evolution over time of a given system. The related models can be formulated in terms of evolution equations, mathematical structures in which the dependence on time plays an essential role. Such equations have been studied intensively in theoretical research and are the source of an enormous number of applications.

A typical class of problems that has been investigated over the years concerns the well-posedness of an evolution equation with the given initial and boundary conditions, possibly with some degeneration (the so-called DIRECT problems). However, in several situations, initial conditions are difficult to determine exactly, while measurements of the solution at different stages of its evolution might be available.Many techniques have been developed to recover, from such pieces of information, important parameters governing the evolution, such as forcing terms or diffusion coefficients. This avenue of investigation is usually referred to as INVERSE problems.

A third way to study evolution equations is to try to influence the evolution of a given system through various kinds of external action called CONTROL. Of course, control problems may vary in nature, ranging from a given system to a desired configuration in finite or infinite time, to trying to optimize a performance criterion.

Although for some time direct, inverse and control problems for evolution equations were viewed as almost independent issues, in recent years it has become clear that they can profit enormously from a strong interaction with each other. For instance, a priori estimates for solutions of partial differential equations that were originally developed to study unique continuation problems, i.e. Carleman's estimates, have been proved to be extremely useful in studying exact controllability and inverse problems.

For these reasons, one of the main cultural goals of our initiative was to bring together experts in the above fields to speed up interaction and stimulate the development of new ideas. To achieve this aim, several conferences were organized, the most recent being the meeting *Differential Equations, Inverse Problems and Control Theory*.

This volume assembles the contributions of most of the speakers who participated in the meeting. It provides an overview that reflects the richness and vitality of the subject. All the contributions underwent peer review, in compliance with the standard procedure for the Springer INdAM Collection.

Bologna, Italy   Angelo Favini
Bari, Italy   Genni Fragnelli
Bari, Italy   Rosa Maria Mininni
June 2014

Acknowledgements

We express our heartfelt thanks to the following sponsors: the Mathematics Department of the University of Bologna and Istituto Nazionale di Alta Matematica. We would particularly like to thank Scuola Normale Superiore di Pisa for the exceptional hospitality offered in the Palazzone in Cortona. Last but not least, we are sincerely grateful to all the authors for their efforts in writing excellent papers and to the editors of the series for their interest in this publication.

Contents

Chapter 1
Exponential Stability of the Wave Equation with Memory and Time Delay

Fatiha Alabau-Boussouira, Serge Nicaise, and Cristina Pignotti

In memory of Alfredo Lorenzi

Abstract We study the asymptotic behaviour of the wave equation with viscoelastic damping in presence of a time-delayed damping. We prove exponential stability if the amplitude of the time delay term is small enough.

1.1 Introduction

This paper is devoted to the stability analysis of a viscoelastic model. In particular, we consider a model combining viscoelastic damping and time-delayed damping. We prove an exponential stability result provided that the amplitude of time-delayed damping is small enough. Moreover, we give a precise estimate on this smallness condition. This shows that even if delay effects usually generate instability (see e.g. [6, 7, 14, 20]), the damping due to viscoelasticity can counterbalance them.

F. Alabau-Boussouira
LMAM, Université de Lorraine and CNRS (UMR 7122), 57045 Metz Cedex 1, France
e-mail: fatiha.alabau@univ-lorraine.fr

S. Nicaise
LAMAV, FR CNRS 2956, Institut des Sciences et Techniques de Valenciennes, Université de Valenciennes et du Hainaut Cambrésis, 59313 Valenciennes Cedex 9, France
e-mail: serge.nicaise@univ-valenciennes.fr

C. Pignotti (✉)
Dipartimento di Ingegneria e Scienze dell'Informazione e Matematica, Università di L'Aquila, 67010 L'Aquila, Italy
e-mail: pignotti@univaq.it

© Springer International Publishing Switzerland 2014
A. Favini et al. (eds.), *New Prospects in Direct, Inverse and Control Problems for Evolution Equations*, Springer INdAM Series 10,
DOI 10.1007/978-3-319-11406-4_1

Let $\Omega \subset \mathbb{R}^n$ be an open bounded set with a smooth boundary. Let us consider the following problem:

$$u_{tt}(x,t) - \Delta u(x,t) + \int_0^\infty \mu(s)\Delta u(x,t-s)ds + ku_t(x,t-\tau) = 0$$

$$\text{in } \Omega \times (0,+\infty) \qquad (1.1)$$

$$u(x,t) = 0 \quad \text{on } \partial\Omega \times (0,+\infty) \qquad\qquad\qquad (1.2)$$

$$u(x,t) = u_0(x,t) \quad \text{in } \Omega \times (-\infty, 0] \qquad\qquad\quad (1.3)$$

where the initial datum u_0 belongs to a suitable space, the constant $\tau > 0$ is the time delay, k is a real number and the memory kernel $\mu : [0,+\infty) \to [0,+\infty)$ is a locally absolutely continuous function satisfying

(i) $\mu(0) = \mu_0 > 0$;
(ii) $\int_0^{+\infty} \mu(t)dt = \tilde{\mu} < 1$;
(iii) $\mu'(t) \leq -\alpha\mu(t)$,   for some $\alpha > 0$.

We know that the above problem is exponentially stable for $k = 0$ (see e.g. [9]).

We will show that an exponential stability result holds if the delay parameter k is *small* with respect to the memory kernel.

Observe that for $\tau = 0$ and $k > 0$ the model (1.1)–(1.3) presents both viscoelastic and standard dissipative damping. Therefore, in that case, under the above assumptions on the kernel μ, the model is exponentially stable.

We will see that exponential stability also occurs for $k < 0$, under a suitable smallness assumption on $|k|$. Note that the term $ku_t(t)$ with $k < 0$ is a so-called anti-damping (see e.g. [8]), namely a damping with an opposite sign with respect to the standard dissipative one, and therefore it induces instability. Indeed, in absence of viscoelastic damping, i.e. for $\mu \equiv 0$, the solutions of the above problem, with $\tau = 0$ and $k < 0$, grow exponentially to infinity.

We will prove our stability results by using a perturbative approach, first introduced in [18] (see also [15] for a more general setting).

The stability properties of the wave equation with memory and time delay have been first studied by Kirane and Said-Houari [11], in the case of finite memory. However, in their model, an extra standard dissipative damping is added in order to contrast the destabilizing effect of the time delay term. The stabilization problem for model (1.1)–(1.3) has been studied also by Guesmia in [10] by using a different approach based on the construction of a suitable Lyapunov functional (see also [5] for the case of finite memory). Our analysis allows to determine an explicit estimate on the constant k_0 (cf. Theorem 1.1). Moreover, our approach can be extended to the case of localized viscoelastic damping (cf. [13]). In fact, we first prove the exponential stability of an auxiliary problem having a decreasing energy and then, regarding the original problem as a perturbation of that one, we extend the exponential decay estimate to it.

The paper is organized as follows. In Sect. 1.2 we study the well-posedness by introducing an appropriate functional setting and we formulate our stability result. In Sect. 1.3 we introduce the auxiliary problem and prove the exponential decay estimate for it. Then, the stability result is extended to the original problem.

1.2 Main Results and Preliminaries

As in [4], let us introduce the new variable

$$\eta^t(x,s) := u(x,t) - u(x,t-s). \tag{1.4}$$

Moreover, as in [14], we define

$$z(x,\rho,t) := u_t(x,t-\tau\rho), \quad x \in \Omega, \ \rho \in (0,1), \ t > 0. \tag{1.5}$$

Using (1.4) and (1.5) we can rewrite (1.1)–(1.3) as

$$u_{tt}(x,t) = (1-\tilde{\mu})\Delta u(x,t) + \int_0^\infty \mu(s)\Delta\eta^t(x,s)ds - kz(x,1,t)$$

$$\text{in } \Omega \times (0,+\infty) \tag{1.6}$$

$$\eta_t^t(x,s) = -\eta_s^t(x,s) + u_t(x,t) \quad \text{in } \Omega \times (0,+\infty) \times (0,+\infty), \tag{1.7}$$

$$\tau z_t(x,\rho,t) + z_\rho(x,\rho,t) = 0 \quad \text{in } \Omega \times (0,1) \times (0,+\infty), \tag{1.8}$$

$$u(x,t) = 0 \quad \text{on } \partial\Omega \times (0,+\infty) \tag{1.9}$$

$$\eta^t(x,s) = 0 \quad \text{in } \partial\Omega \times (0,+\infty), \ t \geq 0, \tag{1.10}$$

$$z(x,0,t) = u_t(x,t) \quad \text{in } \Omega \times (0,+\infty), \tag{1.11}$$

$$u(x,0) = u_0(x) \quad \text{and} \quad u_t(x,0) = u_1(x) \quad \text{in } \Omega, \tag{1.12}$$

$$\eta^0(x,s) = \eta_0(x,s) \quad \text{in } \Omega \times (0,+\infty), \tag{1.13}$$

$$z(x,\rho,0) = z^0(x,-\tau\rho) \quad x \in \Omega, \ \rho \in (0,1), \tag{1.14}$$

where

$$\begin{aligned}
&u_0(x) = u_0(x,0), \quad x \in \Omega,\\
&u_1(x) = \tfrac{\partial u_0}{\partial t}(x,t)|_{t=0}, \quad x \in \Omega,\\
&\eta_0(x,s) = u_0(x,0) - u_0(x,-s), \quad x \in \Omega, \ s \in (0,+\infty),\\
&z^0(x,s) = \tfrac{\partial u_0}{\partial t}(x,s), \quad x \in \Omega, \ s \in (-\tau,0).
\end{aligned} \tag{1.15}$$

Let us denote $\mathscr{U} := (u,u_t,\eta^t,z)^T$. Then we can rewrite problem (1.6)–(1.14) in the abstract form

$$\begin{cases} \mathscr{U}' = \mathscr{A}\mathscr{U}, \\ \mathscr{U}(0) = (u_0,u_1,\eta_0,z^0)^T, \end{cases} \tag{1.16}$$

where the operator $\mathscr{A}$ is defined by

$$\mathscr{A}\begin{pmatrix} u \\ v \\ w \\ z \end{pmatrix} := \begin{pmatrix} v \\ (1-\tilde{\mu})\Delta u + \int_0^\infty \mu(s)\Delta w(s)ds - kz(\cdot,1) \\ -w_s + v \\ -\tau^{-1}z_\rho \end{pmatrix}, \qquad (1.17)$$

with domain (cf. [16])

$$\mathscr{D}(\mathscr{A}) := \Big\{ (u,v,\eta,z)^T \in H_0^1(\Omega) \times H_0^1(\Omega) \times L_\mu^2((0,+\infty); H_0^1(\Omega))$$

$$\times H^1((0,1); L^2(\Omega)) : v = z(\cdot,0), \ (1-\tilde{\mu})u + \int_0^\infty \mu(s)\eta(s)ds \in H^2(\Omega) \cap H_0^1(\Omega),$$

$$\eta_s \in L_\mu^2((0,+\infty); H_0^1(\Omega)) \Big\}, \qquad (1.18)$$

where $L_\mu^2((0,\infty); H_0^1(\Omega))$ is the Hilbert space of H_0^1-valued functions on $(0,+\infty)$, endowed with the inner product

$$\langle \varphi, \psi \rangle_{L_\mu^2((0,\infty); H_0^1(\Omega))} = \int_\Omega \left(\int_0^\infty \mu(s)\nabla\varphi(x,s)\nabla\psi(x,s)ds \right) dx.$$

Denote by $\mathscr{H}$ the Hilbert space

$$\mathscr{H} = H_0^1(\Omega) \times L^2(\Omega) \times L_\mu^2((0,\infty); H_0^1(\Omega)) \times L^2((0,1); L^2(\Omega)),$$

equipped with the inner product

$$\left\langle \begin{pmatrix} u \\ v \\ w \\ z \end{pmatrix}, \begin{pmatrix} \tilde{u} \\ \tilde{v} \\ \tilde{w} \\ \tilde{z} \end{pmatrix} \right\rangle_{\mathscr{H}} := (1-\tilde{\mu})\int_\Omega \nabla u \nabla \tilde{u}dx + \int_\Omega v\tilde{v}dx + \int_\Omega \int_0^\infty \mu(s)\nabla w \nabla \tilde{w}dsdx$$

$$+ \int_0^1 \int_\Omega z(x,\rho)\tilde{z}(x,\rho)\,dxd\rho. \qquad (1.19)$$

Combining the ideas from [19] with the ones from [14] (see also [3]), we can prove that the operator $\mathscr{A}$ generates a strongly continuous semigroup ($\mathscr{A} - cI$ is dissipative for a sufficiently large constant $c > 0$) and therefore the next existence result holds.

Proposition 1.1 *For any initial datum $\mathscr{U}_0 \in \mathscr{H}$ there exists a unique solution $\mathscr{U} \in C([0,+\infty), \mathscr{H})$ of problem (1.16). Moreover, if $\mathscr{U}_0 \in \mathscr{D}(\mathscr{A})$, then*

$$\mathscr{U} \in C([0,+\infty), \mathscr{D}(\mathscr{A})) \cap C^1([0,+\infty), \mathscr{H}).$$

Let us define the energy F of problem (1.1)–(1.3) as

$$F(t) = F(u,t) := \frac{1}{2}\int_\Omega u_t^2(x,t)dx + \frac{1-\tilde{\mu}}{2}\int_\Omega |\nabla u(x,t)|^2 dx$$

$$+\frac{1}{2}\int_0^{+\infty}\int_\Omega \mu(s)|\nabla\eta^t(s)|^2 dsdx + \frac{\theta|k|e^\tau}{2}\int_{t-\tau}^t e^{-(t-s)}\int_\Omega u_t^2(x,s)dsdx,$$
$$\tag{1.20}$$

where θ is any real constant satisfying

$$\theta > 1. \tag{1.21}$$

We will prove the following exponential stability result.

Theorem 1.1 *For any $\theta > 1$ in the definition (1.20), there exists a positive constant k_0 such that for k satisfying $|k| < k_0$ there is $\sigma > 0$ such that*

$$F(t) \le F(0)e^{1-\sigma t}, \quad t \ge 0; \tag{1.22}$$

for every solution of problem (1.1)–(1.3). The constant k_0 depends only on the kernel $\mu(\cdot)$ of the memory term, on the time delay τ and on the domain Ω.

To prove our stability result we will make use of the following result of Pazy (Theorem 1.1 in Chap. 3 of [17]).

Theorem 1.2 *Let X be a Banach space and let A be the infinitesimal generator of a C_0 semigroup $T(t)$ on X, satisfying $\|T(t)\| \le Me^{\omega t}$. If B is a bounded linear operator on X then $A + B$ is the infinitesimal generator of a C_0 semigroup $S(t)$ on X, satisfying $\|S(t)\| \le Me^{(\omega + M\|B\|)t}$.*

Moreover, we will use the following lemma (see Theorem 8.1 of [12]).

Lemma 1.1 *Let $V(\cdot)$ be a non negative decreasing function defined on $[0, +\infty)$. If*

$$\int_S^{+\infty} V(t)dt \le CV(S) \quad \forall S > 0,$$

for some constants $C > 0$, then

$$V(t) \le V(0)\exp\left(1 - \frac{t}{C}\right), \quad \forall t \ge 0.$$

Remark 1.1 Observe that the well-posedness result in the case $\tau = 0$, namely viscoelastic wave equation with standard frictional damping or anti-damping, directly follows from Theorem 1.2. Furthermore, from Theorem 1.2 we can also deduce an exponential stability estimate under a suitable smallness assumption on $|k|$. Indeed, for $|k|$ small, we can look at problem (1.1)–(1.3) (with $\tau = 0$) as a perturbation of the wave equation with only the viscoelastic damping. And it is by now well-known that for the last model an exponential decay estimate is available (see e.g. [9]).

1.3 Stability Results

In this section we will prove Theorem 1.1.
In order to study the stability properties of problem (1.1)–(1.3), we look at an auxiliary problem (cf. [18]) which is *near* to this one and easier to deal with. Then, let us consider the system

$$u_{tt}(x,t) - \Delta u(x,t) + \int_0^\infty \mu(s)\Delta u(x, t - s)ds + \theta|k|e^\tau u_t(x,t)$$

$$+ku_t(x, t - \tau) = 0$$

$$\text{in } \Omega \times (0, +\infty) \qquad (1.23)$$

$$u(x,t) = 0 \quad \text{on } \partial\Omega \times (0, +\infty) \qquad (1.24)$$

$$u(x,t) = u_0(x,t) \quad \text{in } \Omega \times (-\infty, 0]. \qquad (1.25)$$

First of all we show that the energy, defined by (1.20), of any solution of the auxiliary problem is not increasing.

Proposition 1.2 *For every solution of problem (1.23)–(1.25) the energy $F(\cdot)$ is not increasing and the following estimate holds*

$$F'(t) \leq \frac{1}{2}\int_0^\infty \int_\Omega \mu'(s)|\nabla\eta^t(x,s)|^2 dxds$$

$$-\frac{|k|(\theta e^\tau - 1)}{2}\int_\Omega u_t^2(x,t)dx - \frac{|k|(\theta - 1)}{2}\int_\Omega u_t^2(x, t - \tau)dx \qquad (1.26)$$

$$-\frac{\theta|k|e^\tau}{2}\int_{t-\tau}^t e^{-(t-s)}\int_\Omega u_t^2(x,s)dxds.$$

Remark 1.2 Note that the energy $F(\cdot)$ of solutions of the original problem (1.1)–(1.3) is not in general decreasing.

Proof (of Proposition 1.2.) Differentiating (1.20) we have

$$F'(t) = \int_\Omega u_t(x,t)u_{tt}(x,t)dx + (1-\tilde{\mu})\int_\Omega \nabla u(x,t)\nabla u_t(x,t)dx$$
$$+ \int_0^\infty \int_\Omega \mu(s)\nabla\eta^t(x,s)\nabla\eta_t^t(x,s)dxds + \frac{\theta|k|e^\tau}{2}\int_\Omega u_t^2(x,t)dx$$
$$- \frac{\theta|k|}{2}\int_\Omega u_t^2(x,t-\tau)dx - \frac{\theta|k|e^\tau}{2}\int_{t-\tau}^t e^{-(t-s)}\int_\Omega u_t^2(x,s)dxds.$$

Then, integrating by parts and using (1.7) and the boundary condition (1.24),

$$F'(t) = \int_\Omega u_t(x,t)[u_{tt}(x,t) - (1-\tilde{\mu})\Delta u(x,t)]dx$$
$$+ \int_0^\infty \int_\Omega \mu(s)\nabla\eta^t(x,s)(\nabla u_t(x,t) - \nabla\eta_s^t(x,s))dxds + \frac{\theta|k|e^\tau}{2}\int_\Omega u_t^2(x,t)dx$$
$$- \frac{\theta|k|}{2}\int_\Omega u_t^2(x,t-\tau)dx - \frac{\theta|k|e^\tau}{2}\int_{t-\tau}^t e^{-(t-s)}\int_\Omega u_t^2(x,s)dxds.$$

By using Eqs. (1.23), (1.24), after integration by parts, we deduce

$$F'(t) = \int_\Omega u_t(t)\left[-\int_0^\infty \mu(s)\Delta u(x,t-s) + \tilde{\mu}\Delta u(x,t)\right.$$
$$\left. -\theta|k|e^\tau u_t(x,t) - ku_t(x,t-\tau)\right]dx$$
$$+ \int_0^\infty \int_\Omega \mu(s)\nabla\eta^t(x,s)\nabla u_t(x,t)dxds + \frac{1}{2}\int_0^\infty \int_\Omega \mu'(s)|\nabla\eta^t(x,s)|^2dxds$$
$$+ \frac{\theta|k|e^\tau}{2}\int_\Omega u_t^2(x,t)dx - \frac{\theta|k|}{2}\int_\Omega u_t^2(x,t-\tau)dx - \frac{\theta|k|e^\tau}{2}\int_{t-\tau}^t e^{-(t-s)}\int_\Omega u_t^2(x,s)dxds$$
$$= -\theta|k|e^\tau \int_\Omega u_t^2(x,t)dx - k\int_\Omega u_t(x,t)u_t(x,t-\tau)dx + \frac{\theta|k|e^\tau}{2}\int_\Omega u_t^2(x,t)dx$$
$$- \frac{\theta|k|}{2}\int_\Omega u_t^2(x,t-\tau)dx + \frac{1}{2}\int_0^\infty \int_\Omega \mu'(s)|\nabla\eta^t(x,s)|^2dxds$$
$$- \frac{\theta|k|e^\tau}{2}\int_{t-\tau}^t e^{-(t-s)}\int_\Omega u_t^2(x,s)dxds.$$

Now, using Cauchy–Schwarz inequality we obtain (1.26).   □

Corollary 1.1 *For every solution of problem (1.23)–(1.25), we have*

$$-\frac{1}{2}\int_S^T\int_0^\infty\int_\Omega \mu'(s)|\nabla\eta^t(x,s)|^2 dxds \le F(S),\qquad (1.27)$$

and then by the condition $\mu'(t) \le -\alpha\mu(t)$ we directly get

$$\frac{1}{2}\int_S^T\int_0^\infty \mu(s)\int_\Omega |\nabla\eta^t(x,s)|^2 dxdsdt \le \frac{1}{\alpha}F(S).\qquad (1.28)$$

Proof As each term of the right-hand side of (1.26) is non positive, we directly get that

$$-\frac{1}{2}\int_S^T\int_0^\infty\int_\Omega \mu'(s)|\nabla\eta^t(x,s)|^2 dxds \le \int_S^T(-F'(t))dt \le F(S).\qquad \square$$

Theorem 1.3 *For any $\theta > 1$ in the definition (1.20), there exist positive constants C and $\overline{k}$, depending on μ, Ω and τ, such that if $|k| < \overline{k}$ then for any solution of problem (1.23)–(1.25) the following estimate holds*

$$\int_S^{+\infty} F(t)dt \le CF(S)\quad \forall S > 0.\qquad (1.29)$$

In order to prove Theorem 1.3 we need some preliminary results. Our proof relies in many points on [2] but we have to perform all computations because, in order to extend the exponential estimate related to the perturbed problem (1.23)–(1.25) to the original problem (1.1)–(1.3) we need to determine carefully all involved constants. From the definition of the energy we deduce

$$\int_S^T F(t)dt = \frac{1}{2}\int_S^T\int_\Omega u_t^2(x,t)dxdt + \frac{1-\tilde\mu}{2}\int_S^T\int_\Omega |\nabla u(x,t)|^2 dxdt$$

$$+\frac{1}{2}\int_S^T\int_0^\infty\int_\Omega \mu(s)|\nabla\eta^t(x,s)|^2 dxdsdt$$

$$+\frac{\theta|k|e^\tau}{2}\int_S^T\int_{t-\tau}^t e^{-(t-s)}\int_\Omega u_t^2(x,s)dsdxdt.\qquad (1.30)$$

Now, as in [2] we will use multiplier arguments in order to bound the right-hand side of (1.30) (cf. [1]). We note that we could not apply the same arguments directly to our original problem since the energy is not decreasing.

In the following we will denote by C_P the Poincaré constant, namely the smallest positive constant such that

$$\int_\Omega w^2(x)dx \le C_P\int_\Omega |\nabla w(x)|^2 dx,\quad \forall\, w \in H_0^1(\Omega).\qquad (1.31)$$

Lemma 1.2 *Assume*

$$|k| < \frac{1-\tilde{\mu}}{2C_P(\theta e^\tau + 1)}.$$ (1.32)

Then, for any $T \geq S \geq 0$ we have

$$(1-\tilde{\mu})\int_S^T\int_\Omega |\nabla u(x,t)|^2 dxdt \leq C_0 \int_S^T\int_\Omega u_t^2(x,t)dxdt + C_1 F(S),$$ (1.33)

with

$$C_0 = 2 + \theta|k|e^\tau, \qquad C_1 = 4\left(1 + \frac{\tilde{\mu}}{\alpha(1-\tilde{\mu})} + \frac{C_P}{1-\tilde{\mu}} + \frac{1}{2(\theta-1)}\right).$$ (1.34)

Proof Multiplying Eq. (1.23) by u and integrating on $\Omega \times [S,T]$ we have

$$\int_S^T\int_\Omega [u_{tt}(x,t) - \Delta u(x,t) + \int_0^\infty \mu(s)\Delta u(x,t-s)ds$$
$$+\theta|k|e^\tau u_t(x,t) + ku_t(x,t-\tau)]u(x,t)dxdt = 0.$$

So, integrating by parts and using the boundary condition (1.24), we get

$$-\int_S^T\int_\Omega u_t^2(x,t)dxdt + \int_S^T\int_\Omega |\nabla u(x,t)|^2 dxdt + \left[\int_\Omega u(x,t)u_t(x,t)dx\right]_S^T$$

$$+\theta|k|e^\tau \int_S^T\int_\Omega u(x,t)u_t(x,t)dxdt + k\int_S^T\int_\Omega u(x,t)u_t(x,t-\tau)dxdt$$

$$-\tilde{\mu}\int_S^T\int_\Omega |\nabla u(x,t)|^2 dxdt + \int_S^T\int_\Omega\int_0^\infty \mu(s)\nabla u(x,t)\nabla \eta^t(x,s)dsdxdt = 0,$$

where we used (1.4).
   Then,

$$(1-\tilde{\mu})\int_S^T\int_\Omega |\nabla u(x,t)|^2 dxdt = \int_S^T\int_\Omega u_t^2(x,t)dxdt - \left[\int_\Omega u(x,t)u_t(x,t)dx\right]_S^T$$

$$-\theta|k|e^\tau \int_S^T\int_\Omega u(x,t)u_t(x,t)dxdt - k\int_S^T\int_\Omega u(x,t)u_t(x,t-\tau)dxdt$$

$$-\int_S^T\int_\Omega\int_0^\infty \mu(s)\nabla u(x,t)\nabla \eta^t(x,s)dsdxdt.$$ (1.35)

In order to estimate the integral

$$\int_S^T \left| \int_\Omega \int_0^\infty \mu(s) \nabla \eta^t(x,s) \nabla u(x,t) ds dx \right| dt,$$

we note that, for all $\varepsilon > 0$,

$$\int_S^T \left(\int_\Omega |\nabla u(x,t)|^2 dx \right)^{1/2} \int_0^\infty \mu(s) \left(\int_\Omega |\nabla \eta^t(x,s)|^2 dx \right)^{1/2} ds dt$$

$$\leq \frac{\varepsilon}{2} \int_S^T \int_\Omega |\nabla u(x,t)|^2 dx dt + \frac{1}{2\varepsilon} \int_S^T \left[\int_0^\infty \mu(s) \left(\int_\Omega |\nabla \eta^t(x,s)|^2 dx \right)^{1/2} ds \right]^2 dt. \tag{1.36}$$

We have

$$\int_S^T \left[\int_0^\infty \mu(s) \left(\int_\Omega |\nabla \eta^t(x,s)|^2 dx \right)^{1/2} ds \right]^2 dt$$

$$\leq \int_S^T \left(\int_0^\infty \mu(s) ds \right) \left(\int_0^\infty \mu(s) \int_\Omega |\nabla \eta^t(x,s)|^2 dx ds \right) dt$$

$$= \tilde{\mu} \int_S^T \int_0^\infty \mu(s) \int_\Omega |\nabla \eta^t(x,s)|^2 dx ds dt.$$

Therefore, recalling the estimate (1.28), we obtain

$$\int_S^T \left[\int_0^\infty \mu(s) \left(\int_\Omega |\nabla \eta^t(x,s)|^2 dx \right)^{1/2} ds \right]^2 dt \leq \frac{2\tilde{\mu}}{\alpha} F(S). \tag{1.37}$$

Then, (1.36) and (1.37) give

$$\int_S^T \left| \int_\Omega \int_0^\infty \mu(s)(\nabla u(x,t-s) - \nabla u(x,t)) \cdot \nabla u(x,t) ds dx \right| dt$$

$$\leq \frac{\varepsilon}{2} \int_S^T \int_\Omega |\nabla u(x,t)|^2 dx dt + \frac{\tilde{\mu}}{\alpha\varepsilon} F(S). \tag{1.38}$$

Now observe that

$$F(t) \geq \frac{1}{2} \int_\Omega u_t^2(x,t) dx + \frac{1-\tilde{\mu}}{2} \int_\Omega |\nabla u(x,t)|^2 dx. \tag{1.39}$$

Then, from (1.39),

$$\frac{1}{2} \int_\Omega |\nabla u(x,t)|^2 dx \leq \frac{F(t)}{1-\tilde{\mu}}, \tag{1.40}$$

and also, from Poincaré's inequality,

$$\frac{1}{2}\int_{\Omega}|u(x,t)|^2dx \le \frac{C_P}{2}\int_{\Omega}|\nabla u(x,t)|^2dx \le \frac{C_P}{1-\tilde{\mu}}F(t). \qquad (1.41)$$

Using the above inequalities

$$\left|\int_{\Omega}u_t(x,t)u(x,t)dx\right| \le \frac{1}{2}\int_{\Omega}u_t^2(x,t)dx + \frac{1}{2}\int_{\Omega}u^2(x,t)dx \le F(t)\left(1+\frac{C_P}{1-\tilde{\mu}}\right).$$
$$(1.42)$$

Therefore,

$$-\left[\int_{\Omega}u_t(x,t)u(x,t)dx\right]_S^T \le 2F(S)\left(1+\frac{C_P}{1-\tilde{\mu}}\right), \qquad (1.43)$$

where we used also the fact that F is decreasing. Using (1.38), (1.43) and Cauchy–Schwarz's inequality in order to bound the terms in the right-hand side of (1.35) we have that for any $\varepsilon > 0$,

$$(1-\tilde{\mu})\int_S^T\int_{\Omega}|\nabla u(x,t)|^2dxdt \le \int_S^T\int_{\Omega}u_t^2(x,t)dxdt + \frac{\varepsilon}{2}\int_S^T\int_{\Omega}|\nabla u(x,t)|^2dxdt$$

$$+\frac{\tilde{\mu}}{\alpha\varepsilon}F(S) + 2\left(1+\frac{C_P}{1-\tilde{\mu}}\right)F(S) + \frac{\theta|k|e^{\tau}}{2}\int_S^T\int_{\Omega}u^2(x,t)dxdt$$

$$+\frac{\theta|k|e^{\tau}}{2}\int_S^T\int_{\Omega}u_t^2(x,t)dxdt + \frac{|k|}{2}\int_S^T\int_{\Omega}u^2(x,t)dxdt$$

$$+\frac{|k|}{2}\int_S^T\int_{\Omega}u_t^2(x,t-\tau)dxdt.$$

Therefore, from Poincaré's inequality,

$$(1-\tilde{\mu})\int_S^T\int_{\Omega}|\nabla u(x,t)|^2dxdt \le \left(1+\frac{\theta|k|e^{\tau}}{2}\right)\int_S^T\int_{\Omega}u_t^2(x,t)dxdt$$

$$+\frac{\varepsilon+(\theta e^{\tau}+1)|k|C_P}{2}\int_S^T\int_{\Omega}|\nabla u(x,t)|^2dxdt + \frac{\tilde{\mu}}{\alpha\varepsilon}F(S)$$

$$+2\left(1+\frac{C_P}{1-\tilde{\mu}}\right)F(S) + \frac{|k|}{2}\int_S^T\int_{\Omega}u_t^2(x,t-\tau)dxdt.$$

Now, observe that from (1.26),

$$\frac{|k|}{2}\int_S^T\int_\Omega u_t^2(x,t-\tau)dxdt = \frac{1}{\theta-1}\frac{|k|(\theta-1)}{2}\int_S^T\int_\Omega u_t^2(x,t-\tau)dxdt$$

$$\le \frac{1}{\theta-1}\int_S^T(-F'(t))dt \le \frac{1}{\theta-1}F(S).$$

(1.44)

Now, choose $\varepsilon = \frac{1-\tilde\mu}{2}$. Thus, using (1.32) and also (1.44) we obtain

$$(1-\tilde\mu)\int_S^T\int_\Omega |\nabla u(x,t)|^2 dxdt \le 2\Big(1+\frac{\theta|k|e^\tau}{2}\Big)\int_S^T\int_\Omega u_t^2(x,t)dxdt$$

$$+4\Big(1+\frac{\tilde\mu}{\alpha(1-\tilde\mu)}+\frac{C_P}{1-\tilde\mu}+\frac{1}{2(\theta-1)}\Big)F(S),$$

that is (1.33) with constants C_0, C_1 given by (1.34).   □

Lemma 1.3 *For any $T \ge S \ge 0$, the following identity holds:*

$$\tilde\mu\int_S^T\int_\Omega u_t^2(x,t)dxdt = \Big[\int_\Omega u_t(x,t)\int_0^\infty \mu(s)\eta^t(x,s)dsdx\Big]_S^T$$

$$-\int_S^T\int_\Omega u_t(x,t)\int_0^\infty \mu'(s)\eta^t(x,s)dsdxdt$$

$$+(1-\tilde\mu)\int_S^T\int_\Omega \nabla u(x,t)\int_0^\infty \mu(s)\nabla\eta^t(x,s)dsdxdt$$

$$+\int_S^T\int_\Omega \Big|\int_0^\infty \mu(s)\nabla\eta^t(x,s)ds\Big|^2 dxdt$$

$$+\theta|k|e^\tau\int_S^T\int_\Omega u_t(x,t)\int_0^\infty \mu(s)\eta^t(x,s)dsdxdt$$

$$+k\int_S^T\int_\Omega u_t(x,t-\tau)\int_0^\infty \mu(s)\eta^t(x,s)dsdxdt.$$

(1.45)

Proof We multiply Eq. (1.23) by $\int_0^\infty \mu(s)\eta^t(x,s)ds$ and integrate by parts on $[S, T] \times \Omega$. We obtain

$$\int_S^T \int_\Omega \left\{ u_{tt}(x,t) - \Delta u(x,t) + \int_0^\infty \mu(s)\Delta u(x,t-s)ds + ku_t(x,t-\tau) \right.$$

$$\left. +\theta|k|e^\tau u_t(x,t) \right\} \times \left\{ \int_0^\infty \mu(s)\eta^t(x,s)ds \right\} dxdt = 0.$$

(1.46)

Integrating by parts, we have

$$\int_S^T \int_\Omega u_{tt}(x,t) \int_0^\infty \mu(s)\eta^t(x,s)dsdxdt$$

$$= \left[\int_\Omega u_t(x,t) \int_0^\infty \mu(s)\eta^t(x,s)dsdx \right]_S^T$$

$$- \int_S^T \int_\Omega u_t(x,t) \int_0^\infty \mu(s)(u_t(x,t) - \eta_s^t(x,s))dsdxdt$$

$$= \left[\int_\Omega u_t(x,t) \int_0^\infty \mu(s)\eta^t(x,s)dsdx \right]_S^T$$

$$-\tilde{\mu} \int_S^T \int_\Omega u_t^2(x,t)dxdt - \int_S^T \int_\Omega u_t(x,t) \int_0^\infty \mu'(s)\eta^t(x,s)dsdxdt.$$

(1.47)

Moreover,

$$\int_S^T \int_\Omega \left(-\Delta u(x,t) + \int_0^\infty \mu(s)\Delta u(x,t-s)ds \right) \int_0^\infty \mu(s)\eta^t(x,s)dsdxdt$$

$$= \int_S^T \int_\Omega \nabla u(x,t) \int_0^\infty \mu(s)\nabla \eta^t(x,s)dsdxdt$$

$$- \int_S^T \int_\Omega \int_0^\infty \mu(s)\nabla u(x,t-s)ds \int_0^\infty \mu(s)\nabla \eta^t(x,s)dsdxdt$$

$$= \int_S^T \int_\Omega \nabla u(x,t) \int_0^\infty \mu(s)\nabla \eta^t(x,s)dsdxdt$$

$$+ \int_S^T \int_\Omega \int_0^\infty \mu(s)(\nabla u(x,t) - \nabla u(x,t-s))ds \int_0^\infty \mu(s)\nabla \eta^t(x,s)dsdxdt$$

$$-\tilde{\mu} \int_S^T \int_\Omega \nabla u(x,t) \int_0^\infty \mu(s)\nabla \eta^t(x,s)dsdxdt$$

$$= (1-\tilde{\mu}) \int_S^T \int_\Omega \nabla u(x,t) \int_0^\infty \mu(s)\nabla \eta^t(x,s)dsdxdt$$

$$+ \int_S^T \int_\Omega \left| \int_0^\infty \mu(s)\nabla \eta^t(x,s)ds \right|^2 dxdt.$$

(1.48)

Using (1.47) and (1.48) in (1.46) we obtain (1.45). $\qquad \square$

Lemma 1.4 *Assume*

$$|k| < \frac{\tilde{\mu}}{2\theta} e^{-\tau} . \tag{1.49}$$

Then, for any $T \geq S > 0$ and for any $\varepsilon > 0$ we have

$$\int_S^T \int_\Omega u_t^2(x,t) dx dt \leq \varepsilon \int_S^T \int_\Omega |\nabla u(x,t)|^2 dx dt + C_2 F(S) , \tag{1.50}$$

where the constant $C_2 := C_2(\varepsilon)$ is defined by

$$C_2 = \frac{4}{\tilde{\mu}} \left(1 + \frac{1}{2} \frac{1}{\theta - 1} + \frac{\mu(0)}{\tilde{\mu}} C_P \right) + 4 C_P + \frac{2}{\alpha} \left(2 + \frac{(1-\tilde{\mu})^2}{\tilde{\mu} \varepsilon} + C_P |k| (\theta e^\tau + 1) \right). \tag{1.51}$$

Proof In order to prove Lemma 1.4 we have to estimate the terms of the right-hand side of (1.45). First we have,

$$\left| \int_\Omega u_t(x,t) \int_0^\infty \mu(s) \eta^t(x,s) ds dx \right|$$

$$\leq \int_0^\infty \mu(s) \left(\int_\Omega |u_t(x,t)| |\eta^t(x,s)| dx \right) ds$$

$$\leq \int_0^\infty \mu(s) \left(\int_\Omega u_t^2(x,t) dx \right)^{1/2} \left(\int_\Omega (\eta^t(x,s))^2 dx \right)^{1/2} ds$$

$$\leq \frac{1}{2} \int_\Omega u_t^2(x,t) dx + \frac{1}{2} \left(\int_0^\infty \mu(s) \left(\int_\Omega (\eta^t(x,s))^2 dx \right)^{1/2} ds \right)^2 .$$

Then, recalling (1.20) and using Hölder's inequality, we deduce

$$\left| \int_\Omega u_t(x,t) \int_0^\infty \mu(s) \eta^t(x,s) ds dx \right|$$

$$\leq F(t) + \frac{C_P}{2} \left(\int_0^\infty \mu(s) \left(\int_\Omega |\nabla \eta^t(x,s)|^2 dx \right)^{1/2} ds \right)^2 \tag{1.52}$$

$$\leq F(t) + \frac{C_P}{2} \tilde{\mu} \int_0^\infty \mu(s) \int_\Omega |\nabla \eta^t(x,s)|^2 dx ds \leq F(t)(1 + C_P \tilde{\mu}) .$$

Therefore,

$$\left[\int_\Omega u_t(x,t) \int_0^\infty \mu(s) \eta^t(x,s) ds dx \right]_S^T \leq 2(1 + C_P \tilde{\mu}) F(S) . \tag{1.53}$$

Now we proceed to estimate the second term in the right-hand side of (1.45). For any $\delta > 0$ we have

$$\left| \int_S^T \int_\Omega u_t(x,t) \int_0^\infty \mu'(s)\eta^t(x,s)ds dx dt \right|$$

$$\leq \int_S^T \left(\int_\Omega u_t^2(x,t)dx \right)^{1/2} \left(\int_\Omega \left(\int_0^\infty \mu'(s)\eta^t(x,s)ds \right)^2 dx \right)^{1/2} dt$$

$$\leq \frac{\delta}{2} \int_S^T \int_\Omega u_t^2(x,t)dx dt + \frac{1}{2\delta} \int_S^T \int_\Omega \left(\int_0^\infty \mu'(s)\eta^t(x,s)ds \right)^2 dx dt$$

$$\leq \frac{\delta}{2} \int_S^T \int_\Omega u_t^2(x,t)dx dt$$

$$+ \frac{1}{2\delta} \int_S^T \int_\Omega \int_0^\infty (-\mu'(s))ds \int_0^\infty |\mu'(s)|(\eta^t(x,s))^2 ds dx dt,$$

and then by Corollary 1.1

$$\left| \int_S^T \int_\Omega u_t(x,t) \int_0^\infty \mu'(s)\eta^t(x,s)ds dx dt \right|$$

$$\leq \frac{\delta}{2} \int_S^T \int_\Omega u_t^2(x,t)dx dt - \frac{\mu(0)}{2\delta} C_P \int_S^T \int_0^\infty \mu'(s) \int_\Omega |\nabla\eta^t(x,s)|^2 dx ds dt$$

$$\leq \frac{\delta}{2} \int_S^T \int_\Omega u_t^2(x,t)dx dt + \frac{\mu(0)}{\delta} C_P F(S).$$

$$(1.54)$$

Moreover, by (1.28) we have

$$\int_S^T \int_\Omega \left| \int_0^\infty \mu(s)\nabla\eta^t(x,s)ds \right|^2 dx dt$$

$$\leq \int_S^T \int_\Omega \tilde{\mu} \int_0^\infty \mu(s)|\nabla\eta^t(x,s)|^2 ds dx dt \qquad (1.55)$$

$$\leq \frac{2\tilde{\mu}}{\alpha} F(S).$$

Then, it results also

$$\int_S^T \int_\Omega \nabla u(x,t) \int_0^\infty \mu(s) \nabla \eta^t(x,s) ds dx dt$$

$$\leq \int_S^T \left| \int_\Omega \nabla u(x,t) \int_0^\infty \mu(s) \nabla \eta^t(x,s) ds dx \right| dt \qquad (1.56)$$

$$\leq \frac{\varepsilon}{2} \int_S^T \int_\Omega |\nabla u(x,t)|^2 dx dt + \frac{\tilde{\mu}}{\alpha \varepsilon} F(S).$$

Now we estimate the last two integrals in the right-hand side of (1.45).

$$\theta |k| e^\tau \int_S^T \int_\Omega u_t(x,t) \int_0^\infty \mu(s) \eta^t(x,s) ds dx dt$$

$$+ k \int_S^T \int_\Omega u_t(x,t-\tau) \int_0^\infty \mu(s) \eta^t(x,s) ds dx dt$$

$$\leq \frac{|k|}{2} \int_S^T \int_\Omega u_t^2(x,t-\tau) dx dt + \frac{\theta |k| e^\tau}{2} \int_S^T \int_\Omega u_t^2(x,t) dx dt$$

$$+ \frac{|k|(1+\theta e^\tau)}{2} \int_S^T \int_\Omega \left(\int_0^\infty \mu(s) \eta^t(x,s) ds \right)^2 dx dt$$

$$\leq \frac{|k|}{2} \int_S^T \int_\Omega u_t^2(x,t-\tau) dx dt + \frac{\theta |k| e^\tau}{2} \int_S^T \int_\Omega u_t^2(x,t) dx dt$$

$$+ \frac{|k|(1+\theta e^\tau)}{2} C_P \tilde{\mu} \int_S^T \int_\Omega \int_0^\infty \mu(s) |\nabla \eta^t(x,s)|^2 ds dx dt.$$

Therefore, recalling (1.28) and (1.44), we have

$$\theta |k| e^\tau \int_S^T \int_\Omega u_t(x,t) \int_0^\infty \mu(s) \eta^t(x,s) ds dx dt$$

$$+ k \int_S^T \int_\Omega u_t(x,t-\tau) \int_0^\infty \mu(s) \eta^t(x,s) ds dx dt \leq \frac{|k|}{2} \int_S^T \int_\Omega u_t^2(x,t-\tau) dx dt$$

$$+ \frac{\theta |k| e^\tau}{2} \int_S^T \int_\Omega u_t^2(x,t) dx dt + C_P(|k|(\theta e^\tau + 1)) \frac{\tilde{\mu}}{\alpha} F(S)$$

$$\leq \frac{1}{\theta - 1} F(S) + \frac{\theta |k| e^\tau}{2} \int_S^T \int_\Omega u_t^2(x,t) dx dt + \frac{C_P \tilde{\mu}}{\alpha} |k|(\theta e^\tau + 1) F(S).$$

$$(1.57)$$

Using (1.53)–(1.57) in (1.45) we obtain

$$\left(\tilde{\mu} - \frac{\theta|k|e^{\tau}}{2} - \frac{\delta}{2}\right) \int_S^T \int_\Omega u_t^2(x,t)dx \le \frac{\varepsilon}{2}(1 - \tilde{\mu}) \int_S^T \int_\Omega |\nabla u(x,t)|^2 dxdt$$

$$+ \frac{1}{\theta - 1} F(S) + 2(1 + C_P \tilde{\mu})F(S) + \frac{\mu(0)}{\delta} C_P F(S)$$

$$+ \frac{\tilde{\mu}}{\alpha}\left(\frac{1 - \tilde{\mu}}{\varepsilon} + 2\right)F(S) + C_P \tilde{\mu}\frac{|k|(\theta e^{\tau} + 1)}{\alpha}F(S).$$

$$(1.58)$$

Now, fix $\delta = \frac{\tilde{\mu}}{2}$. Then, from (1.49), for any $T \ge S > 0$, we have

$$\int_S^T \int_\Omega u_t^2(x,t)dxdt \le \frac{\varepsilon}{\tilde{\mu}}(1 - \tilde{\mu}) \int_S^T \int_\Omega |\nabla u(x,t)|^2 dxdt + \frac{2}{\tilde{\mu}}\left(2(1 + C_P \tilde{\mu})\right.$$

$$+ \frac{1}{\theta - 1} + 2\frac{\mu(0)}{\tilde{\mu}}C_P + \frac{\tilde{\mu}}{\alpha}(2 + \frac{1-\tilde{\mu}}{\varepsilon} + C_P|k|(\theta e^{\tau} + 1))\Big)F(S),$$

$$(1.59)$$

that is (1.50) with constant C_2 as in (1.51). □

Lemma 1.5 *Assume*

$$|k| < \min\left\{\frac{1 - \tilde{\mu}}{2C_P(\theta e^{\tau} + 1)}, \frac{\tilde{\mu}}{2\theta}e^{-\tau}\right\}. \tag{1.60}$$

Then, for any $T \ge S > 0$,

$$\frac{1 - \tilde{\mu}}{2} \int_S^T \int_\Omega |\nabla u(x,t)|^2 dxdt + \frac{1}{2} \int_S^T \int_\Omega u_t^2(x,t)dxdt \le C^* F(S), \tag{1.61}$$

with

$$C^* = C_0 C_2 + C_1 + C_2, \tag{1.62}$$

where C_0 and C_1 are the constants defined by (1.34) and

$$C_2 := C_2\left(\frac{1 - \tilde{\mu}}{2(C_0 + 1)}\right) = \frac{4}{\tilde{\mu}}\left(1 + \frac{1}{2}\frac{1}{\theta - 1} + \frac{\mu(0)}{\tilde{\mu}}C_P\right) + 4C_P$$

$$+ \frac{2}{\alpha}\left(2 + (6 + 2\theta|k|e^{\tau})\frac{(1 - \tilde{\mu})}{\tilde{\mu}} + C_P|k|(\theta e^{\tau} + 1)\right). \tag{1.63}$$

Proof The assumptions of previous lemmas are verified. Thus, we can use (1.50) in (1.33). Then,

$$(1 - \tilde{\mu}) \int_S^T \int_\Omega |\nabla u(x,t)|^2 dx$$

$$\leq C_0 \varepsilon \int_S^T \int_\Omega |\nabla u(x,t)|^2 dx dt + (C_0 C_2 + C_1) F(S). \tag{1.64}$$

Therefore, from (1.50) and (1.64), we obtain

$$\frac{1 - \tilde{\mu}}{2} \int_S^T \int_\Omega |\nabla u(x,t)|^2 dx + \frac{1}{2} \int_S^T \int_\Omega u_t^2(x,t) dx dt$$

$$\leq \frac{\varepsilon}{2}(C_0 + 1) \int_S^T \int_\Omega |\nabla u(x,t)|^2 dx dt + \frac{1}{2}(C_0 C_2 + C_1 + C_2) F(S). \tag{1.65}$$

Now, fix

$$\varepsilon = \frac{1 - \tilde{\mu}}{2(C_0 + 1)}.$$

Then, from (1.65) we deduce

$$\frac{1 - \tilde{\mu}}{4} \int_S^T \int_\Omega |\nabla u(x,t)|^2 dx + \frac{1}{2} \int_S^T \int_\Omega u_t^2(x,t) dx dt$$

$$\leq \frac{1}{2}(C_0 C_2 + C_1 + C_2) F(S),$$

where, from (1.51) with the above choice of ε, C_2 is as in (1.63). This clearly implies (1.61) with C^* as in (1.62). □

Proof (of Theorem 1.3.) Notice also that (1.27) directly implies that

$$\frac{\theta |k| e^\tau}{2} \int_S^T \int_{t-\tau}^t e^{-(t-s)} \int_\Omega u_t^2(x,s) dx ds dt \leq - \int_S^T F'(t) dt \leq F(S). \tag{1.66}$$

Let us define $\overline{k}$ as

$$\overline{k} := \min \left\{ \frac{1 - \tilde{\mu}}{2C_P(\theta e^\tau + 1)}, \frac{\tilde{\mu}}{2\theta} e^{-\tau} \right\}. \tag{1.67}$$

Then, if $|k| < \overline{k}$, using (1.61), (1.28) and (1.66) in (1.30), we obtain

$$\int_S^T F(t)dt \le C^* F(S) + \frac{1}{\alpha} F(S) + F(S).$$

Therefore (1.29) is verified with

$$C = C^* + 1 + \frac{1}{\alpha}, \tag{1.68}$$

where C^* is as in (1.62) with C_0, C_1 and C_2 defined in (1.34) and (1.63). $\square$

Proof (of Theorem 1.1) From Theorem 1.3 and Lemma 1.1, it follows that for any solution of the auxiliary problem (1.23)–(1.25) if $|k| < \overline{k}$, we have

$$F(t) \le F(0)e^{1-\tilde{\sigma}t}, \quad t \ge 0, \tag{1.69}$$

with

$$\tilde{\sigma} := \frac{1}{C}, \tag{1.70}$$

where C is as in (1.68).

From this and Theorem 1.2 we deduce that Theorem 1.1 holds, with $\sigma := \tilde{\sigma} - e\theta|k|e^\tau$, if

$$-\tilde{\sigma} + e\theta|k|e^\tau < 0,$$

that is if the delay parameter k satisfies

$$|k| < g(|k|) := \frac{1}{Ce\theta e^\tau}, \tag{1.71}$$

with $C := C(|k|)$ defined in (1.68). Now observe that (1.71) is satisfied for $k = 0$ because $g(0) > 0$. Moreover, by recalling the definitions of the constants C_0, C_1, C_2 and C^*, used to define C, we note that $g : [0, \infty) \to (0, \infty)$ is a continuous decreasing function satisfying

$$g(|k|) \to 0 \quad \text{for} \quad |k| \to \infty.$$

Thus, there exists a unique constant $\hat{k} > 0$ such that $\hat{k} = g(\hat{k})$. We can then conclude that for any θ in the definition (1.20) of the energy $F(\cdot)$, inequality (1.71) is satisfied for every k with

$$|k| < k_0 = \min\{\hat{k}, \overline{k}\}. \tag*{$\square$}$$

Remark 1.3 We can compute an explicit lower bound for k_0. Indeed (1.71) may be rewritten as

$$|k|\theta e^{\tau+1}\left(C^* + 1 + \frac{1}{\alpha}\right) < 1.$$

Then, from (1.62), we have

$$[1 + 1/\alpha + C_2(C_0 + 1) + C_1]\theta e^{\tau+1}|k| < 1, \qquad (1.72)$$

that is

$$h(|k|) := \left\{1 + \frac{1}{\alpha} + \left[\frac{4}{\tilde{\mu}}\left(1 + \frac{1}{2}\frac{1}{\theta-1} + \frac{\mu(0)}{\tilde{\mu}}C_P\right) + 4C_P\right.\right.$$

$$\left.+\frac{2}{\alpha}\left(2 + (6 + 2\theta|k|e^{\tau})\frac{(1-\tilde{\mu})}{\tilde{\mu}} + C_P|k|(\theta e^{\tau} + 1)\right)\right](3 + \theta|k|e^{\tau})$$

$$\left.+4\left(1 + \frac{\tilde{\mu}}{\alpha(1-\tilde{\mu})} + \frac{C_P}{1-\tilde{\mu}} + \frac{1}{2(\theta-1)}\right)\right\}\theta e^{\tau+1}|k| < 1.$$
$$(1.73)$$

Now, we use the assumption $|k| < \overline{k}$ with $\overline{k}$ defined in (1.67) in order to majorize the left-hand side of (1.73), $h(|k|)$, with a linear function. We have

$$h(|k|) \leq \left\{1 + \frac{1}{\alpha} + \left[\frac{4}{\tilde{\mu}}\left(1 + \frac{1}{2}\frac{1}{\theta-1} + \frac{\mu(0)}{\tilde{\mu}}C_P\right) + 4C_P\right.\right.$$

$$\left.+\frac{2}{\alpha}\left(2 + (6 + \tilde{\mu})\frac{(1-\tilde{\mu})}{\tilde{\mu}} + \frac{1-\tilde{\mu}}{2}\right)\right](3 + \tilde{\mu}/2)$$

$$\left.+4\left(1 + \frac{\tilde{\mu}}{\alpha(1-\tilde{\mu})} + \frac{C_P}{1-\tilde{\mu}} + \frac{1}{2(\theta-1)}\right)\right\}\theta e^{\tau+1}|k|,$$
$$(1.74)$$

from which follows

$$h(|k|) \leq \left(1 + \frac{1}{\alpha}\gamma_1 + \gamma_2\right)\theta|k|e^{\tau+1},$$

with

$$\gamma_1 = \gamma_1(\tilde{\mu}) = 4\frac{\tilde{\mu}}{1-\tilde{\mu}} - 8 + \frac{36}{\tilde{\mu}} - \frac{23}{2}\tilde{\mu} - \frac{3}{2}\tilde{\mu}^2,$$

$$\gamma_2 = \gamma_2(\mu(0), \tilde{\mu}, \theta, C_P).$$

$$= 6 + 12C_P + \frac{3}{\theta - 1} + \frac{12}{\tilde{\mu}} + \frac{6}{\tilde{\mu}(\theta - 1)}$$

$$+ 12\frac{\mu(0)}{\tilde{\mu}^2}C_P + 2\frac{\mu(0)}{\tilde{\mu}}C_P + 2C_P\tilde{\mu} + \frac{4C_P}{1 - \tilde{\mu}}.$$

Then, we deduce the following explicit lower bound

$$k_0 \geq \frac{e^{-(\tau+1)}}{\theta(1 + \frac{1}{\alpha}\gamma_1 + \gamma_2)}, \tag{1.75}$$

with γ_1, γ_2 as before. For example, if we take

$$\mu(t) = e^{-2t},$$

then $\tilde{\mu} = 1/2$ and so, fixing $\theta = 2$, we can compute $\gamma_1 = \frac{495}{8}$, $\gamma_2 = 45 + 73C_P$. Hence, for this particular choice of the memory kernel, we obtain

$$k_0 \geq \frac{8e^{-(\tau+1)}}{1231 + 1168C_P}.$$

Remark 1.4 In the case $\tau = 0$ and $k < 0$, namely viscoelastic wave equation with anti-damping, we can simplify previous arguments. Indeed, the absence of time delay allows us to take $\theta = 1$ obtaining an exponential stability estimate under the condition

$$|k| < \left(C_1 + 3C_2 + \frac{1}{\alpha}\right)^{-1}\frac{1}{e},$$

where

$$C_1 = 4\left(1 + \frac{\tilde{\mu}}{\alpha(1 - \tilde{\mu})} + \frac{C_P}{1 - \tilde{\mu}}\right)$$

and

$$C_2 = \frac{2}{\tilde{\mu}}\left(2 + \frac{\mu(0)}{\tilde{\mu}}C_P\right) + 4C_P + \frac{2}{\alpha}\left(2 + 6\frac{1 - \tilde{\mu}}{\tilde{\mu}}\right).$$

References

1. Alabau-Boussouira, F., Cannarsa, P.: A new method for proving sharp energy decay rates for memory-dissipative evolution equations for a quasi-optimal class of kernels. C. R. Acad. Sci. Paris, Sér. I **347**, 867–872 (2009)

2. Alabau-Boussouira, F., Cannarsa, P., Sforza, D.: Decay estimates for second order evolution equations with memory. J. Funct. Anal. **254**, 1342–1372 (2008)
3. Ammari, K., Nicaise, S., Pignotti, C.: Feedback boundary stabilization of wave equations with interior delay. Syst. Control Lett. **59**, 623–628 (2010)
4. Dafermos, C.M.: Asymptotic stability in viscoelasticity. Arch. Ration. Mech. Anal. **37**, 297–308 (1970)
5. Dai, Q., Yang, Z.: Global existence and exponential decay of the solution for a viscoelastic wave equation with a delay. Z. Angew. Math. Phys. (2013). doi:10.1007/s00033-013-0365-6
6. Datko, R.: Not all feedback stabilized hyperbolic systems are robust with respect to small time delays in their feedbacks. SIAM J. Control Optim. **26**, 697–713 (1988)
7. Datko, R., Lagnese, J., Polis, M.P.: An example on the effect of time delays in boundary feedback stabilization of wave equations. SIAM J. Control Optim. **24**, 152–156 (1986)
8. Freitas, P., Zuazua, E.: Stability results for the wave equation with indefinite damping. J. Differ. Equ. **132**, 338–352 (1996)
9. Giorgi, C., Muñoz Rivera, J.E., Pata, V.: Global attractors for a semilinear hyperbolic equation in viscoelasticity. J. Math. Anal. Appl. **260**, 83–99 (2001)
10. Guesmia, A.: Well-posedness and exponential stability of an abstract evolution equation with infinite memory and time delay. IMA J. Math. Control Inf. **30**, 507–526 (2013)
11. Kirane, M., Said-Houari, B.: Existence and asymptotic stability of a viscoelastic wave equation with a delay. Z. Angew. Math. Phys. **62**, 1065–1082 (2011)
12. Komornik, V.: Exact Controllability and Stabilization, the Multiplier Method. RMA, vol. 36. Masson, Paris (1994)
13. Munõz Rivera, J.E., Peres Salvatierra, A.: Asymptotic behaviour of the energy in partially viscoelastic materials. Q. Appl. Math. **59**, 557–578 (2001)
14. Nicaise, S., Pignotti, C.: Stability and instability results of the wave equation with a delay term in the boundary or internal feedbacks. SIAM J. Control Optim. **45**, 1561–1585 (2006)
15. Nicaise, S., Pignotti, C.: Stabilization of second-order evolution equations with time delay. Math. Control Signals Syst. (2014). doi:10.1007/s00498-014-0130-1
16. Pata, V.: Exponential stability in linear viscoelasticity with almost flat memory kernels. Commun. Pure Appl. Anal. **9**, 721–730 (2010)
17. Pazy, A.: Semigroups of Linear Operators and Applications to Partial Differential Equations. Applied Mathematical Sciences, vol. 44. Springer, New York (1983)
18. Pignotti, C.: A note on stabilization of locally damped wave equations with time delay. Syst. Control Lett. **61**, 92–97 (2012)
19. Prüss, J.: Evolutionary Integral Equations and Applications. Monographs in Mathematics, vol. 87. Birkhäuser, Basel (1993)
20. Xu, G.Q., Yung, S.P., Li, L.K.: Stabilization of wave systems with input delay in the boundary control. ESAIM Control Optim. Calc. Var. **12**(4), 770–785 (2006)

Chapter 2
Existence of Global Weak Solutions to a Generalized Hyperelastic-Rod Wave Equation with Source

Fabio Ancona and Giuseppe Maria Coclite

In memory of Alfredo with a vivid remembering of his enthusiasm and dedication in fostering various activities of the inverse and control theoretic community

Abstract We consider a weakly dissipative hyperelastic-rod wave equation describing nonlinear dispersive dissipative waves in compressible hyperelastic rods. We endow it with a nonlinear source and establish the existence of global weak solutions for any initial condition in $H^1(\mathbb{R})$.

2.1 Introduction

We are interested in the Cauchy problem for the nonlinear equation

$$\begin{cases} \partial_t u - \partial_{txx}^3 u + \partial_x \left(\frac{g(u)}{2} \right) = \gamma \left(2\partial_x u \partial_{xx}^2 u + u \partial_{xxx}^3 u \right) + f(t, x, u), & t > 0, \ x \in \mathbb{R}, \\ u(0, x) = u_0(x), & x \in \mathbb{R}, \end{cases} \tag{2.1}$$

where the functions $g : \mathbb{R} \to \mathbb{R}$, $f : [0, \infty) \times \mathbb{R} \times \mathbb{R} \to \mathbb{R}$, and the constant $\gamma \in \mathbb{R}$ are given. Observe that if $g(u) = 2\kappa u + 3u^2$, $f \equiv 0$, and $\gamma = 1$, then (2.1) is the classical Camassa–Holm equation [6, 20]. With $g(u) = 3u^2$, Dai [13–15] derived (2.1) as an equation describing finite length, small amplitude radial deformation waves in cylindrical compressible hyperelastic rods, and the equation

F. Ancona
Department of Pure and Applied Mathematics, University of Padova, Via Trieste 63, 35121 Padova, Italy
e-mail: ancona@math.unipd.it

G.M. Coclite (✉)
Department of Mathematics, University of Bari, Via E. Orabona 4, 70125 Bari, Italy
e-mail: giuseppemaria.coclite@uniba.it

© Springer International Publishing Switzerland 2014
A. Favini et al. (eds.), *New Prospects in Direct, Inverse and Control Problems for Evolution Equations*, Springer INdAM Series 10,
DOI 10.1007/978-3-319-11406-4_2

23

is often referred to as the hyperelastic-rod wave equation. The constant γ is given in terms of the material constants and the prestress of the rod.

We shall assume

$$
\begin{aligned}
&u_0 \in H^1(\mathbb{R}), \ g \in C^\infty(\mathbb{R}), \ |g(u)| \le Mu^2, \quad \gamma > 0, \\
&f \in C^\infty([0,\infty) \times \mathbb{R} \times \mathbb{R}), \ |f(\cdot,\cdot,u)|, \ |\partial_t f(\cdot,\cdot,u)| \le L|u|, \ |\partial_u f(\cdot,\cdot,u)| \le L,
\end{aligned}
\tag{2.2}
$$

for some constants $L, M > 0$ and every $u \in \mathbb{R}$. Observe that the case $\gamma = 0$ is much simpler than the one we are considering. Moreover, if $\gamma < 0$, peakons become antipeakons, so we can use a similar argument. The assumptions of infinite differentiability and subquadratic growth of g is made just for convenience. In fact, locally Lipschitz continuity would be sufficient. Define

$$
h(\xi) := \frac{1}{2}\left(g(\xi) - \gamma\xi^2\right)
\tag{2.3}
$$

for $\xi \in \mathbb{R}$. Rewriting Eq. (2.1) as

$$
(1 - \partial_x^2)\partial_t u + \gamma(1 - \partial_x^2)(u\partial_x u) + \partial_x\left(h(u) + \frac{\gamma}{2}(\partial_x u)^2\right) = f(t, x, u),
\tag{2.4}
$$

we see that Eq. (2.1) formally is equivalent to the elliptic-hyperbolic system

$$
\partial_t u + \gamma u \partial_x u + \partial_x P = F, \ -\partial_{xx}^2 P + P = h(u) + \frac{\gamma}{2}(\partial_x u)^2, \ -\partial_{xx}^2 F + F = f(t, x, u).
\tag{2.5}
$$

Moreover, since $e^{-|x|}/2$ is the Green's function of the operator $-\partial_{xx}^2 + 1$, we have that

$$
\begin{aligned}
P(t, x) &= \frac{1}{2}\int_{\mathbb{R}} e^{-|x-y|}\left(h(u(t, y)) + \frac{\gamma}{2}(\partial_x u(t, y))^2\right) dy, \\
F(t, x) &= \frac{1}{2}\int_{\mathbb{R}} e^{-|x-y|} f(t, y, u(t, y)) dy.
\end{aligned}
\tag{2.6}
$$

Motivated by this, we shall use the following definition of weak solution.

Definition 2.1 Let $u\colon [0,\infty) \times \mathbb{R} \to \mathbb{R}$ be a function. We say that u is a weak dissipative solution of the Cauchy problem (2.1) if

(i) $u \in C([0,\infty) \times \mathbb{R}) \cap L^\infty((0, T); H^1(\mathbb{R}))$, $T > 0$;
(ii) u satisfies (2.5) in the sense of distributions for some $P, F \in L^\infty([0,\infty);$ $W^{1,\infty}(\mathbb{R}))$, that is

$$
\begin{aligned}
&\int_0^\infty \int_{\mathbb{R}}\left(u\partial_t\varphi + \gamma\frac{u^2}{2}\partial_x\varphi + P\partial_x\varphi + F\varphi\right) dt dx + \int_{\mathbb{R}} u_0(x)\varphi(0, x) dx = 0, \\
&\int_0^\infty \int_{\mathbb{R}}\left(-P\partial_{xx}^2\varphi + P\varphi - h(u)\varphi - \frac{\gamma}{2}(\partial_x u)^2\varphi\right) dt dx = 0, \\
&\int_0^\infty \int_{\mathbb{R}}\left(-F\partial_{xx}^2\varphi + F\varphi - f(t, x, u)\varphi\right) dt dx = 0,
\end{aligned}
$$

for every test function $\varphi \in C^\infty(\mathbb{R}^2)$ with compact support;
(iii) $u(0, x) = u_0(x)$, for every $x \in \mathbb{R}$;

(iv) (Oleĭnik type Estimate) for each $T > 0$ there exists a positive constant C_T depending on u_0, γ, g, L, and T such that

$$\partial_x u(t, x) \leq \frac{2}{\gamma t} + C_T,$$

for each $x \in \mathbb{R}$, $0 < t \leq T$.

This definition is inspired by the definition of dissipative solution introduced in [3,5,7,9–12,17,18,23] and differs from the one of conservative solution introduced in [4, 19] for the presence of the Oleĭnik type Estimate. The motivation behind the existence of these two definitions can be easily understood considering the traveling waves of the Camassa–Holm equations. The following peakon like function is a travelling wave solution:

$$\frac{1}{2}\left(1 - \frac{1}{\gamma}\right)c + \frac{c}{2}\left(\frac{3}{\gamma} - 1\right)e^{-\frac{|x-ct-\xi|}{\sqrt{\gamma}}}.$$

When two of these solitary waves interact may happen that at a time t_0 they are completely neglected and we have $u(t_0, x) = 0$. In the conservative solutions the shape of these waves is retained after the interactions, in particular there is the conservation of the total energy for all the times $t \neq t_0$. On the contrary in the dissipative solutions there is no reconstruction and then we have the trivial solution for $t \geq t_0$. One would expect that the Oleĭnik type condition (iv) of Definition 2.1 would yield uniqueness of dissipative solutions to (2.1). However, the more general uniqueness result available in literature is [24] and it does not apply to weak solutions, therefore is not clear if the dissipative solutions of [1–3,5,7,9–12,17,18,23] are the same.

Remark 2.1 An alternative definition of weak dissipative solution can be found in [1,2], where the condition (ii) is replaced by the requirement that

$$\frac{d}{dt}u = -\gamma u \partial_x u - \partial_x P + F$$

holds in the L^2-sense for almost every t. One can easily verify that a function that is a solution of (2.1) according with Definition 2.1 turns out to be also a solution of (2.1) in the sense of the definition given in [1,2].

We have to remark that the Lagrangian approach used in [1–5, 17–19] gives the existence of semigroups of solutions in both cases. The vanishing viscosity one used here and in [3,7,9–12,23] gives only the existence of weak solutions without any semigroup property. The interest for this approach is motivated by the analysis of numerical schemes, where for instance the convergence proof is usually based on similar arguments to the ones used for the viscosity approximation (see [11,12]).

Remark 2.2 The only results available in literature that treat (2.1) in the case $f \not\equiv 0$ are [1, 2]. There we prove the existence of a semigroup of asymptotically stable dissipative solutions to (2.1) when $f = -\lambda(u - \partial_{xx}^2 u)$ and $\lambda > 0$ is a constant. Clearly the source term $f = -\lambda(u - \partial_{xx}^2 u)$ does not satisfies (2.2) because it depends on $\partial_{xx}^2 u$. However, the arguments in the present paper can be easily adapted to the case in which (2.5) is replaced by

$$\partial_t u + \gamma u \partial_x u + \partial_x P = -\lambda u, \qquad -\partial_{xx}^2 P + P = h(u) + \frac{\gamma}{2}(\partial_x u)^2 .$$

Moreover, in this case arguing as in Lemma 2.1 we gain the estimate

$$\|u_\varepsilon(t, \cdot)\|_{H^1(\mathbb{R})}^2 + 2\varepsilon e^{-2\lambda t} \int_0^t e^{2\lambda s} \|\partial_x u_\varepsilon(s, \cdot)\|_{H^1(\mathbb{R})}^2 \, ds \le e^{-2t} \|u_0\|_{H^1(\mathbb{R})}^2 ,$$

which in turn letting $\varepsilon \to 0$, yields the decay of the energy

$$\|u(t, \cdot)\|_{H^1(\mathbb{R})} \le e^{-\lambda t} \|u_0\|_{H^1(\mathbb{R})} .$$

Our existence results are collected in the following theorem:

Theorem 2.1 *Let u_0, γ, g, f satisfy (2.2). The initial value problem (2.1) has a weak dissipative solution $u : [0, \infty) \times \mathbb{R} \to \mathbb{R}$ in the sense of Definition 2.1. Moreover, u satisfies the following property:*

$$\partial_x u \in L^p((0, T) \times (a, b)), \tag{2.7}$$

for each $1 \le p < 3$, $T > 0$, $a < b$.

The paper is organized as follows. In Sect. 2.2 we state the viscous problem and establish an Oleinik type estimate and a higher integrability estimate for the viscous approximants. Section 2.3 is devoted to proving basic compactness properties for the viscous approximants. In Sect. 2.4 we get the strong compactness of the derivative of the viscous approximants and prove Theorem 2.1.

2.2 Viscous Approximants: Existence and Estimates

We will prove existence of a weak dissipative solution to the Cauchy problem for (2.1) by proving compactness of a sequence of smooth functions $\{u_\varepsilon\}_{\varepsilon > 0}$ solving the following viscous problems:

$$\begin{cases} \partial_t u_\varepsilon + \gamma u_\varepsilon \partial_x u_\varepsilon + \partial_x P_\varepsilon = F_\varepsilon + \varepsilon \partial_{xx}^2 u_\varepsilon, & t > 0, \ x \in \mathbb{R}, \\ -\partial_{xx}^2 P_\varepsilon + P_\varepsilon = h(u_\varepsilon) + \frac{\gamma}{2}(\partial_x u_\varepsilon)^2, & t > 0, \ x \in \mathbb{R}, \\ -\partial_{xx}^2 F_\varepsilon + F_\varepsilon = f(t, x, u_\varepsilon), & t > 0, \ x \in \mathbb{R}, \\ u_\varepsilon(0, x) = u_{\varepsilon,0}(x), & x \in \mathbb{R}. \end{cases} \tag{2.8}$$

We shall assume that

$$\{u_{\varepsilon,0}\}_{\varepsilon>0} \subset C^\infty(\mathbb{R}), \quad \|u_{\varepsilon,0}\|_{H^1(\mathbb{R})} \le \|u_0\|_{H^1(\mathbb{R})}, \quad \varepsilon > 0, \quad u_{\varepsilon,0} \to u_0 \text{ in } H^1(\mathbb{R}). \tag{2.9}$$

The well-posedness of smooth solutions for (2.8) has been proved in [8]. Due to the smoothness of the solutions of (2.8), it is (rigorously) equivalent to the fourth order equation

$$\partial_t u_\varepsilon - \partial_{txx}^3 u_\varepsilon + \partial_x \left(\frac{g(u_\varepsilon)}{2} \right)$$
$$= \gamma \left(2\partial_x u_\varepsilon \partial_{xx}^2 u_\varepsilon + u_\varepsilon \partial_{xxx}^3 u_\varepsilon \right) + f(t, x, u_\varepsilon) + \varepsilon \partial_{xx}^2 u_\varepsilon - \varepsilon \partial_{xxxx}^4 u_\varepsilon. \tag{2.10}$$

In our estimates we will frequently use the embedding $H^1(\mathbb{R}) \subset L^\infty(\mathbb{R})$. More precisely we have that [21, Theorem 8.5]

$$\|\varphi\|_{L^\infty(\mathbb{R})} \le \frac{1}{\sqrt{2}} \|\varphi\|_{H^1(\mathbb{R})}, \qquad \varphi \in H^1(\mathbb{R}). \tag{2.11}$$

Finally, we introduce the notation

$$q_\varepsilon = \partial_x u_\varepsilon, \qquad \varepsilon > 0.$$

Differentiating the first equation in (2.8) we get

$$\partial_t q_\varepsilon + \gamma u_\varepsilon \partial_x q_\varepsilon - \varepsilon \partial_{xx}^2 q_\varepsilon + \frac{\gamma}{2} q_\varepsilon^2 = h(u_\varepsilon) - P_\varepsilon + \partial_x F_\varepsilon. \tag{2.12}$$

The remaining part of this section is dedicated to several a priori estimates that will play a key role in the proof of our main result.

2.2.1 Energy Estimate

Lemma 2.1 *The following inequality holds*

$$\|u_\varepsilon(t, \cdot)\|_{H^1(\mathbb{R})}^2 + 2\varepsilon e^{2Lt} \int_0^t e^{-2Ls} \|\partial_x u_\varepsilon(s, \cdot)\|_{H^1(\mathbb{R})}^2 \, ds \le e^{2Lt} \|u_0\|_{H^1(\mathbb{R})}^2, \tag{2.13}$$

for every $t \ge 0$ and $\varepsilon > 0$. In particular

$$\|u_\varepsilon(t, \cdot)\|_{H^1(\mathbb{R})} \le e^{Lt} \|u_0\|_{H^1(\mathbb{R})}, \tag{2.14}$$

$$\sqrt{\varepsilon} \|\partial_x u_\varepsilon\|_{L^2((0,t);H^1(\mathbb{R}))} \le \frac{e^{Lt}}{\sqrt{2}} \|u_0\|_{H^1(\mathbb{R})}, \tag{2.15}$$

$$\|u_\varepsilon(t, \cdot)\|_{L^\infty(\mathbb{R})} \le \frac{e^{Lt}}{\sqrt{2}} \|u_0\|_{H^1(\mathbb{R})}, \tag{2.16}$$

for every $t \ge 0$ and $\varepsilon > 0$.

Proof We (2.2) and (2.10) give

$$\frac{d}{dt} \int_{\mathbb{R}} \frac{u_\varepsilon^2 + (\partial_x u_\varepsilon)^2}{2} dx = \int_{\mathbb{R}} \left(u_\varepsilon \partial_t u_\varepsilon + \partial_x u_\varepsilon \partial_{tx}^2 u_\varepsilon \right) dx = \int_{\mathbb{R}} u_\varepsilon \left(\partial_t u_\varepsilon - \partial_{txx}^3 u_\varepsilon \right) dx$$

$$= - \int_{\mathbb{R}} u_\varepsilon \partial_x \left(\frac{g(u_\varepsilon)}{2} \right) dx + 2\gamma \int_{\mathbb{R}} u_\varepsilon \partial_x u_\varepsilon \partial_{xx}^2 u_\varepsilon dx + \gamma \int_{\mathbb{R}} u_\varepsilon^2 \partial_{xxx}^3 u_\varepsilon dx$$

$$+ \int_{\mathbb{R}} u_\varepsilon f(t, x, u_\varepsilon) dx + \varepsilon \int_{\mathbb{R}} u_\varepsilon \left(\partial_{xx}^2 u_\varepsilon - \partial_{xxxx}^4 u_\varepsilon \right) dx$$

$$= \frac{1}{2} \underbrace{\int_{\mathbb{R}} g(u_\varepsilon) \partial_x u_\varepsilon dx}_{=0} + \underbrace{2\gamma \int_{\mathbb{R}} u_\varepsilon \partial_x u_\varepsilon \partial_{xx}^2 u_\varepsilon dx - 2\gamma \int_{\mathbb{R}} u_\varepsilon \partial_x u_\varepsilon \partial_{xx}^2 u_\varepsilon dx}_{=0}$$

$$+ \int_{\mathbb{R}} u_\varepsilon f(t, x, u_\varepsilon) dx - \varepsilon \int_{\mathbb{R}} \left((\partial_x u_\varepsilon)^2 + (\partial_{xx}^2 u_\varepsilon)^2 \right) dx$$

$$= \int_{\mathbb{R}} u_\varepsilon f(t, x, u_\varepsilon) dx - \varepsilon \int_{\mathbb{R}} \left((\partial_x u_\varepsilon)^2 + (\partial_{xx}^2 u_\varepsilon)^2 \right) dx$$

$$\leq L \int_{\mathbb{R}} u_\varepsilon^2 dx - \varepsilon \int_{\mathbb{R}} \left((\partial_x u_\varepsilon)^2 + (\partial_{xx}^2 u_\varepsilon)^2 \right) dx$$

$$\leq 2L \int_{\mathbb{R}} \frac{u_\varepsilon^2 + (\partial_x u_\varepsilon)^2}{2} dx - \varepsilon \int_{\mathbb{R}} \left((\partial_x u_\varepsilon)^2 + (\partial_{xx}^2 u_\varepsilon)^2 \right) dx.$$

Therefore (2.13) follows from the Gronwall's Lemma.

Clearly, (2.14) follows directly from (2.13), and (2.11) follows from (2.11) and (2.13). Finally, from (2.13) we have

$$\varepsilon \| \partial_x u_\varepsilon \|_{L^2((0,t);H^1(\mathbb{R}))}^2 = \varepsilon \int_0^t \| \partial_x u_\varepsilon(s, \cdot) \|_{H^1(\mathbb{R})}^2 ds$$

$$\leq \varepsilon e^{2Lt} \int_0^t e^{-2Ls} \| \partial_x u_\varepsilon(s, \cdot) \|_{H^1(\mathbb{R})}^2 ds \leq \frac{e^{2Lt}}{2} \| u_0 \|_{H^1(\mathbb{R})}^2,$$

that give (2.15).   □

We continue this subsection with some a priori bounds that come directly from the energy estimate stated in Lemma 2.1.

Lemma 2.2 *The family* $\{P_\varepsilon\}_{\varepsilon>0}$ *is uniformly bounded in* $L^\infty(0, T; W^{2,1}(\mathbb{R}))$ *and in* $L^\infty(0, T; W^{1,\infty}(\mathbb{R}))$, *for every* $T > 0$. *More precisely, we have*

$$\| P_\varepsilon(t, \cdot) \|_{L^\infty(\mathbb{R})}, \| \partial_x P_\varepsilon(t, \cdot) \|_{L^\infty(\mathbb{R})} \leq \frac{M+\gamma}{2} e^{2Lt} \| u_0 \|_{H^1(\mathbb{R})}^2, \qquad (2.17)$$

$$\| P_\varepsilon(t, \cdot) \|_{L^1(\mathbb{R})}, \| \partial_x P_\varepsilon(t, \cdot) \|_{L^1(\mathbb{R})} \leq \frac{M+\gamma}{4} e^{2Lt} \| u_0 \|_{H^1(\mathbb{R})}^2, \qquad (2.18)$$

$$\left\| \partial_{xx}^2 P_\varepsilon(t, \cdot) \right\|_{L^1(\mathbb{R})} \leq 3\frac{M+\gamma}{4} e^{2Lt} \| u_0 \|_{H^1(\mathbb{R})}^2, \qquad (2.19)$$

for every $t \geq 0$ *and* $\varepsilon > 0$.

Proof From (2.2), we know

$$\left| h(u_\varepsilon) + \frac{\gamma}{2} (\partial_x u_\varepsilon)^2 \right| \leq \frac{M+\gamma}{2} \left(u_\varepsilon^2 + (\partial_x u_\varepsilon)^2 \right). \qquad (2.20)$$

In addition, since $e^{-|x|}/2$ is the Green's function of the operator $-\partial_{xx}^2 + 1$, we have

$$P_\varepsilon(t, x) = \tfrac{1}{2} \int_\mathbb{R} e^{-|x-y|} \left(h(u_\varepsilon(t, y)) + \tfrac{\gamma}{2} (\partial_x u_\varepsilon(t, y))^2 \right) dy, \qquad (2.21)$$

$$\partial_x P_\varepsilon(t, x) = \tfrac{1}{2} \int_\mathbb{R} e^{-|x-y|} \operatorname{sign}(y - x) \left(h(u_\varepsilon(t, y)) + \tfrac{\gamma}{2} (\partial_x u_\varepsilon(t, y))^2 \right) dy. \quad (2.22)$$

Since

$$\int_\mathbb{R} \frac{e^{-|x|}}{2} dx = 1, \qquad (2.23)$$

we have

$$|P_\varepsilon(t, x)|, |\partial_x P_\varepsilon(t, x)| \leq \tfrac{1}{2} \int_\mathbb{R} e^{-|x-y|} \left| h(u_\varepsilon(t, y)) + \tfrac{\gamma}{2} (\partial_x u_\varepsilon(t, y))^2 \right| dy$$

$$\leq \tfrac{1}{2} \int_\mathbb{R} \left| h(u_\varepsilon(t, y)) + \tfrac{\gamma}{2} (\partial_x u_\varepsilon(t, y))^2 \right| dy,$$

$$\int_\mathbb{R} |P_\varepsilon(t, x)| dx, \int_\mathbb{R} |\partial_x P_\varepsilon(t, x)| dx \leq \tfrac{1}{2} \int_{\mathbb{R} \times \mathbb{R}} e^{-|x-y|} \left| h(u_\varepsilon(t, y)) \right.$$

$$\left. + \tfrac{\gamma}{2} (\partial_x u_\varepsilon(t, y))^2 \right| dy dx$$

$$\leq \int_\mathbb{R} \left| h(u_\varepsilon(t, y)) + \tfrac{\gamma}{2} (\partial_x u_\varepsilon(t, y))^2 \right| dy,$$

therefore (2.17) and (2.18) follow from (2.13) and (2.20).

Finally, since

$$\partial_{xx}^2 P_\varepsilon = P_\varepsilon - h(u_\varepsilon) - \frac{\gamma}{2} (\partial_x u_\varepsilon)^2$$

(2.19) follows from (2.13), (2.18), and (2.20). $\qquad\qquad\qquad\qquad\qquad \square$

Lemma 2.3 *The family $\{F_\varepsilon\}_{\varepsilon>0}$ is uniformly bounded in $L^\infty(0, T; H^2(\mathbb{R}))$ and in $L^\infty(0, T; W^{2,\infty}(\mathbb{R}))$, for every $T > 0$. More precisely, we have*

$$\|F_\varepsilon(t, \cdot)\|_{L^\infty(\mathbb{R})}, \|\partial_x F_\varepsilon(t, \cdot)\|_{L^\infty(\mathbb{R})} \leq \frac{L}{\sqrt{2}} e^{Lt} \|u_0\|_{H^1(\mathbb{R})}, \qquad (2.24)$$

$$\|F_\varepsilon(t, \cdot)\|_{L^2(\mathbb{R})}, \|\partial_x F_\varepsilon(t, \cdot)\|_{L^2(\mathbb{R})} \leq L e^{Lt} \|u_0\|_{H^1(\mathbb{R})}, \qquad (2.25)$$

$$\|\partial_{xx}^2 F_\varepsilon(t, \cdot)\|_{L^\infty(\mathbb{R})} \leq \frac{L+1}{\sqrt{2}} e^{Lt} \|u_0\|_{H^1(\mathbb{R})}, \qquad (2.26)$$

$$\|\partial_{xx}^2 F_\varepsilon(t, \cdot)\|_{L^2(\mathbb{R})} \leq \left(\frac{L}{\sqrt{2}} + 1 \right) e^{Lt} \|u_0\|_{H^1(\mathbb{R})}, \qquad (2.27)$$

for every $t \geq 0$ and $\varepsilon > 0$.

Proof From (2.2), we know

$$|f(t, x, u_\varepsilon)| \leq L|u_\varepsilon|. \qquad (2.28)$$

In addition, since $e^{-|x|}/2$ is the Green's function of the operator $-\partial_{xx}^2 + 1$, we have

$$F_\varepsilon(t, x) = \tfrac{1}{2} \int_{\mathbb{R}} e^{-|x-y|} f(t, y, u_\varepsilon(t, y)) dy, \tag{2.29}$$

$$\partial_x F_\varepsilon(t, x) = \tfrac{1}{2} \int_{\mathbb{R}} e^{-|x-y|} \operatorname{sign}(y - x) f(t, x, u_\varepsilon(t, y)) dy. \tag{2.30}$$

Using (2.23) we have

$$
\begin{aligned}
|F_\varepsilon(t, x)|\,, |\partial_x F_\varepsilon(t, x)| &\le \tfrac{1}{2} \int_{\mathbb{R}} e^{-|x-y|} |f(t, y, u_\varepsilon(t, y))|\, dy \\
&\le \|f(t, \cdot, u_\varepsilon(t, \cdot))\|_{L^\infty(\mathbb{R})}, \\
\int_{\mathbb{R}} |F_\varepsilon(t, x)|^2 dx\,, \int_{\mathbb{R}} |\partial_x F_\varepsilon(t, x)|^2 dx &\le \tfrac{1}{4} \int_{\mathbb{R}} \left(\int_{\mathbb{R}} e^{-|x-y|} |f(t, y, (u_\varepsilon(t, y))| dy \right)^2 dx \\
&\le \tfrac{1}{4} \left(\int_{\mathbb{R}} e^{-|x-y|} dy \right) \left(\int_{\mathbb{R} \times \mathbb{R}} e^{-|x-y|} f^2(t, y, u_\varepsilon(t, y)) dy dx \right) \\
&= \int_{\mathbb{R}} f^2(t, y, u_\varepsilon(t, y)) dy,
\end{aligned}
$$

therefore (2.24) and (2.25) follow from (2.14), (2.11) and (2.20).

Finally, since

$$\partial_{xx}^2 F_\varepsilon = F_\varepsilon - f(t, x, u_\varepsilon)$$

(2.26) and (2.27) follow from (2.14), (2.11), (2.24), (2.25), and (2.28).   □

Lemma 2.4 *The family* $\{u_\varepsilon\}_{\varepsilon>0}$ *is uniformly bounded in* $H^1((0, T) \times \mathbb{R})$, *for each* $T > 0$.

Proof From (2.8) we have

$$\partial_t u_\varepsilon = -\gamma u_\varepsilon \partial_x u_\varepsilon - \partial_x P_\varepsilon + F_\varepsilon + \varepsilon \partial_{xx}^2 u_\varepsilon,$$

that gives

$$
\begin{aligned}
\|\partial_t u_\varepsilon\|_{L^2((0,T)\times\mathbb{R})} &\le \gamma \|u_\varepsilon\|_{L^\infty((0,T)\times\mathbb{R})} \|\partial_x u_\varepsilon\|_{L^2((0,T)\times\mathbb{R})} \\
&\quad + \sqrt{\|\partial_x P_\varepsilon\|_{L^\infty((0,T)\times\mathbb{R})} \|\partial_x P_\varepsilon\|_{L^1((0,T)\times\mathbb{R})}} \\
&\quad + \|F_\varepsilon\|_{L^2((0,T)\times\mathbb{R})} + \varepsilon \|\partial_{xx}^2 u_\varepsilon\|_{L^2((0,T)\times\mathbb{R})}.
\end{aligned}
$$

Therefore the claim follows from Lemmas 2.1–2.3.   □

Lemma 2.5 *The family* $\{P_\varepsilon\}_{\varepsilon>0}$ *is uniformly bounded in* $W^{1,1}_{loc}((0, \infty) \times \mathbb{R})$.

Proof Thanks to Lemma 2.2 we have to prove that $\{\partial_t P_\varepsilon\}_{\varepsilon>0}$ is uniformly bounded in $L^1_{loc}((0, \infty) \times \mathbb{R})$. We split P_ε in the following way

$$P_\varepsilon = P_{\varepsilon,1} + P_{\varepsilon,2},$$

where [see (2.6)]

$$P_{\varepsilon,1}(t,x) = \frac{1}{2}\int_{\mathbb{R}} e^{-|x-y|} h(u_{\varepsilon}(t,y)) dy, \quad P_{\varepsilon,2}(t,x) = \frac{1}{2}\int_{\mathbb{R}} e^{-|x-y|} \frac{\gamma}{2}(\partial_x u_{\varepsilon}(t,y))^2 dy.$$

Since

$$\partial_t P_{\varepsilon} = \partial_t P_{\varepsilon,1} + \partial_t P_{\varepsilon,2},$$
$$\partial_t P_{\varepsilon,1}(t,x) = \frac{1}{2}\int_{\mathbb{R}} e^{-|x-y|} h'(u(t,y)) \partial_t u_{\varepsilon}(t,y) dy,$$
$$\partial_t P_{\varepsilon,2}(t,x) = \frac{1}{2}\int_{\mathbb{R}} e^{-|x-y|} \gamma \partial_x u(t,y) \partial_{tx}^2 u(t,y) dy.$$

We claim that

$$\{\partial_t P_{1,\varepsilon}\}_{\varepsilon} \text{ is uniformly bounded in } L^2((0,T)\times\mathbb{R}), \tag{2.31}$$

$$\{\partial_t P_{2,\varepsilon}\}_{\varepsilon} \text{ is uniformly bounded in } L^1((0,T)\times\mathbb{R}). \tag{2.32}$$

We begin by proving (2.31). Using (2.2), (2.11), the Tonelli theorem, and the Hölder inequality,

$$\|\partial_t P_{1,\varepsilon}\|_{L^2((0,T)\times\mathbb{R})}^2 \leq \max_{|\xi|\leq e^{Lt}\|u_0\|_{H^1(\mathbb{R})}/\sqrt{2}} (h'(\xi))^2 \|\partial_t u_{\varepsilon}\|_{L^2((0,T)\times\mathbb{R})}^2. \tag{2.33}$$

Then (2.31) is a direct consequence of Lemma 2.4.

We continue by proving (2.32). Observe that, from (2.12),

$$\begin{aligned}
\partial_t P_{2,\varepsilon}(t,x) &= \frac{\gamma}{2}\int_{\mathbb{R}} e^{-|x-y|} q_{\varepsilon} \partial_t q_{\varepsilon} dy \\
&= \frac{\gamma}{2}\int_{\mathbb{R}} e^{-|x-y|} \Big(-\gamma q_{\varepsilon} u_{\varepsilon} \partial_x q_{\varepsilon} + \varepsilon q_{\varepsilon} \partial_{xx}^2 q_{\varepsilon} \\
&\quad -\frac{\gamma}{2}q_{\varepsilon}^3 + q_{\varepsilon}(h(u_{\varepsilon}) - P_{\varepsilon} + \partial_x F_{\varepsilon}) \Big) dy.
\end{aligned} \tag{2.34}$$

Using

$$\frac{\gamma}{2}\partial_x(u_{\varepsilon}q_{\varepsilon}^2) = \frac{\gamma}{2}q_{\varepsilon}^3 + \gamma q_{\varepsilon} u_{\varepsilon} \partial_x q_{\varepsilon}, \qquad \partial_x(q_{\varepsilon}\partial_x q_{\varepsilon}) = q_{\varepsilon}\partial_{xx}^2 q_{\varepsilon} + (\partial_x q_{\varepsilon})^2,$$

(2.34), and integration by parts, we get

$$\begin{aligned}
\partial_t P_{2,\varepsilon}(t,x) &= \frac{\gamma}{4}\int_{\mathbb{R}} e^{-|x-y|} \Big(-\frac{\gamma}{2}\partial_x(u_{\varepsilon}q_{\varepsilon}^2) + \varepsilon\partial_x(q_{\varepsilon}\partial_x q_{\varepsilon}) \\
&\quad -\varepsilon(\partial_x q_{\varepsilon})^2 + q_{\varepsilon}(h(u_{\varepsilon}) - P_{\varepsilon} + \partial_x F_{\varepsilon}) \Big) dy \\
&= \frac{\gamma}{4}\int_{\mathbb{R}} e^{-|x-y|} \Big(\text{sign}(y-x)[\frac{\gamma}{2}u_{\varepsilon}q_{\varepsilon}^2 - \varepsilon q_{\varepsilon}\partial_x q_{\varepsilon}] \\
&\quad -\varepsilon(\partial_x q_{\varepsilon})^2 + q_{\varepsilon}(h(u_{\varepsilon}) - P_{\varepsilon} + \partial_x F_{\varepsilon}) \Big) dy.
\end{aligned}$$

Using (2.2), (2.13), (2.11), Lemma 2.2, the Tonelli theorem, and the Hölder inequality,

$$\int_{\mathbb{R}\times\mathbb{R}} e^{-|x-y|} |u_\varepsilon| q_\varepsilon^2 dxdy \le \frac{e^{Lt}}{\sqrt{2}} \|u_0\|_{H^1(\mathbb{R})} \|u_\varepsilon(t,\cdot)\|_{H^1(\mathbb{R})}^2$$
$$\le \frac{e^{3Lt}}{\sqrt{2}} \|u_0\|_{H^1(\mathbb{R})}^3,$$

$$\varepsilon \int_0^T \int_{\mathbb{R}} \int_{\mathbb{R}} e^{-|x-y|} |q_\varepsilon| |\partial_x q_\varepsilon| dtdxdy$$
$$\le \frac{\varepsilon}{2} \int_0^T \|u_\varepsilon(t,\cdot)\|_{H^1(\mathbb{R})}^2 dt$$
$$+ \frac{\varepsilon}{2} \int_0^T \|\partial_x u_\varepsilon(t,\cdot)\|_{H^1(\mathbb{R})}^2 dt$$
$$\le \frac{e^{2Lt}}{2} \left(\varepsilon T + \frac{1}{2}\right) \|u_0\|_{H^1(\mathbb{R})}^2,$$

$$\varepsilon \int_0^T \int_{\mathbb{R}} \int_{\mathbb{R}} e^{-|x-y|} (\partial_x q_\varepsilon)^2 dtdxdy \le 2\varepsilon \int_0^T \|\partial_x u_\varepsilon(t,\cdot)\|_{H^1(\mathbb{R})}^2 dt \le e^{2Lt} \|u_0\|_{H^1(\mathbb{R})}^2,$$

$$\int_{\mathbb{R}\times\mathbb{R}} e^{-|x-y|} |q_\varepsilon| |h(u_\varepsilon)| dxdy \le \frac{1}{2} \int_{\mathbb{R}} q_\varepsilon^2 dy + \frac{1}{2} \max_{|\xi| \le e^{2Lt} \|u_0\|_{H^1(\mathbb{R})}/\sqrt{2}} (h'(\xi))^2 \int_{\mathbb{R}} u_\varepsilon^2 dy$$
$$\le \frac{e^{2Lt}}{2} \left(1 + \max_{|\xi| \le e^{Lt} \|u_0\|_{H^1(\mathbb{R})}/\sqrt{2}} (h'(\xi))^2\right) \|u_0\|_{H^1(\mathbb{R})}^2,$$

$$\int_{\mathbb{R}\times\mathbb{R}} e^{-|x-y|} |q_\varepsilon| (|P_\varepsilon| + |\partial_x F_\varepsilon|) dxdy$$
$$\le \frac{1}{2} \|u_\varepsilon(t,\cdot)\|_{H^1(\mathbb{R})}^2 + \|P_\varepsilon(t,\cdot)\|_{L^\infty(\mathbb{R})} \|P_\varepsilon(t,\cdot)\|_{L^1(\mathbb{R})} + \|\partial_x F_\varepsilon(t,\cdot)\|_{L^2(\mathbb{R})}^2$$
$$\le \frac{e^{2Lt}}{2} \left(1 + \frac{(M+\gamma)^2}{4} e^{2Lt} \|u_0\|_{H^1(\mathbb{R})}^2 + L^2\right) \|u_0\|_{H^1(\mathbb{R})}^2.$$

It follows from these estimates that (2.32) holds.

Since the bound on $\{\partial_t P_\varepsilon\}_\varepsilon$ is a consequence of (2.31) and (2.32), the family $\{P_\varepsilon\}_\varepsilon$ is bounded in $W_{loc}^{1,1}([0,\infty) \times \mathbb{R})$. $\qquad\square$

Lemma 2.6 *The family $\{F_\varepsilon\}_{\varepsilon>0}$ is uniformly bounded in $H^1((0,T); H^1(\mathbb{R}))$, $T > 0$.*

Proof Thanks to Lemma 2.2 we have to prove that $\{\partial_t F_\varepsilon\}_{\varepsilon>0}$ and $\{\partial_{tx}^2 F_\varepsilon\}_{\varepsilon>0}$ are uniformly bounded in $L^\infty((0,T); L^2(\mathbb{R}))$.

We know

$$\partial_t F_\varepsilon(t,x) = \frac{1}{2} \int_{\mathbb{R}} e^{-|x-y|} \left(\partial_t f(t,x,u_\varepsilon(t,y)) + \partial_u f(t,x,u_\varepsilon(t,y))\partial_t u_\varepsilon(t,y)\right) dy,$$

$$\partial_{tx}^2 F_\varepsilon(t,x) = \frac{1}{2} \int_{\mathbb{R}} e^{-|x-y|} \text{sign}\,(y-x) \left(\partial_t f(t,x,u_\varepsilon(t,y))\right.$$
$$\left. + \partial_u f(t,x,u_\varepsilon(t,y))\partial_t u_\varepsilon(t,y)\right) dy,$$

and then

$$|\partial_t F_\varepsilon(t,x)|, |\partial_{tx}^2 F_\varepsilon(t,x)| \le \frac{L}{2} \int_{\mathbb{R}} e^{-|x-y|} \left(|u_\varepsilon(t,y)| + |\partial_t u_\varepsilon(t,y)|\right) dy.$$

Since

$$
\begin{aligned}
& \int_{\mathbb{R}} |\partial_t\, F_\varepsilon(t,x)|^2 dx,\ \int_{\mathbb{R}} |\partial_{tx}^2 F_\varepsilon(t,x)|^2 dx \\
& \quad \le \tfrac{1}{4} \int_{\mathbb{R}} \left(\int_{\mathbb{R}} e^{-|x-y|} \big(|u_\varepsilon(t,y)| + |\partial_t u_\varepsilon(t,y)| \big) dy \right)^2 dx \\
& \quad \le \tfrac{1}{4} \left(\int_{\mathbb{R}} e^{-|x-y|} dy \right) \left(\int_{\mathbb{R}\times\mathbb{R}} e^{-|x-y|} \big(|u_\varepsilon(t,y)| + |\partial_t u_\varepsilon(t,y)| \big)^2 dy\, dx \right) \\
& \quad \le 2 \int_{\mathbb{R}} |u_\varepsilon(t,y)|^2 dy + 2 \int_{\mathbb{R}} |\partial_t u_\varepsilon(t,y)|^2 dy,
\end{aligned}
$$

the claim follows from Lemmas 2.1 and 2.4.   □

2.2.2 Oleĭnik Estimate

The main result of this subsection is the following one side estimate.

Lemma 2.7 *For each $0 < t \le T$ and $x \in \mathbb{R}$,*

$$
\partial_x u_\varepsilon(t,x) \le \frac{2}{\gamma t} + C_T, \tag{2.35}
$$

where

$$
C_T := \sqrt{\frac{2}{\gamma}} \left(\max_{|\xi| \le \frac{e^{LT}}{\sqrt{2}} \|u_0\|_{H^1(\mathbb{R})}} |h(\xi)| + \frac{M+\gamma}{2} e^{2LT} \|u_0\|_{H^1(\mathbb{R})}^2 + \frac{L}{\sqrt{2}} e^{LT} \|u_0\|_{H^1(\mathbb{R})} \right)^{1/2}.
$$

Proof From (2.11), (2.17), and (2.24)

$$
\begin{aligned}
\|h(u_\varepsilon) & - P_\varepsilon + \partial_x F_\varepsilon\|_{L^\infty((0,T)\times\mathbb{R})} \\
& \le \max_{|\xi| \le \frac{e^{LT}}{\sqrt{2}} \|u_0\|_{H^1(\mathbb{R})}} |h(\xi)| \\
& \quad + \frac{M+\gamma}{2} e^{2LT} \|u_0\|_{H^1(\mathbb{R})}^2 + \frac{L}{\sqrt{2}} e^{LT} \|u_0\|_{H^1(\mathbb{R})} =: C_T'.
\end{aligned} \tag{2.36}
$$

So, from (2.12), we have

$$
\partial_t q_\varepsilon + \gamma u_\varepsilon \partial_x q_\varepsilon - \varepsilon \partial_{xx}^2 q_\varepsilon + \frac{\gamma}{2} q_\varepsilon^2 \le C_T'. \tag{2.37}
$$

Let $U = U(t)$ be the solution of

$$
\frac{dU}{dt} + \frac{\gamma}{2} U^2 = C_T', \quad 0 < t \le T, \qquad U(0) = \|\partial_x u_{\varepsilon,0}\|_{L^\infty(\mathbb{R})}. \tag{2.38}
$$

Since, $U = U(t)$ is a super-solution of the parabolic initial value problem (2.12), due to the comparison principle for parabolic equations, we get

$$q_\varepsilon(t, x) \leq U(t), \qquad 0 < t \leq T, \ x \in \mathbb{R}. \qquad (2.39)$$

Finally, consider the map

$$\overline{U}(t) := \frac{2}{\gamma t} + \sqrt{\frac{2}{\gamma} C_T'}, \qquad 0 < t \leq T.$$

Observe that

$$\frac{d\overline{U}}{dt}(t) + \frac{\gamma}{2}\overline{U}^2(t) - C_T' = \frac{2}{t}\sqrt{2\frac{C_T'}{\gamma}} > 0, \qquad 0 < t \leq T,$$

so that $\overline{U} = \overline{U}(t)$ is a super-solution of (2.38). Due to the comparison principle for ordinary differential equations, we get $U(t) \leq \overline{U}(t)$ for all $0 < t \leq T$. Therefore, by this and (2.39), the estimate (2.35) is proved.   □

2.2.3 Higher Integrability Estimate

The main result of this subsection is the following higher integrability estimate.

Lemma 2.8 *Let* $0 < \alpha < 1$, $T > 0$, *and* $a, b \in \mathbb{R}$, $a < b$. *Then there exists a positive constant* C *depending only on* $\|u_0\|_{H^1(\mathbb{R})}$, α, T, a, *and* b, *but independent on* ε, *such that*

$$\int_0^T \int_a^b |\partial_x u_\varepsilon(t, x)|^{2+\alpha} \, dt dx \leq C, \qquad (2.40)$$

where $u_\varepsilon = u_\varepsilon(t, x)$ *is the unique solution of (2.8).*

Proof The proof is a variant of the proof found in Xin and Zhang [23]. Let $\chi \in C^\infty(\mathbb{R})$ be a cut-off function such that

$$0 \leq \chi \leq 1, \qquad \chi(x) = \begin{cases} 1, & \text{if } x \in [a, b], \\ 0, & \text{if } x \in (-\infty, a - 1] \cup [b + 1, \infty). \end{cases}$$

Consider also the map

$$\theta(\xi) := \xi(|\xi| + 1)^\alpha, \qquad \xi \in \mathbb{R},$$

and observe that, since $0 < \alpha < 1$,

$$
\begin{aligned}
\theta'(\xi) &= ((\alpha + 1)|\xi| + 1)(|\xi| + 1)^{\alpha-1}, \\
\theta''(\xi) &= \alpha \, \text{sign}\,(\xi) \left(|\xi| + 1\right)^{\alpha-2}((\alpha + 1)|\xi| + 2) \\
&= \alpha(\alpha+1)\,\text{sign}\,(\xi)\left(|\xi|+1\right)^{\alpha-1} + (1-\alpha)\alpha\,\text{sign}\,(\xi)\left(|\xi|+1\right)^{\alpha-2}, \\
|\theta(\xi)| &\le |\xi|^{\alpha+1} + |\xi|, \quad |\theta'(\xi)| \le (\alpha + 1)|\xi| + 1, \quad |\theta''(\xi)| \le 2\alpha, \\
\xi\theta(\xi) - \tfrac{1}{2}\xi^2\theta'(\xi) &= \tfrac{1-\alpha}{2}\xi^2\left(|\xi| + 1\right)^{\alpha} + \tfrac{\alpha}{2}\xi^2\left(|\xi| + 1\right)^{\alpha-1} \\
&\ge \tfrac{1-\alpha}{2}\xi^2\left(|\xi| + 1\right)^{\alpha}.
\end{aligned}
\tag{2.41}
$$

Multiplying (2.12) by $\chi\theta'(q_\varepsilon)$, using the chain rule, and integrating over $(0, T) \times \mathbb{R}$, we get

$$
\begin{aligned}
\int_0^T \int_{\mathbb{R}} \gamma\chi(x)q_\varepsilon\theta(q_\varepsilon)dtdx &- \tfrac{\gamma}{2}\int_0^T\int_{\mathbb{R}} q_\varepsilon^2\chi(x)\theta'(q_\varepsilon)dtdx \\
= \int_{\mathbb{R}}\chi(x)\big(\theta(q_\varepsilon(T,x))&-\theta(q_\varepsilon(0,x))\big)\,dx - \int_0^T\int_{\mathbb{R}}\gamma u_\varepsilon\chi'(x)\theta(q_\varepsilon)dtdx \\
+\varepsilon\int_0^T\int_{\mathbb{R}}\partial_x q_\varepsilon\chi'(x)&\theta'(q_\varepsilon)dtdx + \varepsilon\int_0^T\int_{\mathbb{R}}(\partial_x q_\varepsilon)^2\,\chi(x)\theta''(q_\varepsilon)dtdx \\
- \int_0^T\int_{\mathbb{R}}(h(u_\varepsilon)&- P_\varepsilon + \partial_x F_\varepsilon)\,\chi(x)\theta'(q_\varepsilon)dtdx.
\end{aligned}
\tag{2.42}
$$

Observe that, by (2.41),

$$
\begin{aligned}
\int_0^T\int_{\mathbb{R}}\gamma\chi(x)q_\varepsilon\theta(q_\varepsilon)dtdx &- \tfrac{\gamma}{2}\int_0^T\int_{\mathbb{R}}q_\varepsilon^2\chi(x)\theta'(q_\varepsilon)dtdx \\
&= \int_0^T\int_{\mathbb{R}}\gamma\chi(x)\Big(q_\varepsilon\theta(q_\varepsilon) - \tfrac{1}{2}q_\varepsilon^2\theta'(q_\varepsilon)\Big)dtdx \\
&\ge \tfrac{\gamma(1-\alpha)}{2}\int_0^T\int_{\mathbb{R}}\chi(x)q_\varepsilon^2\left(|q_\varepsilon| + 1\right)^{\alpha}dtdx.
\end{aligned}
\tag{2.43}
$$

Let $t \ge 0$, since $0 < \alpha < 1$, using the Hölder inequality, (2.11) and the first part of (2.41),

$$
\begin{aligned}
\left|\int_{\mathbb{R}}\chi(x)\theta(q_\varepsilon)dx\right| &\le \int_{\mathbb{R}}\chi(x)\left(|q_\varepsilon|^{\alpha+1} + |q_\varepsilon|\right)dx \\
&\le \|\chi\|_{L^{2/(1-\alpha)}(\mathbb{R})}\|q_\varepsilon(t,\cdot)\|_{L^2(\mathbb{R})}^{\alpha+1} + \|\chi\|_{L^2(\mathbb{R})}\|q_\varepsilon(t,\cdot)\|_{L^2(\mathbb{R})} \\
&\le (b-a+2)^{(1-\alpha)/2}\|u_0\|_{H^1(\mathbb{R})}^{\alpha+1}\,e^{L(\alpha+1)t} \\
&\qquad\qquad\qquad\qquad + (b-a+2)^{1/2}e^{Lt}\|u_0\|_{H^1(\mathbb{R})},
\end{aligned}
\tag{2.44}
$$

and

$$
\begin{aligned}
\left|\int_0^T\int_{\mathbb{R}}\gamma u_\varepsilon\chi'(x)\theta(q_\varepsilon)dtdx\right| &\le \int_0^T\int_{\mathbb{R}}\gamma|u_\varepsilon||\chi'(x)|\left(|q_\varepsilon|^{\alpha+1} + |q_\varepsilon|\right)dtdx \\
&\le \int_0^T\int_{\mathbb{R}}\gamma\,\|u_\varepsilon(t,\cdot)\|_{L^\infty(\mathbb{R})}|\chi'(x)|\left(|q_\varepsilon|^{\alpha+1} + |q_\varepsilon|\right)dtdx \\
&\le \gamma\frac{\|u_0\|_{H^1(\mathbb{R})}}{\sqrt{2}}e^{Lt}\int_0^T\left(\|\chi'\|_{L^{2/(1-\alpha)}(\mathbb{R})}\|q_\varepsilon(t,\cdot)\|_{L^2(\mathbb{R})}^{\alpha+1}\right. \\
&\qquad\qquad\qquad\qquad \left.+ \|\chi'\|_{L^2(\mathbb{R})}\|q_\varepsilon(t,\cdot)\|_{L^2(\mathbb{R})}\right)dt \\
&\le \gamma T\frac{\|u_0\|_{H^1(\mathbb{R})}}{\sqrt{2}}e^{Lt}\big(\|\chi'\|_{L^{2/(1-\alpha)}(\mathbb{R})}\|u_0\|_{H^1(\mathbb{R})}^{\alpha+1}\,e^{L(\alpha+1)t} \\
&\qquad\qquad\qquad\qquad + \|\chi'\|_{L^2(\mathbb{R})}\|u_0\|_{H^1(\mathbb{R})}\,e^{Lt}\big).
\end{aligned}
\tag{2.45}
$$

Moreover, observe that

$$\varepsilon \int_0^T \int_{\mathbb{R}} \partial_x q_\varepsilon \chi'(x)\theta'(q_\varepsilon)dtdx = -\varepsilon \int_0^T \int_{\mathbb{R}} \theta(q_\varepsilon)\chi''(x)dtdx,$$

so, again by the Hölder inequality, (2.11) and the first part of (2.41),

$$
\begin{aligned}
\left| \varepsilon \int_0^T \int_{\mathbb{R}} \frac{\partial q_\varepsilon}{\partial x} \chi'(x)\theta(q_\varepsilon)dtdx \right| &\leq \varepsilon \int_0^T \int_{\mathbb{R}} |\theta(q_\varepsilon)||\chi''(x)|dtdx \\
&\leq \varepsilon \int_0^T \int_{\mathbb{R}} \left(|q_\varepsilon|^{\alpha+1} + |q_\varepsilon| \right) |\chi''(x)|dtdx \\
&\leq \varepsilon \int_0^T \left(\|\chi''\|_{L^{2/(1-\alpha)}(\mathbb{R})} \|q_\varepsilon(t,\cdot)\|_{L^2(\mathbb{R})}^{\alpha+1} + \|\chi''\|_{L^2(\mathbb{R})} \|q_\varepsilon(t,\cdot)\|_{L^2(\mathbb{R})} \right) dt \\
&\leq \varepsilon T \left(\|\chi''\|_{L^{2/(1-\alpha)}(\mathbb{R})} \|u_0\|_{H^1(\mathbb{R})}^{\alpha+1} e^{(\alpha+1)Lt} + \|\chi''\|_{L^2(\mathbb{R})} \|u_0\|_{H^1(\mathbb{R})} e^{Lt} \right).
\end{aligned}
$$
(2.46)

Since $0 < \alpha < 1$, using (2.13) and the third part of (2.41),

$$\varepsilon \left| \int_0^T \int_{\mathbb{R}} \left(\frac{\partial q_\varepsilon}{\partial x} \right)^2 \chi(x)\theta''(q_\varepsilon)dtdx \right| \leq 2\alpha\varepsilon \int_0^T \int_{\mathbb{R}} (\partial_x q_\varepsilon)^2 \, dtdx \\ \leq \alpha \|u_0\|_{H^1(\mathbb{R})}^2 e^{2Lt}.$$
(2.47)

Thanks to (2.36)

$$
\begin{aligned}
\left| \int_0^T \int_{\mathbb{R}} (h(u_\varepsilon) - P_\varepsilon + \partial_x F_\varepsilon) \chi(x)\theta'(q_\varepsilon)dtdx \right| \\
\leq C_T' \int_0^T \int_{\mathbb{R}} \chi(x) ((\alpha+1)|q_\varepsilon| + 1) \, dtdx \\
\leq C_T' \int_0^T \left((\alpha+1) \|\chi\|_{L^2(\mathbb{R})} \|q_\varepsilon(t,\cdot)\|_{L^2(\mathbb{R})} + \|\chi\|_{L^1(\mathbb{R})} \right) dt \\
\leq C_T' T \left((\alpha+1)(b-a+2)^{1/2}e^{Lt} \|u_0\|_{H^1(\mathbb{R})} + (b-a+2) \right).
\end{aligned}
$$
(2.48)

From (2.42), (2.43), (2.44), (2.45), (2.46), (2.47), and (2.48), there exists a constant $c > 0$ depending only on $\|u_0\|_{H^1(\mathbb{R})}$, α, $T > 0$, a, and b, but independent of ε, such that

$$\frac{\gamma(1-\alpha)}{2} \int_0^T \int_{\mathbb{R}} |q_\varepsilon|^2 \chi(x)(|q_\varepsilon| + 1)^\alpha dtdx \leq c.$$
(2.49)

Then

$$\int_0^T \int_a^b |\partial_x u_\varepsilon(t,x)|^{2+\alpha} \, dtdx \leq \int_0^T \int_{\mathbb{R}} |q_\varepsilon|\chi(x)(|q_\varepsilon| + 1)^{\alpha+1} \, dtdx \leq \frac{2c}{\gamma(1-\alpha)},$$

hence estimate (2.40) is proved. □

2.3 Basic Compactness

Lemma 2.9 *There exists a sequence $\{\varepsilon_j\}_{j\in\mathbb{N}}$ tending to zero and three functions*

$$
\begin{aligned}
u &\in L^\infty((0,T);H^1(\mathbb{R}))\cap H^1((0,T)\times\mathbb{R}), && \text{for each } T\geq 0,\\
P &\in L^\infty((0,T);W^{1,\infty}(\mathbb{R})), && \text{for each } T\geq 0,\\
F &\in L^\infty(0,T;H^2(\mathbb{R}))\cap L^\infty(0,T;W^{2,\infty}(\mathbb{R}))\cap H^1((0,T)\times\mathbb{R}), && \text{for each } T\geq 0,
\end{aligned}
$$

such that

$$
u_{\varepsilon_j} \rightharpoonup u \quad \text{in } H^1((0,T)\times\mathbb{R}), \text{ for each } T\geq 0, \tag{2.50}
$$

$$
u_{\varepsilon_j} \to u \quad \text{in } L^\infty_{\mathrm{loc}}((0,\infty)\times\mathbb{R}), \tag{2.51}
$$

$$
P_{\varepsilon_j} \to P \quad \text{strongly in } L^p_{\mathrm{loc}}((0,\infty)\times\mathbb{R}),\ 1\leq p<\infty, \tag{2.52}
$$

$$
F_{\varepsilon_j} \to F \quad \text{strongly in } L^p_{\mathrm{loc}}((0,\infty);W^{1,p}_{loc}(\mathbb{R})),\ 1\leq p<\infty. \tag{2.53}
$$

Proof Fix $T>0$. Lemmas 2.1 and 2.4 say that $\{u_\varepsilon\}_\varepsilon$ is uniformly bounded in $H^1((0,T)\times\mathbb{R})\cap L^\infty((0,T);H^1(\mathbb{R}))$, $T>0$, and (2.50) follows.

Observe that, for each $0\leq s,\,t\leq T$,

$$
\begin{aligned}
\|u_\varepsilon(t,\cdot)-u_\varepsilon(s,\cdot)\|^2_{L^2(\mathbb{R})} &= \int_{\mathbb{R}}\left(\int_s^t \partial_t u_\varepsilon(\tau,x)d\tau\right)^2 dx\\
&\leq \sqrt{|t-s|}\int_0^T\int_{\mathbb{R}}\left(\partial_t u_\varepsilon(\tau,x)\right)^2 d\tau dx.
\end{aligned}
$$

Moreover, $\{u_\varepsilon\}_\varepsilon$ is uniformly bounded in $L^\infty((0,T);H^1(\mathbb{R}))$ and $H^1(\mathbb{R})\subset\subset L^\infty_{\mathrm{loc}}(\mathbb{R})\subset L^2_{\mathrm{loc}}(\mathbb{R})$, then (2.51) is consequence of [22, Theorem 5].

Lemmas 2.2, 2.5, 2.3, and 2.6 give (2.52) and (2.53). □

Throughout this paper we use overbars to denote weak limits (the spaces in which these weak limits are taken should be clear from the context and thus they are not always explicitly stated).

Lemma 2.10 *There exists a sequence $\{\varepsilon_j\}_{j\in\mathbb{N}}$ tending to zero and two functions*

$$
q\in L^p_{\mathrm{loc}}((0,\infty)\times\mathbb{R}), \qquad \overline{q^2}\in L^r_{\mathrm{loc}}((0,\infty)\times\mathbb{R})
$$

such that

$$
q_{\varepsilon_j} \rightharpoonup q \quad \text{in } L^p_{\mathrm{loc}}((0,\infty)\times\mathbb{R}), \qquad q_{\varepsilon_j} \overset{*}{\rightharpoonup} q \quad \text{in } L^\infty_{\mathrm{loc}}((0,\infty);L^2(\mathbb{R})), \tag{2.54}
$$

$$
q^2_{\varepsilon_j} \rightharpoonup \overline{q^2} \quad \text{in } L^r_{\mathrm{loc}}((0,\infty)\times\mathbb{R}), \tag{2.55}
$$

for each $1 < p < 3$ *and* $1 < r < \frac{3}{2}$. *Moreover,*

$$q^2(t,x) \leq \overline{q^2}(t,x) \quad \text{for almost every } (t,x) \in [0,\infty) \times \mathbb{R} \qquad (2.56)$$

$$\partial_x u = q \quad \text{in the sense of distributions on } [0,\infty) \times \mathbb{R}. \qquad (2.57)$$

Proof Formulas (2.54) and (2.55) are direct consequences of Lemmas 2.1 and 2.8. Inequality (2.56) is true thanks to the weak convergence in (2.55). Finally, (2.57) is a consequence of the definition of q_ε, Lemma 2.9, and (2.54).   □

In the following, for notational convenience, we replace the sequences $\{u_{\varepsilon_j}\}_{j \in \mathbb{N}}$, $\{q_{\varepsilon_j}\}_{j \in \mathbb{N}}$, $\{P_{\varepsilon_j}\}_{j \in \mathbb{N}}$, $\{F_{\varepsilon_j}\}_{j \in \mathbb{N}}$ by $\{u_\varepsilon\}_{\varepsilon > 0}$, $\{q_\varepsilon\}_{\varepsilon > 0}$, $\{P_\varepsilon\}_{\varepsilon > 0}$, $\{F_\varepsilon\}_{\varepsilon > 0}$, respectively.

In view of (2.54), we conclude that for any $\eta \in C^1(\mathbb{R})$ with η' bounded, Lipschitz continuous on $\mathbb{R}$ and any $1 < p < 3$ we have

$$\begin{aligned}
\eta(q_\varepsilon) &\rightharpoonup \overline{\eta(q)}, \quad \text{in } L^p_{\text{loc}}((0,\infty) \times \mathbb{R}), \\
\eta(q_\varepsilon) &\rightharpoonup^* \overline{\eta(q)} \quad \text{in } L^\infty_{\text{loc}}((0,\infty); L^2(\mathbb{R})).
\end{aligned} \qquad (2.58)$$

Multiplying the equation in (2.12) by $\eta'(q_\varepsilon)$, we get

$$\begin{aligned}
&\partial_t \eta(q_\varepsilon) + \partial_x \left(\gamma u_\varepsilon \eta(q_\varepsilon)\right) - \varepsilon \partial^2_{xx} \eta(q_\varepsilon) - \varepsilon \eta''(q_\varepsilon) \left(\partial_x \eta(q_\varepsilon)\right)^2 \\
&= \gamma q_\varepsilon \eta(q_\varepsilon) - \frac{\gamma}{2} \eta'(q_\varepsilon) q_\varepsilon^2 + (h(u_\varepsilon) - P_\varepsilon + \partial_x F_\varepsilon) \, \eta'(q_\varepsilon).
\end{aligned} \qquad (2.59)$$

Lemma 2.11 *For any convex* $\eta \in C^1(\mathbb{R})$ *with* η' *bounded, Lipschitz continuous on* $\mathbb{R}$, *we have*

$$\partial_t \overline{\eta(q)} + \partial_x \left(\gamma u \overline{\eta(q)}\right) \leq \gamma \overline{q\eta(q)} - \frac{\gamma}{2} \overline{\eta'(q)q^2} + (h(u) - P + \partial_x F) \overline{\eta'(q)}, \qquad (2.60)$$

in the sense of distributions on $[0,\infty) \times \mathbb{R}$. *Here* $\overline{q\eta(q)}$ *and* $\overline{\eta'(q)q^2}$ *denote the weak limits of* $q_\varepsilon \eta(q_\varepsilon)$ *and* $\eta'(q_\varepsilon) q_\varepsilon^2$ *in* $L^r_{\text{loc}}((0,\infty) \times \mathbb{R})$, $1 < r < \frac{3}{2}$, *respectively.*

Proof In (2.59), by convexity of η, (2.2), (2.51), (2.54), and (2.55), sending $\varepsilon \to 0$ yields (2.60).   □

Remark 2.3 From (2.54) and (2.55), it is clear that

$$q = q_+ + q_- = \overline{q_+} + \overline{q_-}, \quad q^2 = (q_+)^2 + (q_-)^2, \quad \overline{q^2} = \overline{(q_+)^2} + \overline{(q_-)^2},$$

almost everywhere in $[0,\infty) \times \mathbb{R}$, where $\xi_+ := \xi \chi_{[0,+\infty)}(\xi)$, $\xi_- := \xi \chi_{(-\infty,0]}(\xi)$, $\xi \in \mathbb{R}$. Moreover, by (2.35) and (2.54),

$$q_\varepsilon(t,x), q(t,x) \leq \frac{2}{\gamma t} + C_T, \qquad 0 < t \leq T, \ x \in \mathbb{R}. \qquad (2.61)$$

Lemma 2.12 *There holds*

$$\partial_t q + \partial_x (\gamma u q) = \frac{\gamma}{2}\overline{q^2} + h(u) - P + \partial_x F \tag{2.62}$$

in the sense of distributions on $[0, \infty) \times \mathbb{R}$.

Proof Using (2.12), (2.51), (2.52), (2.54), and (2.55), the result (2.62) follows by $\varepsilon \to 0$ in (2.12).   □

The next lemma contains a renormalized formulation of (2.62).

Lemma 2.13 *For any* $\eta \in C^1(\mathbb{R})$ *with* $\eta' \in L^\infty(\mathbb{R})$,

$$
\begin{aligned}
\partial_t \eta(q) &+ \partial_x (\gamma u \eta(q)) \\
&= \gamma q \eta(q) + \left(\tfrac{\gamma}{2}\overline{q^2} - \gamma q^2 \right) \eta'(q) + (h(u) - P + \partial_x F)\, \eta'(q),
\end{aligned} \tag{2.63}
$$

in the sense of distributions on $[0, \infty) \times \mathbb{R}$.

Proof Let $\{\omega_\delta\}_\delta$ be a family of mollifiers defined on $\mathbb{R}$. Denote $q_\delta(t, x) := (q(t, \cdot) \star \omega_\delta)(x)$. Here and in the following all convolutions are with respect to the x variable. According to Lemma II.1 of [16], it follows from (2.62) that q_δ solves

$$\partial_t q_\delta + \gamma u \partial_x q_\delta = \frac{\gamma}{2}\overline{q^2} \star \omega_\delta - \gamma q^2 \star \omega_\delta + h(u) \star \omega_\delta - P \star \omega_\delta + \partial_x F \star \omega_\delta + \rho_\delta, \tag{2.64}$$

where the error ρ_δ tends to zero in $L^1_{\mathrm{loc}}([0, \infty) \times \mathbb{R})$. Multiplying (2.64) by $\eta'(q_\delta)$, we get

$$
\begin{aligned}
\partial_t\, \eta(q_\delta) &+ \partial_x (\gamma u \eta(q_\delta)) \\
&= q \eta(q_\delta) + \tfrac{\gamma}{2} \left(\overline{q^2} \star \omega_\delta \right) \eta'(q_\delta) - \gamma \left(q^2 \star \omega_\delta \right) \eta'(q_\delta) \\
&\quad + (h(u) \star \omega_\delta)\, \eta'(q_\delta) - (P \star \omega_\delta)\, \eta'(q_\delta) + (\partial_x F \star \omega_\delta)\, \eta'(q_\delta) + \rho_\delta \eta'(q_\delta).
\end{aligned} \tag{2.65}
$$

Using the boundedness of η, η', we can send $\delta \to 0$ in (2.65) to obtain (2.63). The weak time continuity is standard.   □

2.4   Proof of Theorem 2.1

Following [23], in this section we wish to improve the weak convergence of q_ε in (2.54) to strong convergence (and then we have an existence result for (2.1)). Roughly speaking, the idea is to derive a "transport equation" for the evolution of the defect measure $\left(\overline{q^2} - q^2 \right)(t, \cdot) \geq 0$, so that if it is zero initially then it will continue to be zero at all later times $t > 0$. The proof is complicated by the fact that we do not have a uniform bound on q_ε from below but merely (2.61) and that in Lemma 2.8 we have only $\alpha < 1$.

Lemma 2.14 *There holds*

$$\lim_{t\to 0+} \int_{\mathbb{R}} q^2(t,x)dx = \lim_{t\to 0+} \int_{\mathbb{R}} \overline{q^2}(t,x)dx = \int_{\mathbb{R}} (\partial_x u_0)^2\,dx. \tag{2.66}$$

Proof Since $u \in C(\mathbb{R}_+ \times \mathbb{R})$ (see Lemma 2.9), from (2.57),

$$\lim_{t\to 0} \int_{\mathbb{R}} q(t,x)\varphi(x)dx = -\lim_{t\to 0} \int_{\mathbb{R}} u(t,x)\partial_x\varphi(x)dx =$$
$$= -\int_{\mathbb{R}} u_0(x)\partial_x\varphi(x)dx = \int_{\mathbb{R}} \partial_x u_0(x)\varphi(x)dx,$$

for each test function $\varphi \in C^\infty(\mathbb{R})$ with compact support. Due to the boundedness of $\{q_\varepsilon\}_{\varepsilon>0}$ in $L^\infty((0,\infty); L^2(\mathbb{R}))$ we get

$$q(t,\cdot) \rightharpoonup \partial_x u_0 \qquad \text{weakly in } L^2(\mathbb{R}) \text{ as } t\to 0+,$$

so

$$\liminf_{t\to 0+} \int_{\mathbb{R}} q^2(t,x)dx \geq \int_{\mathbb{R}} \left(\partial_x u_0(x)\right)^2 dx. \tag{2.67}$$

Moreover, from (2.9), (2.13), (2.51), and (2.55),

$$\int_{\mathbb{R}} u^2(t,x)dx + \int_{\mathbb{R}} \overline{q^2}(t,x)dx \leq \int_{\mathbb{R}} u_0^2(x)dx + \int_{\mathbb{R}} (\partial_x u_0)^2\,dx,$$

and, again using the continuity of u (see Lemma 2.9),

$$\lim_{t\to 0+} \int_{\mathbb{R}} u^2(t,x)dx = \int_{\mathbb{R}} u_0^2 dx.$$

Hence

$$\limsup_{t\to 0+} \int_{\mathbb{R}} \overline{q^2}(t,x)dx \leq \int_{\mathbb{R}} (\partial_x u_0)^2\,dx. \tag{2.68}$$

Clearly, (2.56), (2.67), and (2.68) imply (2.66). □

Lemma 2.15 *For each $R > 0$,*

$$\lim_{t\to 0+} \int_{\mathbb{R}} \left(\eta_R^\mp(q)(t,x) - \eta_R^\pm(q(t,x)) \right) dx = 0, \tag{2.69}$$

where

$$\eta_R(\xi) := \begin{cases} \dfrac{1}{2}\xi^2, & \text{if } |\xi| \leq R, \\[2mm] R|\xi| - \dfrac{1}{2}R^2, & \text{if } |\xi| > R, \end{cases} \qquad (2.70)$$

and

$$\eta_R^+(\xi) := \eta_R(\xi)\chi_{[0,+\infty)}(\xi), \quad \eta_R^-(\xi) := \eta_R(\xi)\chi_{(-\infty,0]}(\xi),$$

for every $\xi \in \mathbb{R}$.

Proof Let $R > 0$. Observe that

$$\overline{\eta_R(q)} - \eta_R(q) = \frac{1}{2}(\overline{q^2} - q^2) - \left(\overline{f_R(q)} - f_R(q)\right),$$

where $f_R(\xi) := \frac{1}{2}\xi^2 - \eta_R(\xi)$, $\xi \in \mathbb{R}$. Since η_R and f_R are convex,

$$0 \leq \overline{\eta_R(q)} - \eta_R(q) = \frac{1}{2}\left(\overline{q^2} - q^2\right) - \left(\overline{f_R(q)} - f_R(q)\right) \leq \frac{1}{2}\left(\overline{q^2} - q^2\right).$$

Then, from (2.66), $\displaystyle\lim_{t \to 0+} \int_{\mathbb{R}} \left(\overline{\eta_R(q)}(t, x) - \eta_R(q(t, x))\right) dx = 0$. Since, $\overline{\eta_R^\pm(q)} - \eta_R^\pm(q) \leq \overline{\eta_R(q)} - \eta_R(q)$, the proof is done. $\qquad\square$

Remark 2.4 Let $R > 0$. Then for each $\xi \in \mathbb{R}$

$$\eta_R(\xi) = \tfrac{1}{2}\xi^2 - \tfrac{1}{2}(R - |\xi|)^2\chi_{(-\infty,-R)\cup(R,\infty)}(\xi),$$
$$\eta_R'(\xi) = \xi + (R - |\xi|)\,\text{sign}\,(\xi)\,\chi_{(-\infty,-R)\cup(R,\infty)}(\xi),$$
$$\eta_R^+(\xi) = \tfrac{1}{2}(\xi_+)^2 - \tfrac{1}{2}(R - \xi)^2\chi_{(R,\infty)}(\xi),$$
$$(\eta_R^+)'(\xi) = \xi_+ + (R - \xi)\chi_{(R,\infty)}(\xi),$$
$$\eta_R^-(\xi) = \tfrac{1}{2}(\xi_-)^2 - \tfrac{1}{2}(R + \xi)^2\chi_{(-\infty,-R)}(\xi),$$
$$(\eta_R^-)'(\xi) = \xi_- - (R + \xi)\chi_{(-\infty,-R)}(\xi).$$

Lemma 2.16 *Assume (2.2) and (2.9). Then for almost all $t \geq 0$*

$$\int_{\mathbb{R}} \left(\overline{(q_+)^2} - (q_+)^2\right)(t, x)dx \leq 2\int_0^t \int_{\mathbb{R}} S(s, x)\,[\overline{q_+}(s, x) - q_+(s, x)]\,dsdx, \qquad (2.71)$$

where $S(s, x) := h\big(u(s, x)\big) - P(s, x) + \partial_x F(s, x)$.

Proof Let $0 < t \leq T$ and $R > C_T$ (see Lemma 2.7). Subtract (2.63) from (2.60) using the entropy η_R^+ (see Lemma 2.15). The result is

$$\partial_t \left(\overline{\eta_R^+(q)} - \eta_R^+(q)\right) + \partial_x \left(\gamma u \left[\overline{\eta_R^+(q)} - \eta_R^+(q)\right]\right)$$
$$\leq \gamma \left[\overline{q\eta_R^+(q)} - q\eta_R^+(q)\right] - \tfrac{\gamma}{2}\left[\overline{q^2(\eta_R^+)'(q)} - q^2(\eta_R^+)'(q)\right] \qquad (2.72)$$
$$- \tfrac{\gamma}{2}\left(\overline{q^2} - q^2\right)(\eta_R^+)'(q) + S(t,x)\left[\overline{(\eta_R^+)'(q)} - (\eta_R^+)'(q)\right].$$

Since η_R^+ is increasing and $\gamma \geq 0$, by (2.56),

$$-\frac{\gamma}{2}\left(\overline{q^2} - q^2\right)(\eta_R^+)'(q) \leq 0. \qquad (2.73)$$

Moreover, from Remark 2.4,

$$\gamma q\eta_R^+(q) - \tfrac{\gamma}{2}q^2(\eta_R^+)'(q) = -\tfrac{\gamma R}{2}q(R - q)\chi_{(R,\infty)}(q),$$
$$\overline{\gamma q\eta_R^+(q) - \tfrac{\gamma}{2}q^2(\eta_R^+)'(q)} = -\tfrac{\gamma R}{2}\overline{q(R - q)\chi_{(R,\infty)}(q)}.$$

Therefore, due to (2.61),

$$\gamma q\eta_R^+(q) - \tfrac{\gamma}{2}q^2(\eta_R^+)'(q) = \overline{q\eta_R^+(q)} - \tfrac{1}{2}\overline{q^2(\eta_R^+)'(q)} = 0,$$
$$\text{in } \Omega_R := \left(\tfrac{2}{R-C_T}, \infty\right) \times \mathbb{R}. \qquad (2.74)$$

Then from (2.72)–(2.74) the following inequality holds in Ω_R:

$$\partial_t \left(\overline{\eta_R^+(q)} - \eta_R^+(q)\right) + \partial_x \left(\gamma u \left[\overline{\eta_R^+(q)} - \eta_R^+(q)\right]\right)$$
$$\leq S(t,x)\left[\overline{(\eta_R^+)'(q)} - (\eta_R^+)'(q)\right]. \qquad (2.75)$$

In view of Remark 2.3 and due to (2.61),

$$\eta_R^+(q) = \frac{1}{2}(q_+)^2, \quad (\eta_R^+)'(q) = q_+, \quad \overline{\eta_R^+(q)} = \frac{1}{2}\overline{(q_+)^2}, \quad \overline{(\eta_R^+)'(q)} = \overline{q_+}, \quad \text{in } \Omega_R.$$

Inserting this into (2.75) and integrating the result over $(\frac{2}{R-C_T}, t) \times \mathbb{R}$ gives

$$\frac{1}{2}\int_{\mathbb{R}} \left[\overline{(q_+)^2}(t,x) - (q_+(t,x))^2)\right] dx$$
$$\leq \int_{\mathbb{R}} \left[\overline{\eta_R^+(q)}(\frac{2}{R-C_T}, x) - \eta_R^+(q)(\frac{2}{R-C_T}, x)\right] dx$$
$$+ \int_{\frac{2}{R-C_T}}^{t} \int_{\mathbb{R}} S(s,x)\left[\overline{q_+}(s,x) - q_+(s,x)\right] ds\,dx,$$

for almost all $\frac{2}{R-C_T} < t \le T$. Sending $R \to \infty$ and using Lemma 2.15, we get (2.71). $\qquad\square$

Lemma 2.17 *For any $t \ge 0$ and any $R > 0$,*

$$\int_{\mathbb{R}} \left[\overline{\eta_R^-(q)} - \eta_R^-(q) \right](t,x)dx$$

$$\le \frac{\gamma R^2}{2} \int_0^t \int_{\mathbb{R}} \overline{(R+q)\chi_{(-\infty,-R)}(q)}dsdx$$

$$- \frac{\gamma R^2}{2} \int_0^t \int_{\mathbb{R}} (R+q)\chi_{(-\infty,-R)}(q)dsdx + \gamma R \int_0^t \int_{\mathbb{R}} \left[\overline{\eta_R^-(q)} - \eta_R^-(q) \right] dsdx$$

$$+ \frac{\gamma R}{2} \int_0^t \int_{\mathbb{R}} \left[\overline{(q_+)^2} - q_+^2 \right] dsdx + \int_0^t \int_{\mathbb{R}} S(s,x) \left[\overline{(\eta_R^-)'(q)} - (\eta_R^-)'(q) \right] dsdx.$$

Proof Let $R > 0$. By subtracting (2.63) from (2.60), using the entropy η_R^- (see Lemma 2.15), we deduce

$$\frac{\partial}{\partial t}\left(\overline{\eta_R^-(q)} - \eta_R^-(q) \right) + \partial_x \left(\gamma u \left[\overline{\eta_R^-(q)} - \eta_R^-(q) \right] \right)$$

$$\le \gamma \left[\overline{q\eta_R^-(q)} - q\eta_R^-(q) \right] - \frac{\gamma}{2} \left[\overline{q^2(\eta_R^-)'(q)} - q^2(\eta_R^-)'(q) \right] \qquad (2.76)$$

$$- \frac{\gamma}{2}(\overline{q^2} - q^2)(\eta_R^-)'(q) + S(t,x) \left[\overline{(\eta_R^-)'(q)} - (\eta_R^-)'(q) \right].$$

Since $-R \le (\eta_R^-)' \le 0$ and $\gamma \ge 0$, by (2.56),

$$- \frac{\gamma}{2}\left(\overline{q^2} - q^2 \right) (\eta_R^-)'(q) \le \frac{\gamma R}{2} \left(\overline{q^2} - q^2 \right). \qquad (2.77)$$

Using Remarks 2.3 and 2.4

$$\gamma q\eta_R^-(q) - \frac{\gamma}{2}q^2(\eta_R^-)'(q) = -\frac{\gamma R}{2}q(R+q)\chi_{(-\infty,-R)}(q), \qquad (2.78)$$

$$\overline{\gamma q\eta_R^-(q)} - \frac{\gamma}{2}\overline{q^2(\eta_R^-)'(q)} = -\frac{\gamma R}{2}\overline{q(R+q)\chi_{(-\infty,-R)}(q)}. \qquad (2.79)$$

Inserting (2.77)–(2.79) into (2.76) gives

$$\frac{\partial}{\partial t}\left(\overline{\eta_R^-(q)} - \eta_R^-(q) \right) + \partial_x \left(\gamma u \left[\overline{\eta_R^-(q)} - \eta_R^-(q) \right] \right)$$

$$\le -\frac{\gamma R}{2}\overline{q(R+q)\chi_{(-\infty,-R)}(q)} + \frac{\gamma R}{2}q(R+q)\chi_{(-\infty,-R)}(q)$$

$$+ \frac{\gamma R}{2}\left(\overline{q^2} - q^2 \right) + S(t,x)\left[\overline{(\eta_R^-)'(q)} - (\eta_R^-)'(q) \right].$$

Integrating this inequality over $(0, t) \times \mathbb{R}$ yields

$$\int_{\mathbb{R}} \left[\overline{\eta_R^-(q)} - \eta_R^-(q) \right](t, x) dx$$
$$\leq -\frac{\gamma R}{2} \int_0^t \int_{\mathbb{R}} \overline{q(R+q)\chi_{(-\infty,-R)}(q)} ds dx$$
$$+\frac{\gamma R}{2} \int_0^t \int_{\mathbb{R}} q(R+q)\chi_{(-\infty,-R)}(q) ds dx + \frac{R}{2} \int_0^t \int_{\mathbb{R}} \left[\overline{q^2} - q^2 \right] ds dx$$
$$+ \int_0^t \int_{\mathbb{R}} S(s, x) \left[\overline{(\eta_R^-)'(q)} - (\eta_R^-)'(q) \right] ds dx.$$

Using Remark 2.4,

$$\overline{\eta_R^-(q)} - \eta_R^-(q)$$
$$= \frac{1}{2} \left(\overline{(q_-)^2} - (q_-)^2 \right) + \frac{1}{2}(R+q)^2 \chi_{(-\infty,-R)}(q) - \frac{1}{2} \overline{(R+q)^2 \chi_{(-\infty,-R)}(q)}.$$

Hence, from Remark 2.3 and (2.80),

$$\int_{\mathbb{R}} \left[\overline{\eta_R^-(q)} - \eta_R^-(q) \right](t, x) dx$$
$$\leq -\frac{\gamma R}{2} \int_0^t \int_{\mathbb{R}} \overline{q(R+q)\chi_{(-\infty,-R)}(q)} ds dx$$
$$+\frac{\gamma R}{2} \int_0^t \int_{\mathbb{R}} q(R+q)\chi_{(-\infty,-R)}(q) ds dx + \gamma R \int_0^t \int_{\mathbb{R}} \left[\overline{\eta_R^-(q)} - \eta_R^-(q) \right] ds dx$$
$$-\frac{\gamma R}{2} \int_0^t \int_{\mathbb{R}} (R+q)^2 \chi_{(-\infty,-R)}(q) ds dx + \frac{\gamma R}{2} \int_0^t \int_{\mathbb{R}} \overline{(R+q)^2 \chi_{(-\infty,-R)}(q)} ds dx$$
$$+\frac{\gamma R}{2} \int_0^t \int_{\mathbb{R}} \left[\overline{(q_+)^2} - q_+^2 \right] ds dx + \int_0^t \int_{\mathbb{R}} S(s, x) \left[\overline{(\eta_R^-)'(q)} - (\eta_R^-)'(q) \right] ds dx,$$

and applying twice the identity

$$\frac{R}{2}(R+q)^2 - \frac{R}{2} q(R+q) = \frac{R^2}{2}(R+q),$$

we deduce (2.76). $\qquad\qquad\qquad\qquad\qquad\qquad\qquad\qquad\qquad\qquad\qquad\qquad$ □

Lemma 2.18 *There holds $\overline{q^2} = q^2$ almost everywhere in $[0, \infty) \times \mathbb{R}$.*

Proof Let $0 < t \leq T$. Adding (2.71) and (2.76) yields

$$\int_{\mathbb{R}} \left(\frac{1}{2} \left[\overline{(q_+)^2} - (q_+)^2 \right] + \left[\overline{\eta_R^-(q)} - \eta_R^-(q) \right] \right)(t, x) dx$$
$$\leq \frac{\gamma R^2}{2} \int_0^t \int_{\mathbb{R}} \overline{(R+q)\chi_{(-\infty,-R)}(q)} ds dx - \frac{\gamma R^2}{2} \int_0^t \int_{\mathbb{R}} (R+q)\chi_{(-\infty,-R)}(q) ds dx$$
$$+\gamma R \int_0^t \int_{\mathbb{R}} \left[\overline{\eta_R^-(q)} - \eta_R^-(q) \right] ds dx + \frac{\gamma R}{2} \int_0^t \int_{\mathbb{R}} \left[\overline{(q_+)^2} - q_+^2 \right] ds dx$$
$$+ \int_0^t \int_{\mathbb{R}} S(s, x) \left(\left[\overline{q_+} - q_+ \right] + \left[\overline{(\eta_R^-)'(q)} - (\eta_R^-)'(q) \right] \right) ds dx.$$

Arguing as in the proof of Lemma 2.7, there exists a constant $C_T' > 0$, depending only on T and on $\|u_0\|_{H^1(\mathbb{R})}$, such that

$$\|S\|_{L^\infty((0,T)\times\mathbb{R})} = \|h(u) - P + \partial_x F\|_{L^\infty((0,T)\times\mathbb{R})} \leq C_T'. \qquad (2.80)$$

By Remarks 2.3 and 2.4,

$$q_+ + (\eta_R^-)'(q) = q - (R+q)\chi_{(-\infty,-R)}(q),$$
$$\overline{q_+} + \overline{(\eta_R^-)'(q)} = q - (R+q)\chi_{(-\infty,-R)}(q),$$

so by the convexity of the map $\xi \mapsto \xi_+ + (\eta_R^-)'(\xi)$,

$$0 \le [\overline{q_+} - q_+] + \left[\overline{(\eta_R^-)'(q)} - (\eta_R^-)'(q)\right]$$
$$= (R+q)\chi_{(-\infty,-R)}(q) - \overline{(R+q)\chi_{(-\infty,-R)}(q)},$$

and, by (2.80),

$$S(s,x)\left(\left[\overline{q_+}(s,x) - q_+(s,x)\right] + \left[\overline{(\eta_R^-)'(q)} - (\eta_R^-)'(q)\right]\right)$$
$$\le -C_T'\left(\overline{(R+q)\chi_{(-\infty,-R)}(q)} - (R+q)\chi_{(-\infty,-R)}(q)\right).$$

Since $\xi \mapsto (R+\xi)\chi_{(-\infty,-R)}(\xi)$ is concave and choosing R large enough,

$$\frac{\gamma R^2}{2}\overline{(R+q)\chi_{(-\infty,-R)}(q)} - \frac{\gamma R^2}{2}(R+q)\chi_{(-\infty,-R)}(q)$$
$$+ S(s,x)\left(\left[\overline{q_+}(s,x) - q_+(s,x)\right] + \left[\overline{(\eta_R^-)'(q)} - (\eta_R^-)'(q)\right]\right)$$
$$\le \left(\frac{\gamma R^2}{2} - C_T'\right)\left(\overline{(R+q)\chi_{(-\infty,-R)}(q)} - (R+q)\chi_{(-\infty,-R)}(q)\right) \le 0.$$

Then, from (2.80) and (2.81),

$$0 \le \int_{\mathbb{R}}\left(\frac{1}{2}\left[\overline{(q_+)^2} - (q_+)^2\right] + \left[\overline{\eta_R(q)} - \eta_R(q)\right]\right)(t,x)dx$$
$$\le \gamma R \int_0^t \int_{\mathbb{R}}\left(\frac{1}{2}\left[\overline{(q_+)^2} - q_+^2\right] + \left[\overline{\eta_R(q)} - \eta_R(q)\right]\right)dsdx,$$

and using the Gronwall's inequality and Lemmas 2.14 and 2.15 we conclude that

$$\int_{\mathbb{R}}\left(\frac{1}{2}\left[\overline{(q_+)^2} - (q_+)^2\right] + \left[\overline{\eta_R(q)} - \eta_R(q)\right]\right)(t,x)dx = 0, \quad \text{for each } t > 0.$$

By the Fatou's lemma, Remark 2.3, and (2.56), sending $R \to \infty$ yields

$$0 \le \int_{\mathbb{R}}\left(\overline{q^2} - q^2\right)(t,x)dx \le 0, \qquad t > 0, \qquad (2.81)$$

and we see that the claim holds. $\qquad\qquad\qquad\qquad\qquad\qquad\qquad\qquad\qquad\qquad\square$

Proof (Proof of Theorem 2.1) The conditions (i), (iii) of Definition 2.1 are satisfied, due to (2.9), (2.13), and Lemma 2.9. We have to verify (ii). Due to Lemma 2.18, we have

$$q_\varepsilon \to q \qquad \text{in } L^2_{\text{loc}}((0, \infty) \times \mathbb{R}). \tag{2.82}$$

Clearly (2.51)–(2.53), and (2.82) imply that u is a distributional solution of (2.5). Finally, (iv) and (2.7) are consequence of Lemmas 2.7 and 2.8, respectively. □

References

1. Ancona, F., Coclite, G.M.: Asymptotic stabilization of weak solutions to a generalized hyperelastic-rod wave equation: dissipative semigroup (submitted)
2. Ancona, F., Coclite, G.M.: Hyperbolic Problems. In: Ancona, F., Bressan, A., Marcati, P., Marson, A. (eds.) Theory, Numerics, Applications. Proceedings of the 14th International Conference of Hyperbolic Problems (HYP2012) held at University of Padova, June 24–29, 2012. AIMS, pp. 447–454. Springerfield, MO (2014)
3. Bendahmane, M., Coclite, G.M., Karlsen, K.H.: H^1-perturbations of smooth solutions for a weakly dissipative hyperelastic-rod wave equation. Mediterr. J. Math. **3**, 417–430 (2006)
4. Bressan, A., Constantin, A.: Global conservative solutions of the Camassa–Holm equation. Arch. Ration. Mech. Anal. **183**, 215–239 (2007)
5. Bressan A., Constantin, A.: Global dissipative solutions of the Camassa–Holm equation. Anal. Appl. **5**, 1–27 (2007)
6. Camassa, R., Holm, D.D.: An integrable shallow water equation with peaked solitons. Phys. Rev. Lett. **71**, 1661–1664 (1993)
7. Coclite, G.M., Karlsen, K.H.: On an initial-boundary value problem for the hyperelastic rod wave equation. Adv. Differ. Equ. **17**, 37–74 (2012)
8. Coclite, G.M., Holden, H., Karlsen, K.H.: Wellposedness of solutions of a parabolic-elliptic system. Discrete Contin. Dyn. Syst. **13**, 659–682 (2005)
9. Coclite, G.M., Holden, H., Karlsen, K.H.: Global weak solutions to a generalized hyperelastic-rod wave equation. SIAM J. Math. Anal. **37**, 1044–1069 (2006)
10. Coclite, G.M., Holden, H., Karlsen, K.H.: Well-posedness of higher-order Camassa–Holm equations. J. Differ. Equ. **246**, 929–963 (2009)
11. Coclite G.M., Karlsen K.H., Risebro N.H.: A convergent finite difference scheme for the Camassa–Holm equation with general H^1 initial data. SIAM J. Numer. Anal. **46**, 1554–1579 (2008)
12. Coclite, G.M., Karlsen, K.H., Risebro, N.H.: An explicit finite difference scheme for the Camassa–Holm equation. Adv. Differ. Equ. **13**, 681–732 (2008)
13. Dai, H.H.: Exact travelling-wave solutions of an integrable equation arising in hyperelastic rods. Wave Motion **28**, 367–381 (1998)
14. Dai, H.H.: Model equations for nonlinear dispersive waves in a compressible Mooney–Rivlin rod. Acta Mech. **127**, 193–207 (1998)
15. Dai, H.H., Huo, Y.: Solitary shock waves and other travelling waves in a general compressible hyperelastic rod. R. Soc. Lond. Proc. Ser. A **456**, 331–363 (2000)
16. DiPerna, R.J., Lions, P.L.: Ordinary differential equations, transport theory and Sobolev spaces. Invent. Math. **98**, 511–547 (1989)
17. Holden, H., Raynaud, X.: Global conservative solutions of the generalized hyperelastic-rod wave equation. J. Differ. Equ. **233**, 448–484 (2007)

18. Holden, H., Raynaud, X.: Global conservative solutions of the Camassa–Holm equation—a Lagrangian point of view. Commun. Partial Differ. Equ. **32**, 1511–1549 (2007)
19. Holden, H., Raynaud, X.: Dissipative solutions for the Camassa–Holm equation. Discrete Contin. Dyn. Syst. **24**, 1047–1112 (2009)
20. Johnson, R.S.: Camassa–Holm, Korteweg–de Vries and related models for water waves. J. Fluid Mech. **455**, 63–82 (2002)
21. Lieb, E.H., Loss, M.: Analysis. Graduate Studies in Mathematics, vol. 14, 2nd edn. American Mathematical Society, Providence (2001)
22. Simon, J.: Compact sets in the space $L^p(0, T; B)$. Ann. Mat. Pura Appl. **146**, 65–96 (1987)
23. Xin, Z., Zhang P.: On the weak solutions to a shallow water equation. Commun. Pure Appl. Math. **53**, 1411–1433 (2000)
24. Xin, Z., Zhang, P.: On the uniqueness and large time behavior of the weak solutions to a shallow water equation. Commun. Partial Differ. Equ. **27**, 1815–1844 (2002)

Chapter 3
Exponential Decay Properties of a Mathematical Model for a Certain Fluid-Structure Interaction

George Avalos and Francesca Bucci

With sincere respect and a deep sense of loss, we dedicate this work to the memory of our colleague Alfredo Lorenzi

Abstract In this work, we derive a result of exponential stability for a coupled system of partial differential equations (PDEs) which governs a certain fluid-structure interaction. In particular, a three-dimensional Stokes flow interacts across a boundary interface with a two-dimensional mechanical plate equation. In the case that the PDE plate component is rotational inertia-free, one will have that solutions of this fluid-structure PDE system exhibit an exponential rate of decay. By way of proving this decay, an estimate is obtained for the resolvent of the associated semigroup generator, an estimate which is uniform for frequency domain values along the imaginary axis. Subsequently, we proceed to discuss relevant point control and boundary control scenarios for this fluid-structure PDE model, with an ultimate view to optimal control studies on both finite and infinite horizon. (Because of said exponential stability result, optimal control of the PDE on time interval $(0, \infty)$ becomes a reasonable problem for contemplation.)

3.1 Introduction

In this work, we undertake a stability analysis of a certain partial differential equation (PDE) system—(3.2)–(3.3) below—which has been previously studied in [16] and [14], among other works, inasmuch as it simultaneously constitutes a mathematically interesting and physically relevant model of a particular fluid-structure

G. Avalos
University of Nebraska-Lincoln, Lincoln, NE, USA
e-mail: gavalos@math.unl.edu

F. Bucci (✉)
Università degli Studi di Firenze, Firenze, Italy
e-mail: francesca.bucci@unifi.it

© Springer International Publishing Switzerland 2014   49
A. Favini et al. (eds.), *New Prospects in Direct, Inverse and Control Problems for Evolution Equations*, Springer INdAM Series 10,
DOI 10.1007/978-3-319-11406-4_3

(F-S) dynamics. This PDE model comprises a Stokes flow, evolving within a three-dimensional cavity $\mathcal{O}$, coupled via a boundary interface, to a two dimensional Euler–Bernoulli or Kirchhoff plate which displaces upon a sufficiently smooth bounded open set Ω, which is taken to be a portion of the cavity boundary $\partial\mathcal{O}$. Our main result here (Theorem 3.2 below) is the derivation of *exponential decay* rates for the composite fluid-structure dynamics, in the case that the Euler–Bernoulli plate PDE model—i.e. the one corresponding to "rotational inertia" parameter $\rho = 0$ in (3.2e)—is used to describe the mechanical displacements along Ω. With particular regard to the mechanical PDE component: Such *thin plate* models have been carefully derived in [25] and [24] (in the latter reference, stability of linear plate dynamics is also considered, under appropriate feedback, as well as situations where the von Karman nonlinear system is in play). It is shown in these works that the rotational inertia parameter ρ in the Kirchhoff plate model is proportional to the square of the thickness of the plate. So in the case of the Euler–Bernoulli PDE—viz., $\rho = 0$—rotational forces are essentially neglected (See also the recent monograph [15].)

In the case $\rho = 0$, the stability result was originally given in [16] (but the primary focus of this work was on global attractors for the given fluid-structure PDE system, in the presence of nonlinearities and forcing terms). The real novelty in the present work lies in the method of proof: whereas in [16], the exponential decay of the given fluid-structure dynamics is obtained via a Lyapunov functional approach, with the authors of [16] operating strictly within the *time domain*, the present work is centered upon working instead in the *frequency domain*. In particular, we work to attain a uniform estimate for the resolvent operator of the generator of the associated fluid-structure semigroup, as it assumes values along the imaginary axis. With such resolvent estimate in hand, we can then appeal to a well-known resolvent criterion—the Gearhart–Herbst–Prüss–Huang Theorem, whose variants by Prüss and Huang are posted here as Theorems 3.3 and 3.4, respectively—so as to ultimately infer exponential decay. The virtue of the frequency domain approach which is employed here, is that it can eventually be adapted so as to treat the case $\rho > 0$ (Kirchhoff plate). Indeed, the frequency domain methodology outlined here is invoked and refined in [3], so as to provide rational decay rates for Stokes–Kirchhoff plate dynamics. (The higher topology for the mechanical velocity component in Kirchhoff plate equation—viz., $H^1(\Omega)$ for the Kirchhoff plate, as opposed to $L^2(\Omega)$ for Euler–Bernoulli—prevents the attainment of exponential decay in [3]. Hence, weaker polynomial rates of decay are attained for $\rho > 0$.)

We should also state that our estimate of said resolvent on the imaginary axis is direct and explicit, in the style of what was attained in [8]; previous exponential stability works which are geared so as to eventually invoke said resolvent criterion (by Prüss) tend to obtain the requisite resolvent estimates via an argument by contradiction (see, e.g., [33]).

In addition, in the final Section of this work we offer some insight into a further analysis which is needed to pursue the solvability of natural/appropriate optimal control problems (with quadratic functionals) associated with the PDE system under investigation. We note that a full understanding of the stability properties of this F-S interaction is not only of intrinsic interest, but indeed a prerequisite step in the study of optimal control problems over an *infinite time* horizon. In this respect, the uniform (exponential) stability result established for the composite PDE system (3.2) in the presence of an elastic equation of Euler–Bernoulli type ensures that *both finite and infinite* time horizon problems are equally valid objects of investigation.

Instead, when the elastic equation is of Kirchhoff type, the (rational) decay rate $O(1/t)$ for *smooth* solutions, which is shown in [3], will not allow us to consider associated optimal control problems with quadratic functionals on an infinite time interval. In particular, the lack of exponential decay for $\rho > 0$ implies that the so-called *finite cost condition* is not necessarily satisfied for *finite energy* solutions of (3.2)–(3.3); see e.g., [28].

(On the other hand, the aforeasid (weaker, though) stability property of the *linear* dynamics might be utilized—just like in the case $\rho = 0$ (see [16])—in order to establish a soughtafter *quasi-stability* property for the dynamical system corresponding to *nonlinear* variants of the PDE system (3.2)–(3.3), in particular the ones which include physical nonlinearities in the structural PDE component.)

We also point out here that the introduction of boundary or point control actions into the model, will necessitate a careful technical analysis of the regularity properties of the (so-called) "input-to-state map" for the abstract equation corresponding to the controlled boundary value problem (3.2). Such a (PDE) analysis will be unavoidable, inasmuch as the control operator (or, operators) which models (model) the physically relevant control action will be *intrinsically* unbounded from the control space into the state space. In consequence—and which has been in the past for other PDE control problems; see e.g., [29]—sharp PDE regularity estimates for the solutions to the "free" (or uncontrolled) system should be instrumental in bringing about sought-after regularity properties for the input-to-state map.

A brief description of a couple of relevant scenarios for the placement of control functions in the model is given in Sect. 3.4, along with some remarks about the technical challenges which are expected. A natural question which arises is whether the recent results on the LQ-problem and Riccati equations for abstract dynamics—inspired by and tailored for coupled PDE systems of hyperbolic/parabolic type (such as [1] and [2])—are applicable, or whether novel theories need to be devised.

3.2 The PDE Model, Statement of the Main Result

In what follows, the geometrical situation which prevailed in [16] will obtain here. Namely: (fluid) domain $\mathscr{O}$ will be a bounded subset of $\mathbb{R}^3$, with boundary $\partial\mathscr{O}$. Moreover, $\partial\mathscr{O} = \overline{S} \cup \overline{\Omega}$, with $S \cap \Omega = \emptyset$, and with (structure) domain Ω being a *flat* portion of $\partial\mathscr{O}$. In particular, $\partial\mathscr{O}$ has the following specific configuration:

$$\Omega \subset \{x = (x_1, x_2, 0)\}, \quad S \subset \{x = (x_1, x_2, x_3) : x_3 \leq 0\}.$$

So if $\nu(x)$ denotes the unit normal vector to $\partial\mathscr{O}$, pointing outward, then

$$\nu|_{\Omega} = [0, 0, 1]. \tag{3.1}$$

The relatively low regularity, allowed for the boundary of the container $\mathscr{O}$, is made rigorous and explicit in the following Geometric Assumption.

Assumption 3.1 (Geometric Assumption) $[\mathscr{O}, \Omega]$ *is assumed to fall within one of the following classes:*

(G.1) $\mathscr{O}$ *is a convex domain with wedge angles* $\leq \frac{2\pi}{3}$. *Moreover,* Ω *has smooth boundary, and* S *is a piecewise smooth surface;*

(G.2) $\mathscr{O}$ *is a convex polyhedron having angles* $\leq \frac{2\pi}{3}$, *and so then* Ω *is a convex polygon with angles* $\leq \frac{2\pi}{3}$.

The picture below illustrates a geometrical configuration which is consistent with the case (G.2).

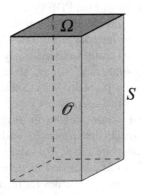

On such geometry, the PDE model is as follows, with rotational inertia parameter $\rho \geq 0$, and in solution variables $u(x,t) = [u^1(x,t), u^2(x,t), u^3(x,t)]$ and $[w(x,t), w_t(x,)]$:

$$u_t - \Delta u + \nabla p = 0 \qquad\qquad \text{in } \mathscr{O} \times (0,\infty) \qquad\qquad (3.2\text{a})$$

$$\text{div}(u) = 0 \qquad\qquad \text{in } \mathscr{O} \times (0,\infty) \qquad\qquad (3.2\text{b})$$

$$u = 0 \qquad\qquad \text{on } S \times (0,\infty) \qquad\qquad (3.2\text{c})$$

$$u = [u^1, u^2, u^3] = [0, 0, w_t] \qquad\qquad \text{on } \Omega \times (0,\infty), \qquad\qquad (3.2\text{d})$$

$$w_{tt} - \rho\Delta w_{tt} + \Delta^2 w = p|_{\Omega} \qquad\qquad \text{in } \Omega \times (0,\infty) \qquad\qquad (3.2\text{e})$$

$$w = \frac{\partial w}{\partial \nu} = 0 \qquad\qquad \text{on } \partial\Omega \times (0,\infty) \qquad\qquad (3.2\text{f})$$

with initial conditions

$$[u(0), w(0), w_t(0)] = [u_0, w_0, w_1] \in \mathbf{H}_\rho . \qquad\qquad (3.3)$$

Here, the space of initial data $\mathbf{H}_\rho$ is defined as follows: Let the (fluid) space $\mathscr{H}_{fluid}$ ($\mathscr{H}_f$, in short) be defined by

$$\mathscr{H}_f := \{f \in \mathbf{L}^2(\mathscr{O}) : \mathrm{div}(f) = 0; \ f \cdot v|_S = 0\}, \qquad (3.4)$$

and let

$$V_\rho = \begin{cases} \hat{L}^2(\Omega) & \text{if } \rho = 0 \\ H_0^1(\Omega) \cap \hat{L}^2(\Omega) & \text{if } \rho > 0, \end{cases} \qquad (3.5)$$

where

$$\hat{L}^2(\Omega) \equiv \left\{ \varpi \in L^2(\Omega) : \int_\Omega \varpi \, d\Omega = 0 \right\}, \qquad (3.6)$$

after invoking the notation of [16]. (We note that the matching of fluid and structure velocities in (3.2d) ultimately dictates this average mean zero constraint; see [16] and [4].) Therewith, we then set

$$\mathbf{H}_\rho = \Big\{ [f, h_0, h_1] \in \mathscr{H}_f \times [H_0^2(\Omega) \cap \hat{L}^2(\Omega)] \times V_\rho,$$
$$\text{with } f \cdot v|_\Omega = [0, 0, f^3] \cdot [0, 0, 1] = h_1 \Big\}. \qquad (3.7)$$

In this paper, we shall focus on the case $\rho = 0$.

In addition: By way of constructing an abstract operator $A_\rho : \mathscr{D}(A_\rho) \subset \mathbf{H}_\rho \to \mathbf{H}_\rho$ which describes the PDE dynamics (3.2)–(3.3), we denote $A_D : L^2(\Omega) \to L^2(\Omega)$ by

$$A_D g = -\Delta g, \qquad \mathscr{D}(A_D) = H^2(\Omega) \cap H_0^1(\Omega). \qquad (3.8)$$

If we subsequently make the denotation for all $\rho \geq 0$

$$P_\rho = I + \rho A_D, \qquad \mathscr{D}(P_\rho) = \begin{cases} L^2(\Omega) & \text{if } \rho = 0 \\ \mathscr{D}(A_D) & \text{if } \rho > 0, \end{cases} \qquad (3.9)$$

then the mechanical PDE component (3.2e) of the system (3.2) can be written as

$$P_\rho w_{tt} + \Delta^2 w = p|_\Omega \qquad \text{on } (0, T).$$

Using the fact from [20] that

$$\mathscr{D}(P_\rho^{1/2}) = \begin{cases} L^2(\Omega) & \text{if } \rho = 0 \\ H_0^1(\Omega) & \text{if } \rho > 0, \end{cases}$$

then we can endow the Hilbert space $\mathbf{H}_\rho$ with the norm-inducing inner product

$$([\mu_0, \omega_1, \omega_2], [\tilde{\mu}_0, \tilde{\omega}_1, \tilde{\omega}_2])_{\mathbf{H}_\rho} = (\mu_0, \tilde{\mu}_0)_\mathscr{O} + (\Delta\omega_1, \Delta\tilde{\omega}_1)_\Omega + (P_\rho^{1/2}\omega_2, P_\rho^{1/2}\tilde{\omega}_2)_\Omega,$$

where $(\cdot, \cdot)_\mathscr{O}$ and $(\cdot, \cdot)_\Omega$ are the L^2-inner products on their respective geometries.

Moreover, as was done in [4] and [3, Lemma 1.1], so as to eliminate the pressure term p in (3.2)–(3.3) (see also [5] for an analogous elimination for a different fluid-structure PDE model), we recognize the pressure term as the solution of the following BVP, pointwise in time:

$$\begin{cases} \Delta p = 0 & \text{in } \mathscr{O} \\ \frac{\partial p}{\partial v} = \Delta u \cdot v\big|_S & \text{on } S \\ \frac{\partial p}{\partial v} + P_\rho^{-1} p = P_\rho^{-1}\Delta^2 w + \Delta u^3\big|_\Omega & \text{on } \Omega. \end{cases} \tag{3.10}$$

To 'solve' for the pressure term, we then invoke appropriate 'Neumann–Robin' maps R_ρ and $\tilde{R}_\rho$ defined as follows:

$$R_\rho g = f \Longleftrightarrow \left\{ \Delta f = 0 \text{ in } \mathscr{O}, \; \frac{\partial f}{\partial v} = 0 \text{ on } S, \; \frac{\partial f}{\partial v} + P_\rho^{-1} f = g \text{ on } \Omega \right\};$$

$$\tilde{R}_\rho g = f \Longleftrightarrow \left\{ \Delta f = 0 \text{ in } \mathscr{O}, \; \frac{\partial f}{\partial v} = g \text{ on } S, \; \frac{\partial f}{\partial v} + P_\rho^{-1} f = 0 \text{ on } \Omega \right\}.$$

By Lax–Milgram Theorem, we then have,

$$R_\rho \in \mathscr{L}\big(H^{-1/2}(\Omega), H^1(\mathscr{O})\big); \quad \tilde{R}_\rho \in \mathscr{L}\big(H^{-1/2}(S), H^1(\mathscr{O})\big).$$

(We are also using implicity the fact that P_ρ^{-1} is positive definite and self-adjoint on Ω.) Consequently, the pressure variable $p(t)$, as necessarily the solution of (3.10)— that is, $p(t)$ is an appropriate *harmonic extension* from the boundary of $\mathscr{O}$ into the interior—can be written pointwise in time as

$$p(t) = G_{\rho,1}(w(t)) + G_{\rho,2}(u(t)),$$

where

$$G_{\rho,1}(w) = R_\rho(P_\rho^{-1}\Delta^2 w); \tag{3.11a}$$

$$G_{\rho,2}(u) = R_\rho(\Delta u^3\big|_\Omega) + \tilde{R}_\rho(\Delta u \cdot v\big|_S). \tag{3.11b}$$

These relations suggest the following choice for the generator $A_\rho : \mathbf{H}_\rho \to \mathbf{H}_\rho$. We set

$$
A_\rho \equiv
\begin{bmatrix}
\Delta - \nabla G_{\rho,2} & -\nabla G_{\rho,1} & 0 \\
0 & 0 & I \\
P_\rho^{-1} G_{\rho,2}\big|_\Omega & -P_\rho^{-1}\Delta^2 + P_\rho^{-1} G_{\rho,1}\big|_\Omega & 0
\end{bmatrix}
\tag{3.12}
$$

with domain

$$
\mathscr{D}(A_\rho) = \Big\{ [u, w_1, w_2] \in \mathbf{H}_\rho : \ u \in \mathbf{H}^2(\mathcal{O}); \quad w_1 \in \mathscr{S}_\rho, \ w_2 \in H_0^2(\Omega),
$$

$$
\tag{3.13}
$$

$$
u = 0 \text{ on } S, \quad u = (0, 0, w_2) \text{ on } \Omega \Big\},
$$

where the mechanical displacement space, denoted by $\mathscr{S}_\rho$, changes with ρ, as follows:

$$
\mathscr{S}_\rho :=
\begin{cases}
H^4(\Omega) \cap H_0^2(\Omega) & \rho = 0 \\
H^3(\Omega) \cap H_0^2(\Omega) & \rho > 0.
\end{cases}
\tag{3.14}
$$

We note also, from the definition of $\mathscr{D}(A_\rho)$ that $[u, w_1, w_2] \in \mathscr{D}(A_\rho)$ implies $\Delta u \in \mathbf{L}^2(\mathcal{O})$ and $\operatorname{div}\Delta u = 0$. Consequently, from elementary Stokes Theory (see, e.g., [17, Proposition 1.4, p. 5], we have

$$
\|\Delta u \cdot \nu\|_{H^{-1/2}(\partial\mathcal{O})} \le C \|u\|_{\mathbf{H}^2(\mathcal{O})} \le C \|[u, w_1, w_2]\|_{\mathscr{D}(A_\rho)}
\tag{3.15}
$$

and so associated pressure π_0 satisfies

$$
\pi_0 \equiv G_{\rho,1}(w_1) + G_{\rho,2}(u) \in H^1(\mathcal{O})
\tag{3.16}
$$

(recall that $\Delta u^3\big|_\Omega = \Delta u \cdot \nu\big|_\Omega$).

Well-posedness of the (*linear*) coupled system (3.2)–(3.3) when $\rho = 0$—namely, when the elastic equation is of Euler–Bernoulli type, of specific concern in the present investigation—was originally established in [16], by using Galerkin approximations. An alternative proof of well-posedness which encompasses both cases $\rho = 0$ and $\rho > 0$ has been recently given in [4]. It is important to emphasize that the proof appeals to the Lumer–Phillips Theorem within classical semigroup theory, and yet also utilizes in a crucial and nontrivial way the Babuška–Brezzi Theorem (see, e.g., [22, p. 116]). The precise statement is given below.

Theorem 3.1 ([4]) *For $\rho \geq 0$ the operator $A_\rho : \mathbf{H}_\rho \rightarrow \mathbf{H}_\rho$ defined by (3.12)–(3.13) generates a C_0-semigroup of contractions $\{e^{A_\rho t}\}_{t \geq 0}$ on $\mathbf{H}_\rho$. Thus, for any $[u_0, w_0, w_1] \in \mathbf{H}_\rho$, the (unique) weak solution to the initial/boundary value problem (3.2)–(3.3) is given by*

$$
\begin{bmatrix} u(t) \\ w(t) \\ w_t(t) \end{bmatrix} = e^{A_\rho t} \begin{bmatrix} u_0 \\ w_0 \\ w_1 \end{bmatrix} \in C([0, T]; \mathbf{H}_\rho). \tag{3.17}
$$

Remark 3.1 (Well-posedness) In a preliminary version of the aforementioned [16]—in the Appendix of the *e-Print* arXiv:1109.4324 [math.AP] (September 2011), to be precise—the authors thereof provide an alternative description of a modelling generator, for the fluid-structure dynamics (3.2)–(3.3). But this earlier semigroup setup, based upon a coupled variational formulation, is fairly unwieldy, and not easily used in any stability analysis and/or control theoretic investigation of (3.2)–(3.3). In this sense, the semigroup e^{At} which is being presently used, and which was formulated in [4], may be said to constitute an improvement.

Remark 3.2 (Higher Regularity for Fluid-Structure Variables) In the Appendix, we briefly expand on the wellposedness paper in [4], to show that indeed, despite the "edge" induced by the boundary interface of $\mathscr{O}$ with Ω, one has the higher regularity specified in $\mathscr{D}(A_\rho)$. (The driving force here is really Lemma 3.1 below.) In particular, with $[\mathscr{O}, \Omega]$ having one of the geometrical configurations in (G.1)–(G.2), then if $[u, w_1, w_2] \in \mathscr{D}(A_\rho)$, one has $u \in \mathbf{H}^2(\mathscr{O})$, $w_1 \in \mathscr{S}_\rho$, as defined in (3.14), and associated pressure variable p is in $H^1(\mathscr{O})$).

To the authors' knowledge, the stability properties of solutions to the *linear* model (3.2) (again, when $\rho = 0$) have been explored for the first time in [16]—which again is primarily concerned with attracting sets for nonlinear fluid-structure dynamics with forcing term—in the course of an analysis of the long-term behaviour of a *nonlinear* coupled dynamics, which comprise a 3D linearized Navier–Stokes system for the fluid velocity field in a bounded domain, and a *nonlinear* elastic plate equation for the transversal displacement of a flat flexible part of the boundary. Along with the various results established in [16] for attracting sets pertinent to the nonlinear model, is the result of exponential stability for the linear dynamics, attained by using Lyapunov function arguments; see [16, Section 3].

We aim here at presenting a different proof of exponential stability, based instead on a (by now classical) resolvent criterion; see Theorem 3.3 and Theorem 3.4 in the next Section. The adoption of a "frequency domain" approach—in contrast with the more commonly invoked "time domain" analysis—is not only of intrinsic interest, but also proves to be very effective in establishing the decay rates of solutions, *even when exponential stability fails*. Indeed, in the case $\rho > 0$, the very same frequency domain approach enables us to establish that the energy of strong solutions decays at the rate of $O(1/t)$, as $t \rightarrow +\infty$; see [3].

The main result of the present work is stated below.

Theorem 3.2 (Exponential Decay Rates) *Let the rotational inertia parameter* $\rho = 0$ *in (3.2e). Then all finite energy solutions of (3.2)–(3.3) decay at an exponential rate. Namely, there exist constants* $\omega > 0$ *and* $M \geq 1$ *such that for arbitrary initial data* $[u_0, w_0, w_1] \in \mathbf{H_0}$, *the corresponding solutions* $[u, w, w_t]$ *of (3.2)–(3.3) satisfy*

$$\|[u(t), w(t), w_t(t)]\|_{\mathbf{H_0}} \leq M\, e^{-\omega t}\, \|[u_0, w_0, w_1]\|_{\mathbf{H_0}}\,.$$

3.3 Exponential Stability

To show that the semigroup defined by (3.17) is exponentially stable, we appeal to a celebrated result of semigroup theory which we recall explicitly—in two variants—for the reader's convenience.

Theorem 3.3 ([36]) *Let* X *be a Hilbert space and* e^{At} *be a semigroup in* X. *Then* e^{At} *decays exponentially iff*

(i) $\{\lambda \in \mathbb{C} : Re\,\lambda \geq 0\} \subset \varrho(A)$, *and*
(ii) there is $M \geq 1$ *such that* $\left\|(\lambda - A)^{-1}\right\| \leq M$ *for all* λ *with* $Re\,\lambda \geq 0$.

Theorem 3.4 ([21]) *Let* e^{At} *be a* C_0-*semigroup generated by* A *in a Hilbert space* H, *satisfying*

$$\|e^{At}\| \leq K_0 \qquad \forall t \geq 0, \quad \text{for some } K_0 > 0.$$

Then e^{At} *decays exponentially iff*

(i) $\{\lambda \in \mathbb{C} : \lambda = i\omega,\ \omega \in \mathbb{R}\} \subset \varrho(A)$, *and*
(ii) $\displaystyle\sup_{\omega \in \mathbb{R}} \left\|(i\omega - A)^{-1}\right\| < \infty$.

In order to invoke the above resolvent criterion, we need, as a preliminary step, to show that the imaginary axis belongs to the resolvent set of the dynamics operator A. This property cannot be freely taken for granted: in the context of other fluid-structure interactions, it is known that certain geometrical configurations will give rise to eigenvalues on the imaginary axis; see, e.g., [5] and [6] (and also [8], where examples of "non-pathological geometries" are given).

3.3.1 Preliminary Results, I: Higher Regularity

In order to justify the smoothness implicitly implied by our multiplier method below—see Appendix—and by way of showing that $\lambda = 0$ is in the resolvent set of $A_\rho : \mathscr{D}(A_\rho) \subset \mathbf{H}_\rho \rightarrow \mathbf{H}_\rho$, we must first establish a higher regularity result for Stokes flow on $\mathscr{O}$ (3-D container), under the aforesaid specifications on $[\mathscr{O}, \Omega]$.

Lemma 3.1 *With $[\mathscr{O}, \Omega]$ obeying the assumptions above—including the flatness of Ω and (G.1)–(G.2)—we consider the following inhomogeneous Stokes problem, with parameter $\lambda \geq 0$:*

$$
\begin{aligned}
\lambda u - \Delta u + \nabla p = u^* && \text{in } \mathscr{O} \\
\text{div}(u) = 0 && \text{in } \mathscr{O} \\
u|_S = [0,0,0] && \text{on } S \\
u|_\Omega = [0,0,w] && \text{in } \Omega,
\end{aligned}
\tag{3.18}
$$

where data $[u^*, w] \in \mathbf{L}^2(\mathscr{O}) \times H_0^{3/2+\varepsilon}(\Omega)$, with $\varepsilon > 0$, and w satisfying the compatibility condition $\int_\Omega w\, d\Omega = 0$. Then one has the following regularity estimate for the solution pair $[u, p]$:

$$
\|u\|_{\mathbf{H}^2(\mathscr{O})} + \|p\|_{H^1(\mathscr{O}) \cap [L^2(\mathscr{O})/\mathbb{R}]} \leq C_\lambda \left\| [u^*, w] \right\|_{\mathbf{L}^2(\mathscr{O}) \times H_0^{3/2+\varepsilon}(\Omega)}.
\tag{3.19}
$$

Proof   **Step 1** We start by recalling the following trace theorem for three dimensional Lipschitz domains:

Theorem 3.5 ([13], Theorem 5, p. 702) *Let $\mathscr{O}$ be a bounded subset of $\mathbb{R}^3$, with Lipschitz boundary. Let $(\gamma_0, \gamma_1) : H^2(\mathscr{O}) \rightarrow H^1(\partial\mathscr{O}) \times L^2(\partial\mathscr{O})$ be the standard continuous trace operator*

$$
\mu \rightarrow \begin{bmatrix} \gamma_0(\mu) = \mu|_{\partial\mathscr{O}} \\ \gamma_1(\mu) = \frac{\partial\mu}{\partial\nu}|_{\partial\mathscr{O}} \end{bmatrix}
$$

(which is generally not surjective on nonsmooth domains; see [35, Theorem 4.11]). Then the range of (γ_0, γ_1) is characterized by the following subspace of $H^1(\partial\mathscr{O}) \times L^2(\partial\mathscr{O})$:

$$
Range\,[(\gamma_0, \gamma_1)] = \left\{ (g_0, g_1) \in H^1(\partial\mathscr{O}) \times L^2(\mathscr{O}): \nabla_{\partial\mathscr{O}}(g_0) + g_1\nu \in \mathbf{H}^{1/2}(\partial\mathscr{O}) \right\}.
\tag{3.20}
$$

(Here, $\nabla_{\partial\mathscr{O}}$ denotes the tangential gradient operator on $\partial\mathscr{O}$.)

To use this result: since $w \in H_0^{3/2+\varepsilon}(\Omega)$, then a continuous extension by zero will allow us to take g_0 in (3.20) to be

$$
g_0 = \begin{cases} w & \text{on } \Omega \\ 0 & \text{on } S \end{cases},
\tag{3.21}
$$

with extension $g_0 \in H^{3/2+\varepsilon}(\partial\mathcal{O})$ (see e.g., [32, Theorem 11.4], if Ω is smooth; see [34, Theorem 3.33], if Ω is polygonal). Thus, with $g_1 \equiv 0$ in (3.20), we can appeal to Theorem 3.5 to have the existence of a function $v^3 \in H^2(\mathcal{O})$, such that

$$v^3\big|_{\partial\mathcal{O}} = \begin{cases} w & \text{on } \Omega \\ 0 & \text{on } S \end{cases} \quad \text{and} \quad \frac{\partial v^3}{\partial v}\bigg|_{\partial\mathcal{O}} = 0 \text{ on } \partial\mathcal{O}, \tag{3.22}$$

with the estimate

$$\left\|v^3\right\|_{H^2(\mathcal{O})} \leq C \left\|\nabla_{\partial\mathcal{O}} g_0\right\|_{\mathbf{H}^{1/2}(\partial\mathcal{O})} \leq C \left\|w\right\|_{H_0^{3/2+\varepsilon}(\Omega)}. \tag{3.23}$$

Therewith we set

$$v \equiv \begin{bmatrix} 0 \\ 0 \\ v^3 \end{bmatrix} \in \mathbf{H}^2(\mathcal{O}). \tag{3.24}$$

Step 2 We use variable $v = v(w)$ as data in the following inhomogeneous Stokes problem: Find $[\tilde{u}, \tilde{p}] \in \mathbf{H}_0^1(\mathcal{O}) \times L^2(\mathcal{O})/\mathbb{R}$ which solves

$$\lambda\tilde{u} - \Delta\tilde{u} + \nabla\tilde{p} = \Delta v - \lambda v + u^* \quad \text{in } \mathcal{O}$$

$$\text{div}(\tilde{u}) = -\text{div}(v) \quad \text{in } \mathcal{O}$$

$$\tilde{u} = 0 \quad \text{on } \partial\mathcal{O}.$$

Since the compatibility condition $\int \text{div}(\tilde{u}) \, d\mathcal{O} = -\int \text{div}(v) \, d\mathcal{O} = 0$ is satisfied, then there exists a unique solution $[\tilde{u}, \tilde{p}] \in \mathbf{H}_0^1(\mathcal{O}) \times L^2(\mathcal{O})/\mathbb{R}$, which depends continuously on the data (see, e.g., Theorem 2.4 and Remark 2.5 in [37]).

To justify additional regularity for this solution, we appeal to the paper [18], from which the geometric configurations (G.1)–(G.2) spring forth; in particular, in [18] we cite (1.9) and (1.10) on p. 75 with $s \equiv 1$ (which is an inference of main Theorem 9.20 therein). From this statement in [18, p. 75], we will have that fluid pair $[\tilde{u}, \tilde{p}]$ of (3.25) is in $\left[\mathbf{H}^2(\mathcal{O}) \cap \mathbf{H}_0^1(\mathcal{O})\right] \times \left[H^1(\mathcal{O}) \cap L^2(\mathcal{O})/\mathbb{R}\right]$ continuously, if it can be shown that divergence data $\text{div}(v) \in H_0^1(\mathcal{O})$. In other words, we must show that $\text{div}(v)$ is *zero at the singular points* of $\mathcal{O}$, for Sobolev index $s = 1$. (See Definition 9.17 (1) and Remark 9.18 (1) on p. 94 of [18]. See also Theorem 8.4 of [23], p. 53). Since $v = [0, 0, v^3]$, with $v^3 \in \mathbf{H}^2(\mathcal{O})$, it is left to establish that Dirichlet trace $\text{div}(v)|_{\partial\mathcal{O}} = 0$.

Step 3 By way of showing that $\text{div}(v) \in H_0^1(\mathcal{O})$, we consider the respective geometric scenarios, (G.1) and (G.2).

Case 1: $[\mathcal{O}, \Omega]$ satisfies (G.1). We proceed somewhat, but not completely, as in the proof of Proposition A.1 (i) of [7, Appendix A], where the domain therein was smooth. Here, $\partial\mathcal{O} = \overline{S} \cup \overline{\Omega}$, where again Ω is flat, and $S = \bigcup_i S_i$, with each S_i being a smooth surface. Set $\tilde{S} \equiv S_i$, for given i. Then $\tilde{S}$ can be described by a smooth parametrization $\mathbf{r}(s, t)$, defined on a rectangle $a \le s \le b$, $c \le t \le d$. Then for this surface, two linear independent unit tangent vectors may be given by

$$\mathbf{e} = \frac{\mathbf{r}_s(s, t)}{|\mathbf{r}_s(s, t)|} \quad \text{and} \quad \boldsymbol{\tau} = \frac{\mathbf{r}_t(s, t)}{|\mathbf{r}_t(s, t)|}. \tag{3.25}$$

In turn, the corresponding unit normal ν is

$$\nu = \frac{\mathbf{e} \times \boldsymbol{\tau}}{|\mathbf{e} \times \boldsymbol{\tau}|}. \tag{3.26}$$

With these vectors $\{\nu, \mathbf{e}, \boldsymbol{\tau}\}$ being a linearly independent but generally not orthonormal set, we proceed to consider the following system, for sufficiently smooth scalar valued function ϕ on $\mathcal{O}$:

$$\begin{bmatrix} \frac{\partial\phi}{\partial\nu} \\ \frac{\partial\phi}{\partial e} \\ \frac{\partial\phi}{\partial\tau} \end{bmatrix} = \begin{bmatrix} \nabla\phi|_{\partial\mathcal{O}} \cdot \nu \\ \nabla\phi|_{\partial\mathcal{O}} \cdot e \\ \nabla\phi|_{\partial\mathcal{O}} \cdot \tau \end{bmatrix} = L \begin{bmatrix} \frac{\partial\phi}{\partial x_1}\big|_{\partial\mathcal{O}} \\ \frac{\partial\phi}{\partial x_2}\big|_{\partial\mathcal{O}} \\ \frac{\partial\phi}{\partial x_3}\big|_{\partial\mathcal{O}} \end{bmatrix}, \tag{3.27}$$

where

$$L = \begin{bmatrix} \nu_1 & \nu_2 & \nu_3 \\ e_1 & e_2 & e_3 \\ \tau_1 & \tau_2 & \tau_3 \end{bmatrix}. \tag{3.28}$$

Inverting this matrix, we have then on $\tilde{S}$

$$\begin{bmatrix} \frac{\partial\phi}{\partial x_1}\big|_{\partial\mathcal{O}} \\ \frac{\partial\phi}{\partial x_2}\big|_{\partial\mathcal{O}} \\ \frac{\partial\phi}{\partial x_3}\big|_{\partial\mathcal{O}} \end{bmatrix} = L^{-1} \begin{bmatrix} \frac{\partial\phi}{\partial\nu} \\ \frac{\partial\phi}{\partial e} \\ \frac{\partial\phi}{\partial\tau} \end{bmatrix} = \mathbf{a}\frac{\partial\phi}{\partial\nu} + \mathbf{b}\frac{\partial\phi}{\partial e} + \mathbf{c}\frac{\partial\phi}{\partial\tau}, \tag{3.29}$$

where $L^{-1} = [\mathbf{a}, \mathbf{b}, \mathbf{c}]$.

We use the above relation as follows. Given $f \in [\mathscr{D}(\overline{\mathcal{O}})]^3$, we now replace $\frac{\partial\phi}{\partial x_i}$ with $\frac{\partial f_i}{\partial x_i}$, $i = 1, 2, 3$. This gives then from (3.29) that, on $\tilde{S}$,

$$\nabla f|_{\tilde{S}} = \mathscr{D}f, \tag{3.30}$$

where for $j = 1, 2, 3$,

$$[\nabla f|_{\partial\mathscr{O}}]_{i,j} = \left.\frac{\partial f_j}{\partial x_i}\right|_{\partial\mathscr{O}} \quad \text{and} \quad [\mathscr{D}f]_{i,j} = a_i\frac{\partial f_j}{\partial v} + b_i\frac{\partial f_j}{\partial e} + c_i\frac{\partial f_j}{\partial \tau}. \tag{3.31}$$

We read off then from (3.30) and (3.31), that on given $S_i = \tilde{S}$,

$$\mathrm{div}(f)|_{\tilde{S}} = \frac{\partial f}{\partial v}\cdot\mathbf{a} + \frac{\partial f}{\partial e}\cdot\mathbf{b} + \frac{\partial f}{\partial \tau}\cdot\mathbf{c}. \tag{3.32}$$

This relation holds true for $f \in [\mathscr{D}(\overline{\mathscr{O}})]^3$. But since $\mathscr{D}(\overline{\mathscr{O}})$ is dense in $H^2(\mathscr{O})$—see e.g., [34, Theorem 3.29]—and since moreover the map $\phi \longmapsto [\phi|_{\partial\mathscr{O}}, \frac{\partial\phi}{\partial v}|_{\partial\mathscr{O}}]$ is in $\mathscr{L}(H^2(\mathscr{O}), H^1(\partial\mathscr{O}) \times L^2(\partial\mathscr{O}))$ and $f \longmapsto \mathrm{div}(f)|_{\partial\mathscr{O}}$ in $\mathscr{L}(\mathbf{H}^2(\mathscr{O}), H^{1/2}(\partial\mathscr{O}))$—see [35, Theorem 4.11] and [34, Theorem 3.38]—we can extend by density the relation (3.32) to all $f \in \mathbf{H}^2(\mathscr{O})$. Thus, for the boundary trace $\mathrm{div}(v)|_S$, where v is given by (3.24) and (3.22), we have, upon taking $f \equiv v$ in (3.32),

$$\mathrm{div}(v)|_S = \mathrm{div}(v)|_{\bigcup_i S_i} = 0. \tag{3.33}$$

Likewise on the smooth boundary segment Ω (flat), we will have a parametrization $\mathbf{r} = \langle g(s,t), h(s,t), 0\rangle$, on some closed rectangle $a \le s \le b, c \le t \le d$.

Accordingly, we can take linearly independent unit tangent and normal vectors to be

$$\mathbf{e} = \frac{1}{\sqrt{g_s^2 + h_s^2}}\begin{bmatrix} g_s \\ h_s \\ 0 \end{bmatrix}, \quad \mathbf{\tau} = \frac{1}{\sqrt{g_t^2 + h_t^2}}\begin{bmatrix} g_t \\ h_t \\ 0 \end{bmatrix}, \quad v = \begin{bmatrix} 0 \\ 0 \\ 1 \end{bmatrix}. \tag{3.34}$$

By considering the system totally analogous to (3.27), we can proceed as we did on $\tilde{S}$, so as to have for all $f \in \mathbf{H}^2(\mathscr{O})$ (cfr. (3.32))

$$\mathrm{div}(f)|_{\Omega} = \frac{\partial f}{\partial v}\cdot\begin{bmatrix} 0 \\ 0 \\ 1 \end{bmatrix} + \frac{\partial f}{\partial e}\cdot\begin{bmatrix} \frac{-\tau_2}{\tau_1 e_2 - \tau_2 e_1} \\ \frac{\tau_1}{\tau_1 e_2 - \tau_2 e_1} \\ 0 \end{bmatrix} + \frac{\partial f}{\partial \tau}\cdot\begin{bmatrix} \frac{e_2}{\tau_1 e_2 - \tau_2 e_1} \\ \frac{-e_1}{\tau_1 e_2 - \tau_2 e_1} \\ 0 \end{bmatrix}. \tag{3.35}$$

Setting $f = v$ in (3.35), where again v is given by (3.24) and (3.22), we have

$$\mathrm{div}(v)|_{\Omega} = 0 + \begin{bmatrix} 0 \\ 0 \\ \frac{\partial w}{\partial e} \end{bmatrix}\cdot\begin{bmatrix} \frac{-\tau_2}{\tau_1 e_2 - \tau_2 e_1} \\ \frac{\tau_1}{\tau_1 e_2 - \tau_2 e_1} \\ 0 \end{bmatrix} + \begin{bmatrix} 0 \\ 0 \\ \frac{\partial w}{\partial \tau} \end{bmatrix}\cdot\begin{bmatrix} \frac{e_2}{\tau_1 e_2 - \tau_2 e_1} \\ \frac{-e_1}{\tau_1 e_2 - \tau_2 e_1} \\ 0 \end{bmatrix} = 0. \tag{3.36}$$

To conclude Case 1: the fact that $\mathrm{div}(v) \in H^1(\mathscr{O})$, and relations (3.33), (3.36) yield now that the divergence data $\mathrm{div}(v)$ in (3.25) is in $H_0^1(\mathscr{O})$.

Case 2: $[\mathcal{O}, \Omega]$ satisfies **(G.2)**. We let Γ_j, $j = 1, \ldots, N$, denote the jth faces of the polyhedron $\mathcal{O}$, with the last boundary face $\Gamma_N \equiv \Omega$. Then as noted in [12], one will have, on each face Γ_j, a set of tangent vectors $\{\mathbf{e}, \boldsymbol{\tau}\}$ and normal vector $\boldsymbol{v}$, which are *orthonormal* in $\mathbb{R}^3$ (just as we construct for Ω below). This being the case, one can use *verbatim* the proof of Proposition A.1(i) of [7, Appendix A], so as to have, on any $f \in \mathbf{H}^2(\mathcal{O})$,

$$\operatorname{div}(f)|_{\Gamma_j} = \frac{\partial f}{\partial v} \cdot \boldsymbol{v} + \frac{\partial f}{\partial e} \cdot \mathbf{e} + \frac{\partial f}{\partial \tau} \cdot \boldsymbol{\tau}, \quad j = 1, \ldots, N. \tag{3.37}$$

In particular, if $f = v$, where again v is given by (3.24) and (3.22), we have

$$\operatorname{div}(v)|_{\Gamma_j} = 0 \cdot \boldsymbol{v} + 0 \cdot \mathbf{e} + 0 \cdot \boldsymbol{\tau}, \quad j = 1, \ldots, N-1. \tag{3.38}$$

On $\Gamma_N = \Omega$ (which we recall lies on the $x_1 x_2$-plane) we take one tangent vector $\mathbf{e} = \langle e_1, e_2, 0 \rangle$ to be any unit vector which is parallel to an edge of Ω; we take unit normal $\boldsymbol{v} = \langle 0, 0, 1 \rangle$. Then second unit tangent vector is given by $\boldsymbol{\tau} = \mathbf{e} \times \boldsymbol{v} = \langle e_2, -e_1, 0 \rangle$. Therewith, we have from (3.37), with $f = v$, where again v is given by (3.24) and (3.22),

$$\operatorname{div}(v)|_{\Gamma_N} = 0 \cdot \boldsymbol{v} + \begin{bmatrix} 0 \\ 0 \\ \frac{\partial w}{\partial e} \end{bmatrix} \cdot \begin{bmatrix} e_1 \\ e_2 \\ 0 \end{bmatrix} + \begin{bmatrix} 0 \\ 0 \\ \frac{\partial w}{\partial e} \end{bmatrix} \cdot \begin{bmatrix} e_2 \\ -e_1 \\ 0 \end{bmatrix} = 0. \tag{3.39}$$

The relations (3.38) and (3.39) now give the conclusion that in Case 2, divergence data in $\operatorname{div}(v)$ in (3.25) is in $H_0^1(\mathcal{O})$.

Step 4 To conclude the proof of Lemma 3.1: Since $\operatorname{div}(v) \in H_0^1(\mathcal{O})$, then as we noted above, the regularity results in [18] then allow the conclusion that Stokes solution $[\tilde{\mu}, \tilde{p}]$ of (3.25) satisfies

$$[\tilde{\mu}, \tilde{p}] \in \left[\mathbf{H}^2(\mathcal{O}) \cap \mathbf{H}_0^1(\mathcal{O})\right] \times \left[H^1(\mathcal{O}) \cap L^2(\mathcal{O})/\mathbb{R}\right], \quad \text{continuously.} \tag{3.40}$$

In turn, we recover the solution u of (3.18) via

$$u = \tilde{u} + v. \tag{3.41}$$

Combining (3.41) with (3.24), (3.22) and (3.25), we see that $[u, p]$ indeed solves the Stokes system in (3.18), and moreover from (3.24) and (3.40) it follows

$$[u, p] \in \left[\mathbf{H}^2(\mathcal{O}) \cap \mathscr{H}_f\right] \times H^1(\mathcal{O}). \tag{3.42}$$

Finally, the estimate (3.19) follows from (3.40) and (3.23). This concludes the proof of Lemma 3.1. $\qquad\square$

3.3.2 Preliminary Results, II: Spectral Analysis

Here, we limit ourselves to showing that $\lambda = 0$ belongs to the resolvent set $\varrho(A)$; in other words, the resolvent operator is boundedly invertible on the state space $\mathbf{H}_\rho$. The reader is referred to [3, Section 2] for a detailed analysis and proof of the fact that the spectrum has empty intersection with the whole imaginary axis, in the more challenging case $\rho > 0$. The arguments used therein can be easily adapted to the case $\rho = 0$. In particular: as in [3, Section 2], one can invoke a simple energy argument to quickly rule out the possibility of eigenvalues of A on $i\mathbb{R}$. This fact, and the particular form of the adjoint $A^* : \mathbf{H}_0 \to \mathbf{H}_0$, will in turn yield the conclusion that there is no residual spectrum on the imaginary axis. Lastly, to eliminate the possibility that there be continuous spectrum on the imaginary axis, a certain argument by contradiction—which ultimately uses the average mean zero constraint in (3.6) for the structural displacement—can be introduced.

As the parameter ρ equals 0 throughout, in order to simplify the notation we set $\mathbf{H}_0 =: \mathbf{H}$, as well as $A_0 =: A$. (We note that P_0 then coincides with the identity operator I throughout.)

Proposition 3.1 *The generator $A : \mathscr{D}(A) \subset \mathbf{H} \to \mathbf{H}$ is boundedly invertible on $\mathbf{H}$. Namely, $\lambda = 0$ is in the resolvent set of A.*

Proof Given data $[u^*, w_1^*, w_2^*] \in \mathbf{H}$, we look for $[u, w_1, w_2] \in \mathscr{D}(A)$ which solves

$$
A \begin{bmatrix} u \\ w_1 \\ w_2 \end{bmatrix} = \begin{bmatrix} u^* \\ w_1^* \\ w_2^* \end{bmatrix}.
\tag{3.43}
$$

To this end, we must seek for $[u, w_1, w_2]$ in $\mathscr{D}(A)$ and associated pressure $\pi_0 \in H^1(\mathcal{O})$ which uniquely solve

$$
\begin{array}{lll}
\Delta u - \nabla \pi_0 = u^* & \text{in } \mathcal{O} & \text{(3.44a)} \\[4pt]
\text{div}(u) = 0 & \text{in } \mathcal{O} & \text{(3.44b)} \\[4pt]
u = 0 & \text{on } S & \text{(3.44c)} \\[4pt]
u = (0, 0, w_2) & \text{on } \Omega & \text{(3.44d)} \\[4pt]
w_2 = w_1^* & \text{in } \Omega & \text{(3.44e)} \\[4pt]
\Delta^2 w_1 - \pi_0 \big|_\Omega = -w_2^* & \text{in } \Omega & \text{(3.44f)} \\[4pt]
w_1 = \dfrac{\partial w_1}{\partial \nu} = 0 & \text{on } \partial \Omega. & \text{(3.44g)}
\end{array}
$$

Moreover, we must justify that the pressure variable π_0 above is given by the expression

$$
\pi_0 = G_1(w_1) + G_2(u),
\tag{3.45}
$$

where we simply denoted by G_i, $i = 1, 2$, the operators $G_{0,i}$ defined in (3.11) (in line with the appearance of A in (3.12)).

1. *The Plate Velocity.* From (3.44e), the velocity component w_2 is immediately resolved.

2. *The Fluid Velocity.* We next consider the Stokes system (3.44a)–(3.44d). From (3.44e) and (3.44c)–(3.44d) we have $u|_{\partial \mathcal{O}}$ satisfies

$$\int_{\partial \mathcal{O}} u \cdot v \, d\sigma = \int_{\Omega} [0, 0, u^3] \cdot v \, d\Omega = \int_{\Omega} w_2 \, d\Omega = \int_{\Omega} w_1^* \, d\Omega = 0, \quad (3.46)$$

where the last equality follows by the membership $[u^*, w_1^*, w_2^*] \in \mathbf{H}$. (This last equality demonstrates the intrinsic nature of the zero average value constraint on structural displacement data.) Since this compatibility condition is satisfied and data $\{u^*, w_1^*\} \in \mathbf{L}^2(\mathcal{O}) \times H_0^2(\Omega)$, then by Lemma 3.1 we can find a unique (fluid and pressure) pair $(u, q_0) \in [\mathbf{H}^2(\mathcal{O}) \cap \mathscr{H}_f] \times H^1(\mathcal{O})/\mathbb{R}$ which solve

$$\Delta u - \nabla q_0 = u^* \qquad \text{in } \mathcal{O} \qquad (3.47a)$$

$$\text{div}(u) = 0 \qquad \text{in } \mathcal{O} \qquad (3.47b)$$

$$u = 0 \qquad \text{on } S \qquad (3.47c)$$

$$u = (0, 0, w_1^*) \qquad \text{on } \Omega. \qquad (3.47d)$$

Moreover, one has the estimate

$$\|u\|_{\mathbf{H}^2(\mathcal{O}) \cap \mathscr{H}_f} + \|q_0\|_{H^1(\mathcal{O})/\mathbb{R}} \leq C \big[\|u^*\|_{\mathscr{H}_f} + \|w_1^*\|_{H_0^2(\Omega)} \big]. \qquad (3.48)$$

3. *The Mechanical Displacement.* Subsequently, we consider the plate component boundary value problem (BVP) (3.44f)–(3.44g). By ellipticity and elliptic regularity (see [32] and [9]) there exists a solution $\hat{w}_1 \in H^4(\Omega) \cap H_0^2(\Omega)$ to the problem

$$\begin{cases} \Delta^2 \hat{w}_1 = q_0|_{\Omega} - w_2^* & \text{in } \Omega \\ \hat{w}_1 = \dfrac{\partial \hat{w}_1}{\partial v} = 0 & \text{on } \partial \Omega \end{cases}$$

where q_0 is the pressure variable in (3.47a). Moreover, we have the estimate

$$\|\hat{w}_1\|_{H^4(\Omega) \cap H_0^2(\Omega)} \leq C \|q_0|_{\Omega} + w_2^*\|_{L^2(\Omega)}$$

$$\leq C \|q_0|_{\Omega}\|_{H^{1/2}(\Omega)} + \|w_2^*\|_{L^2(\Omega)}$$

$$\leq C \|[u^*, w_1^*, w_2^*]\|_{\mathbf{H}} \qquad (3.49)$$

(in the last inequality we have also invoked Sobolev Trace Theory and (3.48)).

Now if, as in [16], we let $\mathbb{P}$ denote the orthogonal projection of $H_0^2(\Omega)$ onto $H_0^2(\Omega) \cap \hat{L}^2(\Omega)$—orthogonal with respect to the inner product $[\omega, \tilde{\omega}] \rightarrow (\Delta\omega, \Delta\tilde{\omega})_\Omega$—, then one can readily show that its orthogonal complement $I - \mathbb{P}$ can be characterized as

$$(I - \mathbb{P})H_0^2(\Omega) = \text{Span}\{\varphi\}, \quad \text{where}$$

$$\left\{\Delta^2\varphi = 1 \text{ in } \Omega, \ \varphi = \frac{\partial\varphi}{\partial\nu} = 0 \text{ on } \partial\Omega\right\}$$
(3.50)

(see [16, Remark 2.1, p. 6]). With these projections, we then set

$$w_1 = \mathbb{P}\hat{w}_1$$

$$\pi_0 = q_0 - \Delta^2(I - \mathbb{P})\hat{w}_1 .$$
(3.51)

With this assignment of variables, then by (3.49) and $\hat{w}_1 = \mathbb{P}\hat{w}_1 + (I - \mathbb{P})\hat{w}_1$, we will have that w_1 solves (3.44f)–(3.44g). (And of course since π_0 and q_0 differ only by a constant, then the pair (u, π_0) also solves (3.44a)–(3.44d).)

Moreover, from elliptic theory, (3.48) and (3.49), we have the estimate

$$\|w_1\|_{H^4(\Omega)\cap H_0^2(\Omega)\cap \hat{L}^2(\Omega)} + \|\pi_0\|_{H^1(\mathcal{O})}$$

$$\leq C \left(\|\Delta^2(I - \mathbb{P})\hat{w}_1\|_{L^2(\Omega)} + \|q_0\|_{H^1(\mathcal{O})/\mathbb{R}} + \|w_2^*\|_{L^2(\Omega)}\right)$$

$$\leq C \|[u^*, w_1^*, w_2^*]\|_{\mathbf{H}} ,$$
(3.52)

where implicitly we are also using the fact that $\Delta^2(I - \mathbb{P}) \in \mathcal{L}(H_0^2(\Omega), \mathbb{R})$, by the Closed Graph Theorem.

4. *Resolution of the Pressure.* As we noted in (3.16) we have $\Delta u \cdot \nu \in H^{-1/2}(\partial\mathcal{O})$, with the estimate

$$\|\Delta u \cdot \nu|_\Omega\|_{H^{-1/2}(\partial\mathcal{O})} \leq C \|u\|_{\mathbf{H}^2(\mathcal{O})} \leq C \left[\|u^*\|_{\mathscr{H}_f} + \|w_1^*\|_{H_0^2(\Omega)}\right],$$
(3.53)

where for the second inequality we have also used (3.48).

We will apply this estimate to the pressure variable π_0 in (3.44)—given explicitly in (3.51)—which solves *a fortiori*

$$\begin{cases} \Delta\pi_0 = 0 & \text{in } \mathcal{O} \\ \frac{\partial\pi_0}{\partial\nu} = \Delta u \cdot \nu|_S & \text{on } S \\ \frac{\partial\pi_0}{\partial\nu} + \pi_0 = \Delta^2 w_1 + \Delta u^3|_\Omega & \text{on } \Omega \end{cases} .$$

In fact: Applying the divergence operator to both sides of (3.44a) and using $\operatorname{div}(\Delta u) = \operatorname{div}(u^*) = 0$, we obtain that π_0 is harmonic in $\mathcal{O}$. Moreover, dotting both sides of (3.44a) with respect to the normal vector, and subsequently taking the boundary trace to the portion S, we get the boundary condition on S (implicitly we are also using $u^* \cdot v|_S = 0$, as $[u^*, w_1^*, w_2^*] \in \mathbf{H}$). Finally, as $u^* \cdot v|_\Omega = w_2^*$, and as $[u^*, w_1^*, w_2^*] \in \mathbf{H}$, we have from (3.44f)

$$\pi_0|_\Omega = w_2^* + \Delta^2 w_1 = \Delta u \cdot v|_\Omega - \nabla \pi_0 \cdot v|_\Omega + \Delta^2 w_1 \,,$$

which gives the asserted Robin boundary condition on Ω. Necessarily then, the pressure term must be given by the expression

$$\pi_0 = G_1(w_1) + G_2(u) \in H^1(\mathcal{O}) \tag{3.54}$$

(with the well-definition of right hand side assured by (3.53)).

Finally, we collect: (i) (3.47a)–(3.47d) and (3.48) (for the fluid variable u); (ii) (3.49) and (3.44e) and (3.51) (for the respective structure and pressure variables w_1, w_2 and π_0); (iii) (3.52) and (3.54) (for the characterization of the pressure term π_0). In this way we have obtained the solution of (3.44)–(3.45) in $\mathscr{D}(A)$, for given $[u^*, w_1^*, w_2^*] \in \mathbf{H}$. (Because of (3.44e), (3.48) and (3.52), the solution is unique.) We conclude now that $0 \in \varrho(A)$.   □

3.3.3 Proof of Main Result Theorem 3.2

Proof of Theorem 3.2 By Theorem 3.4, the fluid structure semigroup $\{e^{At}\}_{t \geq 0}$ will be uniformly stable provided its associated resolvent operator $R(\lambda; A)$ is bounded on the imaginary axis; viz.,

$$\|R(i\beta; A)\|_{\mathbf{H}} \leq C \qquad \text{for all } \beta \in \mathbb{R}. \tag{3.55}$$

By way of establishing (3.55), we consider the following resolvent equation, for $\beta \in \mathbb{R} \setminus \{0\}$ (recall that we have already established that $0 \in \varrho(A)$): Given data $[u^*, w_1^*, w_2^*] \in \mathbf{H}$, we look for $[w_1, w_2, u] \in \mathscr{D}(A)$ which solves

$$(i\beta - A) \begin{bmatrix} u \\ w_1 \\ w_2 \end{bmatrix} = \begin{bmatrix} u^* \\ w_1^* \\ w_2^* \end{bmatrix}. \tag{3.56}$$

From (3.12) and (3.13) we consider solution variables $[u, w_1, w_2] \in \mathscr{D}(A)$ satisfies the following PDE system

$$i\beta u - \Delta u + \nabla p = u^* \in \mathscr{H}_f \tag{3.57a}$$

$$\operatorname{div}(u) = 0 \qquad\qquad\qquad\qquad \text{in } \mathcal{O} \tag{3.57b}$$

$$u = 0 \qquad\qquad\qquad\qquad\qquad\qquad \text{on } S \qquad (3.57c)$$

$$u = (u^1, u^2, u^3) = (0, 0, i\beta w_1 - w_1^*) \qquad\qquad \text{on } \Omega \qquad (3.57d)$$

$$i\beta w_1 - w_2 = w_1^* \in \left[H_0^2(\Omega) \cap \hat{L}^2(\Omega) \right] \qquad (3.57e)$$

$$-\beta^2 w_1 + \Delta^2 w_1 - p|_\Omega = w_2^* + i\beta w_1^* \in \hat{L}^2(\Omega) \qquad (3.57f)$$

$$w_1\big|_{\partial\Omega} = \frac{\partial w_1}{\partial n}\bigg|_{\partial\Omega} = 0. \qquad (3.57g)$$

Here, associated pressure variable p is given by

$$p = G_1(w_1) + G_2(u) \in H^1(\mathcal{O}). \qquad (3.58)$$

Step 1 *(A Relation for the Fluid Gradient).* We start by taking the **H**-inner product of both sides of (3.56), with respect to $[u, w_1, w_2]$. This gives

$$i\beta \left\| \begin{bmatrix} u \\ w_1 \\ w_2 \end{bmatrix} \right\|_{\mathbf{H}}^2 - \left(A \begin{bmatrix} u \\ w_1 \\ w_2 \end{bmatrix}, \begin{bmatrix} u \\ w_1 \\ w_2 \end{bmatrix} \right)_{\mathbf{H}} = \left(\begin{bmatrix} u^* \\ w_1^* \\ w_2^* \end{bmatrix}, \begin{bmatrix} u \\ w_1 \\ w_2 \end{bmatrix} \right)_{\mathbf{H}}.$$

Combining this with the readily derivable relation

$$\left(A \begin{bmatrix} u \\ w_1 \\ w_2 \end{bmatrix}, \begin{bmatrix} u \\ w_1 \\ w_2 \end{bmatrix} \right)_{\mathbf{H}} = -\|\nabla u\|_{\mathcal{O}}^2 - 2i\,\mathrm{Im}(\Delta w_1, \Delta w_2)_\Omega, \qquad (3.59)$$

(see [3]) we then will have the following "static dissipation":

$$\|\nabla u\|_{L^2(\mathcal{O})}^2 = \mathrm{Re}\left(\begin{bmatrix} u^* \\ w_1^* \\ w_2^* \end{bmatrix}, \begin{bmatrix} u \\ w_1 \\ w_2 \end{bmatrix} \right)_{\mathbf{H}}. \qquad (3.60)$$

This gives then, for arbitrary $\varepsilon > 0$,

$$\|\nabla u\|_{L^2(\mathcal{O})} \leq \varepsilon \left\| \begin{bmatrix} u \\ w_1 \\ w_2 \end{bmatrix} \right\|_{\mathbf{H}} + C_\varepsilon \left\| \begin{bmatrix} u^* \\ w_1^* \\ w_2^* \end{bmatrix} \right\|_{\mathbf{H}}. \qquad (3.61)$$

Step 2 *(Control of the Mechanical Velocity).* This comes quickly: using the fluid Dirichlet boundary condition in (3.57) we have

$$i\beta w_1 = u^3\big|_\Omega + w_1^*.$$

We estimate this expression by invoking in sequence, the Sobolev Embedding Theorem, Poincaré's Inequality and (3.61). In this way, we then obtain

$$
\begin{aligned}
\|\beta w_1\|_{H^{1/2}(\Omega)} &\leq \| u^3 \|_\Omega + \|w_1^*\|_{H^{1/2}(\Omega)} \\
&\leq C \left(\|\nabla u\|_{L^2(\mathcal{O})} + \|w_1^*\|_{H_0^2(\Omega)} \right) \\
&\leq \varepsilon C \left\| \begin{bmatrix} u \\ w_1 \\ w_2 \end{bmatrix} \right\|_{\mathbf{H}} + C_\varepsilon \left\| \begin{bmatrix} u^* \\ w_1^* \\ w_2^* \end{bmatrix} \right\|_{\mathbf{H}}.
\end{aligned}
\tag{3.62}
$$

Using subsequently the resolvent relation $w_2 = i\beta w_1 - w_1^*$ now gives

$$
\|\beta w_1\|_{H^{1/2}(\Omega)} + \|w_2\|_{H^{1/2}(\Omega)} \leq \varepsilon C \left\| \begin{bmatrix} u \\ w_1 \\ w_2 \end{bmatrix} \right\|_{\mathbf{H}} + C_\varepsilon \left\| \begin{bmatrix} u^* \\ w_1^* \\ w_2^* \end{bmatrix} \right\|_{\mathbf{H}}.
\tag{3.63}
$$

Step 3 *(Control of the Mechanical Displacement).* We multiply both sides of the mechanical equation (3.57f) by w_1 and integrate. This gives the relation

$$
\left(\Delta^2 w_1, w_1\right)_{L^2(\Omega)} = \left(p|_\Omega, w_1\right)_\Omega + \beta^2 \|w_1\|_{L^2(\Omega)}^2 + \left(w_2^* + i\beta w_1^*, w_1\right)_{L^2(\Omega)}.
\tag{3.64}
$$

To handle the first term we use the fact that since $[u, w_1, w_2] \in \mathbf{H}$, then in particular

$$
\int_\Omega w_1 \, d\Omega = 0.
$$

In consequence, one can extend w_1 by zero—see e.g., Theorem 3.3 of [34]— so as to have well-posedness of the following boundary value problem (see [37, Theorem 2.4 and Remark 2.5]):

$$
\begin{cases}
-\Delta \psi + \nabla q = 0 & \text{in } \mathcal{O} \\
\operatorname{div} \psi = 0 & \text{in } \mathcal{O} \\
\psi|_S = 0 & \text{on } S \\
\psi|_\Omega = (\psi^1, \psi^2, \psi^3)|_\Omega = (0, 0, w_1) & \text{on } \Omega
\end{cases}
\tag{3.65}
$$

with the estimate

$$
\|\nabla \psi\|_{\mathbf{L}^2(\mathcal{O})} + \|q\|_{L^2(\mathcal{O})} \leq C \|w_1\|_{H^{1/2+\varepsilon}(\Omega)}.
\tag{3.66}
$$

Implicitly, we are also using Poincaré Inequality, and the fact that the Dirichlet trace map $\gamma_0 : H^1(\mathcal{O}) \to H^{1/2}(\partial\mathcal{O})$ is bounded and surjective on Lipschitz domains; see [19]. (See also [26, 27] for similar "Dirichlet maneuvering", used therein to show decays for an altogether different fluid-structure PDE.)

With this solution variable ψ in hand, we now address the first term on the right hand side of (3.64): since normal vector ν equals $(0, 0, 1)$ on Ω, and $u|_{\Omega} = \langle 0, 0, u^3 \rangle$ on Ω, and moreover $\mathrm{div}(u) = 0$ in $\mathcal{O}$, we have

$$
(p|_{\Omega}, w_1)_{\Omega} = 0 + \left(p\,\nu, \begin{bmatrix} 0 \\ 0 \\ w_1 \end{bmatrix} \right)_{L^2(\Omega)}
$$

$$
= -\left(\begin{bmatrix} D_3 u^1 \\ D_3 u^2 \\ D_3 u^2 \end{bmatrix}, \begin{bmatrix} 0 \\ 0 \\ w_1 \end{bmatrix} \right)_{\Omega} + \left(p\,\nu, \begin{bmatrix} 0 \\ 0 \\ w_1 \end{bmatrix} \right)_{L^2(\Omega)}
$$

$$
= -\left(\frac{\partial u}{\partial \nu}, \begin{bmatrix} 0 \\ 0 \\ w_1 \end{bmatrix} \right)_{L^2(\Omega)} + \left(p\,\nu, \begin{bmatrix} 0 \\ 0 \\ w_1 \end{bmatrix} \right)_{L^2(\Omega)}
$$

$$
= -\left(\frac{\partial u}{\partial \nu}, \psi \right)_{L^2(\partial\mathcal{O})} + (p\,\nu, \psi)_{L^2(\partial\mathcal{O})}, \tag{3.67}
$$

after invoking the boundary conditions in (3.65).

The use of Green's Identities and the fluid equation in (3.67) then gives

$$
(p|_{\Omega}, w_1)_{\Omega} = -\left(\frac{\partial u}{\partial \nu}, \psi \right)_{L^2(\partial\mathcal{O})} + (p\,\nu, \psi)_{L^2(\partial\mathcal{O})}
$$

$$
= -(\Delta u, \psi)_{L^2(\mathcal{O})} - (\nabla u, \nabla \psi)_{L^2(\mathcal{O})} + (\nabla p, \psi)_{L^2(\mathcal{O})}
$$

$$
= -i\beta(u, \psi)_{L^2(\mathcal{O})} - (\nabla u, \nabla \psi)_{L^2(\mathcal{O})} + (u^*, \psi)_{L^2(\mathcal{O})}.
$$

Estimating the latter right hand side by means of (3.61), (3.62) and (3.66) (and a rescaling of parameter $\varepsilon > 0$), we get for $\beta \in \mathbb{R} \setminus \{0\}$,

$$
|(p|_{\Omega}, w_1)_{\Omega}| \leq \left(|\beta| \, \|\nabla u\|_{L^2(\mathcal{O})} + \|\nabla u\|_{L^2(\mathcal{O})} + \|u^*\|_{L^2(\mathcal{O})} \right) \|\nabla \psi\|_{L^2(\mathcal{O})}
$$

$$
\leq \varepsilon\, C \left\| \begin{bmatrix} u \\ w_1 \\ w_2 \end{bmatrix} \right\|_{\mathbf{H}}^2 + C_{\varepsilon} \left\| \begin{bmatrix} u^* \\ w_1^* \\ w_2^* \end{bmatrix} \right\|_{\mathbf{H}}^2 + \left(\|\nabla u\|_{L^2(\mathcal{O})} + \|u^*\|_{L^2(\mathcal{O})} \right) \|\nabla \psi\|_{L^2(\mathcal{O})}
$$

$$
\leq 2\varepsilon\, C \left\| \begin{bmatrix} u \\ w_1 \\ w_2 \end{bmatrix} \right\|_{\mathbf{H}}^2 + C_{\varepsilon} \left\| \begin{bmatrix} u^* \\ w_1^* \\ w_2^* \end{bmatrix} \right\|_{\mathbf{H}}^2. \tag{3.68}
$$

Applying the obtained estimate (3.68) to the right hand side of (3.64), and using once more (3.62), we get

$$
\left(\Delta^2 w_1, w_1 \right)_{L^2(\Omega)} \leq 3\varepsilon C \left\| \begin{bmatrix} u \\ w_1 \\ w_2 \end{bmatrix} \right\|_{\mathbf{H}}^2 + C_\varepsilon \left\| \begin{bmatrix} u^* \\ w_1^* \\ w_2^* \end{bmatrix} \right\|_{\mathbf{H}}^2 .
$$

Another application of Green's formula, along the fact that w_1 satisfies clamped boundary conditions, gives

$$
\| \Delta w_1 \|_{L^2(\Omega)}^2 \leq 3\varepsilon C \left\| \begin{bmatrix} u \\ w_1 \\ w_2 \end{bmatrix} \right\|_{\mathbf{H}}^2 + C_\varepsilon \left\| \begin{bmatrix} u^* \\ w_1^* \\ w_2^* \end{bmatrix} \right\|_{\mathbf{H}}^2 . \tag{3.69}
$$

Thus, combining (3.61), (3.63) and (3.69) yields now

$$
\| [u, w_1, w_2] \|_{\mathbf{H}}^2 \leq 3\varepsilon C \, \| [u, w_1, w_2] \|_{\mathbf{H}}^2 + C_\varepsilon \, \| [u^*, w_1^*, w_2^*] \|_{\mathbf{H}}^2 .
$$

This yields the required uniform norm estimate (3.55), upon taking $\varepsilon > 0$ small enough. The proof of exponential decay of finite energy solutions of (3.2)–(3.3) is now complete.   □

Remark 3.3 In the course of our "frequency domain" proof of the exponential stability result Theorem 3.2, we have seen that, analogous to "time dependent" stability proofs of PDE systems under dissipative feedback, one must essentially estimate the finite energy norms of the structural solution variables in terms of the fluid dissipation. When $\rho = 0$ (the case under present consideration), control of the structural velocity solution variable βw_1 comes about very quickly; in particular, the boundary condition (3.57d) is invoked to control structural velocity variable βw_1 in $L^2(\Omega)$-norm (even the $H^{1/2}(\Omega)$-norm; see (3.62)).

However, for $\rho > 0$ the corresponding finite energy topology of the static structural velocity is instead $H_0^1(\Omega) \cap \hat{L}^2(\Omega)$ (see (3.5)). Because of this higher topology, the matching velocity boundary condition (3.57d) is of limited use for $\rho > 0$: indeed, the dissipation of the fluid velocity is measured in $\mathbf{H}^1(\mathcal{O})$-norm; thus by the Sobolev Trace Theorem, an immediate application of the boundary condition (3.57d) gives only a measurement of βw_1 in $H^{1/2}(\Omega)$-norm, at best. As we show in [3], for $\rho > 0$ a measurement of βw_1 in $H^1(\Omega)$-norm is ultimately accomplished, but at the price of 'penalizing' the majorizing constant by parameter $|\beta|$. This 'β-corrupted' estimate is ultimately what gives rise to the rational decay rate $O(1/t)$, as opposed to exponential decay. From what we have seen, an improvement in this rate of decay does not seem likely.

If, however, for $\rho > 0$ one were to incorporate *structural damping* into the PDE component (3.2e), so as to have instead

$$
w_{tt} - \rho \Delta w_{tt} + \Delta^2 w - \eta \Delta w_t = p|_\Omega \qquad \text{in } \Omega \times (0, \infty), \text{ where parameter } \eta > 0,
$$

then the frequency domain approach we have illustrated here would allow for a recovery of exponential decay for the fluid-structure solution variables. Indeed, very likely the resulting—structurally *and* fluid—damped C_0-semigroup would be *analytic*.

3.4 The Associated Optimal Control Problems: Relevant Scenarios, Expected Difficulties

In this section we briefly discuss a couple of possible implementations for the placement of control actions into the PDE system (3.2)–(3.3); these are complemented with some remarks about the technical challenges which are expected in the forthcoming study of the associated optimal control problems (with quadratic functionals).

3.4.1 A First Setup: Point Control on the Mechanical Component

A classical scenario worth studying is the case of *point control* exerted on the elastic wall Ω. The control action may be mathematically described by

$$\mathscr{B}g = \sum_{j=1}^{J} a_j \, g_j \delta_{\xi_j} \, ,$$

where ξ_j are points in Ω, and δ_{ξ_j} denote the corresponding *delta functions*. The control space here is $U = \mathbb{R}^J$ and

$$\mathscr{B} : U \to H^{-1-\sigma}(\Omega) \, , \ \sigma > 0 \, ;$$

accordingly, the initial/boundary value problem (IBVP) is as follows:

$$
\begin{cases}
u_t - \Delta u + \nabla p = 0 & \text{in } \mathscr{O} \times (0, T) \\
\operatorname{div} u = 0 & \text{in } \mathscr{O} \times (0, T) \\
u = 0 & \text{on } S \times (0, T) \\
u = (u^1, u^2, u^3) = (0, 0, w_t) & \text{on } \Omega \times (0, T) \\
w_{tt} - \rho \Delta w_{tt} + \Delta^2 w = p|_\Omega + \mathscr{B}g & \text{in } \Omega \times (0, T) \\
w = \frac{\partial w}{\partial \nu} = 0 & \text{on } \partial \Omega \times (0, T) \\
u(\cdot, 0) = u_0 & \text{in } \mathscr{O} \\
w(\cdot, 0) = w_0 \, , \ w_t(\cdot, 0) = w_1 & \text{in } \Omega \, .
\end{cases}
\tag{3.70}
$$

To deal with the IBVP problem (3.70), with the natural state space Y given by $\mathbf{H}_\rho$ and the control space U defined above, we will appeal to the corresponding abstract formulation

$$\begin{cases} y'(t) = Ay(t) + Bg(t), & 0 < t < T \\ y(0) = y_0 \in Y, \end{cases} \tag{3.71}$$

where

- $A : \mathscr{D}(A) \subset Y \to Y$ is the infinitesimal generator of a C_0-semigroup e^{At} on Y, $t \geq 0$;
- $B \in \mathscr{L}(U, [\mathscr{D}(A^*)]')$; equivalently, $A^{-1}B \in \mathscr{L}(U, Y)$.

The most prominent features of the controlled dynamics come are that (i) the control operator B is *not bounded* from U into Y; (ii) the semigroup e^{At} is *not analytic*.

Remark 3.4 We elaborate on the second feature: the (uncontrolled) PDE system (3.70) appears similar in structure to certain "rotational-inertia free and non-free" hyperbolic-parabolic PDE's, which describe elastic plate dynamics subjected to a thermal damping (see [24] and [29]). When $\rho = 0$, such thermoelastic systems are in fact associated with analytic semigroups; when rotational inertia parameter is present, these thermoelastic PDE systems are not. However, for all $\rho \geq 0$ the distinct PDE components of these systems of thermoelasticity evolve on the very *same* geometry, and hence the strong coupling of hyperbolic-parabolic dynamics is accomplished via *interior* differential terms. By contrast, the coupling in the present fluid-structure PDE system (3.70) occurs only on the boundary interface Ω. Thus, the dissipation manifested by the three dimensional (analytic) Stokes component is propagated to the two dimensional hyperbolic plate component in restricted fashion; namely, via boundary traces of the fluid and pressure.

To the control system (3.71) we associate a quadratic functional:

$$J(g) = \int_0^T \left(\|Ry(t)\|_Z^2 + \|g(t)\|_U^2 \right) dt, \tag{3.72}$$

where $R \in \mathscr{L}(Y, Z)$ denotes the *observation operator* and Z the observation space; thus, the optimal control problem is formulated as follows.

Problem 3.1 (The Optimal Control Problem) Given $y_0 \in Y$, we seek a control function $g \in L^2(0, T; U)$ which minimizes the cost functional (3.72), where $y(\cdot) = y(\cdot; y_0, g)$ is the solution to (3.71) corresponding to $g(\cdot)$.

As is well known, the core of the work is to pinpoint the regularity of the input-to-state map

$$L : g(\cdot) \longmapsto (Lg)(t) := \int_0^t e^{A(t-s)} Bg(s) \, ds, \tag{3.73}$$

which, in turn, is related to the *regularity properties* of the kernel $e^{At}B$ (or, equivalently, of $B^* e^{A^* t}$).

We expect to make use of the sharp regularity theory for the uncoupled plate equation in the presence of point control; see e.g., [38]. We also expect that, with the possible exception of the one-dimensional case for Ω, the presence of point control acting on the *hyperbolic* component of the PDE system will prevent one from establishing that the *gain operator* is bounded—unbounded even on a dense subset of Y—unless the observation operator R possesses appropriate smoothing properties.

3.4.2 A Different Setup: Boundary Control on the Fluid Component

Another interesting scenario is the case of *boundary* control acting on some part Σ of S. A tentative condition[1] to be taken into consideration is $u = g\,v$ on Σ. The IBVP becomes as follows:

$$
\begin{cases}
u_t - \Delta u + \nabla p = 0 & \text{in } \mathcal{O} \times (0, T) \\
\operatorname{div} u = 0 & \text{in } \mathcal{O} \times (0, T) \\
u = 0 & \text{on } S \setminus \Sigma \times (0, T) \\
u = g\,v & \text{on } \Sigma \times (0, T) \\
u = (u^1, u^2, u^3) = (0, 0, w_t) & \text{on } \Omega \times (0, T) \\
w_{tt} - \rho \Delta w_{tt} + \Delta^2 w = p|_\Omega & \text{in } \Omega \times (0, T) \\
w = \frac{\partial w}{\partial v} = 0 & \text{on } \partial\Omega \times (0, T) \\
u(\cdot, 0) = u_0 & \text{in } \mathcal{O} \\
w(\cdot, 0) = w_0, \ w_t(\cdot, 0) = w_1 & \text{in } \Omega .
\end{cases}
\tag{3.74}
$$

A first task to be accomplished is to derive the proper abstract formulation of the IBVP (3.74), having chosen—beside the state space $Y \equiv \mathbf{H}_\rho$—a natural control space U, as well as a class of control functions (such as, e.g., $L^2(0, T; U)$). At the outset, an important question which arises is whether the PDE problem (3.74) can in fact be modeled by an appropriate abstract ODE control system of the form (3.71); then $y(t) = e^{At} y_0 + (Lg)(t)$, with the operator L given by (3.73). If this is the case, the presence of boundary control acting on the *parabolic* component of the PDE system would suggest that we investigate whether the optimal control

---

[1] This condition was suggested by Giovanna Guidoboni (Indiana University and Purdue University at Indianapolis; also Acting Co-Director of the *School of Science Institute of Mathematical Modeling and Computational Science*), in connection with the modeling of ocular blood flow and specifically with the issue of reducing the ocular pressure.

theory devised in [1] ($T < \infty$) and [2] ($T = \infty$) is applicable. It is known that the analysis to be performed—in order to prove that the PDE problem actually fits into the abstract class of control systems formerly introduced in [1]—would then require that suitable regularity results for the boundary traces of the fluid component on Σ are established. This kind of "hidden regularity" has been shown to hold true in the case of a different F-S interaction (body immersed in a fluid); see [10] and [11]. (See [30] and [31] for a study of a Bolza problem associated with the very same PDE model.)

Remark 3.5 It is important to remind the reader (and to be emphasized) that a major feature of the aforesaid theory of [1,2], inspired by and tailored for certain boundary control problems for significant coupled PDE systems of *hyperbolic-parabolic* type, guarantees well-posedness of Riccati equations with *bounded* gains (on a dense subset of Y), *without* requiring smoothing effects of the observation operator R.

3.5   Appendix: Proof of Higher Regularity for $\mathscr{D}(A_\rho)$

Here, with Lemma 3.1 in hand, we infer the higher regularity for the fluid-structure generator, as stated in (3.13). Namely, if $[u_0, w_0, w_1] \in \mathscr{D}(A_\rho)$, then corresponding $[u_0, w_1, w_2] \in \mathbf{H}^2(\mathscr{O}) \times \mathscr{S}_\rho \times H_0^2(\Omega)$, and associated pressure variable π_0, as given in (3.16), is in $H^1(\mathscr{O})$.

In [4], by way of demonstrating the maximality of $A_\rho : \mathscr{D}(A_\rho) \subset \mathbf{H}_\rho \to \mathbf{H}_\rho$, we undertook the task of finding $[u, w_1, w_2] \in \mathscr{D}(A_\rho) \subset \mathbf{H}_\rho$ which solves the resolvent equation

$$(\lambda I - A_\rho) \begin{bmatrix} u \\ w_1 \\ w_2 \end{bmatrix} = \begin{bmatrix} u^* \\ w_1^* \\ w_2^* \end{bmatrix}$$

for given $\lambda > 0$, and data $[w_1^*, w_2^*, u^*] \in \mathbf{H}_\rho$. In PDE terms, this is the following system:

$$\lambda w_1 - w_2 = w_1^* \qquad\qquad \text{in } \Omega \qquad\qquad (3.75\text{a})$$

$$\lambda w_2 + P_\rho^{-1} \Delta^2 w_1 - + P_\rho^{-1} p\big|_\Omega = w_2^* \qquad\qquad \text{in } \Omega \qquad\qquad (3.75\text{b})$$

$$w_1 = \frac{\partial w_1}{\partial \nu} = 0 \qquad\qquad \text{on } \partial\Omega \qquad\qquad (3.75\text{c})$$

$$\lambda u - \Delta u + \nabla p = u^* \qquad\qquad \text{in } \mathscr{O} \qquad\qquad (3.75\text{d})$$

$$\text{div}(u) = 0 \qquad\qquad \text{in } \mathscr{O} \qquad\qquad (3.75\text{e})$$

$$u|_S = [0, 0, 0] \qquad\qquad \text{on } S \qquad\qquad (3.75\text{f})$$

$$u|_\Omega = [0, 0, w_2] \qquad\qquad \text{in } \Omega, \qquad\qquad (3.75\text{g})$$

where again, pressure term

$$p = G_{\rho,1}(w_1) + G_{\rho,2}(u) . \tag{3.76}$$

We start by briefly recalling what was done in [4] to recover the structural component w_1: in [4], the existence of a unique pair

$$[w_1, \tilde{c}] \in \left[H_0^2(\Omega) \cap \hat{L}(\Omega) \right] \times \mathbb{R} \tag{3.77}$$

which solves the following mixed variational formulation

$$\begin{cases} a_\lambda(w_1, \phi) + b(\phi, \tilde{c}) = \mathbb{F}(\phi) & \forall \phi \in H_0^2(\Omega) \\ b(w_1, r) = 0 & \forall r \in \mathbb{R} \end{cases} \tag{3.78}$$

was established. Here,

(i) $a_\lambda(\psi, \phi) = \lambda^2 (P_\rho^{1/2} \psi, P_\rho^{1/2} \phi)_\Omega + (\Delta \psi, \Delta \phi)_\Omega + \lambda (\nabla \tilde{f}(\psi), \nabla \tilde{f}(\phi))_\mathscr{O}$

$$+ \lambda^2 (\tilde{f}(\psi), \tilde{f}(\phi))_\mathscr{O}, \qquad \forall \psi, \phi \in H_0^2(\Omega);$$

(ii) $b(\phi, r) = -r \int_\Omega \phi \, d\Omega, \qquad \forall \phi \in H_0^2(\Omega) \text{ and } r \in \mathbb{R};$

(iii) $\mathbb{F}(\phi) = (\nabla \tilde{f}(w_1^*), \nabla \tilde{f}(\phi))_\mathscr{O} + \lambda (\tilde{f}(w_1^*), \tilde{f}(\phi))_\mathscr{O} - (\nabla \tilde{\mu}(u^*), \nabla \tilde{f}(\phi))_\mathscr{O}$

$$- \lambda (\tilde{\mu}(u^*), \tilde{f}(\phi))_\mathscr{O} + (u^*, \tilde{f}(\phi))_\mathscr{O} + (P_\rho(\lambda w_1^* + w_2^*), \phi)_\Omega, \qquad \forall \phi \in H_0^2(\Omega).$$

Also, $[\tilde{f}, \tilde{\pi}]$ and $[\tilde{\mu}, \tilde{q}]$ are the following solution maps (the solvability of each of which is assured by [37, Theorem 2.4 and Remark 2.5]): For given $\phi \in H_0^{1/2+\varepsilon}(\Omega)$, $\varepsilon > 0$, let $[\tilde{f}(\phi), \tilde{\pi}(\phi)] \in \mathbf{H}^1(\mathscr{O}) \times L^2(\mathscr{O})/\mathbb{R}$ solve

$$\begin{cases} \lambda \tilde{f} - \Delta \tilde{f} + \nabla \tilde{\pi} = 0 & \text{in } \mathscr{O} \\ \operatorname{div}(\tilde{f}) = \dfrac{1 \cdot \int_\Omega \phi \, d\Omega}{meas(\mathscr{O})} & \text{in } \mathscr{O} \\ \tilde{f}|_S = [0, 0, 0] & \text{in } S \\ \tilde{f}|_\Omega = [0, 0, \phi] & \text{in } \Omega. \end{cases} \tag{3.79}$$

(We note that the classic compatibility condition for solvability is satisfied, and that pressure variable $\tilde{\pi}$ is uniquely defined up to a constant. Also, the fact that $\varepsilon > 0$ allows us to extend Dirichlet boundary ϕ, by zero, so to have a function in $\mathbf{H}^{1/2+\varepsilon}(\partial\mathscr{O})$; see e.g., [34, Theorem 3.33].)

Likewise, for given fluid data $u^* \in \mathbf{L}^2(\mathcal{O})$ the solution variables $[\tilde{\mu}(u^*), \tilde{q}(u^*)] \in H^1(\mathcal{O}) \times L^2(\mathcal{O})/\mathbb{R}$ solve

$$\begin{cases} \lambda \tilde{\mu} - \Delta \tilde{\mu} + \nabla \tilde{q} = u^* & \text{in } \mathcal{O} \\ \text{div}(\tilde{\mu}) = 0 & \text{in } \mathcal{O} \\ \tilde{\mu}\big|_{\partial \mathcal{O}} = \mathbf{0} & \text{on } \partial \mathcal{O}. \end{cases} \qquad (3.80)$$

With solution component w_1 of (3.78) in hand, we then have from (3.75a),

$$w_2 \equiv \lambda w_1 - w_1^* \in \left[H_0^2(\Omega) \cap \hat{L}(\Omega) \right]. \qquad (3.81)$$

As was done in [4], we can in turn use $\tilde{c}$ of (3.77), w_2 of (3.81), and the maps (3.79) and (3.80), to recover the fluid component variable $[u, p]$ of (3.75d)–(3.75g) via the identification

$$\begin{cases} u = \tilde{f}(w_2) + \tilde{\mu}(u^*) \in \mathbf{H}^1(\mathcal{O}) \cap \mathcal{H}_f \\ p = \tilde{\pi}(w_2) + \tilde{q}(u^*) + \tilde{c} \in L^2(\mathcal{O}) \end{cases} \qquad (3.82)$$

where $[\tilde{f}, \tilde{\pi}]$ and $[\tilde{\mu}, \tilde{q}]$ are the solution maps defined respectively in (3.79) and (3.80).

But from Lemma 3.1, $[u, p]$ as the solution of (3.75d)–(3.75g)—with Dirichlet data $w_2 \in \left[H_0^2(\Omega) \cap \hat{L}(\Omega) \right]$—in fact enjoys the additional regularity

$$[u, p] \in \left[\mathbf{H}^2(\mathcal{O}) \cap \mathcal{H}_f \right] \times H^1(\mathcal{O}). \qquad (3.83)$$

Using (3.81) and (3.82), it was shown in [4] that the structural solution component w_1 of the variational system (3.78) indeed solves the biharmonic BVP (3.75b)–(3.75c). Consequently, as "forcing term" $p|_\Omega$ in (3.75b) is in $L^2(\Omega)$ (conservatively) from (3.83), then by elliptic regularity w_1 resides in the space $\mathcal{S}_\rho$, as defined in (3.14). (In the case that Ω is polygonal—i.e., $[\mathcal{O}, \Omega]$ obeys (G.2)—we appeal to [9, Theorem 2] to justify this extra regularity for w_1.)

Finally, the characterization of pressure term in (3.82) via the relation (3.76) was shown in [4].

Acknowledgements We would like to thank the two anonymous referees for their extremely useful comments and suggestions, which ultimately helped to enhance the strength of this work.

The research of G. Avalos was partially supported by the NSF Grants DMS-0908476 and DMS-1211232.

The research of F. Bucci was partially supported by the Italian MIUR under the PRIN 2009KNZ5FK Project (*Metodi di viscosità, geometrici e di controllo per modelli diffusivi nonlineari*), by the GDRE (Groupement De Recherche Européen) CONEDP (*Control of PDEs*), and also by the Università degli Studi di Firenze under the Project *Calcolo delle variazioni e teoria del controllo*.

F. Bucci is member of the Gruppo Nazionale per l'Analisi Matematica, la Probabilità e le loro Applicazioni (GNAMPA) of the Istituto Nazionale di Alta Matematica (INdAM).

References

1. Acquistapace, P., Bucci, F., Lasiecka, I.: Optimal boundary control and Riccati theory for abstract dynamics motivated by hybrid systems of PDEs. Adv. Differ. Equ. **10**(12), 1389–1436 (2005)
2. Acquistapace, P., Bucci F., Lasiecka, I.: A theory of the infinite horizon LQ-problem for composite systems of PDEs with boundary control. SIAM J. Math. Anal. **45**(3), 1825–1870 (2013)
3. Avalos, G., Bucci, F.: Spectral analysis and rational decay rates of strong solutions to a fluid-structure PDE system. arXiv:1312.4812 [math.AP] (2013)
4. Avalos, G., Clark, T.: A mixed variational formulation for the wellposedness and numerical approximation of a PDE model arising in a 3-D fluid-structure interaction. Evol. Equ. Control Theory **3**(4), 557–578 (2014)
5. Avalos, G., Dvorak, M.: A new maximality argument for a coupled fluid-structure interaction, with implications for a divergence-free finite element method. Appl. Math. (Warsaw) **35**(3), 259–280 (2008)
6. Avalos, G., Triggiani, R.: The coupled PDE system arising in fluid-structure interaction, Part I: explicit semigroup generator and its spectral properties. Contemp. Math. **440**, 15–54 (2007)
7. Avalos, G., Triggiani, R.: Boundary feedback stabilization of a coupled parabolic-hyperbolic Stokes–Lamé PDE System. J. Evol. Eq. **9**(2), 341–370 (2009)
8. Avalos, G., Triggiani, R.: Fluid structure interaction with and without internal dissipation of the structure: a contrast study in stability. Evol. Equ. Control Theory **2**(4), 563–598 (2013)
9. Blum, H., Rannacher, R.: On the boundary value problem of the biharmonic operator on domains with angular corners. Math. Methods Appl. Sci. **2**(4), 556–581 (1980)
10. Bucci, F., Lasiecka, I.: Optimal boundary control with critical penalization for a PDE model of fluid-solid interactions. Calc. Var. Part. Differ. Equ. **37**(1–2), 217–235 (2010)
11. Bucci, F., Lasiecka, I.: Regularity of boundary traces for a fluid-solid interaction model. Discrete Contin. Dyn. Syst. Ser. S **4**(3), 505–521 (2011)
12. Buffa, A.: Trace theorems on non-smooth boundaries for functional spaces related to Maxwell equations: an overview. In: Computational Electromagnetics (Kiel, 2001). Lecture Notes in Computational Science and Engineering, vol. 28, pp. 23–34. Springer, Berlin (2003)
13. Buffa, A., Geymonat, G.: On traces of functions in $W^{2,p}(\Omega)$ for Lipschitz domains in $\mathbb{R}^3$. C. R. Acad. Sci. Paris Sér. I Math. **332**(8), 699–704 (2001)
14. Chambolle, A., Desjardins, B., Esteban, M.J., Grandmont, C.: Existence of weak solutions for the unsteady interaction of a viscous fluid with an elastic plate. J. Math. Fluid Mech. **7**(3), 368–404 (2005)
15. Chueshov, I., Lasiecka, I.: Von Karman Evolution Equations. Well-Posedness and Long-Time Dynamics. Springer Monographs in Mathematics. Springer, New York (2010)
16. Chueshov, I., Ryzhkova, I.: A global attractor for a fluid-plate interaction model. Commun. Pure Appl. Anal. **12**(4), 1635–1656 (2013)
17. Constantin, P., Foias, C.: Navier–Stokes Equations. The University of Chicago Press, Chicago (1988)
18. Dauge, M.: Stationary Stokes and Navier–Stokes systems on two- or three-dimensional domains with corners, I. Linearized equations. SIAM J. Math. Anal. **20**(1), 74–97 (1989)
19. Gagliardo, E.: Caratterizzazioni delle tracce sulla frontiera relative ad alcune classi di funzioni in n variabili. Ren. Sem. Mat. Univ. Padova **27**, 284–305 (1957)
20. Grisvard, P.: Caractérisation de quelques espaces d'interpolation. Arch. Ration. Mech. Anal. **25**, 40–63 (1967)
21. Huang, F.L.: Characteristic conditions for exponential stability of linear dynamical systems in Hilbert spaces. Ann. Differ. Equ. **1**(1), 43–53 (1985)
22. Kesavan, S.: Topics in Functional Analysis and Applications. Wiley, New York (1989)
23. Kufner, A.: Weighted Sobolev Spaces. Wiley, New York (1985)

24. Lagnese, J.E.: Boundary Stabilization of Thin Plates. SIAM Studies in Applied Mathematics, vol. 10. Society for Industrial and Applied Mathematics, Philadelphia (1989)
25. Lagnese, J.E., Lions, J.L.: Modelling, Analysis and Control of Thin Plates. Recherches en Mathématiques Appliquées, vol. 6. Masson, Paris (1988)
26. Lasiecka, I, Lu, Y.: Asymptotic stability of finite energy in Navier–Stokes-elastic wave interaction. Semigroup Forum **82**, 61–82 (2011)
27. Lasiecka, I., Lu, Y.: Stabilization of a fluid structure interaction with nonlinear damping. Control Cybern. **42**(1), 155–181 (2013)
28. Lasiecka, I., Triggiani, R.: Differential and Algebraic Riccati Equations with Applications to Boundary/Point Control Problems: Continuous Theory and Approximation Theory. Lecture Notes in Control and Information Sciences, vol. 164. Springer, New York (1991)
29. Lasiecka, I., Triggiani, R.: Control Theory for Partial Differential Equations: Abstract Parabolic Systems: Continuous and Approximation Theories. Encyclopedia of Mathematics and its Applications, vol. 1. Cambridge University Press, Cambridge (2000)
30. Lasiecka, I. Tuffaha, A.: A Bolza optimal synthesis problem for singular estimate control systems. Control Cybern. **38**(4B), 1429–1460 (2009)
31. Lasiecka, I. Tuffaha, A.: Riccati theory and singular estimates for a Bolza control problem arising in linearized fluid-structure interaction. Syst. Control Lett. **58**(7), 499–509 (2009)
32. Lions, J.L., Magenes, E.: Non-Homogeneous Boundary Value Problems and Applications, vol. I. Springer, New York (1972)
33. Liu, Z., Zheng, S.: Exponential stability of the Kirchhoff plate with thermal or viscoelastic damping. Q. Appl. Math. **55**(3), 551–564 (1997)
34. McLean, W.: Strongly Elliptic Systems and Boundary Integral Equations. Cambridge University Press, New York (2000)
35. Nečas, J.: Direct Methods in the Theory of Elliptic Equations (translated by Gerard Tronel and Alois Kufner). Springer, New York (2012)
36. Prüss, J.: On the spectrum of C_0-semigroups. Trans. Am. Math. Soc. **284**(2), 847–857 (1984)
37. Temam, R.: Navier–Stokes Equations. Theory and Numerical Analysis. AMS Chelsea Publishing, Providence (2001)
38. Triggiani, R.: Interior and boundary regularity with point control, Part I: wave and Euler–Bernoulli equations. Diff. Int. Eqn. **6**, 111–129 (1993)

Chapter 4
Inverse Coefficient Problem for Grushin-Type Parabolic Operators

Karine Beauchard and Piermarco Cannarsa

To the memory of Alfredo Lorenzi, to his enthusiasm for mathematics and human warmth

Abstract The approach to Lipschitz stability for uniformly parabolic equations introduced by Imanuvilov and Yamamoto in 1998 based on Carleman estimates, seems hard to apply to the case of Grushin-type operators studied in this paper. Indeed, such estimates are still missing for parabolic operators degenerating in the interior of the space domain. Nevertheless, we are able to prove Lipschitz stability results for inverse coefficient problems for such operators, with locally distributed measurements in arbitrary space dimension. For this purpose, we follow a strategy that combines Fourier decomposition and Carleman inequalities for certain heat equations with nonsmooth coefficients (solved by the Fourier modes).

4.1 Introduction

4.1.1 Model

The relevance of the Heisenberg group to quantum mechanics has long been acknowledged. Indeed, it was recognized by Weyl [13] that the Heisenberg algebra generated by the momentum and position operators comes from a Lie algebra representation associated with a corresponding group—namely the Heisenberg group (Weyl group in the traditional language of physicists). In such a group, the role played by the so-called Heisenberg laplacian is absolutely central, being analogous

K. Beauchard
CMLS, Ecole Polytechnique, 91128 Palaiseau Cedex, France
e-mail: Karine.Beauchard@math.polytechnique.fr

P. Cannarsa (✉)
Dipartimento di Matematica, Università di Roma Tor Vergata, 00133 Roma, Italy
e-mail: cannarsa@mat.uniroma2.it

© Springer International Publishing Switzerland 2014
A. Favini et al. (eds.), *New Prospects in Direct, Inverse and Control Problems for Evolution Equations*, Springer INdAM Series 10,
DOI 10.1007/978-3-319-11406-4_4

to the standard laplacian in Euclidean spaces, see [11]. On an even larger scale, deep connections have been pointed out between the properties of subriemannian operators, like the Heisenberg laplacian, and other topics of interest to current mathematical research such as isoperimetric problems and systems theory, see, for instance [9].

Another important example of sublaplacian is the Grushin operator which takes the form

$$Gu = -(\partial_x^2 u + x^2 \partial_y^2 u) \tag{4.1}$$

on the plane. As a matter of fact, the Heisenberg laplacian and the Grushin operator are deeply related: the former can be transformed into the latter, and the corresponding heat kernels are connected by an integral map, see [12].

This paper is a part of a general project we are pursuing, which consists of investigating the possibility of extending the known controllability, observability, and Lipschitz stability properties of the heat equation, to degenerate parabolic problems. On all such topics, several results are available for parabolic operators which degenerate at the boundary of the space domain in low dimension, see, for instance [1,4–8].

In two space dimensions, a fairly complete analysis of Grushin operator is presented in [2] as far as controllability and observability are concerned, and generalized to the multidimensional case in [3]. The inverse source problem is treated in [3]. To the best of our knowledge, there are no results on inverse coefficient problems for Grushin-type equations. The goal of this article is to prove a Lipschitz stability estimate for the inverse coefficient problem, by adapting the techniques developed in [3] for the inverse source problem.

We consider Grushin-type equations of the form

$$\begin{cases} \partial_t u - \Delta_x u - |x|^{2\gamma} b(x) \Delta_y u = 0, & (t,x,y) \in (0,T) \times \Omega, \\ u(t,x,y) = 0, & (t,x,y) \in (0,T) \times \partial\Omega, \\ u(0,x,y) = u^0(x,y), & (x,y) \in \Omega, \end{cases} \tag{4.2}$$

where $T > 0$, $\Omega := \Omega_1 \times \Omega_2$, Ω_1 is a bounded open subset of $\mathbb{R}^{N_1}$, with C^4 boundary, such that $0 \in \Omega_1$, Ω_2 is a bounded open subset of $\mathbb{R}^{N_2}$, with C^2 boundary, $N_1, N_2 \in \mathbb{N}^* := \{1,2,3,\ldots\}$, $b \in C^1(\overline{\Omega}_1; (0,\infty))$, $\gamma \in (0,1]$ and $|.|$ is the Euclidean norm on $\mathbb{R}^{N_1}$.

Specifically, we are interested in the inverse coefficient problem: is it possible to recover the coefficient $b(x)$ from the knowledge of an observation $\partial_t u|_{(T_0,T_1) \times \omega}$, where ω is a nonempty open subset of Ω?

First, we recall well-posedness and regularity results for such equations. To this aim, we introduce the space $H_\gamma^1(\Omega)$, which is the closure of $C_0^\infty(\Omega)$ for the topology defined by the norm

$$\|f\|_{H_\gamma^1} := \left(\int_\Omega \left(|\nabla_x f|^2 + |x|^{2\gamma} |\nabla_y f|^2 \right) dx dy \right)^{1/2},$$

and the Grushin operator G_γ defined by

$$G_\gamma u := -\Delta_x u - |x|^{2\gamma} b(x) \Delta_y u \qquad \forall u \in D(G_\gamma),$$

where

$$D(G_\gamma) = \left\{ f \in H^1_\gamma(\Omega) \ : \ \exists c > 0 \text{ s.t. } \left| \int_\Omega \left(\nabla_x f \cdot \nabla_x g + |x|^{2\gamma} \nabla_y f \cdot \nabla_y g \right) dx dy \right| \right.$$
$$\left. \leqslant c \|g\|_{L^2(\Omega)} \quad \forall g \in H^1_\gamma(\Omega) \right\},$$

Theorem 4.1 *Let $\gamma > 0$. For every $u_0 \in L^2(\Omega)$ and $g \in L^2((0, T) \times \Omega)$, there exists a unique weak solution $u \in C^0([0, T]; L^2(\Omega)) \cap L^2(0, T; H^1_\gamma(\Omega))$ of (4.2). Moreover, $u \in C^0((0, T]; D(G_\gamma))$.*

We refer to [2] for the proof with $N_1 = N_2 = 1$; the general case can be treated similarly.

4.1.2   Hypotheses and Notations

We introduce an open subset $\Omega'_1 \subset\subset \Omega_1$ such that $0 \notin \Omega'_1$ and $\delta > 0$ such that

$$x \in \Omega'_1 \qquad \Rightarrow \qquad |x| \geqslant \delta.$$

The function b is a priori assumed to satisfy

$$b \in \mathcal{M} := \{b \in C^1(\overline{\Omega}_1; [m, M]) \ : \ b \equiv 1 \text{ on } \Omega_1 \setminus \Omega'_1\}$$

for some positive constants m, M with $0 < m \leqslant 1 \leqslant M$ that are fixed in the whole article. In particular, $b \equiv 1$ on a neighborhood of $x = 0$ and $\partial\Omega_1$.

In order to introduce the hypotheses on the initial data u^0 of system (4.2), under which we prove Lipschitz stability estimate, the following notation is required. Let $\mathscr{A}$ be the operator defined by

$$D(\mathscr{A}) := H^2 \cap H^1_0(\Omega_2), \qquad \mathscr{A}\varphi := -\Delta_y\varphi$$

and let $(\mu_n)_{n\in\mathbb{N}^*}$ be the nondecreasing sequence of its eigenvalues, with associated eigenvectors $(\varphi_n)_{n\in\mathbb{N}^*}$, so that

$$\begin{cases} -\Delta_y\varphi_n(y) = \mu_n\varphi_n(y), & y \in \Omega_2, \\ \varphi_n(y) = 0, & y \in \partial\Omega_2. \end{cases} \qquad (4.3)$$

When $v = v(x, y) \in L^2(\Omega)$, then, we denote by $v_n = v_n(x)$ its Fourier components (with respect to variable y)

$$v_n(x) := \int_{\Omega_2} v(x, y)\varphi_n(y) dy, \forall n \in \mathbb{N}^*.$$

To prove the Lipschitz stability estimate, the initial data u^0 of system (4.2) will be assumed to belong to the class

$$\mathscr{D}_{N,K_1,T_1} := \Big\{ u^0 \in D(G_\gamma^{s/2}); \ u_N^0 \geq 0 \ \text{on} \ \Omega_1 \ \text{and}$$

$$\sup_{x \in \Omega_1'} \Big(e^{\frac{T_1}{2} \Delta_{\Omega_1'} u_N^0} \Big)(x) \geq K_1 e^{\delta^{2\gamma} m T_1 \mu_N} \| u^0 \|_{D(G_\gamma^{s/2})} \Big\}$$

where $s > N_1/2$ is fixed in the whole article, $K_1, T_1 > 0$ and $N \in \mathbb{N}^*$. Here, $e^{\tau \Delta_{\Omega_1'}}$ denotes the heat flow on Ω_1': for $\phi^0 \in L^2(\Omega_1')$, the function $\phi(\tau, x) := \Big(e^{\tau \Delta_{\Omega_1'}} \phi^0 \Big)(x)$ is the solution of

$$\begin{cases} \partial_\tau \phi(\tau, x) - \Delta_x \phi(\tau, x) = 0, & (\tau, x) \in (0, +\infty) \times \Omega_1', \\ \phi(\tau, x) = 0, & (\tau, x) \in (0, +\infty) \times \partial \Omega_1', \\ \phi(0, x) = \phi^0(x), & x \in \Omega_1'. \end{cases}$$

Note that

$$\Big\{ u^0 \in D(G_\gamma^{s/2}); \ u_N^0 \geq 0 \ \text{on} \ \Omega_1 \Big\} = \cup_{j=1}^\infty \mathscr{D}_{N,1/j,T_1}.$$

In particular if $u^0 \in D(G_\gamma^{s/2})$ is ≥ 0 on Ω_1 then $u_1^0 \geq 0$ thus $u^0 \in \mathscr{D}_{1,1/j}$ for j large enough. Thus, this class of functions is quite general.

Finally, we denote by C a constant which may change from line to line.

4.1.3   Main Results

Our main result consists of a Lipschitz stability estimate, with observation on a vertical strip $\omega := \omega_1 \times \Omega_2$, for appropriate initial conditions. When $\gamma \in (0, 1)$, the Lipschitz stability estimate holds in any positive time.

Theorem 4.2 *Let $\gamma \in (0, 1)$, ω_1 be a nonempty open subset of Ω_1, $\omega := \omega_1 \times \Omega_2$ a vertical strip, $T \in (0, \infty)$, $T_1 \in (0, T)$ and $K_1 > 0$. Then there exists a constant $\mathscr{C} = \mathscr{C}(K_1, T_1) > 0$ such that, for every $b, \tilde{b} \in \mathscr{M}$, $N \in \mathbb{N}$, $u^0 \in L^2(\Omega)$, and $\tilde{u}^0 \in \mathscr{D}_{N,K_1,T_1}$, the associated solutions u and $\tilde{u}$ of (4.2) satisfy*

$$\int_{\Omega_1'} (b - \tilde{b})(x)^2 dx \leq \frac{\mathscr{C}}{\| \tilde{u}^0 \|_{D(G_\gamma^{s/2})}^2} \Bigg[\int_0^T \int_\omega |\partial_t(u - \tilde{u})(t, x, y)|^2 dxdydt$$

$$+ \int_{\Omega_1' \times \Omega_2} |G_\gamma(u - \tilde{u})(T_1, x, y)|^2 dxdy \Bigg]. \tag{4.4}$$

Note that the constant $\mathscr{C}$ above does not depend on N.

When $\gamma = 1$, the Lipschitz stability estimate holds for a sufficiently large time, as stated below.

Theorem 4.3 *We assume* $\gamma = 1$, ω_1 *is a nonempty open subset of* Ω_1 *and* $\omega :=$ $\omega_1 \times \Omega_2$ *is a vertical strip. There exists* $T_1^* > 0$ *such that, for every* $T_1, T \in (0, \infty)$ *with* $T_1^* < T_1 < T$ *there exists* $\mathscr{C} = \mathscr{C}(K_1, T_1) > 0$ *such that, for every* $b, \tilde{b} \in \mathscr{M}$, $N \in \mathbb{N}$, $u^0 \in L^2(\Omega)$, $\tilde{u}^0 \in \mathscr{D}_{N, K_1, T_1}$, *the associated solutions* u *and* $\tilde{u}$ *of (4.2) satisfy (4.4).*

The assumption on the initial data ($\tilde{u}^0 \in \mathscr{D}_{N, K_1, T_1}$) is an important restriction, essentially related to technical difficulty. The validity of the Lipschitz stability estimate under more general assumptions is an interesting open problem that will be investigated in future works.

4.1.4 Structure of the Article

This article is organized as follows. Section 4.2 is devoted to preliminary results concerning the well posedness of (4.2), the Fourier decomposition of its solutions, the dissipation speed of the Fourier modes, embeddings between spaces related to the Grushin operator, and Harnack's inequality. In Sect. 4.3, we prove our main results, namely, Theorems 4.2 and 4.3.

4.2 Preliminaries

4.2.1 Well Posedness

In this section, we recall a known regularity result (see, e.g., [3]) for the solution of problem (4.2) that will be used in what follows.

Theorem 4.4 *Let* $\gamma \in (0, 1]$, $u^0 \in D(G_\gamma)$, $g \in H^1((0, T), L^2(\Omega))$, *and*

$$u \in C^0([0, T]; L^2(\Omega)) \cap L^2(0, T; H^1_\gamma(\Omega))$$

be the solution of

$$\begin{cases} \partial_t u - \Delta_x u - |x|^{2\gamma} b(x) \Delta_y u = g(t, x, y), & (t, x, y) \in (0, \infty) \times \Omega, \\ u(t, x, y) = 0, & (t, x, y) \in (0, \infty) \times \partial\Omega, \\ u(0, x, y) = u^0(x, y), & (x, y) \in \Omega. \end{cases}$$

Then the function $v := \partial_t u$ *belongs to* $L^2(0, T; H^1_\gamma(\Omega))$ *and is a weak solution of*

$$\begin{cases} \partial_t v - \Delta_x v - b(x)|x|^{2\gamma} \Delta_y v = \partial_t g(t, x, y), & (t, x, y) \in (0, \infty) \times \Omega, \\ v(t, x, y) = 0, & (t, x, y) \in (0, \infty) \times \partial\Omega, \\ v(0, x, y) = -G_\gamma u_0(x, y) + g(0, x, y), & (x, y) \in \Omega. \end{cases} \quad (4.5)$$

4.2.2 Fourier Decomposition

The following known result is the starting point of our analysis (see [2] for the proof).

Theorem 4.5 *Let* $u^0 \in L^2(\Omega)$, $g \in L^2((0, T) \times \Omega)$ *and* u *be the weak solution of*

$$\begin{cases} \partial_t u - \Delta_x u - |x|^{2\gamma} b(x) \Delta_y u = g(t, x, y), & (t, x, y) \in (0, \infty) \times \Omega, \\ u(t, x, y) = 0, & (t, x, y) \in (0, \infty) \times \partial\Omega, \\ u(0, x, y) = u^0(x, y), & (x, y) \in \Omega. \end{cases}$$

For every $n \in \mathbb{N}^*$, *the function*

$$u_n(t, x) := \int_{\Omega_2} u(t, x, y) \varphi_n(y) dy$$

belongs to $C^0([0, T]; L^2(\Omega))$ *and is the unique weak solution of*

$$\begin{cases} \partial_t u_n - \Delta_x u_n + \mu_n |x|^{2\gamma} b(x) u_n = g_n(t, x), & (t, x) \in (0, T) \times \Omega_1, \\ u_n(t, x) = 0, & t \in (0, T) \times \partial\Omega_1, \\ u_n(0, x) = u_{n,0}(x), & x \in \Omega_1, \end{cases} \quad (4.6)$$

where

$$g_n(t, x) := \int_{\Omega_2} g(t, x, y) \varphi_n(y) dy \quad and \quad u_{0,n}(x) = \int_{\Omega_2} u_0(x, y) \varphi_n(y) dy.$$

4.2.3 Dissipation Speed

We introduce, for every $n \in \mathbb{N}^*$, $\gamma > 0$, the operator $G_{n,\gamma}$ defined on $L^2(\Omega_1)$ by

$$D(G_{n,\gamma}) := H^2 \cap H^1_0(\Omega_1), \ G_{n,\gamma} u := -\Delta_x u + \mu_n |x|^{2\gamma} b(x) u. \quad (4.7)$$

The smallest eigenvalue of $G_{n,\gamma}$ is given by

$$\lambda_{n,\gamma} = \min\left\{ \frac{\int_{\Omega_1} \left[|\nabla v(x)|^2 + \mu_n |x|^{2\gamma} b(x) v(x)^2\right] dx}{\int_{\Omega_1} v(x)^2 dx} : v \in H_0^1(\Omega_1), \, v \neq 0 \right\}.$$

The asymptotic behavior of $\lambda_{n,\gamma}$ as $n \to \infty$, which quantifies the dissipation speed of the solutions of (4.2), is described below (see [3] for a proof).

Theorem 4.6 *For every $\gamma > 0$, there exists constants $c_*, c^* > 0$ such that*

$$c_* \mu_n^{\frac{1}{1+\gamma}} \leq \lambda_{n,\gamma} \leq c^* \mu_n^{\frac{1}{1+\gamma}}, \qquad \forall n \in \mathbb{N}^*.$$

4.2.4   Continuous Embeddings

In this section, we obtain a continuous embedding result for the domains of powers of $G_{N,\gamma}$, which is used in the proof of our main theorem.

Theorem 4.7 *For every $s > N_1/2$ we have that $D(G_{N,\gamma}^{s/2}) \subset L^\infty(\Omega_1)$, with continuous embedding.*

We prove the conclusion just when s is an even positive integer. In this case, setting $k = s/2$, it suffices to show that

$$k \in \mathbb{N}, \; k > \frac{N_1}{4} \quad \Longrightarrow \quad D(G_{N,\gamma}^k) \subset L^\infty(\Omega_1). \tag{4.8}$$

Let $k = 1$. We have that

$$u \in D(G_{N,\gamma}) \quad \Longleftrightarrow \quad u \in H^2 \cap H_0^1(\Omega_1). \tag{4.9}$$

Therefore, u is continuous for $N_1 = 1$. Moreover,

$$u \in \begin{cases} W^{1,p}(\Omega_1) \; \forall p > 1 \text{ if } N_1 = 2 \\ W^{1,2^*}(\Omega_1) \qquad \text{if } N_1 > 2 \end{cases} \tag{4.10}$$

where

$$\frac{1}{2^*} = \frac{1}{2} - \frac{1}{N_1}. \tag{4.11}$$

So, u is Hölder continuous in $\overline{\Omega}_1$ by Sobolev's embedding provided that $2^* > N_1$, that is, $N_1 < 4$. We have thus checked (4.8) for $k = 1$.

Now, suppose $N_1 \geq 4$ (so that $2^* \leq N_1$), let $k = 2$, and take $u \in D(G_{N,\gamma}^2)$. Set $v := G_{N,\gamma} u$, and observe that u satisfies the boundary value problem

$$\begin{cases} -\Delta_x u + \mu_N |x|^{2\gamma} b(x) u = v(x) & x \in \Omega_1 \\ u(x) = 0 & x \in \partial\Omega_1. \end{cases}$$

Moreover, since

$$v \in W^{1,2^*}(\Omega_1),$$

it follows that

$$v \in \begin{cases} L^p(\Omega_1) & \forall p > 1 \text{ if } 2^* = N_1 = 4 \\ L^{2^{**}}(\Omega_1) & \text{if } N_1 > 4. \end{cases}$$

Thus, owing to the L^p-regularity of solutions to elliptic equations with Hölder continuous coefficients, we have that

$$u \in \begin{cases} W^{2,p}(\Omega_1) & \forall p > 1 \text{ if } 2^* = N_1 = 4 \\ W^{2,2^{**}}(\Omega_1) & \text{if } N_1 > 4. \end{cases} \tag{4.12}$$

The above inclusions imply that u is smooth right away if $2^{**} > N_1$, that is, $N_1 < 6$. By a refinement of the above argument one obtains the embedding in (4.8). Indeed, for $N_1 \geq 6$, (4.12) yields

$$u \in \begin{cases} W^{1,p}(\Omega_1) & \forall p > 1 \text{ if } 2^{**} = N_1 = 6 \\ W^{1,2^{***}}(\Omega_1) & \text{if } N_1 > 6. \end{cases} \tag{4.13}$$

This implies that u is Hölder continuous for $N_1 < 2^{***}$, that is, $N_1 < 8$. We have thus checked (4.8) for $k = 1, 2$. The general result follows by iteration.

4.2.5 Harnack's Inequality

In our next proposition, we recall the well-known Harnack inequality for the heat equation (see [10]).

Theorem 4.8 *Let U be an open subset of Ω_1 and let $V \subset\subset U$ be connected. Let $T > 0$, $0 < t_1 < T_1 < T$, and set $U_T := (0, T) \times U$. Then there exists $C_H > 0$ such that, for every solution $u \in C^2(U_T)$ of*

$$\partial_t u - \Delta u = 0 \quad \text{in } U_T$$

with $u \geqslant 0$ on U_T, one has that

$$\inf_{x \in V} u(T_1, x) \geqslant C_H \sup_{x \in V} u(t_1, x).$$

4.3 Proof of Lipschitz Stability

In this section, first, we prove Theorem 4.2, then we explain how to adapt the reasoning to the proof of Theorem 4.3.

The function $v(t, x, y) := (u - \tilde{u})(t, x, y)$ satisfies

$$\begin{cases} \partial_t v_N - \Delta_x v_N + \mu_N |x|^{2\gamma} b(x) v_N = \mu_N |x|^{2\gamma} (b - \tilde{b})(x) \tilde{u}_N, & (t, x) \in (0, T) \times \Omega_1, \\ v_N(t, x) = 0, & (t, x) \in (0, T) \times \partial \Omega_1, \\ v_N(0, x) = v_N^0(x), & x \in \Omega_1. \end{cases}$$

Step 1: **Use of Harnack inequality and assumption** $\tilde{u}^0 \in \mathscr{D}_{N,K_1,T_1}$. We recall that

$$\begin{cases} \partial_t \tilde{u}_N - \Delta_x \tilde{u}_N + \mu_N |x|^{2\gamma} b(x) \tilde{u}_N = 0, & (t, x) \in (0, +\infty) \times \Omega_1, \\ \tilde{u}_N(t, x) = 0, & t \in (0, +\infty) \times \partial \Omega_1, \\ \tilde{u}_N(0, x) = \tilde{u}_N^0(x), & x \in \Omega_1. \end{cases}$$

Let us introduce the solution $v_N(t, x)$ of

$$\begin{cases} \partial_t v_N - \Delta_x v_N + \mu_N \delta^{2\gamma} m v_N = 0, & (t, x) \in (0, T) \times \Omega_1', \\ v_N(t, x) = 0, & (t, x) \in (0, T) \times \partial \Omega_1', \\ v_N(0, x) = \tilde{u}_N^0(x), & x \in \Omega_1'. \end{cases}$$

Then

$$\mu_N |x|^{2\gamma} b(x) \geqslant \mu_N \delta^{2\gamma} m, \quad \forall x \in \Omega_1',$$
$$\tilde{u}_N(t, x) \geqslant 0 = v_N(t, x), \quad \forall (t, x) \in (0, +\infty) \times \partial \Omega_1',$$
$$\tilde{u}_N(0, x) = v_N(0, x), \quad \forall x \in \Omega_1'.$$

By the maximum principle, we deduce that

$$\tilde{u}_N(t, x) \geqslant v_N(t, x), \quad \forall (t, x) \in (0, +\infty) \times \Omega_1'.$$

Note that

$$v_N(t, x) = e^{-\mu_N \delta^{2\gamma} mt} \left(e^{t \Delta_{\Omega_1'}} u_N^0 \right)(x).$$

Thus

$$\inf_{z \in \Omega_1'} |\tilde{u}_N(T_1, z)| \geq \inf_{z \in \Omega_1'} |v_N(T_1, z)|$$

$$\geq e^{-\mu_N \delta^{2\gamma} m T_1} \inf_{z \in \Omega_1'} \left| \left(e^{T_1 \Delta_{\Omega_1'}} \tilde{u}_N^0 \right)(z) \right|$$

$$\geq e^{-\mu_N \delta^{2\gamma} m T_1} C_H \sup_{z \in \Omega_1'} \left| \left(e^{\frac{T_1}{2} \Delta_{\Omega_1'}} \tilde{u}_N^0 \right)(z) \right| \text{ by Proposition 4.8}$$

$$\geq C_H K_1 \|\tilde{u}^0\|_{D(G_\gamma^{s/2})}$$

because $\tilde{u}^0 \in \mathscr{D}_{N,K_1,T_1}$. In particular, $\inf_{z \in \Omega_1'} |\tilde{u}_N(T_1, z)|$ is positive thus

$$\int_{\Omega_1'} (b - \tilde{b})(x)^2 dx \leq \frac{1}{\mu_N^2 \delta^{4\gamma} \inf_{z \in \Omega_1'} |\tilde{u}_N(T_1, z)|^2} \int_{\Omega_1'} \left| \mu_N |x|^{2\gamma} (b - \tilde{b})(x) \tilde{u}_N(T_1, x) \right|^2 dx$$

$$\leq \frac{1}{\mu_N^2 \delta^{4\gamma} C_H^2 K_1^2 \|\tilde{u}^0\|_{D(G_\gamma^{s/2})}^2} \int_{\Omega_1'} \left| \mu_N |x|^{2\gamma} (b - \tilde{b})(x) \tilde{u}_N(T_1, x) \right|^2 dx.$$

Therefore

$$\int_{\Omega_1'} (b - \tilde{b})(x)^2 dx \leq \frac{C}{2\|\tilde{u}^0\|_{D(G_\gamma^{s/2})}^2} \int_{\Omega_1'} \left| \mu_N |x|^{2\gamma} (b - \tilde{b})(x) \tilde{u}_N(T_1, x) \right|^2 dx$$

$$\leq \frac{C}{\|\tilde{u}^0\|_{D(G_\gamma^{s/2})}^2} \int_{\Omega_1'} \left(|\partial_t v_N(T_1, x)|^2 + |G_{N,\gamma} v_N(T_1, x)|^2 \right) dx$$

$$(4.14)$$

where

$$C := \frac{1}{\mu_1^2 \delta^{4\gamma} C_H^2 K_1^2}.$$

In order to dominate properly the first term of the right hand side, we revisit the proof of Proposition 6 of [3].

Step 2:   Duhamel formula reads as

$$\partial_t v_N(T_1) = e^{-G_{N,\gamma}(T_1 - t)} \partial_t v_N(t) + \int_t^{T_1} e^{-G_{N,\gamma}(T_1 - \tau)} g_N(\tau) d\tau, \quad \forall t \in (0, T_1)$$

where

$$g_N(\tau, x) = \mu_N |x|^{2\gamma} (b - \tilde{b})(x) \partial_t \tilde{u}_N(\tau, x).$$

Thus,

$$\|\partial_t v_N(T_1)\|_{L^2(\Omega_1)} \leq e^{-\lambda_{N,\gamma}(T_1 - t)} \|\partial_t v_N(t)\|_{L^2(\Omega_1)}$$

$$+ \int_t^{T_1} e^{-\lambda_{N,\gamma}(T_1 - \tau)} \|g_N(\tau)\|_{L^2(\Omega_1)} d\tau.$$

Moreover,

$$\|g_N(\tau)\|_{L^2(\Omega_1)} \leqslant C\mu_N \|b - \tilde{b}\|_{L^2(\Omega_1')} \|\partial_t \tilde{u}_N(\tau)\|_{L^\infty(\Omega_1)}.$$

By the continuous embedding proved in Sect. 4.2.4, we have

$$\|\partial_t \tilde{u}_N(\tau)\|_{L^\infty(\Omega_1)} \leqslant C\|\partial_t \tilde{u}_N(\tau)\|_{D(G_{N,\gamma}^{s/2})}$$
$$\leqslant C\|\tilde{u}_N^0\|_{D(G_{N,\gamma}^{s/2})} e^{-\lambda_{N,\gamma}\tau}$$
$$\leqslant C\|\tilde{u}^0\|_{D(G_\gamma^{s/2})} e^{-\lambda_{N,\gamma}\tau}.$$

Therefore,

$$\|g_N(\tau)\|_{L^2(\Omega_1)} \leqslant C\mu_N \|b - \tilde{b}\|_{L^2(\Omega_1')} \|\tilde{u}^0\|_{D(G_\gamma^{s/2})} e^{-\lambda_{N,\gamma}\tau} \tag{4.15}$$

and

$$\|\partial_t v_N(T_1)\|_{L^2(\Omega_1)} \leqslant e^{-\lambda_{N,\gamma}(T_1-t)} \|\partial_t v_N(t)\|_{L^2(\Omega_1)}$$
$$+ C(T_1 - t)\|\tilde{u}^0\|_{D(G_\gamma^{s/2})} \mu_N e^{-\lambda_{N,\gamma}T_1} \|b - \tilde{b}\|_{L^2(\Omega_1')}.$$

Taking the square, we get

$$\int_{\Omega_1} |\partial_t v_N(T_1,x)|^2 dx \leqslant 2e^{-2\lambda_{N,\gamma}(T_1-t)} \int_{\Omega_1} |\partial_t v_N(t,x)|^2 dx$$
$$+ 2C^2(T_1 - t)^2 \|\tilde{u}^0\|_{D(G_\gamma^{s/2})}^2 \mu_N^2 e^{-2\lambda_{N,\gamma}T_1} \int_{\Omega_1'} (b - \tilde{b})(x)^2 dx.$$

Integrating over $t \in (T_1/3, 2T_1/3)$, we obtain

$$\int_{\Omega_1} |\partial_t v_N(T_1)|^2 \leqslant \frac{6}{T_1} e^{-2\lambda_{N,\gamma}T_1/3} \int_{T_1/3}^{2T_1/3} \int_{\Omega_1} |\partial_t v_N(t,x))|^2 dxdt$$
$$+ 2C^2 \left(\frac{2T_1}{3}\right)^2 \|\tilde{u}^0\|_{D(G_\gamma^{s/2})}^2 \mu_N^2 e^{-2\lambda_{N,\gamma}T_1} \int_{\Omega_1'} |x|^{4\gamma} (b - \tilde{b})(x)^2 dx. \tag{4.16}$$

Step 3:   We apply Carleman estimate. Working exactly as in the step 2 of the proof of Proposition 6 of [3], we get, for N large enough

$$\int_{T_1/3}^{2T_1/3} \int_{\Omega_1} |\partial_t v_N(t,x)|^2 dxdt \leqslant Ce^{C\mu_N^{p(\gamma)}} \left(\int_0^{T_1} \int_{\omega_1} |\partial_t v_N(t,x)|^2 dxdt \right.$$
$$\left. + \int_0^{T_1} \int_{\Omega_1} |g_N(t,x)|^2 dxdt \right)$$

where $C = C(T_1) > 0$ and

$$p(\gamma) := \begin{cases} 1/2 & \text{if } \gamma \in [1/2, 1], \\ 2/3 & \text{if } \gamma \in (0, 1/2). \end{cases} \tag{4.17}$$

Moreover, estimate (4.15) justifies that

$$\int_0^{T_1} \int_{\Omega_1} |g_N(t,x)|^2 dxdt \leqslant \frac{C^2}{2\lambda_{N,\gamma}} \mu_N^2 \|\tilde{u}^0\|^2_{D(G_\gamma^{3/2})} \int_{\Omega_1'} (b-\tilde{b})(x)^2 dx.$$

Thus,

$$\int_{T_1/3}^{2T_1/3} \int_{\Omega_1} |\partial_t v_N(t,x)|^2 dxdt \leqslant C e^{C\mu_N^{p(\gamma)}} \int_0^{T_1} \int_{\omega_1} |\partial_t v_N(t,x)|^2 dxdt$$

$$+ \frac{C}{\lambda_{N,\gamma}} \mu_N^2 e^{C\mu_N^{p(\gamma)}} \|\tilde{u}^0\|^2_{D(G_\gamma^{s/2})} \int_{\Omega_1'} (b-\tilde{b})(x)^2 dx.$$
$$(4.18)$$

Step 4:   By combining (4.14), (4.16) and (4.18), we obtain

$$\int_{\Omega_1'} (b-\tilde{b})(x)^2 dx$$

$$\leqslant \frac{C}{\|u^0\|^2_{D(G_\gamma^{s/2})}} \left(\int_{\Omega_1'} |G_{N,\gamma} v_N(T_1,x)|^2 dx + e^{C\mu_N^{p(\gamma)} - 2\lambda_{N,\gamma}T_1/3} \int_0^{T_1} \int_{\omega_1} |\partial_t v_N(t,x)|^2 dxdt \right)$$

$$+ \left(\frac{1}{\lambda_{N,\gamma}} \mu_N^2 e^{C\mu_N^{p(\gamma)} - 2\lambda_{N,\gamma}T_1/3} + \mu_N^2 e^{-2\lambda_{N,\gamma}T_1} \right) \int_{\Omega_1'} (b-\tilde{b})(x)^2 dx.$$

For N large enough, the source term in the right hand side may be absorbed by the left hand side. Indeed, by (4.17), we have $\frac{1}{1+\gamma} > p(\gamma)$ for every $\gamma \in (0,1)$; thus, by Proposition 4.6, $\mu_N^{p(\gamma)} = o(\lambda_{N,\gamma})$ when $N \to \infty$. Thus, we get a constant $\mathscr{C} > 0$ (independent of N) such that, for N large enough (i.e. $N \geqslant N_*$),

$$\int_{\Omega_1'} (b-\tilde{b})(x)^2 |x|^{4\gamma} dx$$

$$\leqslant \frac{\mathscr{C}}{\|u^0\|^2_{D(G_\gamma^{s/2})}} \left(\int_0^{T_1} \int_{\omega_1} |\partial_t v_N(t,x)|^2 dxdt + \int_{\Omega_1'} |G_{N,\gamma} v_N(T_1,x)|^2 dx \right)$$

$$\leqslant \frac{\mathscr{C}}{\|u^0\|^2_{D(G_\gamma^{s/2})}} \left(\int_0^{T_1} \int_{\omega} |\partial_t v(t,x,y)|^2 dxdydt + \int_\Omega |G_\gamma v(T_1,x,y)|^2 dxdy \right).$$

When $\gamma = 1$, then $\lambda_{N,\gamma}$ behaves asymptotically like $C\mu_N^{p(\gamma)}$, thus the time T_1 needs to be taken large enough for the same conclusion to hold.

The previous arguments treat the high frequencies (i.e. $N \geqslant N_*$). For low frequencies (i.e., $N < N_*$), the Lipschitz stability estimate for the inverse source problem in the uniformly parabolic case (see [14]) yields the conclusion.

Acknowledgements The authors are grateful to Masahiro Yamamoto and Philippe Gravejat for fruitful discussions.

This research was performed in the framework of the GDRE CONEDP, and was partially supported by the Istituto Nazionale di Alta Matematica "F. Severi" and Ècole Polytechnique. The authors wish to express their gratitude to the above institutions. The first author was also supported by the "Agence Nationale de la Recherche" (ANR) Projet Blanc EMAQS number ANR-2011-BS01-017-01.

References

1. Alabau-Boussouira, F., Cannarsa, P., Fragnelli, G.: Carleman estimates for degenerate parabolic operators with applications to null controllability. J. Evol. Equ. **6**, 161–204 (2006)
2. Beauchard, K., Cannarsa, P., Guglielmi, R.: Null controllability of Grushin-type equations in dimension two. J. Eur. Math. Soc. **16**(1), 67–101 (2014)
3. Beauchard, K., Cannarsa, P., Yamamoto, M.: Inverse source problem and null controllability for multidimensional parabolic operators of Grushin type. Inverse Probl. **30**(2), 025006, 26 (2014)
4. Cannarsa, P., Martinez, P., Vancostenoble, J.: Null controllability of degenerate heat equations. Adv. Differ. Equ. **10**(2), 153–190 (2005)
5. Cannarsa, P., Fragnelli, G., Rocchetti, D.: Null controllability of degenerate parabolic operators with drift. Netw. Heterog. Media **2**, 695–715 (2007)
6. Cannarsa, P., Martinez, P., Vancostenoble, J.: Carleman estimates for a class of degenerate parabolic operators. SIAM J. Control Optim. **47**(1), 1–19 (2008)
7. Cannarsa, P., Martinez, P., Vancostenoble, J.: Carleman estimates and null controllability for boundary-degenerate parabolic operators. C. R. Math. Acad. Sci. Paris **347**(3–4), 147–152 (2009)
8. Cannarsa, P., Tort, J., Yamamoto, M.: Determination of source terms in a degenerate parabolic equation. Inverse Probl. **26**, 105003, 20 (2010)
9. Capogna, L., Danielli, D., Pauls, S.D., Tyson, J.T.: An Introduction to the Heisenberg Group and the Sub-Riemannian Isoperimetric Problem. Progress in Mathematics, vol. 259. Birkhauser, Basel (2007)
10. Evans, L.: Partial Differential Equations. American Mathematical Society, Providence (2010)
11. Folland, G.B.: Harmonic Analysis in Phase Space. Annals of Mathematics Studies. Princeton University Press, Princeton (1989)
12. Furutani, K., Iwasaki, C., Kagawa, T.: An action function for a higher step Grushin operator. J. Geom. Phys. **62**(9), 1949–1976 (2012)
13. Weyl, H.: The Theory of Groups and Quantum Mechanics. Dover, New York (1931)
14. Yamamoto, M., Zou, J.: Simultaneous reconstruction of the initial temperature and heat radiative coefficient. Inverse Probl. **17**, 1181–1202 (2001)

Chapter 5
Determining the Scalar Potential in a Periodic Quantum Waveguide from the DN Map

Mourad Choulli, Yavar Kian, and Eric Soccorsi

Dedicated to the memory of Alfredo Lorenzi (1944–2013)

Abstract We prove logarithmic stability in the determination of the time-dependent scalar potential in a periodic quantum cylindrical waveguide, from the boundary measurements of the solution to the dynamic Schrödinger equation.

5.1 Introduction

In this short paper we review the main ideas and results developed in [4].

5.1.1 Statement of the Problem

Let ω be a bounded connected open subset of $\mathbb{R}^2$ that contains the origin, with C^2-boundary $\partial\omega$. We put $\Omega = \mathbb{R} \times \omega$ and write $x = (x_1, x')$ with $x' = (x_2, x_3)$ for every $x = (x_1, x_2, x_3) \in \Omega$ throughout this text. Given $T > 0$, we consider the following initial boundary value problem (IBVP in short)

$$\begin{cases} (-i\,\partial_t - \Delta u + V(t, x))u = 0 & \text{in } Q = (0, T) \times \Omega, \\ u(0, \cdot) = u_0 & \text{in } \Omega, \\ u = g & \text{on } \Sigma = (0, T) \times \partial\Omega, \end{cases} \tag{5.1}$$

M. Choulli (✉)
IECL, UMR CNRS 7502, Université de Lorraine, Ile du Saulcy, 57045 Metz cedex 1, France
e-mail: mourad.choulli@univ-lorraine.fr

Y. Kian • E. Soccorsi
Aix-Marseille Université, CNRS, CPT UMR 7332, 13288 Marseille, France

Université de Toulon, CNRS, CPT UMR 7332, 83957 La Garde, France
e-mail: yavar.kian@univ-amu.fr; eric.soccorsi@univ-amu.fr

© Springer International Publishing Switzerland 2014
A. Favini et al. (eds.), *New Prospects in Direct, Inverse and Control Problems for Evolution Equations*, Springer INdAM Series 10,
DOI 10.1007/978-3-319-11406-4_5

where the time dependent electric potential V is 1-periodic with respect to the infinite variable x_1:

$$V(t, x_1 + 1, x') = V(t, x_1, x'), \quad (t, x_1, x') \in Q. \tag{5.2}$$

In the present paper we examine the stability issue in the determination of V from the knowledge of the "boundary" operator

$$\Lambda_V : (u_0, g) \longrightarrow (\partial_\nu u_{|\Sigma}, u(T, \cdot)), \tag{5.3}$$

where the measure of $\partial_\nu u_{|\Sigma}$ (resp., $u(T, \cdot)$) is performed on Σ (resp., in Ω). Here $\nu(x), x \in \partial\Omega$, denotes the outward unit normal to Ω and $\partial_\nu u(t, x) = \nabla u(t, x) \cdot \nu(x)$, $t \in (0, T)$.

5.1.2 What is Known so Far

There are only a few results available in the mathematical literature on the identification of time-dependent coefficients appearing in an IBVP, such as [1, 2, 6, 10]. All these results were obtained in bounded domains. Several authors considered the problem of recovering time independent coefficients in an unbounded domain from boundary measurements. In most of the cases the unbounded domain under consideration is either a half space [15, 16] or an infinite slab [12, 14, 18].

The case of an infinite cylindrical waveguide was addressed in [3, 13]. For inverse problems with time-independent coefficients in unbounded domains we also refer to [8]. In [7], uniqueness modulo gauge invariance was proved in the inverse problem of determining the time-dependent electric and magnetic potentials from the Dirichlet-to-Neumann map for the Schrödinger equation in a simply-connected bounded or unbounded domain. More specifically the inverse problem of determining periodic coefficients in the Helmholtz equation was recently examined in [9].

5.1.3 Boundary Operator

We define the trace operator τ_0 by

$$\tau_0 w = (w_{|\Sigma}, w(0, \cdot)) \text{ for all } w \in C_0^\infty([0, T] \times \mathbb{R}, C^\infty(\overline{\omega})),$$

and extend it to a bounded operator from $H^2(0, T; H^2(\Omega))$ into

$$L^2((0, T) \times \mathbb{R}; H^{3/2}(\partial\omega)) \times L^2(\Omega).$$

Then the space $X_0 = \tau_0(H^2(0, T; H^2(\Omega)))$ is easily seen to be Hilbertian for the norm

$$\|w\|_{X_0} = \inf\{\|W\|_{H^2(0,T;H^2(\Omega))}, W \in H^2(0, T; H^2(\Omega)) \text{ such that } \tau_0 W = w\}$$

and we recall from [4, Corollary 2.1] the following useful existence and uniqueness result:

Proposition 5.1 *Fix $M > 0$ and let $V \in C([0, T], W^{2,\infty}(\Omega))$ be such that*

$$\|V\|_{C([0,T];W^{2,\infty}(\Omega))} \leq M.$$

Then for every $(g, u_0) \in X_0$, the IBVP (5.1) admits a unique solution

$$\mathfrak{s}(g, u_0) \in Z = L^2(0, T; H^2(\Omega)) \cap H^1(0, T; L^2(\Omega))$$

and there is a constant $C > 0$, depending only on ω, T and M, such that we have

$$\|\mathfrak{s}(g, u_0)\|_Z \leq C \|(g, u_0)\|_{X_0}. \tag{5.4}$$

Armed with Proposition 5.1 we turn now to defining the operator Λ_V appearing in (5.3). To do that we introduce the linear bounded operator τ_1 from $L^2((0, T) \times \mathbb{R}; H^2(\omega)) \cap H^1(0, T; L^2(\Omega))$ into $X_1 = L^2(\Sigma) \times L^2(\Omega)$, obeying

$$\tau_1 w = \left(\partial_\nu w_{|\Sigma}, w(T, \cdot)\right) \text{ for } w \in C_0^\infty([0, T] \times \mathbb{R}; C^\infty(\overline{\omega})).$$

In view of (5.4) we have $\|\tau_1\mathfrak{s}(g, u_0)\|_{X_1} \leq C \|\mathfrak{s}(g, u_0)\|_Z \leq C \|(g, u_0)\|_{X_0}$, where, as in the remaining part of this text, C denotes some generic positive constant. As a consequence the operator $\Lambda_V = \tau_1 \circ \mathfrak{s}$ is bounded from X_0 into X_1 and $\|\Lambda_V\| = \|\Lambda_V\|_{\mathscr{B}(X_0,X_1)} \leq C$.

5.1.4 Main Result

The main result of this paper is borrowed from [4, Theorem 1.1] and claims logarithmic stability in the determination of V from Λ_V. Putting $\Omega' = (0, 1) \times \omega$, $Q' = (0, T) \times \Omega'$ and $\Sigma'_* = (0, T) \times (0, 1) \times \partial\omega$, it may be stated as follows.

Theorem 5.1 *For $M > 0$ fixed, let V_1, $V_2 \in W^{2,\infty}(0, T; W^{2,\infty}(\Omega))$ fulfill (5.2) together with the three following conditions:*

$$(V_2 - V_1)(T, \cdot) = (V_2 - V_1)(0, \cdot) = 0 \text{ in } \Omega', \tag{5.5}$$

$$V_2 - V_1 = 0 \text{ in } \Sigma'_*, \tag{5.6}$$

$$\|V_j\|_{W^{2,\infty}(0,T;W^{2,\infty}(\Omega'))} \leq M, \ j = 1, 2. \tag{5.7}$$

Then there are two constants $C > 0$ and $\gamma^* > 0$, depending only on T, ω and M, such that the estimate

$$\|V_2 - V_1\|_{L^2(Q')} \leq C \left(\ln \left(\frac{1}{\|\Lambda_{V_2} - \Lambda_{V_1}\|_{\mathcal{B}(X_0, X_1)}} \right) \right)^{-1/5},$$

holds whenever $0 < \|\Lambda_{V_2} - \Lambda_{V_1}\|_{\mathcal{B}(X_0, X_1)} < \gamma^*$.

5.1.5 Outline

The paper is organized as follows. In Sect. 5.2 we introduce the Floquet–Bloch–Gel'fand transform, that is used to decompose the IBVP (5.1)–(5.2) into a collection of IBVPs in Q', with quasi-periodic boundary conditions on $(0, T) \times \{0, 1\} \times \omega$. Section 5.3 is devoted to building suitable optics geometric solutions (abbreviated as OGS in the sequel) for each of these problems. Finally a sketch of the proof of Theorem 5.1, which is by means of the OGS defined in Sect. 5.3, is given in Sect. 5.4.

5.2 Floquet–Bloch–Gel'fand Analysis

The main tool in the analysis of the periodic system (5.1)–(5.2) is the partial Floquet–Bloch–Gel'fand transform (abbreviated to FBG in the sequel) with respect to the x_1-direction, that is described below.

5.2.1 Partial FBG Transform

For any arbitrary open subset Y of $\mathbb{R}^n$, $n \in \mathbb{N}^*$, we define the partial FBG transform with respect to x_1 of $f \in C_0^\infty(\mathbb{R} \times Y)$ by

$$\check{f}_{Y,\theta}(x_1, y) = (\mathcal{U}_Y f)_\theta(x_1, y) = \sum_{k=-\infty}^{+\infty} e^{-ik\theta} f(x_1 + k, y), \quad x_1 \in \mathbb{R}, \ y \in Y, \ \theta \in [0, 2\pi).$$

$$(5.8)$$

With reference to [17, Sect. XIII.16], $\mathcal{U}_Y$ extends to a unitary operator, still denoted by $\mathcal{U}_Y$, from $L^2(\mathbb{R} \times Y)$ onto the Hilbert space

$$\int_{(0,2\pi)}^{\oplus} L^2((0,1) \times Y) d\theta/(2\pi) = L^2((0, 2\pi) d\theta/(2\pi); L^2((0,1) \times Y)).$$

Let $H^s_{\sharp,loc}(\mathbb{R} \times Y)$, $s = 1, 2$, denote the subspace of distributions f in $\mathbb{R} \times Y$ such that $f_{|I \times Y} \in H^s(I \times Y)$ for any bounded open subset $I \subset \mathbb{R}$. Then a function $f \in H^s_{\sharp,loc}(\mathbb{R} \times Y)$ is said to be 1-periodic with respect to x_1 if it satisfies $f(x_1 + k, y) = f(x_1, y)$ for a.e. $(x_1, y) \in (0, 1) \times Y$ and all $k \in \mathbb{Z}$. The subspace of functions of $H^s_{\sharp,loc}(\mathbb{R} \times Y)$, that are 1-periodic with respect to x_1, is denoted by $H^s_{\sharp,per}(\mathbb{R} \times Y)$. Such a function being obviously determined by its values on $(0, 1) \times Y$, we put $H^s_{\sharp,per}((0, 1) \times Y) = \{u_{|(0,1) \times Y}, u \in H^s_{\sharp,per}((0, 1) \times Y)\}$. Since $\check{f}_{Y,\theta}(x_1 + 1, y) = e^{i\theta} \check{f}_{Y,\theta}(x_1, y)$ for a.e. $(x_1, y) \in \mathbb{R} \times Y$ and all $\theta \in [0, 2\pi)$, by (5.8), we next set $H^s_{\sharp,\theta}((0, 1) \times Y) = \{e^{i\theta x_1}u, u \in H^s_{\sharp,per}((0, 1) \times Y)\}$ and then derive from [5, Chap. II, Sect. 1, Définition 1] that

$$\mathscr{U}_Y H^s(\mathbb{R} \times Y) = \int_{(0,2\pi)}^{\oplus} H^s_{\sharp,\theta}((0, 1) \times Y)\frac{d\theta}{2\pi}, \quad s = 1, 2.$$

For the sake of simplicity we will systematically omit the subscript Y in $\mathscr{U}_Y$ and $\check{f}_{Y,\theta}$ in the remaining part of this text.

5.2.2   FBG Decomposition

Let τ'_0 denote the linear bounded operator from $H^2(0, T; H^2(\Omega'))$ into $L^2((0, T) \times (0, 1); H^{3/2}(\partial\omega)) \times L^2(\Omega')$ such that $\tau'_0 w = (w_{|\Sigma'_*}, w(0, \cdot))$ for $w \in C^\infty_0((0, T) \times (0, 1); C^\infty(\overline{\omega}))$. Thus, putting $\mathscr{X}'_{0,\theta} = \tau'_0(H^2(0, T; H^2_{\sharp,\theta}(\Omega')))$ for all $\theta \in [0, 2\pi)$, it is easy to check that $\mathscr{X}_0 = \mathscr{U} X_0 = \int_{(0,2\pi)}^{\oplus} \mathscr{X}'_{0,\theta} d\theta/(2\pi)$ and

$$\mathscr{U} \tau_0 \mathscr{U}^{-1} = \int_{(0,2\pi)}^{\oplus} \tau'_0 d\theta/(2\pi),$$

where the notation τ'_0 stands for the operator τ'_0 restricted to $H^2(0, T; H^2_{\sharp,\theta}(\Omega'))$. The last identity means that $(\mathscr{U} \tau_0 f)_\theta = \tau'_0 (\mathscr{U} f)_\theta$ for all $f \in H^2(0, T; H^2(\Omega))$ and a. e. $\theta \in (0, 2\pi)$.

Further, we have $\mathscr{Z} = \mathscr{U} Z = \int_{(0,2\pi)}^{\oplus} \mathscr{Z}'_\theta d\theta/(2\pi)$, where

$$\mathscr{Z}'_\theta = L^2(0, T; H^2_{\sharp,\theta}(\Omega')) \cap H^1(0, T; L^2(\Omega')).$$

Thus, applying the transform $\mathscr{U}$ to (5.1), we immediately get the:

Proposition 5.2 *Let $V \in W^{2,\infty}(0, T; W^{2,\infty}(\Omega))$ fulfill (5.2) and let $(g, u_0) \in X_0$. Then u is the solution $\mathfrak{s}(g, u_0) \in Z$ to (5.1) defined in Proposition 5.1 if and only if*

$\mathscr{U} u \in \mathscr{L}$ and each $\check{u}_\theta = (\mathscr{U} u)_\theta \in \mathscr{L}_\theta$, for $\theta \in [0, 2\pi)$, is solution to the following IBVP

$$\begin{cases} (-i\partial_t - \Delta + V)v = 0 & \text{in } Q' = (0, T) \times \Omega', \\ v(0, \cdot) = \check{u}_{0,\theta} & \text{in } \Omega', \\ v = \check{g}_\theta & \text{on } \Sigma'_*, \end{cases} \qquad (5.9)$$

where $\check{g}_\theta$ (resp. $\check{u}_{0,\theta}$) stands for $(\mathscr{U} g)_\theta$ (resp. $(\mathscr{U} u_0)_\theta$), that is

$$(\check{g}_\theta, \check{u}_{0,\theta}) = (\mathscr{U}(g, u_0))_\theta.$$

The existence and uniqueness of solutions to (5.9) for $\theta \in [0, 2\pi)$ is guaranteed by [4, Lemma 2.1]:

Lemma 5.1 *Assume that V obeys the conditions of Proposition 5.2 and satisfies*

$$\|V\|_{W^2(0,T;W^{2,\infty}(\Omega'))} \leq M,$$

for some $M > 0$. Then for every $(\check{g}_\theta, \check{u}_{0,\theta}) \in \mathscr{X}'_{0,\theta}$, $\theta \in [0, 2\pi)$, there exists a unique solution $\mathfrak{s}_\theta(\check{g}_\theta, \check{u}_{0,\theta}) \in \mathscr{L}'_\theta$ to (5.9), such that the estimate

$$\|\mathfrak{s}_\theta(\check{g}_\theta, \check{u}_{0,\theta})\|_{\mathscr{L}'_\theta} \leq C \|(\check{g}_\theta, \check{u}_{0,\theta})\|_{\mathscr{X}'_{0,\theta}}, \qquad (5.10)$$

holds for some constant $C > 0$ depending only on T, ω and M.

5.2.3 Boundary Operators

In view of Lemma 5.1 the linear operator $\mathfrak{s}_\theta$, $\theta \in [0, 2\pi)$, is bounded from $\mathscr{X}'_{0,\theta}$ into $\mathscr{L}'_\theta$, with

$$\|\mathfrak{s}_\theta\| = \|\mathfrak{s}_\theta\|_{\mathscr{B}(\mathscr{X}'_{0,\theta}, \mathscr{L}'_\theta)} \leq C, \ \theta \in [0, 2\pi). \qquad (5.11)$$

Let τ'_1 be the linear bounded operator from

$$L^2((0, T) \times (0, 1); H^2(\Omega')) \cap H^1(0, T; L^2(\Omega'))$$

$$\longrightarrow \mathscr{X}'_1 = L^2((0, T) \times (0, 1) \times \partial\omega) \times L^2(\Omega'),$$

satisfying $\tau'_1 w = (\partial_\nu w_{|\Sigma'_*}, w(T, \cdot))$ for all $w \in C_0^\infty((0, T) \times (0, 1); C^\infty(\overline{\omega}))$, in such a way that $\mathscr{X}_1 = \mathscr{U} X_1 = \int_{(0,2\pi)}^\oplus \mathscr{X}'_1 d\theta/(2\pi)$ and $\mathscr{U} \tau_1 \mathscr{U}^{-1} = \int_{(0,2\pi)}^\oplus \tau'_1 d\theta/(2\pi)$. Then we have $\|\tau'_1 \mathfrak{s}_\theta(\check{g}_\theta, \check{u}_{0,\theta})\|_{\mathscr{X}'_1} \leq C \|\mathfrak{s}_\theta(\check{g}_\theta, \check{u}_{0,\theta})\|_{\mathscr{L}'_\theta} \leq C \|(\check{g}_\theta, \check{u}_{0,\theta})\|_{\mathscr{X}'_{0,\theta}}$ for every $\theta \in [0, 2\pi)$, from (5.10), so the reduced boundary operator $\Lambda_{V,\theta} = \tau'_1 \circ s_\theta \in \mathscr{B}(\mathscr{X}'_{0,\theta}, \mathscr{X}'_1)$. Further, it follows from Proposition 5.2 and Lemma 5.1 that

$$\mathscr{U} \Lambda_V \mathscr{U}^{-1} = \int_{(0,2\pi)}^{\oplus} \Lambda_{V,\theta} d\theta/(2\pi),$$

hence [5, Chap. II, Sect. 2, Proposition 2] yields:

$$\|\Lambda_V\|_{\mathscr{B}(X_0, X_1)} = \sup_{\theta \in (0,2\pi)} \|\Lambda_{V,\theta}\|_{\mathscr{B}(\mathscr{X}'_{0,\theta}, \mathscr{X}'_1)}. \tag{5.12}$$

5.3 Optics Geometric Solutions

For each $\theta \in [0, 2\pi)$ we aim to build solutions to the system

$$\begin{cases} (-i\partial_t - \Delta + V)v = 0 & \text{in } Q', \\ u(\cdot, 1, \cdot) = e^{i\theta} u(\cdot, 0, \cdot) & \text{on } (0, T) \times \omega, \\ \partial_{x_1} u(\cdot, 1, \cdot) = e^{i\theta} \partial_{x_1} u(\cdot, 0, \cdot) & \text{on } (0, T) \times \omega. \end{cases} \tag{5.13}$$

Specifically, for $r > 0$ fixed, we seek solutions $u_{k,\theta}, k \in \mathbb{Z}$, to (5.13) of the form

$$u_{k,\theta}(t, x) = \left(e^{i\theta x_1} + w_{k,\theta}(t, x)\right) e^{-i((\xi \cdot \xi + 4\pi^2 k^2)t + 2\pi k x_1 + x' \cdot \xi)}, \quad (t, x) = (t, x_1, x') \in Q', \tag{5.14}$$

where $w_{k,\theta} \in H^2(0, T; H^2_{\sharp,\theta}(\Omega'))$ obeys

$$\|w_{k,\theta}\|_{H^2(0,T;H^2(\Omega'))} \le \frac{c}{r}(1 + |k|), \tag{5.15}$$

for some constant $c > 0$ independent of r, k and θ, and $\xi \in \mathbb{C}^2 \setminus \mathbb{R}^2$ is such that

$$\Im\xi \cdot \Re\xi = 0. \tag{5.16}$$

The main issue here is the quasi-periodic condition imposed on $w_{k,\theta}$ (through the requirement that $w_{k,\theta}(t, \cdot)$ is in $H^2_{\sharp,\theta}(\Omega')$ for a.e. $t \in (0, T)$). This problem may be overcomed upon adapting the framework introduced in [11] for the definition of OGS in periodic media, giving (see [4, Lemma 3.2]):

Lemma 5.2 *Let $\xi \in \mathbb{C}^2 \setminus \mathbb{R}^2$ obey (5.16) and let $f \in H^2(0, T; H^2(\Omega'))$. Then for all $\theta \in [0, 2\pi)$ and all $k \in \mathbb{Z}$, there exists $E_{k,\theta} \in \mathscr{B}(H^2(0, T; H^2(\Omega')); H^2(0, T; H^2_{\sharp,\theta}(\Omega')))$ such that $\varphi = E_{k,\theta} f$ is solution to the equation*

$$(-i\partial_t - \Delta + 4i\pi k \partial_{x_1} + 2i\xi \cdot \nabla_{x'})\varphi = f \text{ in } Q'. \tag{5.17}$$

Moreover we have

$$\|E_{k,\theta}\|_{\mathscr{B}(H^2(0,T;H^2(\Omega')))} \leq \frac{c_0}{|\Im\xi|}, \tag{5.18}$$

for some constant $c_0 > 0$, which is independent of ξ, k and θ.

The occurrence of (5.17) in Lemma 5.2 follows from a direct calculation showing that $u_{k,\theta}$ fulfills (5.13) if and only if $w_{k,\theta}$ is solution to

$$\begin{cases} (-i\partial_t - \Delta + 4i\pi k\partial_{x_1} + 2i\xi\cdot\nabla_{x'} + V)w + e^{i\theta x_1}W_{k,\theta} = 0 & \text{in } Q', \\ w(\cdot,1,\cdot) = e^{i\theta}w(\cdot,0,\cdot) & \text{on } (0,T)\times\omega, \\ \partial_{x_1}w(\cdot,1,\cdot) = e^{i\theta}\partial_{x_1}w(\cdot,0,\cdot) & \text{on } (0,T)\times\omega, \end{cases} \tag{5.19}$$

with

$$W_{k,\theta} = V + \theta^2 - 4\pi k\theta.$$

Taking $r = |\Im\xi|$ so large (relative to c_0 and $\|V\|_{W^{2,\infty}(0,T;W^{2,\infty}(\Omega))}$) that

$$\begin{aligned} G_{k,\theta} : H^2(0,T;H^2_{\sharp,\theta}(\Omega')) &\longrightarrow \quad H^2(0,T;H^2_{\sharp,\theta}(\Omega')) \\ q &\longmapsto -E_{k,\theta}\left(Vq + e^{i\theta x_1}W_{k,\theta}\right) \end{aligned}$$

is a contraction mapping, we may apply Lemma 5.2 with $f = -(Vw_{k,\theta} + e^{i\theta x_1}W_{k,\theta})$ $\in H^2(0,T;H^2(\Omega'))$. In light of (5.17), $w_{k,\theta} = E_{k,\theta}f$ is thus a solution to (5.19) and fulfills (5.15). As a consequence we have (see [4, Proposition 3.1]) obtained:

Proposition 5.3 *We assume that $V \in W^{2,\infty}(0,T;W^{2,\infty}(\Omega))$ satisfies (5.2) and*

$$\|V\|_{W^{2,\infty}(0,T;W^{2,\infty}(\Omega))} \leq M$$

for some $M \geq 0$. Pick $r \geq r_0 = c_0(1 + M)$, where c_0 is the same as in (5.18), and let $\xi \in \mathbb{C}^2 \setminus \mathbb{R}^2$ fulfill (5.16) and $|\Im\xi| = r$. Then for all $\theta \in [0, 2\pi)$ and $k \in \mathbb{Z}$, there exists $w_{k,\theta} \in H^2(0,T;H^2_{\sharp,\theta}(\Omega'))$ obeying (5.15) such that the function $u_{k,\theta}$ defined by (5.14) is a $H^2(0,T;H^2_{\sharp,\theta}(\Omega'))$-solution to (5.13).

5.4 Stability Estimate

This section contains the proof of Theorem 5.1.

5.4.1 Auxiliary Result

Fix $r > 0$ and let $\zeta = (\eta, \ell) \in \mathbb{R}^2 \times \mathbb{R}$ with $\eta \neq 0_{\mathbb{R}^2}$. Then there exists $\zeta_j = \zeta_j(r, \eta, \ell) = (\xi_j, \tau_j) \in \mathbb{C}^2 \times \mathbb{R}$, $j = 1, 2$, such that we have

$$|\Im \xi_j| = r, \ \tau_j = \xi_j \cdot \xi_j, \ \zeta_1 - \overline{\zeta_2} = \zeta, \ \Re \xi_j \cdot \Im \xi_j = 0, \tag{5.20}$$

and

$$|\xi_j| \leq \frac{1}{2}\left(|\eta| + \frac{|\ell|}{|\eta|}\right) + r, \ |\tau_j| \leq |\eta|^2 + \frac{\ell^2}{|\eta|^2} + 2r^2. \tag{5.21}$$

This can be checked by direct calculation upon setting

$$\xi_j = \frac{1}{2}\left((-1)^{j+1} + \frac{\ell}{|\eta|^2}\right)\eta + (-1)^j i \eta_r^{\perp},$$

$$\tau_j = \frac{1}{4}\left((-1)^{j+1} + \frac{\ell}{|\eta|^2}\right)^2 |\eta|^2 - r^2, \ j = 1, 2,$$

where $\eta^{\perp}$ is any non zero $\mathbb{R}^2$-vector, orthogonal to η and $\eta_r^{\perp} = r\eta^{\perp}/|\eta^{\perp}|$.

This, combined with Proposition 5.3, immediately yields the:

Lemma 5.3 *Assume that $V_j \in W^{2,\infty}(0, T; W^{2,\infty}(\Omega))$, $j = 1, 2$, fulfill (5.2) and fix $r \geq r_0 = c_0(1 + M) > 0$, where $M \geq \max_{j=1,2} \|V_j\|_{W^{2,\infty}(0,T;W^{2,\infty}(\Omega))}$ and c_0 is the same as in (5.18). Pick $\zeta = (\eta, \ell) \in \mathbb{R}^2 \times \mathbb{R}$ with $\eta \neq 0_{\mathbb{R}^2}$, and let $\zeta_j = (\xi_j, \tau_j) \in \mathbb{C}^2 \times \mathbb{R}$, $j = 1, 2$, obey (5.20)–(5.21). Then, there exists a constant $C > 0$ depending only on T, $|\omega|$ and M, such that for every $k \in \mathbb{Z}$ and $\theta \in [0, 2\pi)$, the function $u_{j,k,\theta}$, $j = 1, 2$, defined in Proposition 5.3 by substituting ξ_j for ξ, satisfies the estimate*

$$\|u_{j,k,\theta}\|_{H^2(0,T;H^2(\Omega'))} \leq C(1 + q(\zeta, k))^{\frac{13}{2}} \frac{(1 + r^2)^3}{r} e^{|\omega| r}, \ k \in \mathbb{Z}, \ \theta \in [0, 2\pi), \ r \geq r_0,$$

with

$$q(\zeta, k) = q(\eta, \ell, k) = |\eta|^2 + \frac{|\ell|}{|\eta|} + k^2.$$

5.4.2 Sketch of the Proof

Let $\zeta = (\eta, \ell)$, r and $\zeta_j = (\xi_j, \tau_j)$, $j = 1, 2$, be as in Lemma 5.3, fix $k \in \mathbb{Z}$, and put

$$(k_1, k_2) = \begin{cases} (k/2, -k/2) & \text{if } k \text{ is even,} \\ ((k+1)/2, -(k-1)/2) & \text{if } k \text{ is odd.} \end{cases}$$

Further we pick $\theta \in [0, 2\pi)$ and note u_j, $j = 1, 2$, the OGS $u_{j,k_j,\theta}$, defined by Lemma 5.3. In light of Lemma 5.1 there is a unique solution $v \in L^2(0, T; H^2_{\sharp,\theta}(\Omega')) \cap H^1(0, T; L^2(\Omega'))$ to the IBVP

$$\begin{cases} (-i\partial_t + \Delta + V_2)v = 0 \text{ in } Q' \\ v(0, \cdot) = u_1(0, \cdot) & \text{in } \Omega', \\ v = u_1 & \text{on } \Sigma'_*. \end{cases} \quad (5.22)$$

Hence $u = v - u_1$ is solution to the following system

$$\begin{cases} (-i\partial_t + \Delta + V_2)u = (V_1 - V_2)u_1 \text{ in } Q' \\ u(0, \cdot) = 0 & \text{in } \Omega', \\ u = 0 & \text{on } \Sigma'_*, \\ u(\cdot, 1, \cdot) = e^{i\theta}u(\cdot, 0, \cdot) & \text{on } (0, T) \times \omega \\ \partial_{x_1}u(\cdot, 1, \cdot) = e^{i\theta}\partial_{x_1}u(\cdot, 0, \cdot) & \text{on } (0, T) \times \omega, \end{cases} \quad (5.23)$$

so we get

$$\int_{Q'} (V_1 - V_2)u_1\overline{u_2}\, dtdx = \int_{\Sigma'_*} \partial_\nu u\overline{u_2}\, dtd\sigma(x) - i\int_{\Omega'} u(T, \cdot)\overline{u_2(T, \cdot)}\, dx, \quad (5.24)$$

by integrating by parts and taking into account the quasi-periodic boundary conditions satisfied by u and u_2. Notice from (5.22)–(5.23) that $\partial_\nu u = \left(\Lambda^1_{V_2,\theta} - \Lambda^1_{V_1,\theta}\right)(\mathfrak{g}_1)$ and $u(T, .) = \left(\Lambda^2_{V_2,\theta} - \Lambda^2_{V_1,\theta}\right)(\mathfrak{g}_1)$, where

$$\mathfrak{g}_1 = \left(u_{1|\Sigma'_*}, u_1(0, .)\right) \in \mathscr{X}'_{0,\theta}.$$

Thus, putting

$$\beta_k = \begin{cases} 0 & \text{if } k \text{ is even or } k \in \mathbb{R} \setminus \mathbb{Z} \\ 4\pi^2 & \text{if } k \text{ is odd,} \end{cases}$$

for all $k \in \mathbb{Z}$, and

$$\varrho = \varrho_{k,\theta} = e^{-i\theta x_1}w_1 + e^{i\theta x_1}\overline{w_2} + w_1\overline{w_2}, \quad (5.25)$$

we deduce from (5.14), (5.20) and (5.24) that

$$\int_{Q'} (V_1 - V_2) e^{-i((\ell + \beta_k k)t + 2\pi k x_1 + x' \cdot \eta)} \, dt dx = A + B + C, \tag{5.26}$$

with

$$A = -\int_{Q'} (V_2 - V_1) \varrho(t, x) e^{-i((\ell + \beta_k k)t + 2\pi k x_1 + x' \cdot \eta)} \, dt dx, \tag{5.27}$$

$$B = \int_{\Sigma'_*} \left(\Lambda^1_{V_2,\theta} - \Lambda^1_{V_1,\theta} \right) (\mathfrak{g}_1) \overline{u_2} \, dt d\sigma(x), \tag{5.28}$$

$$C = -i \int_{\Omega'} \left(\Lambda^2_{V_2,\theta} - \Lambda^2_{V_1,\theta} \right) (\mathfrak{g}_1) \overline{u_2(T, \cdot)} \, dx. \tag{5.29}$$

Upon setting

$$V(t, x) = \begin{cases} (V_2 - V_1)(t, x) & \text{if } (t, x) \in Q, \\ 0 & \text{if } (t, x) \in \mathbb{R}^4 \setminus Q, \end{cases} \quad \text{and } \phi_k(x_1) = e^{i 2\pi k x_1}, \ x_1 \in \mathbb{R}, \ k \in \mathbb{Z},$$

we may rewrite (5.26) as

$$\int_{Q'} (V_1 - V_2) e^{-i((\ell + \beta_k k)t + 2\pi k x_1 + x' \cdot \eta)} \, dt dx = \left\langle \hat{V}(\ell + \beta_k k, \eta), \phi_k \right\rangle_{L^2(0,1)}, \tag{5.30}$$

where $\hat{V}$ stands for the partial Fourier transform of V with respect to $t \in \mathbb{R}$ and $x' \in \mathbb{R}^2$. Further, due to (5.15) and (5.25), we have $\|\varrho\|_{L^1(Q')} \leq c'(1 + |k|)^2/r^2$, where the constant $c' > 0$ depends only on T, $|\omega|$ and M. Since $\|V_1 - V_2\|_\infty \leq 2M$, it follows from this and (5.27) (upon substituting c' for $4Mc'$) in the above estimate that

$$|A| \leq \|V_1 - V_2\|_\infty \|\varrho\|_{L^1(Q')} \leq c' \frac{(1 + q(\zeta, k))}{r^2}, \tag{5.31}$$

where q is defined in Lemma 5.3. Moreover, we have

$$|B| + |C| \leq C^2 \|\Lambda^1_{V_2,\theta} - \Lambda^1_{V_1,\theta}\|_{\mathscr{B}(\mathscr{X}'_{0,\theta}, \mathscr{X}'_1)} (1 + q(\zeta, k))^{13} \frac{(1 + r^2)^6}{r^2} e^{2|\omega| r}, \ r \geq r_0, \tag{5.32}$$

from (5.28)–(5.29) and Lemma 5.3. Now, putting (5.26) and (5.30)–(5.32) together, we end up getting that

$$\left| \left\langle \hat{V}(\ell + \beta_k k, \eta), \phi_k \right\rangle_{L^2(0,1)} \right| \leq c'' \frac{(1 + q(\zeta, k))}{r^2} \left(r + \gamma (1 + q(\zeta, k))^{12} (1 + r^2)^6 e^{2|\omega| r} \right)$$

for $r \geq r_0$, where $\gamma = \|\Lambda_{V_2,\theta} - \Lambda_{V_1,\theta}\|_{\mathscr{B}(\mathscr{X}'_{0,\theta}, \mathscr{X}'_1)}$ and the constant $c'' > 0$ is independent of k, r and $\zeta = (\eta, \ell)$. From this and the Parseval–Plancherel theorem, entailing

$$\|V_2 - V_1\|^2_{L^2(Q')} = \|V\|^2_{L^2(\mathbb{R} \times (0,1) \times \mathbb{R}^2)} = \sum_{k \in \mathbb{Z}} \int_{\mathbb{R}^3} |\langle \hat{V}(\ell, \eta), \phi_k \rangle_{L^2(0,1)}|^2 d\zeta,$$

then follows the:

Theorem 5.2 *Let M and V_j, $j = 1, 2$, be the same as in Theorem 5.1. Then we may find two constants $C > 0$ and $\gamma^* > 0$, depending on T, ω and M, such that we have*

$$\|V_2 - V_1\|_{L^2(Q')} \leq C \left(\ln \left(\frac{1}{\|\Lambda_{V_2,\theta} - \Lambda_{V_1,\theta}\|_{\mathscr{B}(\mathscr{X}'_0, \mathscr{X}'_1)}} \right) \right)^{-1/5},$$

for any $\theta \in [0, 2\pi)$, provided $0 < \|\Lambda_{V_2,\theta} - \Lambda_{V_1,\theta}\|_{\mathscr{B}(\mathscr{X}'_{0,\theta}, \mathscr{X}'_1))} < \gamma^*$.

Finally, putting (5.12) together with Theorem 5.2, we obtain Theorem 5.1.

References

1. Choulli, M.: Une introduction aux problèmes inverses elliptiques et paraboliques. Mathématiques et Applications, vol. 65. Springer, Berlin (2009)
2. Choulli, M., Kian, Y.: Stability of the determination of a time-dependent coefficient in parabolic equations. Math. Control Relat. Fields **3**(2), 143–160 (2013)
3. Choulli, M., Soccorsi, E.: Some inverse anisotropic conductivity problem induced by twisting a homogeneous cylindrical domain. arXiv:1209.5662 (2012)
4. Choulli, M., Kian, Y., Soccorsi, E.: Stable determination of time-dependent scalar potential from boundary measurements in a periodic quantum waveguide. arXiv:1306.6601 (2013)
5. Dixmier, J.: Les algèbres d'opérateurs dans l'espace hilbertien (algèbres de Von Neumann). Cahiers scientifiques, vol. 25. Gauthier-Villars, Paris (1957)
6. Eskin, G.: Inverse hyperbolic problems with time-dependent coefficients. Commun. Part. Differ. Equ. **32**(11), 1737–1758 (2007)
7. Eskin, G.: Inverse problem for the Schrödinger equation with time-dependent electromagnetic potentials and the Aharonov-Bohm effect. J. Math. Phys. **49**(2), 1–18 (2008)
8. Ferreira, D.D.S., Kurylev, Y., Lassas, M., Salo, M.: The Calderon problem in transversally anisotropic geometries. arXiv:1305.1273 (2013)
9. Fliss, S.: A Dirichlet-to-Neumann approach for the exact computation of guided modes in photonic crystal waveguides. arXiv:1202.4928 (2012)
10. Gaitan, P., Kian, Y.: Stability result for a time-dependent potential in a cylindrical domain. Inverse Probl. **29**(6), 065006 (18 pp.) (2013)
11. Hähner, P.: A periodic Faddeev-type operator. J. Differ. Equ. **128**, 300–308 (1996)
12. Ikehata, M.: Inverse conductivity problem in the infinite slab. Inverse Probl. **17**, 437–454 (2001)
13. Kian, Y.: Stability of the determination of a coefficient for the wave equation in an infinite wave guide. arXiv:1305.4888 (2013)

14. Li, X., Uhlmann, G.: Inverse problems on a slab. Inverse Probl. Imag. **4**, 449–462 (2010)
15. Nakamura, S.I.: Uniqueness for an inverse problem for the wave equation in the half space. Tokyo J. Math. **19** (1), 187–195 (1996)
16. Rakesh: An inverse problem for the wave equation in the half plane. Inverse Probl. **9**, 433–441 (1993)
17. Reed, M., Simon, B.: Methods of Modern Mathematical Physics IV: Analysis of Operators. Academic, New York (1978)
18. Salo, M., Wang, J.N.: Complex spherical waves and inverse problems in unbounded domains. Inverse Probl. **22**, 2299–2309 (2006)

Chapter 6
A General Approach to Identification Problems

Angelo Favini, Alfredo Lorenzi, and Hiroki Tanabe

To the memory of Alfredo Lorenzi

Abstract Two approaches to a general form of identification problems are described. Some applications to particular inverse problems are also given.

6.1 Introduction

At the beginning of this paper we want to remember Prof. Alfredo Lorenzi, who passed away on November 9th, 2013. He will be greatly missed, both as a dear friend and a brilliant researcher. His enthusiasm, integrity and passion will keep inspiring us throughout our lives.

Very recently identification problems related to equations in a Banach space X of the type

$$\begin{cases} \dfrac{dy}{dt} = y'(t) = Ay(t) + \displaystyle\sum_{j=1}^{n} f_j(t)z_j + h(t), & t \in [0, r], \\ y(0) = y_0, \\ \Phi_j[y(t)] = g_j(t), & j = 1, \dots, n \end{cases} \tag{6.1}$$

A. Favini (✉)
Dipartimento di Matematica, Università di Bologna, Piazza di Porta S. Donato 5,
40126 Bologna, Italy
e-mail: angelo.favini@unibo.it

A. Lorenzi
Dipartimento di Matematica, Università di Milano, via Cesare Saldini 50, 20133 Milano, Italy
e-mail: alfredo.lorenzi@unimi.it

H. Tanabe
Hirai Sanso 12-13, Takarazuka 665-0817, Japan
e-mail: h7tanabe@jttk.zaq.ne.jp

© Springer International Publishing Switzerland 2014
A. Favini et al. (eds.), *New Prospects in Direct, Inverse and Control Problems for Evolution Equations*, Springer INdAM Series 10,
DOI 10.1007/978-3-319-11406-4_6

in the unknown functions $(y, f_1, \ldots, f_n) \in C([0, r]; D(A)) \times \prod_{j=1}^{n} C([0, r]; \mathbb{C})$, $h \in C([0, r]; X)$ have been studied under the assumptions that the closed linear operator A in X satisfies the weak parabolic resolvent estimate

$$\|(\lambda - A)^{-1}\|_{\mathscr{L}(X)} \leq C(1 + |\lambda|)^{-\beta} \tag{6.2}$$

for every complex number λ in the region

$$\Sigma_\alpha := \{\lambda \in \mathbb{C} : \operatorname{Re}\lambda \geq -C(1 + |\operatorname{Im}\lambda|)^\alpha\},$$

where $C > 0, 0 < \beta \leq \alpha \leq 1$. Moreover, $z_j \in X$, $y_0 \in D(A)$, $\Phi_j \in X^*$, $g_j \in C([0, r]; \mathbb{C})$, $j = 1, \ldots, n$ (see [4]). In the paper in preparation, from Favini et al. [6], the authors consider as A a multivalued linear operators and the equation in (6.1) then reads as the inclusion

$$y'(t) - \sum_{j=1}^{n} f_j(t)z_j - h(t) \in Ay(t). \tag{6.3}$$

The main aim in this paper is to present a new approach taking as particular cases some inverse problems discussed in literature. We also note that the approach given in [6] inspired the second approach that we propose. Precisely, in a first step, we will consider the problem to find a pair of functions $u : [0, r] \to X$, $F : [0, r] \to \mathscr{F}$, $X, \mathscr{F}$ two Banach spaces, such that

$$\begin{cases} u'(t) = Au(t) + M(F(t), Z) + g(t), & t \in [0, r], \\ u(0) = u_0, \\ \Phi[u(t)] = H(t), & t \in [0, r], \end{cases}$$

where M, Z, g, u_0, Φ, H will be specified in a moment.

The approach is basically consisting in reducing this inverse problem to a direct problem, following previous papers [4, 5] and [2]. See also [1] and [3].

Furthermore, the same general problem is faced in a direct way. In this sense it is more classic, but it applies to the multivalued case

$$u'(t) - M(F(t), Z) - g(t) \in Au(t)$$

allowing to handle very strong degeneracy in the equation.

6.2   A First Approach

Let $X, \mathscr{F}, \mathscr{Z}$ be three complex Banach spaces with norms $\| \cdot \|, \| \cdot \|_{\mathscr{F}}, \| \cdot \|_{\mathscr{Z}}$, respectively.

We make the following assumptions:

(H1) $A : D(A) \subset X \to X$ is a linear closed operator whose resolvent

$\rho(A)$ contains the set $\Sigma_\alpha := \{\lambda \in \mathbb{C} : \mathrm{Re}\lambda \geq -C_0(1 + |\mathrm{Im}\lambda|)^\alpha\},$

(6.4)

where $C_0, \alpha \in (0, 1]$;

(H2) $\|(\lambda - A)^{-1}\|_{\mathscr{L}(X)} \leq C_0(1 + |\lambda|)^{-\beta},$ (6.5)

where $C_0, \beta \in (0, \alpha]$ and $\lambda \in \Sigma_\alpha$;

(H3) $M \in B(\mathscr{F} \times \mathscr{Z}; X),$

(6.6)

where $B(\mathscr{F} \times \mathscr{Z}; X)$ denotes the Banach space of all bounded bilinear operators from $\mathscr{F} \times \mathscr{Z}$ into X;

(H4) $\Phi \in \mathscr{L}(X; \mathscr{F});$ (6.7)

(H5)   for each fixed $Z \in \mathscr{Z}$ and for all $H \in \mathscr{F}$, the equation

$$\Phi[M(F, Z)] = H$$

is uniquely solvable in $\mathscr{F}$ and its solution can be represented by $F = \Psi[H, Z]$, where the (nonlinear) operator $\Psi : D(\Psi) \subset \mathscr{F} \times \mathscr{Z} \to \mathscr{F}$ is linear continuous as a function of H i.e.

$$\|\Psi[H, Z]\|_{\mathscr{F}} \leq C_1(Z)\|H\|_{\mathscr{F}} \text{ for all } H \in \mathscr{F};$$

(H6)   there exist Banach spaces X_A^θ and $\mathscr{Z}^\theta$ embedded in X and $\mathscr{Z}$, respectively, with $\theta > 1 - \beta$, such that

$$\|M(F, Z)\|_{X_A^\theta} \leq C(\theta)\|F\|_{\mathscr{F}}\|Z\|_{\mathscr{Z}^\theta}.$$

X_A^θ is in fact the space

$$X_A^\theta = \{x \in X; |x|_{X_A^\theta} =: \sup_{t \geq t_0} t^\theta \|A(t - A)^{-1}x\|_X < \infty\},$$

where $t_0 > 0$ and $0 < \theta < 1$. The norm of X_A^θ is defined by

$$\|x\|_{X_A^\theta} = |x|_{X_A^\theta} + \|x\|_X.$$

It is easy to show that $D(A) \subset X_A^\theta$ for $\theta \leq \beta$. The set $D(A)$ makes a Banach space with norm $\| \cdot \|_{D(A)} = \|A \cdot \|_X$. The interpolation space $(X, D(A))_{\theta,\infty}$, $0 < \theta < 1$, is defined by

$$(X, D(A))_{\theta,\infty} = \{ u = u_0(t) + u_1(t) \ \forall t \in (0, \infty);$$

$$\sup_{0<t<\infty} \|t^\theta u_0(t)\|_X < \infty, \ \sup_{0<t<\infty} \|t^{\theta-1} u_1(t)\|_{D(A)} < \infty \},$$

$$\|u\|_{(X,D(A))_{\theta,\infty}} = \inf_{u=u_0(t)+u_1(t)} \{ \sup_{0<t<\infty} \|t^\theta u_0(t)\|_X + \sup_{0<t<\infty} \|t^{\theta-1} u_1(t)\|_{D(A)} \}.$$

The inclusion relation $X_A^\theta \subset (X, D(A))_{\theta,\infty}$ can be shown by decomposing $u \in X_A^\theta$ as

$$u = u_0(t) + u_1(t), \quad u_0(t) = -A(t - A)^{-1} u, \quad u_1(t) = t(t - A)^{-1} u.$$

We then consider the following identification problem: find a pair of functions $u : [0, r] \to X$ and $F : [0, r] \to \mathscr{F}$ such that

$$\begin{cases} u'(t) = Au(t) + M(F(t), Z) + g(t), & t \in [0, r], \\ u(0) = u_0, \\ \Phi[u(t)] = H(t), & t \in [0, r], \end{cases} \tag{6.8}$$

where $u_0 \in D(A)$, $Z \in \mathscr{Z}$, $g \in C([0, r]; X)$, $H \in C^1([0, r]; \mathscr{F})$.

Apply operator Φ to both sides of the differential equation in (6.8). Under assumption (H4) we get the equation

$$H'(t) - \Phi[Au(t)] = \Phi[M(F(t), Z)] + \Phi[g(t)], \quad t \in [0, r]. \tag{6.9}$$

From assumptions (H3) and (H5) we deduce

$$F(t) = \Psi[H'(t), Z] - \Psi[\Phi[g(t)], Z] - \Psi[\Phi[Au(t)], Z], \quad t \in [0, r]. \tag{6.10}$$

Inserting (6.10) into the differential equation of (6.8) we get that the identification problem (6.8) is equivalent to the unusual Cauchy problem

$$\begin{cases} u'(t) - Au(t) + M(\Psi[\Phi[Au(t)], Z], Z) \\ = M(\Psi[H'(t), Z], Z) - M(\Psi[\Phi[g(t)], Z], Z) + g(t), & t \in [0, r], \\ u(0) = u_0. \end{cases} \tag{6.11}$$

Introduce now the linear operator B defined by

$$D(B) = D(A), \quad Bu := -M(\Psi[\Phi[Au], Z], Z), u \in D(B). \tag{6.12}$$

From assumptions (H5) and (H6) we deduce the bounds

$$
\begin{aligned}
\|Bu\|_{X_A^\theta} &\le C(\theta)\|\Psi[\Phi[Au], Z]\|_{\mathscr{F}}\|Z\|_{\mathscr{Z}^\theta} \\
&\le C(\theta)C_1(Z)\|\Phi[Au]\|_{\mathscr{F}}\|Z\|_{\mathscr{Z}^\theta} \\
&\le C(\theta)C_1(Z)\|\Phi\|_{\mathscr{L}(X;\mathscr{F})}\|Au\|_X\|Z\|_{\mathscr{Z}^\theta} \\
&\le C(\theta)C_1(Z)\|\Phi\|_{\mathscr{L}(X;\mathscr{F})}\|Z\|_{\mathscr{Z}^\theta}\|u\|_{D(A)}.
\end{aligned}
\tag{6.13}
$$

We recall the following lemma, generalizing a well known result, see Lunardi [7]. For the proof we refer the readers to Theorem 1 of [5].

Lemma 6.1 *Let A satisfy (H1), (H2) and suppose $B \in \mathscr{L}(D(A), X_A^\theta)$, where $1 - \beta < \theta < 1$. Then $A + B$ generates an infinitely differentiable semigroup in X, too, and*

$$
\|(\lambda - A - B)^{-1}\|_{\mathscr{L}(X)} \le C(1 + |\lambda|)^{-\beta}
$$

for all $\lambda \in \Sigma_\alpha$, $|\lambda|$ large enough.

Therefore, we can apply to (6.11) the uniqueness and existence results in [2,4] and [6]. See, in particular, the following proposition due to [6].

Proposition 6.1 *Let A be a possibly multivalued linear operator satisfying (H1) and (H2) with (α, β) satisfying $2\alpha + \beta > 2$. Suppose that $2\alpha + \beta + \theta > 3$, $u_0 \in D(A)$, $Au_0 \cap (X, D(A))_{\theta,\infty} \ne \varnothing$, $f \in C([0, r]; X) \cap B([0, r]; (X, D(A))_{\theta,\infty})$. Then the problem*

$$
u'(t) - f(t) \in Au(t), \quad t \in [0, r],
$$

$$
u(0) = u_0,
$$

admits a unique solution u such that $u \in C^1([0, r]; X)$,

$$
u' - f \in C^{\frac{2\alpha + \beta - 3 + \theta}{\alpha}}([0, r]; X) \cap B([0, r]; X_A^{\frac{2\alpha + \beta - 3 + \theta}{\alpha}}).
$$

Remark 6.1 Since $3 - 2\alpha - \beta < 1$ under our assumption there exists $\theta \in (0, 1)$ such that $2\alpha + \beta + \theta > 3$.

We need another lemma as follows.

Lemma 6.2 *Let A, B be two operators satisfying assumptions of Lemma 6.1. Then the following inclusions hold:*

(i) $X_{A+B}^{\theta+1-\beta} \hookrightarrow X_A^\theta, 0 < \theta < \beta,$
(ii) $X_A^\theta \hookrightarrow X_{A+B}^{\theta+\beta-1}, 1 - \beta < \theta < 1,$
(iii) $X_A^{\theta+1-\beta} \hookrightarrow X_{A+B}^\theta, 0 < \theta < \beta,$

provided that $0 \in \rho(A + B)$.

Proof The first assertion follows from

$$A(t - A)^{-1} - (A + B)(t - A - B)^{-1}$$
$$= A[(t - A)^{-1} - (t - A - B)^{-1}] - B(t - A - B)^{-1}$$
$$= -A(t - A)^{-1}B(t - A - B)^{-1} - B(t - A - B)^{-1}$$
$$= -[A(t - A)^{-1} + I]B(t - A - B)^{-1}$$
$$= -t(t - A)^{-1}B(A + B)^{-1}(A + B)(t - A - B)^{-1}.$$

By replacing A and B by $A + B$ and $-B$ respectively the second and third assertions are obtained. $\square$

We can thus establish the following theorem (see [4]):

Theorem 6.1 *Suppose that A satisfies (H1) and (H2) with (α, β) such that $\alpha + \beta + \alpha\beta > 2$ and that M and Φ satisfy (H3)–(H6). Let $\theta > 3 - \alpha - \beta - \alpha\beta$, $u_0 \in D(A)$, $Au_0 \in (X, D(A))_{\theta,\infty}$, $Z \in \mathscr{Z}^\theta$, $g \in C([0, r]; X) \cap B([0, r]; X_A^\theta)$, $H \in C^1([0, r]; \mathscr{F})$, $\Phi(u_0) = H(0)$. Then there exists a unique solution (u, F) to problem (6.8) such that*

$$u \in C^1([0, r]; X), \quad u' \in B([0, r]; X_A^{\frac{\alpha+\beta+\alpha\beta-3+\theta}{\alpha}}), \quad F \in C([0, r]; \mathscr{F}),$$
$$Au \in C^{\frac{2\alpha+\beta-\theta+3}{\alpha}}([0, r]; X) \cap B([0, r]; X_A^{\frac{\alpha+\beta+\alpha\beta-3+\theta}{\alpha}}).$$

Notice that if $\alpha = \beta = 1$ we obtain an optimal regularity result.

Remark 6.2 The assumption $\alpha + \beta + \alpha\beta > 2$, which is stronger than $2\alpha + \beta > 2$, implies the existence of $\theta \in (0, 1)$ satisfying $\theta > 3 - \alpha - \beta - \alpha\beta$.

Proof Notice that our operator $A + B$ is possibly non invertible. Thus we must apply a change of variable $u = e^{kt}v$, where k is large enough so that $A + B - k$ has a bounded inverse. Then problem (6.11) reads

$$\begin{cases} v'(t) - (A + B - k)v(t) = e^{-kt}[M(\Psi[H'(t), Z], Z) - M(\Psi[\Phi[g(t)], Z], Z) + g(t)] \\ v(0) = u_0. \end{cases}$$

Denote by $G(t)$ the right hand side of the above equation. Noting that the assumption $\theta > 3 - \alpha - \beta - \alpha\beta$ of the theorem implies $2\alpha + \beta + \theta > 3$ we apply Proposition 6.1 to this new problem

$$\begin{cases} v'(t) = (A + B - k)v(t) + G(t), \quad 0 \le t \le r, \\ v(0) = u_0. \end{cases}$$

In view of Proposition 6.1 and Lemma 6.2 we deduce that if $2\alpha + \beta > 2$, $2\alpha + \beta + \theta > 3$, $G \in C([0,r]; X) \cap B([0,r]; (X, D(A))_{\theta,\infty})$ and $u_0 \in D(A)$, $Au_0 \in (X, D(A))_{\theta,\infty}$, which implies $(A + B)u_0 \in (X, D(A))_{\theta,\infty}$ since $Bu_0 \in X_A^\theta \subset (X, D(A))_{\theta,\infty}$, then such a problem admits a unique solution v such that $v \in C^1([0,r]; X)$,

$$v' - G = (A + B - k)v \in C^{\frac{2\alpha+\beta-3+\theta}{\alpha}}([0,r]; X) \cap B([0,r], X_{A+B-k}^{\frac{2\alpha+\beta-3+\theta}{\alpha}})$$

$$\subseteq C^{\frac{2\alpha+\beta-3+\theta}{\alpha}}([0,r]; X) \cap B([0,r], X_A^{\frac{\alpha+\beta+\alpha\beta-3+\theta}{\alpha}})$$

(6.14)

provided that $\theta > 3 - \alpha - \beta - \alpha\beta$. It is easy to see that the hypotheses of the theorem imply

$$G \in C([0,r]; X) \cap B([0,r]; X_A^\theta).$$

(6.15)

Since

$$\theta - (\alpha + \beta + \alpha\beta - 3 + \theta)/\alpha = [(1 - \alpha)(1 - \theta) + 2 - \beta - \alpha\beta]/\alpha \geq 0,$$

one has

$$X_A^\theta \subset X_A^{\frac{\alpha+\beta+\alpha\beta-3+\theta}{\alpha}}.$$

(6.16)

Hence from (6.15) it follows that

$$G \in C([0,r]; X) \cap B([0,r]; X_A^{\frac{\alpha+\beta+\alpha\beta-3+\theta}{\alpha}}).$$

This and (6.14) imply

$$v' \in C([0,r]; X) \cap B([0,r]; X_A^{\frac{\alpha+\beta+\alpha\beta-3+\theta}{\alpha}}).$$

(6.17)

In view of (6.14) one has

$$Av = A(A + B - k)^{-1}(A + B - k)v \in C^{\frac{2\alpha+\beta-3+\theta}{\alpha}}([0,r]; X).$$

(6.18)

Since $(\alpha + \beta + \alpha\beta - 3 + \theta)/\alpha < \beta$, one observes $D(A) \subset X_A^{\frac{\alpha+\beta+\alpha\beta-3+\theta}{\alpha}}$. Therefore it follows that

$$v \in C^{\frac{2\alpha+\beta-3+\theta}{\alpha}}([0,r]; D(A)) \subset C([0,r]; X_A^{\frac{\alpha+\beta+\alpha\beta-3+\theta}{\alpha}}) \subset B([0,r]; X_A^{\frac{\alpha+\beta+\alpha\beta-3+\theta}{\alpha}}).$$

(6.19)

With the aid of (6.13) one observes

$$\|Bv(t)\|_{X_A^\theta} \le C(\theta)C_1(Z)\|\Phi\|_{\mathscr{L}(X;\mathscr{F})}\|Z\|_{\mathscr{Z}^\theta}\|v(t)\|_{D(A)}.$$

This inequality, (6.18) and (6.16) yield

$$Bv \in B([0,r]; X_A^\theta) \subset B([0,r]; X_A^{\frac{\alpha+\beta+\alpha\beta-3+\theta}{\alpha}}). \tag{6.20}$$

From (6.19), (6.20) and (6.14) it follows that

$$Av \in B([0,r]; X_A^{\frac{\alpha+\beta+\alpha\beta-3+\theta}{\alpha}}). \tag{6.21}$$

Returning to the original notation $u = e^{kt}v$ and defining the function F by (6.10) we complete the proof of the theorem. □

6.3 Application 1

Let A be a linear operator with domain $D(A)$ on a Banach space X and satisfy (H1) and (H2). Let $\mathscr{F} = \mathbb{R}^N$ or $\mathbb{C}^N$, $\mathscr{Z} = X^N = X \times \ldots \times X$, N times, $X_A^\theta = \{x \in X : \sup_{t>0}(1+t)^\theta\|A(t-A)^{-1}x\| < +\infty\}$, $\mathscr{Z}^\theta = (X_A^\theta)^N$. Let $F = (f_1, \ldots, f_N) \in \mathbb{R}^N$, $Z = (z_1, \ldots, z_N) \in \mathscr{Z}^\theta$, $M(F,Z) = \sum_{j=1}^N f_j z_j$, $\Phi = (\Phi_1, \ldots, \Phi_N) \in (X^*)^N$ and let $(\Phi_k[z_j])_{j,k=1}^N$ be an invertible matrix.

Denote by $\tilde{\Psi}(Z) = (\tilde{\Psi}_{j,k}(Z))_{j,k=1}^N$ its inverse.

Observe that for all fixed $Z \in \mathscr{Z}^\theta = (X_A^\theta)^N$ and $H \in \mathbb{R}^N$, equation $\Phi[M(F,Z)] = H$ admits the unique solution $F = \tilde{\Psi}(Z)H := \Psi[H, Z]$. Therefore,

$$M(\Psi[\Phi[Au], Z], Z) = \sum_{j,k=1}^N \tilde{\Psi}_{j,k}(Z)\Phi_k[Au]z_j$$

and

$$\|M(F,Z)\|_{X_A^\theta} \le \sum_{j=1}^N |f_j|\|z_j\|_{X_A^\theta} \le \|F\|_{\mathbb{R}^N} \left(\sum_{j=1}^N \|z_j\|_{X_A^\theta}^2\right)^{\frac{1}{2}}$$

$$= \|F\|_{\mathscr{F}}\|Z\|_{\mathscr{Z}^\theta}, \quad (F,Z) \in \mathbb{R}^N \times \mathscr{Z}^\theta.$$

Moreover, the closed linear operator

$$\tilde{A}u = Au - \sum_{j,k=1}^N (\tilde{\Psi}(Z))_{j,k}\Phi_k[Au]z_j, \quad u \in D(\tilde{A}) = D(A)$$

generates a $C^\infty-$ semigroup if $\beta \in (0, \alpha)$ and $\theta \in (1 - \beta, 1)$ (respectively, an analytic semigroup, if $\beta = 1$ and $\theta \in (0, 1)$.)

Then, it is easy to apply the theorem above.

6.4 Application 2

Let A be a linear operator with domain $D(A)$ on a Banach space X and satisfy (H1) and (H2). Let $\mathscr{F} = l^2(\mathbb{R})$, $\mathscr{Z} = l^2(X)$, $\mathscr{Z}^\theta = l^2(X_A^\theta)$, $F = \{f_j\}_{j=1}^\infty \in l^2(\mathbb{R})$, $Z = \{z_j\}_{j=1}^\infty \in \mathscr{Z}^\theta$, $M(F, Z) = \sum_{j=1}^\infty f_j z_j$. Let $\Phi = \{\Phi_j\}_{j=1}^\infty \in l^2(X^*)$ such that $\sum_{j,k=1}^\infty |\Phi_k[z_j]|^2 < \infty$. Therefore, equation $\Phi[M(F, Z)] = H$, with $H \in l^2(\mathbb{R})$, means

$$\Phi\left(\sum_{j=1}^\infty f_j z_j\right) = \left(\sum_{j=1}^\infty f_j \Phi_k[z_j]\right)_{k\in\mathbb{N}} = H.$$

Suppose that the infinite matrix $\left(\Phi_k[z_j]\right)_{j,k=1}^\infty$ defines an invertible operator in $\mathscr{L}(l^2(\mathbb{R}))$ the inverse of which is denoted by $\tilde{\Psi}(Z)$, so that $F = \tilde{\Psi}(Z)H = \Psi[H, Z]$. One has

$$M(\Psi[\Phi[Au], Z], Z) = \sum_{j=1}^\infty \left(\tilde{\Psi}(Z)\Phi[Au]\right)_j z_j = \sum_{j,k=1}^\infty \tilde{\Psi}_{j,k}(Z)\Phi_k[Au]z_j.$$

Moreover,

$$\|M(F, Z)\|_{X_A^\theta} \le \sum_{j=1}^\infty |f_j|\|z_j\|_{X_A^\theta} \le \|F\|_{l^2(\mathbb{R})} \left(\sum_{j=1}^\infty \|z_j\|_{X_A^\theta}^2\right)^{\frac{1}{2}}$$

$$= \|F\|_{l^2(\mathbb{R})}\|Z\|_{\mathscr{Z}^\theta}, \quad (F, Z) \in l^2(\mathbb{R}) \times \mathscr{Z}^\theta.$$

Therefore, all assumptions (H1)–(H6) are verified. Hence, the identification problem above, i.e.

$$\begin{cases} u'(t) = Au(t) + \sum_{j=1}^\infty f_j(t)z_j, & t \in [0, r], \\ u(0) = u_0, \\ \Phi_k[u(t)] = g_k(t), & k = 1, 2, .. \end{cases}$$

admits a unique solution $(u, \{f_j\}_{j\in\mathbb{N}})$.

6.5 Application 3

Here the identification problem reads: find $(N + 1)n$ functions $u_1, \ldots, u_n : [0, r] \to X$ and $f_{jk} : [0, r] \to \mathbb{C}, j = 1, .., n, k = 1, \ldots, N$ such that

$$
\begin{cases}
u'_j(t) + A_j u_j(t) + B_j(u_1(t), \ldots, u_n(t)) \\
= \sum_{k=1}^{N} f_{j,k}(t) z_k + g_j(t), & t \in [0, r], j = 1, \ldots, n, \\
u_j(0) = u_{0j}, & j = 1, \ldots, n, \\
\Phi_k[u_j(t)] = g_{jk}(t), & t \in [0, r], j = 1, \ldots, n, k = 1, \ldots, N,
\end{cases}
$$

where A_j is an operator satisfying (H1), (H2) and B_j is a continuous multilinear form from $D(A_1) \times \ldots \times D(A_n)$ into $X_{A_j}^{\theta}$ with $1 - \beta < \theta < 1, j = 1, \ldots, n$, $z_k \in \bigcap_{j=1}^{n} X_{A_j}^{\theta}, k = 1, \ldots, N, g_j \in C([0, r]; X_{A_j}^{\theta}), j = 1, \ldots, n, \Phi_k \in X^*, k = 1, \ldots, N$. It is easy to verify that if the data are sufficiently smooth and $(\Phi_k[z_l])_{l,k=1}^{N}$ is an invertible matrix, then all our hypotheses hold and the previous results apply.

6.6 A Second Approach

In this section we want to solve the identification problem (6.8) in a more direct way. This is the idea. First, notice that here A is possibly multivalued and the problem is expressed as

$$
\begin{cases}
y'(t) - g(t) - M(F(t), Z) \in Ay(t), & t \in [0, r], \\
y(0) = y_0 \in D(A), \\
\Phi[y(t)] = H(t), & t \in [0, r].
\end{cases}
\tag{6.22}
$$

Theorem 6.2 *Suppose that A is a possibly multivalued linear operator satisfying (H1) and (H2) with (α, β) satisfying $2\alpha + \beta > 2$ and that M and Φ satisfy (H3)-(H6). Suppose also that $2\alpha + \beta + \theta > 3$, $Z \in \mathcal{Z}^{\theta}$, $y_0 \in D(A)$, $Ay_0 \cap X_A^{\theta} \neq \emptyset$, $g \in C([0, r]; X) \cap B([0, r]; X_A^{\theta})$, $H \in C^1([0, r]; \mathcal{F})$, $\Phi(y_0) = H(0)$. Then, problem (6.22) admits a unique solution (y, F) such that $y \in C^1([0, r]; X)$, $F \in C([0, r]; \mathcal{F})$ and*

$$
y' - M(F(\cdot), Z) - g \in C^{\frac{2\alpha+\beta-3+\theta}{\alpha}}([0, r]; X) \cap B([0, r]; X_A^{\frac{2\alpha+\beta-3+\theta}{\alpha}}).
$$

Clearly, such a general result improves Theorem 6.1.

Proof Necessarily a solution (y, F) to problem (6.22) satisfies

$$y(t) = e^{tA}y_0 + \int_0^t e^{(t-s)A}M(F(s), Z)ds + \int_0^t e^{(t-s)A}g(s)ds,$$

$$\Phi[y(t)] = H(t).$$

This yields

$$H(t) = \Phi[e^{tA}y_0] + \Phi\left[\int_0^t e^{(t-s)A}M(F(s), Z)ds\right] + \Phi\left[\int_0^t e^{(t-s)A}g(s)ds\right].$$
(6.23)

It is shown in [6] that

if $\alpha + \beta + \theta > 2$, then $\lim_{t \to 0} e^{tA}\phi \to \phi$ in X for $\phi \in (X, D(A))_{\theta,\infty}$,   (6.24)

if $g \in C([0, r]; X) \cap B([0, r]; X_A^\theta)$, the function $t \mapsto \int_0^t e^{(t-s)A}g(s)ds$ belongs to $C^1([0, r]; X)$.

In case A is single valued these statements are established in [4].

If $F \in C([0, r]; \mathscr{F})$, then $M(F(\cdot), Z) \in C([0, r]; X_A^\theta) \subset C([0, r]; (X, D(A))_{\theta,\infty})$. Therefore differentiating both members of (6.23), we get

$$H'(t) = \Phi\left[\frac{d}{dt}e^{tA}y_0\right] + \Phi[M(F(t), Z)] + \Phi\left[\int_0^t \frac{\partial}{\partial t}e^{(t-s)A}M(F(s), Z)ds\right]$$

$$+ \frac{d}{dt}\Phi\left[\int_0^t e^{(t-s)A}g(s)ds\right].$$

From (H5) this reads to the following integral equation to be satisfied by F:

$$F(t) = \Psi[H'(t), Z] - \Psi\left[\Phi\left[\frac{d}{dt}e^{tA}y_0\right], Z\right] - \Psi\left[\Phi\left[\int_0^t \frac{\partial}{\partial t}e^{(t-s)A}M(F(s), Z)ds\right], Z\right]$$

$$- \Psi\left[\frac{d}{dt}\Phi\left[\int_0^t e^{(t-s)A}g(s)ds\right], Z\right].$$
(6.25)

By assumption there exists $\phi \in Ay_0 \cap X_A^\theta$. Hence in view of (6.24)

$$\frac{d}{dt}e^{tA}y_0 = \frac{d}{dt}e^{tA}A^{-1}\phi = e^{tA}\phi \to \phi \text{ in } X \text{ as } t \to 0.$$

Therefore, using both (H4) and (H5) we observe that the known part of the integral equation (6.25) belongs to $C([0, r]; \mathscr{F})$.

Let S be the operator in $C([0, r]; \mathscr{F})$ defined by

$$SF(t) = -\Psi\left[\Phi\left[\int_0^t \frac{\partial}{\partial t} e^{(t-s)A} M(F(s), Z) ds\right], Z\right].$$

Observe the estimate

$$\|SF(t)\|_{\mathscr{F}} \le C_1(Z) \left\|\Phi\left[\int_0^t \frac{\partial}{\partial t} e^{(t-s)A} M(F(s), Z) ds\right]\right\|_{\mathscr{F}}$$

$$\le C_1(Z) \int_0^t \left\|\Phi\left[\frac{\partial}{\partial t} e^{(t-s)A} M(F(s), Z)\right]\right\|_{\mathscr{F}} ds$$

$$\le C_1(Z)\|\Phi\|_{\mathscr{L}(X;\mathscr{F})} \int_0^t \left\|\frac{\partial}{\partial t} e^{(t-s)A} M(F(s), Z)\right\|_X ds$$

$$\le C_1(Z)\|\Phi\|_{\mathscr{L}(X;\mathscr{F})} \int_0^t \left\|\frac{\partial}{\partial t} e^{(t-s)A}\right\|_{\mathscr{L}((X,D(A))_{\theta,\infty},X)}$$

$$\times \|M(F(s), Z)\|_{(X,D(A))_{\theta,\infty}} ds$$

$$\le C'(Z) \int_0^t (t - s)^{\frac{\beta-2+\theta}{\alpha}} \|M(F(s), Z)\|_{X_A^\theta} ds$$

$$\le C'(Z)C(\theta) \int_0^t (t - s)^{\frac{\beta-2+\theta}{\alpha}} \|F(s)\|_{\mathscr{F}} \|Z\|_{\mathscr{Z}^\theta} ds.$$

Therefore

$$\|SF(t)\|_{\mathscr{F}} \le C'(Z)C(\theta)\|Z\|_{\mathscr{Z}^\theta} \int_0^t (t - s)^{-1+\theta_0} \|F(s)\|_{\mathscr{F}} ds,$$

where $\theta_0 := \frac{\theta-(2-\alpha-\beta)}{\alpha}$. Hence, repeating the arguments as in [4, Corollary 3.3],

$$\|S^n F(t)\|_{\mathscr{F}} \le [C'(Z)C(\theta)\|Z\|_{\mathscr{Z}^\theta}]^n \frac{\Gamma(\theta_0)^n t^{r\theta_0}}{\Gamma(n\theta_0)n\theta_0} \|F\|_{C([0,r];\mathscr{F})};$$

since $[\Gamma(n\theta_0)]^{\frac{1}{n}} \to +\infty$ as $n \to +\infty$, we conclude that the operator S has spectral radius equals to 0. Consequently the integral equation (6.25) admits a unique solution $F \in C([0, r]; \mathscr{F})$. It is evident that

$$M(F(\cdot), Z) \in C([0, r]; X_A^\theta) \subset C([0, r]; X) \cap B([0, r]; X_A^\theta).$$

Therefore it only remains to apply Proposition 6.1 to conclude the proof of the theorem.   □

Example 6.1 Let L, M two linear operators. $X, D(L) \subseteq D(M)$, $0 \in \rho(L)$, $\|M(\lambda M - L)^{-1}\|_{\mathscr{L}(X)} \leq C(1 + |\lambda|)^{-\beta}$ for all number $\lambda \in \Sigma_\alpha := \{\lambda \in \mathbb{C} : \text{Re}\lambda \geq -C(1 + |\text{Im}\lambda|)^\alpha\}$. Then the operator $A = LM^{-1}$ is a multivalued linear operator with $D(A) = M(D(L))$ and satisfies the estimates described above. Then, the identification problem: to find $(y, f_1(t), \dots f_n(t))$ such that

$$\begin{cases} \frac{d}{dt}(My(y)) = Ly(t) + \sum_{j=1}^n f_j(t)z_j, \\ (My)(0) = My_0, \\ \Phi_j[My(t)] = g_j(t), \qquad\qquad 0 \leq t \leq r, \end{cases}$$

can be easily solved as in Application 1 if we take $A = LM^{-1}$, the trivial verification is left to the reader.

Of course, we could extend all previous applications, too.

Acknowledgements The authors wish to thank the referee for his/her comments which contributed to improve the final version of the manuscript. We wish to express our deepest thanks to Dr. Genni Fragnelli and Dr. Gian Piero Favini for their assistance and help in typesetting the paper.

References

1. Favaron, A., Favini, A.: On the behavior of singular semigroups in intermediate and intepolations spaces and its applications to maximal regularity for degenerate integrodifferential evolution equations. Abstr. Appl. Anal. **2013**(275494) (2013). doi:10.1155/2013/275494
2. Favini, A., Tanabe, H.: Degenerate differential equations of parabolic type and inverse problems. In: Proceedings of Seminar on Partial Differential Equations in Osaka 2012, pp. 89–100. Osaka University, Suita (2013)
3. Favini, A., Yagi, A.: Degenerate Differential Equations in Banach Spaces. Monographs and Textbooks in Pure and Applied Mathematics, vol. 215. Marcel Dekker, New York (1999)
4. Favini, A., Lorenzi, A., Tanabe, H.: Direct and inverse problems for systems of singular differential boudary value problems. Electron. J. Differ. Equ. **2012**, 1–34 (2012)
5. Favini A., Lorenzi A., Marinoschi G., Tanabe H.: Perturbation methods and identification problems for degenerate evolution equations. In: Beznea, V., Brinzanes, V., Iosilescu, M., Marinoschi, G., Purice, R., Timotiu, D. (eds.) Contribution to the Seventh Congress of Romanian Mathematicians, 2011, pp. 88–96. Publishing House of the Romanian Academy of Sciences, Bucarest (2013)
6. Favini A., Lorenzi A., Tanabe, H.: Direct and inverse degenerate parabolic differential and integrodifferential equations with multivalued operators (in preparation)
7. Lunardi, A.: Analytic Semigroups and Optimal Regularity in Parabolic Problems. Modern Birkhäuser Classics. Birkhäuser/Springer Basel AG, Basel (1995)

Chapter 7
A Control Approach for an Identification Problem Associated to a Strongly Degenerate Parabolic System with Interior Degeneracy

Genni Fragnelli, Gabriela Marinoschi, Rosa Maria Mininni, and Silvia Romanelli

In memory of Alfredo Lorenzi, distinguished and tireless mathematician

Abstract We study an identification problem associated with a strongly degenerate parabolic evolution equation of the type

$$y_t - Ay = f(t, x), \quad (t, x) \in Q := (0, T) \times (0, L)$$

equipped with Dirichlet boundary conditions, where $T > 0, L > 0$, and f is in a suitable L^2 space. The operator A has the form $A_1 y = (u y_x)_x$, or $A_2 y = u y_{xx}$, and strong degeneracy means that the diffusion coefficient $u \in W^{1,\infty}(0, L)$ satisfies $u(x) > 0$ except for an interior point of $(0, L)$ and $\frac{1}{u} \notin L^1(0, L)$. Since an identification problem related to A_1 was studied in Fragnelli et al. (J. Evol. Equ., 2014), here we devote more attention to the identification problem of u when $A = A_2$. In this setting new weighted spaces of L^2-type must be considered. Our techniques are based on the minimization problem of a functional depending on u, provided that some observations are known. Optimality conditions are also given.

G. Fragnelli • R.M. Mininni • S. Romanelli (✉)
Dipartimento di Matematica, Università degli Studi di Bari Aldo Moro, Via E. Orabona 4, 70125 Bari, Italy
e-mail: genni.fragnelli@uniba.it; rosamaria.mininni@uniba.it; silvia.romanelli@uniba.it

G. Marinoschi
Institute of Mathematical Statistics and Applied Mathematics of the Romanian Academy, Calea 13 Septembrie 13, Bucharest, Romania

Simion Stoilow Institute of Mathematics, Research Group of the Project PN-II-ID-PCE-2011-3-0045, Bucharest, Romania
e-mail: gabimarinoschi@yahoo.com

© Springer International Publishing Switzerland 2014
A. Favini et al. (eds.), *New Prospects in Direct, Inverse and Control Problems for Evolution Equations*, Springer INdAM Series 10,
DOI 10.1007/978-3-319-11406-4_7

121

7.1   Introduction

The main aim of this paper is to study an identification problem of coefficients of a
parabolic evolution equation of the type

$$y_t - Ay = f(t, x), \quad (t, x) \in Q := (0, T) \times (0, L)$$

where $T > 0$, $L > 0$, and f in a suitable L^2 space, are given. Here A is a second
order differential operator of the type $A_1 y = (uy_x)_x$ (divergence form), or $A_2 y =
uy_{xx}$ (nondivergence form) equipped with Dirichlet boundary conditions. Note that
in our setting the diffusion coefficient u vanishes at an interior point of the interval
$(0, L)$, according to the strongly degenerate case defined as follows.

Definition 7.1 The operators $A_1 y = (uy_x)_x$ and $A_2 y = uy_{xx}$ are called *strongly
degenerate* if there exists $x_0 \in (0, L)$ such that $u(x_0) = 0$, $u > 0$ on $[0, L] \setminus \{x_0\}$,
$u \in W^{1,\infty}(0, L)$ and $\dfrac{1}{u} \notin L^1(0, L)$.

A common example in the choice of u is given by taking $u(x) = |x - x_0|^k$, with
$0 < x_0 < L$ and $k \geq 1$ fixed. The operators A_i, $i = 1, 2$, are assumed to have
suitable domains in Hilbert spaces of L^2 type. In Sect. 7.2 the case of the operator
A_1 is presented and a survey of results is given, based on [9]. This is a useful step
before to focus on the new results concerning the identification problem of u in the
case of the operator A_2 in nondivergence form treated in Sects. 7.3 and 7.4. In order
to face this identification problem which seems completely new, due to the presence
of an interior degeneracy of the coefficient u, we are forced to work in suitable
weighted spaces according to the generation results in [10]. Indeed, in [10] it was
proved that A_2, with a suitable domain, is selfadjoint and nonpositive on $L^2_{\frac{1}{u}}(0, L)$.
Hence, inspired by the methods used in [9], we examine the minimization problem
of the functional

$$J(u) = \frac{\lambda_1}{2} \left(\int_Q u(x) y_x^u(t, x) \, dxdt - M_f \right)^2 + \frac{\lambda_2}{2} \left(\int_0^L y^u(T, x) \, dx - M_T \right)^2$$

$$+ \frac{\lambda_3}{2} \left(\int_Q y^u(t, x) \, dxdt - M \right)^2. \quad (7.1)$$

Note that we take into account the observation of the spatial mean M_T of the state
at a final time T combined with the mean value M and the mean flux M_f over Q.

We point out that our results fit in a very recent literature devoted to identification
and controllability problems involving different types of degeneracy, e.g. [1–9],
motivated by very important applications to concrete models of the real life as
Budyko-Sellers models in climatology, Wright-Fisher and Fleming-Viot models in
genetics, Black-Merton-Scholes models in mathematical finance. In this framework

Fig. 7.1

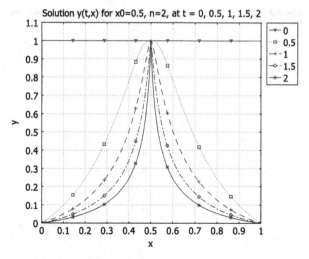

Solution y(t,x) for x0=0.5, n=2, at t = 0, 0.5, 1, 1.5, 2

Fig. 7.2

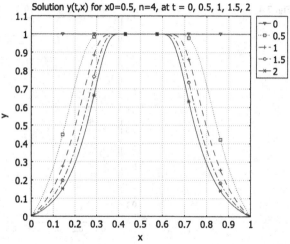

Solution y(t,x) for x0=0.5, n=4, at t = 0, 0.5, 1, 1.5, 2

it may be of interest for the applications to observe that the effects of the vanishing diffusion coefficient upon the solution of the diffusion equation (concentration, temperature, density) change according to the particular form of degeneracy of u. This behavior is illustrated in the graphics realized with Comsol Multiphysics v. 3.5a (FLN License 1025226), e.g., for $u(x) = |x - x_0|^n$, $x \in [0, 1]$, $x_0 = 0.5$, $y_0 = 1$ and $f = 0$. Figure 7.1 represents the values of the solution $y(t, x)$ to the parabolic problem (7.2) introduced in Sect. 7.2 along Ox, for $t = 0; 0.5; 1; 1.5; 2$, for $n = 2$, and Fig. 7.2 shows the distribution of the values of the solution $y(t, x)$ along Ox at the same times for $n = 4$. We observe that if u has a zero of a higher order of multiplicity ($n = 4$) at x_0, then the solution lies at high values in a larger subset of $(0, 1)$.

Fig. 7.3

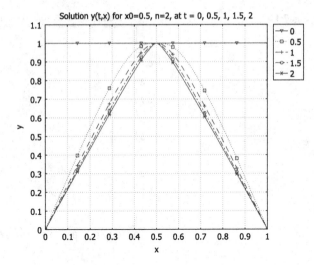

Fig. 7.4

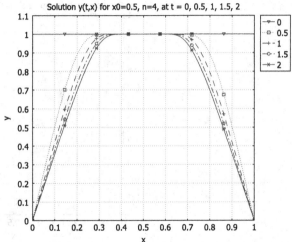

The solution to the parabolic problem in nondivergence case defined by system (7.26) in Sect. 7.3, corresponding to the same data as before, is presented in Figs. 7.3 and 7.4.

7.2   Identification of the Diffusion Coefficient in the Strongly Degenerate Divergence Case

In this section we recall the main results in [9], where the aim is to identify the diffusion coefficient $u(x)$ in the following degenerate parabolic problem written in divergence form

$$\begin{cases} y_t - (uy_x)_x = f(t,x), & (t,x) \in Q, \\ y(t,0) = y(t,L) = 0 & t \in (0,T), \\ y(0,x) = y_0(x) & x \in (0,L), \end{cases} \tag{7.2}$$

where $Q := (0,T) \times (0,L)$, $T, L \in (0,\infty)$. The identification of u is done on the basis of certain available observations, that is the spatial mean M_T of the state y at a final time T, the mean value M of y, or the mean flux M_f over Q, which justifies the choice of the cost functional defined in (7.1). The diffusion coefficient u is identified as a solution to the minimization problem

$$\text{Min}\,\{J(u);\ u \in U\} \tag{7.3}$$

subject to (7.2), where

$$U = \{u \in W^{1,\infty}(0,L);\ u_m(x) \le u(x) \le u_M(x),\ u(0) = u_0,\ u(L) = u_L,$$

$$|u_x(x)| \le u_\infty \text{ a.e. } x \in (0,L)\}. \tag{7.4}$$

In (7.4) we assume

$$x_0 \in (0,L),\ x_0 \text{ fixed and known, } u_\infty \in [0,\infty),$$

$$u_m,\ u_M \in C[0,L],\ u_M(x) \le \alpha(x)u_m(x) \text{ for } x \in [0,L],\ \alpha \in C[0,L], \alpha \ge 1, \tag{7.5}$$

$$0 < u_m(x) < u_M(x) \text{ for } x \in [0,L]\backslash\{x_0\},\ u_m(x_0) = u_M(x_0) = 0,$$

$$\int_0^L \frac{1}{u_M(x)}\,dx = +\infty. \tag{7.6}$$

We also assume that $M_f \in \mathbf{R}$, M_T, M, $\lambda_1, \lambda_2, \lambda_3 \ge 0$ and that there exists at least one $i \in \{1,2,3\}$ such that $\lambda_i > 0$. The constants λ_i, $i = 1,2,3$, are considered for inducing a higher or lower importance to the terms in the functional. The notation

y^u indicates the solution to (7.2) corresponding to u. For simplicity, later we shall drop the superscript ^{u}. We denote by the subscripts t and x the partial derivatives with respect to t and x, respectively.

In [9], the identification of u from the observations previously mentioned was led as a control problem in coefficients, the final goal being the optimality conditions, that is the necessary conditions that u must satisfy, as a solution to (7.3). To this end, the degenerate state system (7.2) was treated by the classical variational framework and the solution existence was proved by the Lions' Theorem (see [11]). Next, the existence of at least one minimizer to (7.3) was proved and a generalized form of the optimality conditions was deduced.

After specifying a few notation we recall the main results in [9], beginning with the existence of the solution to the state system (Theorem 7.1 below), the existence of an optimal pair (Theorem 7.2) and the optimality condition related to (7.3), in Proposition 7.1.

Following [10] we define the weighted spaces

$$H_u^1(0, L) = \{y \in L^2(0, L); \ y \text{ locally absolutely continuous in } [0, L]\setminus\{x_0\},$$

$$\sqrt{u}y_x \in L^2(0, L), \ y(0) = y(L) = 0\}, \qquad (7.7)$$

$$H_u^2(0, L) = \{y \in H_u^1(0, L); \ uy_x \in H^1(0, L)\} \qquad (7.8)$$

and specify that $H_u^1(0, L)$ is a Hilbert space and $H_u^1(0, L) \hookrightarrow L^2(0, L) \hookrightarrow (H_u^1(0, L))'$, where $(H_u^1(0, L))'$ is the dual of $H_u^1(0, L)$ and "$\hookrightarrow$" means a continuous and dense embedding. We denote $H = L^2(0, L)$, $V_u = H_u^1(0, L)$, $V_u' = (H_u^1(0, L))'$.

Definition 7.2 Let $y_0 \in L^2(0, L)$, $f \in L^2(0, T; (H_u^1(0, L))')$ and u with the properties of Definition 7.1. We call a *solution* to (7.2) a function

$$y \in C([0, T]; L^2(0, L)) \cap L^2(0, T; H_u^1(0, L)) \cap W^{1,2}([0, T]; (H_u^1(0, L))'),$$
$$(7.9)$$

which satisfies the equation

$$\int_0^T \left\langle \frac{dy}{dt}(t), \psi(t) \right\rangle_{V_u', V_u} dt + \int_Q uy_x \psi_x dxdt = \int_0^T \langle f(t), \psi(t) \rangle_{V_u', V_u} dt, \qquad (7.10)$$

for any $\psi \in L^2(0, T; H_u^1(0, L))$, and the initial condition $y(0) = y_0$.

Theorem 7.1 *If $y_0 \in L^2(0, L)$ and $f \in L^2(0, T; (H_u^1(0, L))')$, then (7.2) has a unique solution*

$$y \in C([0, T]; L^2(0, L)) \cap L^2(0, T; H_u^1(0, L)) \cap W^{1,2}([0, T]; (H_u^1(0, L))'),$$
$$(7.11)$$

satisfying the estimate

$$\sup_{t\in[0,T]} \|y(t)\|_H^2 + \int_0^T \|y(t)\|_{V_u}^2 \, dt \leq C_T(\|y_0\|_H^2 + \|f\|_{L^2(0,T;V_u')}^2). \tag{7.12}$$

If, in addition, $y_0 \in H_u^1(0,L)$ and $f \in L^2(Q) := L^2(0,T;L^2(0,L))$, then

$$y \in W^{1,2}([0,T];L^2(0,L)) \cap L^2(0,T;H_u^2(0,L)) \cap L^\infty(0,T;H_u^1(0,L)) \tag{7.13}$$

and it satisfies

$$\sup_{t\in[0,T]} \|y(t)\|_{V_u}^2 + \int_0^T \left(\left\| \frac{dy}{dt}(t) \right\|_H^2 + \|(uy_x)_x(t)\|_H^2 \right) dt$$

$$\leq C_T \left(\|y_0\|_{V_u}^2 + \|f\|_{L^2(Q)}^2 \right), \tag{7.14}$$

where C_T denotes several positive constants.

Theorem 7.2 *Let $y_0 \in L^2(0,L)$, $y_0 \geq 0$ on $(0,L)$, $f \in L^2(Q)$, $f \geq 0$ a.e. on Q. Then, the minimization problem (7.3) has at least one solution u with the corresponding state y belonging to the spaces mentioned in (7.11). If $y_0 \in H_u^1(0,L)$, then the state y is more regular, as in (7.13).*

Now, let u^* be a solution to (7.3) and let y^* be the solution to (7.2) corresponding to the coefficient u^*. We introduce the system for the adjoint state p, by

$$\begin{cases} \dfrac{\partial p}{\partial t} + (u^* p_x)_x = -\lambda_1 \left(\int_Q u^* y_x^* dxdt - M_f \right) u_x^* \\ \qquad +\lambda_3 \left(\int_Q y^* dxdt - M \right) & \text{in } Q, \\[2mm] p(T,x) = -\lambda_2 \left(\int_0^L y^*(T,x)dx - M_T \right), & \text{in } (0,L), \\[2mm] p(t,0) = p(t,L) = 0, & \text{in } (0,T), \end{cases} \tag{7.15}$$

and give the following result which describes the property of u^*, as a minimizer in (7.3). Before this, we specify that (7.15) has a unique solution p belonging to the same spaces as y (see (7.11) and (7.13) in Theorem 7.1).

Proposition 7.1 *Let (u^*, y^*) be a solution to (7.3). Then u^* satisfies the necessary condition*

$$\int_Q (u^* - u)y_x^* \left[p_x + \lambda_1 \left(\int_Q u^* y_x^* dxdt - M_f \right) \right] dxdt \leq 0 \tag{7.16}$$

for all $u \in U$, where p is the solution to (7.15).

Since y^* and p do not belong to $H^1(0, L)$ but to $H^1_u(0, L)$, it is not clear whether condition (7.16) can be further handled in order to deduce a more explicit form for u^*. However, an alternative way to characterize it is to compute u^* as the limit of a sequence of more regular solutions u^*_ε to an approximating minimization problem, with a nondegenerate state system. Thus, we introduce the approximating problem

$$\text{Min } \{J(u); \ u \in U_\varepsilon\}, \tag{7.17}$$

$\epsilon > 0$, subject to the state system (7.2), where

$$U_\varepsilon = \{u \in W^{1,\infty}(0, L); \ u_m(x) + \varepsilon \leq u(x) \leq u_M(x) + 2\varepsilon, \ u(0) = u^\varepsilon_0, \ u(L) = u^\varepsilon_L,$$

$$|u_x(x)| \leq u_\infty \text{ a.e. } x \in (0, L)\}. \tag{7.18}$$

For all $u \in U_\varepsilon$, $u(x) \geq u_m(x) + \varepsilon \geq \varepsilon$, and so system (7.2) with $u \in U_\varepsilon$ is nondegenerate, implying that its solution y_ε corresponding to $u \in U_\varepsilon$ belongs to the space (see [11, p. 163])

$$C([0, T]; L^2(0, L)) \cap L^2(0, T; H^1_0(0, L)) \cap W^{1,2}([0, T]; H^{-1}(0, L)). \tag{7.19}$$

Consequently, the control problem (7.17) has at least a solution $(u_\varepsilon, y_\varepsilon)$, with $u_\varepsilon \in U_\varepsilon$ and y_ε (corresponding to u_ε) satisfying (7.19). The next theorem proves the connection between a solution to (7.3) and a solution to (7.17) and shows in fact that a minimizer u^* to the original problem can be retrieved as the limit of a sequence of minimizers corresponding to the approximating problem (7.17).

Theorem 7.3 Let $y_0 \in L^2(0, L)$, $f \in L^2(Q)$, $y_0 \geq 0$ a.e. on $(0, L)$, and $f \geq 0$ a.e. on Q. Let $(u^*_\varepsilon, y^*_\varepsilon)_{\varepsilon>0}$ be a sequence of solutions to (7.17). Then (on subsequences), as $\varepsilon \to 0$ we have

$$u^*_\varepsilon \to u^* \text{ uniformly in } [0, L], \ u^*_{\varepsilon x} \to u^*_x \text{ weak* in } L^\infty(0, L), \tag{7.20}$$

$$y^*_\varepsilon \to y^* \text{ weakly in } L^2(0, T; H^1_{u^*}(0, L)) \cap W^{1,2}([0, T]; (H^1_{u^*}(0, L))'), \tag{7.21}$$

$$y_\varepsilon(T) \to y(T) \text{ weakly in } L^2(0, L), \tag{7.22}$$

where (u^*, y^*) is a solution to (7.3). Moreover,

$$\int_0^L (u^*_\varepsilon - u_\varepsilon)(x)\Phi(x)dx \geq 0, \quad \text{for all } u_\varepsilon \in U_\varepsilon, \tag{7.23}$$

where

$$\Phi(x) := -\int_0^T y^*_{\varepsilon x}(t, x) \left\{ p_{\varepsilon x}(t, x) + \lambda_1 \left(\int_Q u^*_\varepsilon y^*_{\varepsilon x} dx dt - M_f \right) \right\} dt. \tag{7.24}$$

In (7.24) p_ε is the solution to (7.15) in which (u^*, y^*) is replaced by $(u^*_\varepsilon, y^*_\varepsilon)$. Due to the fact that the approximating minimizer is positive and y^*_ε belongs to the spaces mentioned in (7.19), relation (7.23) will further provide more clear information about u_ε^* (see [9]).

Also a very explicit form of u^* has been identified in [9] from a final observation involving a state system with Dirichlet–Neumann boundary conditions.

7.3 Identification of the Diffusion Coefficient in the Strongly Degenerate Nondivergence Case

In this section we consider the minimization problem in the nondivergence case, that is

$$\text{Min}\{J(u); \ u \in U\} \tag{7.25}$$

subject to

$$\begin{cases} y_t - u y_{xx} = f(t,x), & (t,x) \in Q, \\ y(t,0) = y(t,L) = 0 & t \in (0,T), \\ y(0,x) = y_0(x) & x \in (0,L), \end{cases} \tag{7.26}$$

where J and U have been defined in (7.1) and (7.4)–(7.6), respectively.

We introduce the adapted functional framework, defining the following Hilbert spaces:

$$L^2_{\frac{1}{u}}(0,L) := \left\{ y \in L^2(0,L) \ \Big| \ \int_0^L \frac{y^2}{u} dx < \infty \right\},$$

$$\mathscr{H}^1_{\frac{1}{u}}(0,L) := L^2_{\frac{1}{u}}(0,L) \cap H^1_0(0,L),$$

and

$$\mathscr{H}^2_{\frac{1}{u}}(0,L) := \left\{ y \in \mathscr{H}^1_{\frac{1}{u}}(0,L) \cap W^{2,1}_{loc}(0,L) \ \Big| \ u y_{xx} \in L^2_{\frac{1}{u}}(0,L) \right\},$$

endowed with the associated norms

$$\|y\|^2_{L^2_{\frac{1}{u}}(0,L)} := \int_0^L \frac{y^2}{u} dx, \quad y \in L^2_{\frac{1}{u}}(0,L),$$

$$\|y\|^2_{\mathscr{H}^1_{\frac{1}{u}}(0,L)} := \|y\|^2_{L^2_{\frac{1}{u}}(0,L)} + \|y_x\|^2_{L^2(0,L)}, \quad y \in \mathscr{H}^1_{\frac{1}{u}}(0,L),$$

$$\|y\|^2_{\mathscr{H}^2_{\frac{1}{u}}(0,L)} := \|y\|^2_{\mathscr{H}^1_{\frac{1}{u}}(0,L)} + \|u y_{xx}\|^2_{L^2_{\frac{1}{u}}(0,L)}, \quad y \in \mathscr{H}^2_{\frac{1}{u}}(0,L).$$

We recall also the following Green's formula:

Lemma 7.1 (See [10, Lemma 3.2]) *For all* $(y, z) \in \mathcal{H}^2_{\frac{1}{u}}(0, L) \times \mathcal{H}^1_{\frac{1}{u}}(0, L)$ *one has*

$$\int_0^L y_{xx} z \, dx = - \int_0^L y_x z_x \, dx. \tag{7.27}$$

We give the following definition:

Definition 7.3 Assume that $y_0 \in L^2_{\frac{1}{u}}(0, L)$ and $f \in L^2_{\frac{1}{u}}(Q) := L^2(0, T; L^2_{\frac{1}{u}}(0, L))$. A function y is said to be a solution to (7.26) if

$$y \in C([0, T]; L^2_{\frac{1}{u}}(0, L)) \cap L^2(0, T; \mathcal{H}^1_{\frac{1}{u}}(0, L))$$

and satisfies

$$\int_0^L \frac{y(T) \, \varphi(T)}{u} \, dx - \int_0^L \frac{y_0 \, \varphi(0)}{u} \, dx - \int_Q \frac{y \, \varphi_t}{u} \, dx dt$$

$$= - \int_Q y_x \varphi_x \, dx dt + \int_Q \frac{f \, \varphi}{u} \, dx dt \tag{7.28}$$

for all $\varphi \in \mathcal{V} := H^1(0, T; L^2_{\frac{1}{u}}(0, L)) \cap L^2(0, T; \mathcal{H}^1_{\frac{1}{u}}(0, L))$.

Now, as in [10] we consider the strongly degenerate operator A_2 defined by

$$A_2 y := u y_{xx},$$

with domain $D(A_2) = \mathcal{H}^2_{\frac{1}{u}}(0, L)$. Here u satisfies the conditions of Definition 7.1.

Using the fact that the operator $A_2 : D(A_2) \to L^2_{\frac{1}{u}}(0, L)$ is selfadjoint, nonpositive on $L^2_{\frac{1}{u}}(0, L)$ and generates a positivity preserving semigroup (see [10, Theorem 3.4]), one can prove the following existence result:

Theorem 7.4 (See [10, Theorem 4.3 and Remark 4.4]) *For all* $f \in L^2_{\frac{1}{u}}(Q)$ *and* $y_0 \in L^2_{\frac{1}{u}}(0, L)$, *there exists a unique (mild) solution*

$$y \in C([0, T]; L^2_{\frac{1}{u}}(0, L)) \cap L^2(0, T; \mathcal{H}^1_{\frac{1}{u}}(0, L)) \tag{7.29}$$

to (7.26) such that

$$\sup_{t \in [0,T]} \|y(t)\|^2_{L^2_{\frac{1}{u}}(0,L)} + \int_0^T \|y(t)\|^2_{\mathcal{H}^1_{\frac{1}{u}}(0,L)} dt \leq C_T \left(\|y_0\|^2_{L^2_{\frac{1}{u}}(0,L)} + \|f\|^2_{L^2_{\frac{1}{u}}(Q)} \right). \tag{7.30}$$

Moreover, if $y_0 \in \mathscr{H}_{\frac{1}{u}}^1(0, L)$, then

$$y \in H^1\left(0, T; L_{\frac{1}{u}}^2(0, L)\right) \cap L^2\left(0, T; \mathscr{H}_{\frac{1}{u}}^2(0, L)\right) \cap C\left([0, T]; \mathscr{H}_{\frac{1}{u}}^1(0, L)\right) \quad (7.31)$$

such that

$$\sup_{t \in [0,T]} \left(\|y(t)\|_{\mathscr{H}_{\frac{1}{u}}^1(0,L)}^2 \right) + \int_0^T \left(\|y_t(t)\|_{L_{\frac{1}{u}}^2(0,L)}^2 + \|uy_{xx}(t)\|_{L_{\frac{1}{u}}^2(0,L)}^2 \right) dt$$
$$\leq C_T \left(\|y_0\|_{\mathscr{H}_{\frac{1}{u}}^1(0,L)}^2 + \|f\|_{L_{\frac{1}{u}}^2(Q)}^2 \right), \quad (7.32)$$

where C_T denotes a positive constant.

It is clear that if $y_0 \in \mathscr{H}_{\frac{1}{u}}^1(0, L)$, the strong solution to (7.26) satisfies the equivalent relation

$$\int_Q \frac{y_t \varphi}{u} \, dxdt + \int_Q y_x \varphi_x \, dxdt = \int_Q \frac{f\varphi}{u} \, dxdt, \quad (7.33)$$

for all $\varphi \in L^2(0, T; \mathscr{H}_{\frac{1}{u}}^1(0, L))$.

Note that from (7.4) and (7.5) it clearly follows

$$\left| \frac{u(x)}{v(x)} \right| \leq \|\alpha\|_{L^\infty(0,L)}, \quad \text{a.e. in } [0, L], \quad (7.34)$$

for all $u, v \in U$.

Using the previous estimate, one can prove the following result:

Lemma 7.2 *For all $u, v \in U$, one has*

$$L_{\frac{1}{u}}^2(0, L) = L_{\frac{1}{v}}^2(0, L),$$

$$\mathscr{H}_{\frac{1}{u}}^1(0, L) = \mathscr{H}_{\frac{1}{v}}^1(0, L),$$

$$\mathscr{H}_{\frac{1}{u}}^2(0, L) = \mathscr{H}_{\frac{1}{v}}^2(0, L).$$

Proof Let $y \in L_{\frac{1}{u}}^2(0, L)$. Then it follows

$$\int_0^L \frac{y^2}{v} \, dx = \int_0^L \frac{u}{v} \frac{y^2}{u} \, dx \leq K \int_0^L \frac{y^2}{u} \, dx,$$

where $K := \|\alpha\|_{L^\infty(0,L)}$. Hence $y \in L^2_{\frac{1}{v}}(0, L)$. Analogously one can prove that $L^2_{\frac{1}{v}}(0, L) \subset L^2_{\frac{1}{u}}(0, L)$. As an immediate consequence, it follows

$$\mathcal{H}^1_{\frac{1}{u}}(0, L) = \mathcal{H}^1_{\frac{1}{v}}(0, L)$$

for all $u, v \in U$.

Now we prove that

$$\mathcal{H}^2_{\frac{1}{u}}(0, L) = \mathcal{H}^2_{\frac{1}{v}}(0, L).$$

Let $y \in \mathcal{H}^2_{\frac{1}{u}}(0, L)$. Then, by definition of $\mathcal{H}^2_{\frac{1}{u}}(0, L)$, $y \in \mathcal{H}^1_{\frac{1}{u}}(0, L) = \mathcal{H}^1_{\frac{1}{v}}(0, L)$ and

$$\int_0^L \frac{(vy_{xx})^2}{v} dx = \int_0^L v(y_{xx})^2 dx = \int_0^L \frac{v}{u}(y_{xx})^2 u\, dx \leq K \int_0^L (y_{xx})^2 u\, dx.$$

Hence $y \in \mathcal{H}^2_{\frac{1}{v}}(0, L)$. Analogously one can prove that $\mathcal{H}^2_{\frac{1}{v}}(0, L) \subset \mathcal{H}^2_{\frac{1}{u}}(0, L)$. $\square$

Since by Theorem 7.4, in particular $uy_x, y \in L^2(Q)$ and $y(T) \in L^2(0, L)$, it follows that all terms in $J(u)$ are well defined, even in the nondivergence case.

In the next theorem we give the existence result for a minimizer to (7.25).

Theorem 7.5 *Let us take $v \in U$ and $y_0 \in L^2_{\frac{1}{v}}(0, L)$, $f \in L^2_{\frac{1}{v}}(Q)$. Then, the minimization problem (7.25) has at least one solution u. If $y_0 \in \mathcal{H}^1_{\frac{1}{v}}(0, L)$, then (7.25) has at least one solution with the corresponding state y being a strong solution to (7.26).*

Proof Let $y_0 \in L^2_{\frac{1}{v}}(0, L)$ and $f \in L^2_{\frac{1}{v}}(Q)$, with $v \in U$. Then there exists a unique (mild) solution y^v to (7.26) and $J(v) \geq 0$. Let us denote by d the infimum of J on U. For simplicity, in the following, we shall drop the superscript u in y^u.

Now, consider a minimizing sequence $(u_n)_{n\geq 1}$, $u_n \in U$, such that

$$d \leq J(u_n) := \frac{\lambda_1}{2} \left(\int_Q u_n(x)\, (y_n)_x(t, x)\, dxdt - M_f \right)^2$$

$$+ \frac{\lambda_2}{2} \left(\int_0^L y_n(T, x)\, dx - M_T \right)^2$$

$$+ \frac{\lambda_3}{2} \left(\int_Q y_n(t, x)\, dxdt - M \right)^2 \leq d + \frac{1}{n}, \qquad (7.35)$$

where y_n is the mild solution to (7.26) corresponding to u_n, i.e.,

$$y_n \in C\left([0, T]; L^2_{\frac{1}{u}}(0, L)\right) \cap L^2\left(0, T; \mathscr{H}^1_{\frac{1}{u}}(0, L)\right),$$

equivalently written

$$\int_0^L \frac{y_n(T)\,\varphi(T)}{u_n}\,dx - \int_0^L \frac{y_0\,\varphi(0)}{u_n}\,dx - \int_Q \frac{y_n\,\varphi_t}{u_n}\,dxdt$$

$$= -\int_Q y_{nx}\varphi_x\,dxdt + \int_Q \frac{f\,\varphi}{u_n}\,dxdt \qquad (7.36)$$

for all $\varphi \in \mathscr{V}$.

Moreover, y_n satisfies (7.30).

Since $(u_n)_{n \geq 1}$ is equicontinuous and bounded, by the Arzelà Theorem, u_n converges uniformly to $u \in C[0, L]$, and up to subsequence, $u_{nx} \to u_x$ weak* in $L^\infty(0, L)$ as $n \to +\infty$. Clearly,

$$u(0) = u_0, \quad u(L) = u_L, \quad u_m(x) \leq u(x) \leq u_M(x), \quad |u_x(x)| \leq u_\infty \text{ a.e. } x \in (0, L).$$

Hence $u \in U$.

By (7.30) and Lemma 7.2 we deduce that $(y_n)_n$ is bounded in $L^\infty\left(0, T; L^2_{\frac{1}{u}}(0, L)\right) \cap L^2\left(0, T; \mathscr{H}^1_{\frac{1}{u}}(0, L)\right)$. Hence, there exists $y \in L^\infty\left(0, T; L^2_{\frac{1}{u}}(0, L)\right) \cap L^2\left(0, T; \mathscr{H}^1_{\frac{1}{u}}(0, L)\right)$ such that, up to a subsequence, as $n \to \infty$

$$y_n \to y \quad \text{weak* in } L^\infty\left(0, T; L^2_{\frac{1}{u}}(0, L)\right), \qquad (7.37)$$

$$y_n \to y \quad \text{weak* in } L^\infty\left(0, T; L^2(0, L)\right).$$

and

$$y_{nx} \to y_x \quad \text{weakly in } L^2(Q),$$

i.e.

$$\lim_{n \to +\infty} \int_Q y_{nx}(t, x)\varphi_x(t, x)\,dxdt = \int_Q y_x(t, x)\varphi_x(t, x)\,dxdt \qquad (7.38)$$

for all $\varphi \in L^2\left(0, T; \mathscr{H}^1_{\frac{1}{u}}(0, L)\right)$.

By (7.30) we also have that $\left(\dfrac{y_n}{\sqrt{u_n}}\right)_{n \geq 1}$ is bounded in $L^\infty(0, T; L^2(0, L))$ and so, on a subsequence

$$\frac{y_n}{\sqrt{u_n}} \to \zeta \quad \text{weak* in } L^\infty(0, T; L^2(0, L)).$$

Let us denote $\zeta_n = \frac{y_n}{\sqrt{u_n}}$ a.e. on Q. Then, $y_n = \sqrt{u_n}\zeta_n$ a.e. on Q, which converges to $y = \sqrt{u}\zeta$, whence we deduce that $\zeta = \frac{y}{\sqrt{u}}$ a.e. on Q.

Moreover,

$$\lim_{n\to+\infty} \int_Q \frac{f\varphi}{u_n} dxdt = \int_Q \frac{f\varphi}{u} dxdt, \qquad (7.39)$$

for all $\varphi \in L^2(0, T; \mathcal{H}^1_{\frac{1}{u}}(0, L))$. Indeed, setting $g_n := \frac{f\varphi}{u_n}$, it follows that $(g_n)_{n\geq 1}$ converges almost everywhere to $\frac{f\varphi}{u}$ and, by (7.34),

$$\int_Q |g_n| dxdt = \int_Q \left|\frac{f\varphi}{u}\frac{u}{u_n}\right| dxdt \leq \|\alpha\|_{L^\infty(0,L)} \int_0^T \left\|\frac{f\varphi}{u}(t)\right\|_{L^1(0,L)} dt.$$

Hence, (7.39) follows by the Lebesgue dominated convergence Theorem.

We show now that

$$\lim_{n\to\infty} \int_Q \frac{y_n\varphi}{u_n} dxdt = \int_Q \frac{y\varphi}{u} dxdt \qquad (7.40)$$

for all $\varphi \in L^2(0, T; L^2_{\frac{1}{u}}(0, L))$.

To this aim, let us take $\varphi \in L^2_{\frac{1}{u}}(Q)$ and note that $\frac{\varphi}{\sqrt{u_n}}$ converges almost everywhere to $\frac{\varphi}{\sqrt{u}}$ in $L^2(Q)$. Moreover, as before,

$$\int_Q \left|\frac{\varphi}{\sqrt{u_n}}\right| dxdt \leq \|\alpha\|_{L^\infty(0,L)} \int_0^T \left\|\frac{\varphi(t)}{\sqrt{u}}\right\|_{L^1(0,L)} dt$$

and so, by the Lebesgue dominated convergence Theorem we get that

$$\frac{\varphi}{\sqrt{u_n}} \to \frac{\varphi}{\sqrt{u}} \quad \text{strongly in } L^2(Q).$$

Since

$$\frac{y_n}{\sqrt{u_n}} \to \frac{y}{\sqrt{u}} \quad \text{weak* in } L^\infty(0, T; L^2(0, L))$$

we get (7.40) as claimed.

Since A_2 generates a C_0-semigroup on $L^2_{\frac{1}{u}}(0, L)$ and y_n is a mild solution to (7.26), we can write

$$y_n(t) = e^{A_2 t} y_0 + \int_0^t e^{A_2(t-s)} f(s) ds, \quad \text{for all } t \in [0, T]. \qquad (7.41)$$

Then, according to Ball's theorem (see e.g., [12, p. 258]), for every $v^* \in D(A_2^*)$, (A_2^* being the adjoint of A_2), the map $t \to (y(t), v^*)_{L_{\frac{1}{u}}^2 (0,L)}$ is absolutely continuous on $[0, T]$ and

$$\frac{d}{dt}(y_n(t), v^*)_{L_{\frac{1}{u}}^2 (0,L)} = (y_n(t), A_2^* v^*)_{L_{\frac{1}{u}}^2 (0,L)} + (f(t), v^*)_{L_{\frac{1}{u}}^2 (0,L)}, \quad \text{for all } t \in [0, T],$$

where $(\cdot, \cdot)_{L_{\frac{1}{u}}^2 (0,L)}$ denotes the inner product in $L_{\frac{1}{u}}^2 (0, L)$. We also recall that A_2 is self-adjoint and so, integrating the previous equation with respect to t, we get

$$(y_n(t), v^*)_{L_{\frac{1}{u}}^2 (0,L)} - (y_0, v^*)_{L_{\frac{1}{u}}^2 (0,L)} = \int_0^t (y_n(s), u_n v_{xx}^*)_{L_{\frac{1}{u}}^2 (0,L)} \, ds$$

$$+ \int_0^t (f(s), v^*)_{L_{\frac{1}{u}}^2 (0,L)} \, ds$$

for all $t \in [0, T]$. Taking into account the previous convergences we get that

$$y_n(t) \to y(t) \text{ weakly in } L_{\frac{1}{u}}^2 (0, L), \text{ for all } t \in [0, T]$$

and in particular at $t = T$ and $t = 0$.

Then, we get all ingredients to pass to the limit in (7.36) and get (7.28) which shows that y is the solution to (7.26).

Finally, by passing to the limit in (7.35) on the basis of the weak lower semicontinuity of the convex integrands we get

$$\liminf_{n \to \infty} J(u_n) \geq J(u),$$

and, hence, $J(u) = d$, which asserts that u is a solution to (7.25).

Let now take $y_0 \in \mathcal{H}_{\frac{1}{u}}^1 (0, L)$. The proof is led as before, specifying that y_n is in this case the strong solution to (7.26), equivalently written

$$\int_Q \frac{y_{nt} \varphi}{u_n} dxdt + \int_Q y_{nx} \varphi_x dxdt = \int_Q \frac{f \varphi}{u_n} dxdt, \qquad (7.42)$$

for all $\varphi \in L^2(0, T; \mathcal{H}_{\frac{1}{u}}^1 (0, L))$ and it satisfies estimate (7.32). It follows that, on a subsequence

$$\frac{y_{nt}}{\sqrt{u_n}} \to \Psi \text{ weakly in } L^2(Q).$$

Recalling (7.40) and since $\left(\dfrac{y_n}{\sqrt{u_n}}\right)_t$ converges in the sense of distributions to $\left(\dfrac{y}{\sqrt{u}}\right)_t$ we get that $\Psi = \dfrac{y}{\sqrt{u}}$ a.e. on Q.

Passing to the limit in (7.42), we obtain

$$\int_Q \frac{y_t \varphi}{u} dxdt + \int_Q y_x \varphi_x dxdt = \int_Q \frac{f\varphi}{u} dxdt, \tag{7.43}$$

for all $\varphi \in L^2(0, T; \mathcal{H}_1^1(0, L))$, so that y is a strong solution to (7.26). Then we pass to the limit in (7.35) and obtain that u is a minimizer in (7.25), as claimed. $\quad\square$

7.4 Optimality Conditions in the Nondivergence Case

In this section we assume hypotheses from Theorem 7.5, second part, and consider (u^*, y^*) be a solution to (7.25). We recall that y^* is the strong solution to the state system. Let us denote

$$I_f = \int_Q u^* y_x^* \, dxdt - M_f, \quad I_T = \int_0^L y^*(T, x) \, dx - M_T, \quad I_M = \int_Q y^* \, dxdt - M \tag{7.44}$$

and let us introduce the dual system as

$$\begin{cases} \dfrac{\partial p}{\partial t} + u^* p_{xx} = -\lambda_1 I_f u_x^* u^* + \lambda_3 I_M u^*, & \text{in } Q, \\[2mm] p(T, x) = -\lambda_2 I_T u^*, & \text{in } (0, L), \\[2mm] p(t, 0) = p(t, L) = 0, & \text{in } (0, T). \end{cases} \tag{7.45}$$

Theorem 7.6 *Let (u^*, y^*) be a solution to (7.25). Then, u^* satisfies the necessary condition*

$$\int_Q (u^* - u) \left(\frac{y_{xx}^* p}{u^*} - \lambda_1 I_f y_x^*\right) dxdt \geq 0, \tag{7.46}$$

for all $u \in U$, where p is the solution to (7.45).

Proof Let $\lambda \in (0, 1)$ and denote

$$u^\lambda(x) = u^*(x) + \lambda v(x),$$

where

$$v(x) = u(x) - u^*(x), \quad \text{with } u \in U. \tag{7.47}$$

It is obvious that $v \in W^{1,\infty}(0,L)$, $v(x_0) = 0$, $u_m(x) - u_M(x) \le v(x) \le u(x)$, $v(0) = v(L) = 0$ and $\frac{1}{v} \notin L^1(0,L)$. We introduce the system

$$\begin{cases} \dfrac{\partial Y}{\partial t} - u^* Y_{xx} = v y_{xx}^*, & \text{in } Q; \\[2mm] Y(0,x) = 0, & \text{in } (0,L); \\[2mm] Y(t,0) = Y(t,L) = 0, & \text{in } (0,T). \end{cases} \tag{7.48}$$

Since y^* is a strong solution to the state system (7.26), it follows that $u^* y_{xx}^* \in L^2_{\frac{1}{u^*}}(Q)$. Then, again by Theorem 7.4, system (7.48) has a unique solution

$$Y \in H^1(0,T; L^2_{\frac{1}{u^*}}(0,L)) \cap L^2(0,T; \mathcal{H}^1_{\frac{1}{u^*}}(0,L)), \qquad u^* Y_{xx} \in L^2_{\frac{1}{u^*}}(Q).$$

Denoting by $y^\lambda(t,x)$ the solution to (7.26) corresponding to $u^\lambda(x)$, one can prove by a direct computation that

$$Y(t,x) = \lim_{\lambda \to 0} \frac{y^\lambda(t,x) - y^*(t,x)}{\lambda} \qquad \text{weakly in } L^2_{\frac{1}{u^*}}(Q).$$

Since the term on the right-hand side in the equation of the dual system (7.45) belongs to $L^2_{\frac{1}{u^*}}(Q)$, system (7.45) has, still by the second part of Theorem 7.4, a unique solution

$$p \in H^1(0,T; L^2_{\frac{1}{u^*}}(0,L)) \cap L^2(0,T; \mathcal{H}^1_{\frac{1}{u^*}}(0,L)), \qquad p^* Y_{xx} \in L^2_{\frac{1}{u^*}}(Q).$$

Now, we write that (u^*, y^*) is a solution to (7.25), that is

$$J(u^*) \le J(u), \qquad \text{for all } u \in U,$$

which is true, in particular, for $u = u^\lambda$. After some calculations, this inequality yields the following relation

$$\int_Q (-\lambda_1 I_f u_x^* + \lambda_3 I_M) Y \, dx \, dt + \int_Q \lambda_1 I_f v y_x^* \, dx \, dt + \lambda_2 I_T \int_0^L Y(T,x) \, dx \ge 0. \tag{7.49}$$

We test the equation in the system (7.48) by $p(t)$ and integrate over $(0, T)$. We have

$$\int_Q \frac{Y_t p}{u^*} \, dxdt + \int_Q Y_x p_x = \int_Q \frac{vy^*_{xx} p}{u^*} \, dxdt$$

whence, integrating by parts the first term on the left-hand side, and taking into account the conditions in the system (7.45) we get

$$\int_0^L \frac{p(T)Y(T)}{u^*} \, dx - \int_Q \frac{Yp_t}{u^*} \, dxdt + \int_Q Y_x p_x \, dxdt = \int_Q \frac{vy^*_{xx} p}{u^*} \, dxdt. \qquad (7.50)$$

Further, this yields

$$-\int_0^T \left((p_t + u^* p_{xx})(t), Y(t)\right)_{L^2_{\frac{1}{u^*}}(0,L)} \, dt + (p(T), Y(T))_{L^2_{1/u^*}(0,L)}$$

$$= \int_0^T \left(vy^*_{xx}(t), p(t)\right)_{L^2_{\frac{1}{u^*}}(0,L)} \, dt.$$

From (7.45) we obtain

$$\int_Q (-\lambda_1 I_f u^*_x + \lambda_3 I_M) Y dxdt + \lambda_2 I_T \int_0^L Y(T, x) dx = -\int_Q \frac{vy^*_{xx} p}{u^*} dxdt. \qquad (7.51)$$

Comparing with (7.49) it follows that

$$\int_Q \lambda_1 I_f vy^*_x dxdt - \int_Q \frac{vy^*_{xx} p}{u^*} dxdt \geq 0$$

for all $u \in U$, and this implies (7.46), as claimed. □

Acknowledgements G. M. acknowledges the support of the project CNCS-UEFISCDI PN-II-ID-PCE-2011-3-0045 of the Romanian National Authority for Scientific Research. G.F., R.M. M., S.R. acknowledge the support of the GNAMPA project *Equazioni di evoluzione degeneri e singolari: controllo e applicazioni*.

References

1. Ait Ben Hassi, E.M., Ammar Khodja, F., Hajjaj, A., Maniar, L.: Null controllability of degenerate parabolic cascade systems. Port. Math. **68**, 345–367 (2011)
2. Ait Ben Hassi, E.M., Ammar Khodja, F., Hajjaj, A., Maniar, L.: Carleman estimates and null controllability of coupled degenerate systems. Evol. Equ. Control Theory **2**, 441–459 (2013)
3. Cannarsa, P., Fragnelli, G.: Null controllability of semilinear degenerate parabolic equations in bounded domains. Electron. J. Differ. Equ. **2006**, 1–20 (2006)

4. Cannarsa, P., Fragnelli, G., Rocchetti, D.: Null controllability of degenerate parabolic operators with drift. Netw. Heterog. Media **2**, 693–713 (2007)
5. Favini, A., Marinoschi, G.: Identification of the time derivative coefficient in a fast diffusion degenerate equation. J. Optim. Theory Appl. **145**, 249–269 (2010)
6. Flores, C., De Teresa, L.: Carleman estimates for degenerate parabolic equations with first order terms and applications. C. R. Math. Acad. Sci. Paris **348**, 391–396 (2010)
7. Fragnelli, G.: Null controllability of degenerate parabolic equations in non divergence form via Carleman estimates. Discrete Continuous Dyn. Syst. Ser. S **6**, 687–701 (2013)
8. Fragnelli, G., Mugnai, D.: Carleman estimates and observability inequalities for parabolic equations with interior degeneracy. Adv. Nonl. Anal. **2**, 339–378 (2013)
9. Fragnelli, G., Marinoschi, G., Mininni, R.M., Romanelli, S.: Identification of a diffusion coefficient in strongly degenerate parabolic equations with interior degeneracy. J. Evol. Equ. 2014. DOI 10.1007/s00028-14-0247-1
10. Fragnelli, G., Ruiz Goldstein, G., Goldstein, J.A., Romanelli, S.: Generators with interior degeneracy on spaces of L^2 type. Electron. J. Differ. Equ. **2012**, 1–30 (2012)
11. Lions, J.L.: Quelques méthodes de resolution des problèmes aux limites non linéaires. Dunod, Paris (1969)
12. Pazy, A.: Semigroups of Linear Operators and Applications to Partial Differential Equations. Springer, New York (1983)

Chapter 8
On the Nature of the Instability of Radial Power Equilibria of a Semilinear Parabolic Equation

Jerome A. Goldstein and Junqiang Han

Dedicated to the memory of Alfredo Lorenzi

Abstract The semilinear problem

$$\frac{\partial u}{\partial t} = (-1)^{m+1}\Delta^m u + u^p \quad (x \in \mathbb{R}^N, t \geq 0)$$

has positive equilibria of the form

$$u(x,t) = C\,|x|^{-a}$$

for many values of (N, p). Of concern is getting more information on exactly how stable or unstable these solutions are. When $m = 1$, the results have a qualitatively different nature for the two cases $N \leq 10$ and $N \geq 11$.

8.1 Introduction

The semilinear parabolic equation

$$\frac{\partial u}{\partial t} = (-1)^{m+1}\Delta^m u + u^p \quad (t \geq 0, x \in \mathbb{R}^N, m \in \mathbb{N}) \tag{8.1}$$

J.A. Goldstein (✉)
Department of Mathematical Sciences, The University of Memphis, 38152 Memphis, TN, USA
e-mail: jgoldste@memphis.edu

J. Han
Department of Applied Mathematics, Northwestern Polytechnical University, 710072 Xi'an,
Shaanxi, China
e-mail: southhan@163.com

© Springer International Publishing Switzerland 2014
A. Favini et al. (eds.), *New Prospects in Direct, Inverse and Control Problems
for Evolution Equations*, Springer INdAM Series 10,
DOI 10.1007/978-3-319-11406-4_8

has positive time independent solutions of the form

$$u(x,t) = v(r) = Cr^{-a} \tag{8.2}$$

where $p > 1$, $r = |x|$, and $C, a > 0$. At issue is the stability or instability of these solutions.

The simplest case is that of $m = 1$, in which case the Maximum Principle holds; it fails for $m \geq 2$. Some stability/instability results for the positive solutions (8.2) of (8.1) when $m = 1$ were obtained by Gui et al. [5, 6] and further refined by Polacik and Yanagida [9]. These are deep and subtle results. These seem to be no known analogous results for $m \geq 2$.

We shall approach the stability issue from the (specialized) perspective of linearized stability. For $m = 1$, this has been considered already by Mizoguchi [8] in the unstable case, but, as before, nothing seems to have been done for $m \geq 2$.

A simple calculation shows that, formally, the linearization of (8.1) about v (using (8.1) and (8.2)) is

$$\begin{aligned}
\frac{\partial w}{\partial t} &= (-1)^{m+1} \Delta^m w + (pv^{p-1})w \\
&= (-1)^{m+1} \Delta^m w + \frac{c}{|x|^{2m}} w
\end{aligned} \tag{8.3}$$

where

$$c = c(p, m, N).$$

The operator

$$(-1)^m \Delta^m - \frac{c}{|x|^{2m}}$$

on $L^2(\mathbb{R}^N)$ is formally symmetric and is nonnegative iff

$$(-1)^m \langle \Delta^m \psi, \psi \rangle \geq c \left\langle \frac{1}{|x|^{2m}} \psi, \psi \right\rangle, \tag{8.4}$$

i.e.,

$$\int_{\mathbb{R}^N} \left| (-\Delta)^{\frac{m}{2}} \psi(x) \right|^2 dx \geq c \int_{\mathbb{R}^N} \frac{|\psi(x)|^2}{|x|^{2m}} dx \tag{8.5}$$

holds for enough choices of ψ. When $m = 1$, (8.5) reduces to Hardy's inequality

$$\int_{\mathbb{R}^N} |\nabla \psi|^2 dx \geq c \int_{\mathbb{R}^N} \frac{|\psi(x)|^2}{|x|^2} dx.$$

This holds for all $\psi \in C_c^\infty (\mathbb{R}^N \backslash \{0\})$ iff $c \leq \left(\frac{N-2}{2}\right)^2$. For $m = 2$, the analogous result is due to Rellich [10], and for $m \geq 3$, the equality was shown for sufficiently large N by Galaktionov and Kamotski [4]. The corresponding linear PDE has very interesting unusual behavior. Let $C(N, m)$ be the best constant for the Hardy inequality (8.4) (or (8.5)). When $c > C(N, m)$, (8.3) has no positive solutions at all. This was shown for $m = 1$ by Baras and Goldstein [2], who also established "instantaneous blowup". For $m \geq 2$ the absence of positive solutions was shown by Galaktionov and Kamotski [4] (but without instantaneous blowup, which is based on the Maximum Principle).

The standard versions of the Principle of Linearized Stability/Instability do not apply to (8.1), (8.2). A discussion of failures and hopes of this approach is given in Sect. 8.2. Section 8.3 deals with the equilibrium (8.2) of Eq. (8.1) when $m = 1$. Detailed calculations are given and lead to the main theorem, stated near the end of Sect. 8.3. Some concluding results follow. The case of $m \geq 2$ will be treated in detail in a future paper. Section 8.4 discusses a better formulation of the Principle of Linearized Stability/Instability.

8.2 The Principle of Linearized Stability

Consider a "semilinear" equation of the form

$$\frac{dv}{dt} = Lv + N(v) \tag{8.6}$$

where L is linear and N is nonlinear. (Of course, L could be zero, so the equation could be fully nonlinear.) The usual context is: X is a Banach space, $v : \mathbb{R}^+ = [0, \infty) \to X$ and L generates a strongly continuous (or (C_0)) semigroup $\{e^{tL} : t \geq 0\}$ of bounded linear operators on X. Also, N is a nonlinear operator from its domain $\mathscr{D}(N) \subset X$ to X. Typically N is assumed to be locally Lipschitzian from $\mathscr{D}(N) \subset Y$ to X, where Y is densely and continuously embedded in X (For instance, in the quasilinear Navier–Stokes system, X is the Hilbert space of divergence free vector fields in $\left[L^2(\Omega)\right]^3$ for Ω a domain in $\mathbb{R}^3$, L is the negative selfadjoint Stokes operator, and $Y = \mathscr{D}((-L)^\alpha)$ where $\frac{3}{4} < \alpha < 1$.) By a **global mild solution** of (8.6) satisfying $v(0) = v_0$ is meant a continuous solution of the integral equation

$$u(t) = e^{tL}v_0 + \int_0^t e^{(t-s)L} N(v(s))ds$$

for all $t \geq 0$. Local (in t) mild solutions exists and are unique by an application of the Banach Fixed Point Theorem. If $(0, \tau_{max}(v_1))$ is the maximal time interval of existence for the unique (local) mild solution v of

$$\frac{dv}{dt} = Lv + N(v), \quad v(0) = v_1, \tag{8.7}$$

then either $\tau_{max}(v_1) = \infty$ and v is a global solution, or else $\tau_{max}(v_1) < \infty$, in which case we have the blow up result

$$\lim_{t \to \tau_{max}(v_1)^-} \|v(t)\| = \infty.$$

An **equilibrium** is a constant solution. Thus $v(t) \equiv v_0 \in X$ is an equilibrium if and only if

$$Lv_0 + N(v_0) = 0.$$

Typically L is a partial differential equation satisfying some boundary or other conditions and it can be extended by ignoring the latter. Let L_0 be such an extension, and similarly let N_0 be a "natural" extension of N. Suppose

$$v(t) \equiv \psi_0$$

is a solution of

$$L_0\psi_0 + N_0(\psi_0) = 0.$$

We can work in the Banach space

$$Y = span(\psi_0) \oplus X$$

(direct sum or product), or, better, in the associated affine space

$$\psi_0 \oplus X =: X_0.$$

Let u be a solution of (8.7) satisfying $u(0) = v_0$. Then we say that the equilibrium ψ satisfies

(1) v is X_0 **unstable** if given $\epsilon > 0$, there exist a $\delta > 0$ such that if $v_0 \in X_0$ and

$$0 < \|v_0 - \psi_0\|_X < \epsilon,$$

then

$$\|u(t) - \psi_0\|_X \geq \delta$$

for sufficiently large t.

(2) v is X_0 **asymptotically stable** if there exist $\epsilon > 0$ such that if $v_0 \in X_0$ and

$$\|v_0 - \psi_0\|_X < \epsilon,$$

then

$$\|u(t) - \psi_0\|_X \to 0 \text{ as } t \to \infty.$$

Let ψ_0 be an equilibrium, and suppose N is differentiable at ψ_0 in some sense. Let B_0 be the linear operator

$$B_0 = L + N'(\psi_0).$$

Suppose the closure B of B_0 generates a semigroup $S = \{e^{tB} : t \geq 0\}$ of linear operators on X. If e^{tB} is unbounded on X for each $t > 0$, it seems natural to call ψ_0 **strongly linearly unstable**. So suppose S is a (C_0) semigroup. (For semigroups, see, for instance, [4].) We classify the stability types of S using the properties (Pi) defined below. S is called

(P1) **asymptotically exponentially stable** if

$$\left\| e^{tB} \right\| \leq M e^{-at}$$

   holds for all $t > 0$ and some positive constants M and a;
(P2) **exponentially asymptotically unstable** if

$$\left\| e^{tB} \right\| \geq M_0 e^{ta_0}$$

   holds for all $t > 0$ and some positive constants M_0 and a_0;
(P3) **stable** if

$$\sup_{t>0} \left\| e^{tB} \right\| < \infty;$$

(P4) **asymptotically stable** if

$$\left\| e^{tB} f \right\| \to 0$$

   as $t \to \infty$ for all $f \in X$;
(P5) **unstable** if

$$\lim_{t \to \infty} \left\| e^{tB} \right\| = \infty.$$

The equilibrium ψ_0 of (8.6) is called **linearly asymptotically stable** [resp. **linearly unstable**] if (P1) holds [resp., if (P2) holds].

If the equilibrium ψ_0 is linearly asymptotically stable, then we may hope that there is an $\epsilon > 0$ such that if $\|v_1 - \psi_0\|_Z < \epsilon$, where the Banach space Z embeds continuously and densely into Y, then there is a unique global mild solution v of (8.7) and

$$v(t) - v_0 \to 0 \tag{8.8}$$

in X as $t \to \infty$. If the equilibrium is linearly unstable, then we may expect that there exist $\epsilon > 0$, $\delta > 0$ such that if v is the unique global mild solution of (8.7) and if $0 < \|v_1 - v_0\|_Z < \epsilon$, then there is a $T > 0$ such that

$$\|v(t) - v_0\| \geq \delta \tag{8.9}$$

for all $t > T$.

The above paragraph is a generalized version of the Principle of Linearized Stability/Instability and it requires the asserted existence results. If (P3) or (P4) holds, then either of (8.8), (8.9) could hold for solutions of (8.7) with v_1 close to v_0. So the equilibrium may be unstable if (P3) or (P4) holds. But it is "closer to being stable" than in the case of when (P2) holds. We want to give a name to this case. "Almost Stable" fits, in that "almost everywhere" is not everywhere and "almost uniform convergence" is not uniform convergence. Thus almost stability need not be stability. But since v_0 could be unstable in this case, it seems more prudent to use the term **borderline instability** or **borderline linearly unstable** to refer to the equilibrium v_0 when (P3) or (P4) holds. Typically there could be a "stable manifold", that is, initial values v_1 for solutions of (8.7) on certain curves passing through v_0 for which the corresponding solution v satisfies (8.8).

There is an enormous literature on the Principle of Linearized Stability in special cases (see, for example, [3, 11]). But a unifying general form of the theorem has not yet been found. This is discussed further in Sect. 20.4.

8.3   The Case of (8.1)

The popular Eq. (8.1) is studied as a partial differential equation rather than an ordinary differential equation in some particular Banach space X. Indeed, there is not a natural space X associated with (8.1).

So let

$$u(x,t) = v_0(x) = Cr^{-a}$$

be a solution of (8.1) with $r = |x|$ and $C, a > 0$. Then, for $x \in \mathbb{R}^N$,

$$-u_t + \Delta u = \Delta v_0 = (v_0)_{rr} + \frac{N-1}{r}(v_0)_r = -(v_0)^p$$

implies

$$C(a(a+1) - a(N-1))r^{-a-2} = -C^p r^{-ap}.$$

It follows that

$$a + 2 = ap$$

or

$$a = \frac{2}{p-1}. \tag{8.10}$$

Next,

$$-a^2 + (N-2)a = C^{p-1} > 0. \tag{8.11}$$

(8.10) and (8.11) imply

$$N > 2\left(1 + \frac{1}{p-1}\right).$$

Thus we take $N \geq 3$ and

$$p > \frac{N}{N-2}, \tag{8.12}$$

and we calculate

$$C = \left[\frac{2(N-2)}{p-1} - \left(\frac{2}{p-1}\right)^2\right]^{\frac{1}{p-1}}. \tag{8.13}$$

Then we have, by (8.10)–(8.13), a unique equilibrium of (8.1) of the form (8.2) for $N \geq 3$ and $p > \frac{N}{N-2}$. (There are no such solutions if $N < 3$ or if $N = 3$ and $1 < p \leq 3$.)

What about the stability of this equilibrium v_0? This v_0 does not belong to any of the usual $L^q(\mathbb{R}^N)$ spaces or any Sobolev space. It does belong to a unique weak $L^q(\mathbb{R}^N)$ space (or Marcinkiewicz space), but this space is not "natural" for (8.1). So let us avoid choosing X momentarily and consider v_0 as a solution of (8.6) with $L = \Delta$ and $N(v) = v^p$. Formally,

$$N'(v_0) = pv_0^{p-1} \equiv pC^{p-1}r^{-a(p-1)} = pC^{p-1}r^{-2} =: \frac{c}{|x|^2}.$$

This defines c. The operator

$$A = \Delta + \frac{c}{|x|^2}$$

on $L^2(\mathbb{R}^N)$ with domain $C_c^\infty(\mathbb{R}^N)$ is, for $N \geq 3$, symmetric. Let A_c be the Friedrichs extension if A is semibounded, and any selfadjoint extension of A is not semibounded; in all cases, A has a selfadjoint extension, A_c. Then

$$A_c = A_c^* \leq 0$$

if $c \leq \left(\frac{N-2}{2}\right)^2$, while

$$\sigma\left(A_c\right) = \mathbb{R}$$

if $c > \left(\frac{N-2}{2}\right)^2$. Moreover, $\left(\frac{N-2}{2}\right)^2$ is the best constant in the Hardy inequality

$$\left(\frac{N-2}{2}\right)^2 \int_{\mathbb{R}^N} \frac{|u(x)|^2}{|x|^2} dx \leq \int_{\mathbb{R}^N} |\nabla u(x)|^2 dx$$

for all $u \in C_c^\infty(\mathbb{R}^N)$. Note that the Hardy inequality also holds for $N = 1, 2$ for all $u \in C_c^\infty(\mathbb{R}^N \setminus \{0\})$. For $N = 1$ we usually consider only $u \in C_c^\infty(0, \infty)$. See [7].

Remark 8.1 Recall

$$A_c = \Delta + \frac{c}{|x|^2}$$

(with suitable domain) generates a (C_0) contraction semigroup on $L^2\left(\mathbb{R}^N\right)$ iff

$$c \leq \left(\frac{N-2}{2}\right)^2$$

and e^{tA_c} is unbounded for all $t \neq 0$ if $c > \left(\frac{N-2}{2}\right)^2$. Similarly [1],

$$A_{c_q} = \Delta + \frac{c}{|x|^2}$$

(with suitable domain) generates a (C_0) contraction semigroup on $L^q\left(\mathbb{R}^N\right)$ for $1 < q < \infty$ iff

$$c \leq \left(\frac{N-2}{2}\right)^2 \left(\frac{4}{qq'}\right) =: \tilde{C}\left(N, q\right) \equiv: \tilde{C}\left(N, q'\right)$$

where $\frac{1}{q} + \frac{1}{q'} = 1$. The constant $\tilde{C}\left(N, q\right)$ for $N \geq 3$ is a linear increasing function of $\frac{1}{q}$ as $\frac{1}{q}$ goes from 0 to $\frac{1}{2}$. Thus to find a space X (or X_0) so that $\psi_0 + X$ has the best chance to give linearized stability, it seems prudent to take $q = 2$, i.e., to take $X = L^2\left(\mathbb{R}^N\right)$, at least if one restricts to considering $L^q\left(\mathbb{R}^N\right)$ spaces.

In our case,

$$c = pC^{p-1} = p\left((N-2)a - a^2\right). \tag{8.14}$$

Let

$$M := N - 2 \in \mathbb{N}, \quad q = p - 1 > 0.$$

We want to view $c = c_M(q)$ as a function of q with parameter M, and compare $c = c_M(q)$ with the optimal Hardy inequality constant. Then $q > \frac{2}{M}$ and

$$c = c_M(q) = (q+1)\left(\frac{2M}{q} - \frac{4}{q^2}\right) = 2M + \frac{2M-4}{q} - \frac{4}{q^2}, \tag{8.15}$$

$$\frac{dc_M(q)}{dq} = \frac{4-2M}{q^2} + \frac{8}{q^3} = \frac{2}{q^2}\left(2 - M + \frac{4}{q}\right). \tag{8.16}$$

$\frac{dc_M(q)}{dq} > 0$ for all $q > 0$ and $M = 1, 2$. For $M \geq 3$,

$$\begin{cases} \frac{dc_M(q)}{dq} > 0 & \text{for } q < q_M^*, \\ \frac{dc(q)}{dq} < 0 & \text{for } q > q_M^*, \end{cases} \tag{8.17}$$

where

$$q_M^* = \frac{4}{M-2} \text{ for } M \geq 3.$$

Thus $c_M(q)$ is maximized at q_M^*. We want to know whether

$$c_M(q) \leq \frac{M^2}{4} \text{ or } c_M(q) > \frac{M^2}{4},$$

where

$$\frac{M^2}{4} = \left(\frac{N-2}{2}\right)^2$$

is the Hardy constant.

For $M = 1$ [resp. $M = 2$], $\frac{M^2}{4} = \frac{1}{4}$ [resp. 1] and

$$c_1(q) = 2 - \frac{2}{q} - \frac{4}{q^2}$$

$$[\text{resp. } c_2(q) = 4 - \frac{4}{q^2}]$$

increases from 0 at $q = 2$ to 2 at $q = \infty$ [resp. from 0 at $q = 1$ to 4 at $q = \infty$]. Since $p > \frac{N}{N-2}$ implies $q > \frac{2}{M}$ ($= 2$ or 1, according to $M = 1$ or 2), we have

$$c_M(q) > \frac{M^2}{4}, \tag{8.18}$$

and for $M = 1$,

$$c_1(q) > \frac{1}{4}$$

iff

$$2 - \frac{2}{q} - \frac{4}{q^2} > \frac{1}{4}$$

iff

$$q > \frac{4 + 8\sqrt{2}}{7};$$

while if $M = 2$, (8.18) holds iff

$$4 - \frac{4}{q^2} > 1,$$

iff

$$q > \frac{2}{\sqrt{3}}.$$

Furthermore, when $M = 1$,

$$c \leq \frac{M^2}{4} = \frac{1}{4}$$

iff

$$\frac{2}{M} = 2 < q \leq \frac{4 + 8\sqrt{2}}{7} \text{ (which exceeds 2)},$$

and when $M = 2$,

$$c \leq \frac{M^2}{4} = 1$$

iff

$$\frac{2}{M} = 1 < q \leq \frac{2}{\sqrt{3}} \text{ (which exceeds 1)}.$$

For $M \geq 3$ (or $N \geq 5$), by (8.15),

$$\max_q c_M(q) = c_M(q_M^*) = c_M\left(\frac{4}{M-2}\right)$$

$$= 2M + \frac{(M-2)^2}{2} - \frac{(M-2)^2}{4}$$

$$= 2M + \frac{(M-2)^2}{4} = \left(\frac{M+2}{2}\right)^2 = \frac{N^2}{4}.$$

This exceeds $\frac{M^2}{4}$, the Hardy constant. Fix M and let

$$S_M := \left\{q > \frac{2}{M} : c_M(q) \leq \frac{M^2}{4}\right\}, \quad T_M := \left\{q > \frac{2}{M} : c_M(q) > \frac{M^2}{4}\right\}.$$
$$(8.19)$$

Since c_M is increasing up to $q = q_M^*$ and decreasing thereafter, it follows that T_M is an interval whose endpoints (except possibly for $\frac{2}{M}$) are the roots of

$$c_M(q) - \frac{M^2}{4},$$

i.e., by (8.15),

$$2M + \frac{2M-4}{q} - \frac{4}{q^2} - \frac{M^2}{4} = 0$$

iff

$$\left(2M - \frac{M^2}{4}\right)q^2 + (2M-4)q - 4 = 0,$$

iff (for $M \neq 8$)

$$q = q_\pm = \frac{4 - 2M \pm \sqrt{(2M-4)^2 + 16\left(2M - \frac{M^2}{4}\right)}}{4M - \frac{M^2}{2}} = \frac{4 - 2M \pm 4\sqrt{M+1}}{4M - \frac{M^2}{2}}.$$
$$(8.20)$$

For $3 \leq M \leq 7$, we have

$$4 - 2M < 0$$

Table 8.1

M	3	4	5	6	7
$\widehat{q_M}$	$\frac{4}{5}$	$\frac{\sqrt{5}-1}{2}$	$\frac{8\sqrt{6}-12}{15}$	$\frac{2\sqrt{7}-4}{3}$	$\frac{16\sqrt{2}-20}{17}$

and

$$4M - \frac{M^2}{2} = \frac{M}{2}(8 - M) > 0.$$

Hence

$$q_- < 0 < q_+,$$

In this case, therefore

$$c_M(q) > \frac{M^2}{4} \text{ for } 3 \le M \le 7$$

iff

$$q \in (\widehat{q_M}, \infty)$$

where $\widehat{q_M}$ is the q_+, given by (8.20) (Table 8.1).
In each case, for $3 \le M \le 7$,

$$\inf q = \frac{2}{M} < \widehat{q_M}$$

and so (see (8.19))

$$S_M = \left\{ q : q > \frac{2}{M}, c_M(q) \le \frac{M^2}{4} \right\} = \left(\frac{2}{M}, \widehat{q_M} \right].$$

For $M = 8$, by (8.15),

$$\max c_8(q) = c_8(q_8^*) = c_8\left(\frac{2}{3}\right)$$

$$= 16 + \frac{12}{\frac{2}{3}} - \frac{4}{\left(\frac{2}{3}\right)^2} = 16 + 18 - 9$$

$$= 25 > \frac{8^2}{4} = 16,$$

thus

$$q \in T_8 = \left\{ q > \frac{1}{4} : c_8(q) > 16 \right\}$$

iff $q > \frac{1}{4}$ and

$$c_8 (q) = 16 + \frac{12}{q} - \frac{4}{q^2} > 16$$

iff

$$q > \frac{1}{3}.$$

So

$$S_8 = \left(\frac{1}{4}, \frac{1}{3}\right], \quad T_8 = \left(\frac{1}{3}, \infty\right).$$

Now let $M \geq 9$. By (8.20), $c_M(q) = \frac{M^2}{4}$ iff q satisfies

$$\left(2M - \frac{M^2}{4}\right) q^2 + (2M - 4)q - 4 = 0.$$

Here is why. For $M \geq 9$,

$$c_M(q) - \frac{M^2}{4} = 2M + \frac{2M - 4}{q} - \frac{4}{q^2} - \frac{M^2}{4} = 0$$

iff

$$\left(2M - \frac{M^2}{4}\right) q^2 + (2M - 4)q - 4 = 0.$$

This equation has two roots,

$$q_- = \frac{4\left(-2 + M - 2\sqrt{1 + M}\right)}{-8M + M^2}, \quad q_+ = \frac{4\left(-2 + M + 2\sqrt{1 + M}\right)}{-8M + M^2},$$

which satisfy $0 < \frac{2}{M} < q_- < q_M^* < q_+$.

$$q \in S_M = \left\{ q > \frac{2}{M} : c_M(q) \leq \frac{M^2}{4} \right\}$$

iff $q > \frac{2}{M}$ and

$$c_M(q) = 2M + \frac{2M - 4}{q} - \frac{4}{q^2} \leq \frac{M^2}{4}$$

iff

$$\frac{2}{M} < q \le q_- = \frac{4\left(-2 + M - 2\sqrt{1 + M}\right)}{-8M + M^2} \quad \text{or} \quad q \ge q_+$$

$$= \frac{4\left(-2 + M + 2\sqrt{1 + M}\right)}{-8M + M^2}.$$

So

$$S_M = \left(\frac{2}{M}, \frac{4\left(-2 + M - 2\sqrt{1 + M}\right)}{-8M + M^2}\right] \cup \left[\frac{4\left(-2 + M + 2\sqrt{1 + M}\right)}{-8M + M^2}, +\infty\right),$$

i.e.,

$$p \in \left(\frac{2 + M}{M}, \frac{4\left(-2 - M - 2\sqrt{1 + M}\right) + M^2}{-8M + M^2}\right] \cup$$

$$\times \left[\frac{4\left(-2 - M + 2\sqrt{1 + M}\right) + M^2}{-8M + M^2}, +\infty\right). \qquad (8.21)$$

On the other hand,

$$q \in \left\{q > \frac{2}{M} : c_M(q) > \frac{M^2}{4}\right\}$$

iff $q > \frac{2}{M}$ and

$$c_M(q) = 2M + \frac{2M - 4}{q} - \frac{4}{q^2} > \frac{M^2}{4}$$

iff

$$q_- = \frac{4\left(-2 + M - 2\sqrt{1 + M}\right)}{-8M + M^2} < q < q_+ = \frac{4\left(-2 + M + 2\sqrt{1 + M}\right)}{-8M + M^2}.$$

So

$$q \in \left(\frac{4\left(-2 + M - 2\sqrt{1 + M}\right)}{-8M + M^2}, \frac{4\left(-2 + M + 2\sqrt{1 + M}\right)}{-8M + M^2}\right),$$

i.e.,

$$p \in \left(\frac{4\left(-2 - M - 2\sqrt{1 + M}\right) + M^2}{-8M + M^2}, \frac{4\left(-2 - M + 2\sqrt{1 + M}\right) + M^2}{-8M + M^2}\right).$$

We call the equilibrium v_0 **strongly linearly unstable** in case $c > \left(\frac{N-2}{2}\right)^2$, in which case the linearized semigroup consists of unbounded operators. We have the borderline unstable case when $c \leq \left(\frac{N-2}{2}\right)^2$, which corresponds to the linearized semigroup satisfying the stability condition (P3) or the asymptotical stability condition (P4), which it does in this case.

We summarize these results in the following theorem.

Theorem 8.1 *Let* $v_0(x) = C r^{-a}$ *be the unique radial positive equilibrium solution of*

$$\frac{\partial u}{\partial t} = \Delta u + u^p$$

for $x \in \mathbb{R}^N$, $r = |x|$, $N \geq 3$ *and* $p > \frac{N}{N-2}$.
  For $3 \leq N \leq 10$, v_0 *is strongly linearly unstable if*

$$N = 3 \text{ and } p \in \left(3, p_3^* = \frac{11 + 8\sqrt{2}}{7}\right),$$

$$N = 4 \text{ and } p \in \left(2, p_4^* = \frac{3 + 2\sqrt{3}}{3}\right),$$

$$N = 5 \text{ and } p \in \left(1\frac{2}{3}, p_5^* = 1\frac{4}{5}\right),$$

$$N = 6 \text{ and } p \in \left(1\frac{1}{2}, p_6^* = \frac{\sqrt{5}+1}{2}\right),$$

$$N = 7 \text{ and } p \in \left(1\frac{2}{5}, p_7^* = \frac{3 + 8\sqrt{6}}{15}\right),$$

$$N = 8 \text{ and } p \in \left(1\frac{1}{3}, p_8^* = \frac{2\sqrt{7}-1}{4}\right),$$

$$N = 9 \text{ and } p \in \left(1\frac{2}{7}, p_9^* = \frac{16\sqrt{2}-13}{7}\right),$$

$$N = 10 \text{ and } p \in \left(1\frac{1}{4}, p_{10}^* = 1\frac{1}{3}\right).$$

For $N \leq 10$, v_0 *is borderline linearly unstable for* $p \geq p_N^*$. *For* $N \geq 11$, *let*

$$p_-^*(N) = \frac{4\left(-2 - M - 2\sqrt{1 + M}\right) + M^2}{-8M + M^2},$$

$$p_+^*(N) = \frac{4\left(-2 - M + 2\sqrt{1 + M}\right) + M^2}{-8M + M^2}.$$

Then

$$\frac{2}{N-2} < p_-^* (N) < p_+^* (N) < \infty$$

and v_0 is strongly linearly unstable iff

$$p \in \left(p_-^* (N), p_+^* (N) \right),$$

and v_0 is borderline linearly unstable iff

$$p \in \left(\frac{2}{N-2}, p_-^* (N) \right] \cup \left[p_+^* (N), \infty \right).$$

Thus there is a change in the nature of the linearized instability of the power equilibrium v_0 to the nonlinear heat equation when N increases from 10 to 11. For each $N \geq 11$, there are three nonempty pairwise disjoint intervals with union $\left(\frac{2}{N-2}, \infty \right)$, namely

$$\left(\frac{2}{N-2}, p_-^* (N) \right], \quad \left(p_-^* (N), p_+^* (N) \right), \quad \left[p_+^* (N), \infty \right),$$

such that v_0 is strongly linearly unstable in the second interval, but v_0 is borderline linearly unstable in the first and third interval.

We remark that in the borderline linearly unstable case, v_0 is not only stable in the sense of (P3), it is in fact asymptotically stable in the sense of (P4). Here is a proof.

As before, let A_γ be the Friedrichs extension of $A = \Delta + \frac{\gamma}{|x|^2}$ on $L^2 \left(\mathbb{R}^N \right)$ (or $L^2(0, \infty)$ if $N = 1$) if $\gamma \leq \left(\frac{N-2}{2} \right)^2$, and any selfadjoint extension if $\gamma > \left(\frac{N-2}{2} \right)^2$; selfadjoint extensions of $\Delta + \frac{\gamma}{|x|^2}$ always exists.

Lemma 8.1 *If $\gamma \leq \left(\frac{N-2}{2} \right)^2$, then*

$$\| e^{tA_\gamma} f \| \to 0$$

as $t \to \infty$ for all $f \in L^2 \left(\mathbb{R}^N \right)$ (or $L^2(0, \infty)$ if $N = 1$).

Proof By the spectral theorem and the associated functional calculus, A_γ is unitarily equivalent to a multiplication operator. There is a measure space (Ω, Σ, ν) and a Σ-measurable function

$$m : \Omega \to \mathbb{R}$$

(with $m : \Omega \to (-\infty, 0)$ if $\gamma \le \left(\frac{N-2}{2}\right)^2$ such that A_γ is unitarily equivalent (via a unitary operator U) to multiplication by m. Thus for all $f \in L^2\left(\mathbb{R}^N\right)$) (or $L^2(0, \infty)$ if $N = 1$) and for $\hat{f} \in L^2\left(\Omega, \Sigma, \nu\right)$ the corresponding function (i.e., $\hat{f} = Uf$),

$$\left\| e^{tA_\gamma} f \right\|^2 = \int_\Omega e^{2tm(\omega)} \left| \hat{f}(\omega) \right|^2 \mu(d\omega) \to 0$$

as $t \to \infty$ by Lebesgue's Dominated Theorem, since $m(\omega) < 0$ a.e.. $\square$

Consequently (P4) holds with $\gamma \le \left(\frac{N-2}{2}\right)^2$.

Often when studying (8.1), one makes use of the maximum principle or, equivalently, the fact that $f(x) \ge 0$ for $x \in \mathbb{R}^N$ implies $\left(e^{t\Delta} f\right)(x) \ge 0$ for a.e. $x \in \mathbb{R}^N$ and $t > 0$ (provided $e^{t\Delta} f$ exists). But our proof did not make use of this. Thus one can consider extending our results to

$$\frac{\partial u}{\partial t} = -(-\Delta)^m u + u^p$$

for $t > 0$, $x \in \mathbb{R}^N$ and $m = 2, 3, \cdots$. We will present such results in a future paper.

8.4   In Search of a Better Principle of Linearized Stability/Instability

Consider the partial differential equation

$$\frac{\partial u}{\partial t} = F(u), \tag{8.22}$$

where F is given its maximal possible domain. Let $X \subset Y$ be Banach spaces and let G be a suitable restriction of F to Y so that the corresponding initial value problem (possibly with boundary conditions) can be written as

$$\frac{du}{dt} = G(u), \quad u(0) = u_0. \tag{8.23}$$

This problem is *wellposed* in Y if one has the usual existence, uniqueness and continuous dependence. We call (8.23) *wellposed* if it is wellposed in some space Y.

Similarly, an equilibrium solution of (8.22) is *asymptotically stable* in X (even if ψ_0 is not in X) if $f \in X$, u satisfies (8.23) in $Y = span\,(\psi_0) \times X$ with $u_0 = \psi_0 + f$ and $u(t) \to \psi_0$ as $\to \infty$, provided $\|f\|_X$ is small enough. Similarly for stable, unstable, etc. Thus proving stability involves finding a space, perhaps "the right space", in which the equilibrium becomes stable. From another perspective, we study stability and instability in the affine subspace $\psi_0 \times X$ of Y.

These definitions can be extended to linearized stability of equilibria. Let $F(\psi_0) = 0$ so that ψ_0 is an equilibrium of (8.22). One may speak of ψ_0 as being asymptotically stable in X even if ψ_0 is not in X. Namely, ψ_0 is asymptotically stable in the affine space $\psi_0 \times X$ if $f \in X$, (8.22) holds with $u_0 = \psi_0 + f$, and $\|f\|_X$ is small enough implies

$$\|u(t) - \psi_0\|_X \to 0 \text{ as } t \to \infty.$$

Similarly for stable, unstable, etc.

Gui et al., in an interesting series of papers [5, 6, 12], discussed the stability of

$$v_0(x) = Cr^{-a}$$

as an equilibrium of (8.1). Their results are as follows. One says that v_0 is "unstable in any reasonable sense" if $3 \leq N \leq 10$ or if $N \geq 11$ and

$$p < p_+^*(N)$$

(see our Theorem of Sect. 8.3 for the definition of $p_+^*(N)$). However, they showed that v_0 is stable in certain weighted sup norm spaces and certain weighted L^p spaces if $p > p_+^*(N)$ and $N \geq 11$.

But in the unstable cases, one could possibly have stability if one perturbed v_0 by a small vector in, say, a suitable subset of $L^2(\mathbb{R}^N)$, a sort of a stable manifold. This seems a possibility whenever v_0 is borderline linearly unstable. This remains to be explored, especially for $3 \leq N \leq 10$. Furthermore, in case $m \geq 2$, no stability results are known (although we have some borderline linearly unstable results in some cases).

Acknowledgements We thank Peter Polacik for very interesting comments and suggestions which inspired this research. We also thank Gisèle Goldstein for her valuable help in the early stages of this research. This research was finished while the second author was visiting Professor J. A. Goldstein in The University of Memphis. The second author acknowledges partial financial support from Nature Science Fund of Shaanxi Province, No. 2012JM1014.

References

1. Arendt, G., Ruiz Goldstein, G., Goldstein, J.A.: Outgrowths of Hardy's inequality, recent advances in differential equations and mathematical physics. Contemp. Math. **412**, 51–68 (2006)
2. Baras, P., Goldstein, J.A.: The heat equation with a singular potential. Trans. Am. Math. Soc. **284**, 121–139 (1984)
3. Desch, W., Schappacher, W.: Linearized stability for nonlinear semigroups. Lecture Notes in Mathematics, vol. 1223, pp. 61–73. Springer, Berlin (1986)
4. Galaktionov, V.A., Kamotski, I.V.: On nonexistence of Baras-Goldstein type for higher-order parabolic equations with singular potentials. Trans. Am. Math. Soc. **362**, 4117–4136 (2010)

5. Gui, C., Ni, W.N., Wang, X.: On the stability and instability of positive steady states of a semilinear heat equation in $\mathbb{R}^n$. Commun. Pure Appl. Math. **45**, 1153–1181 (1992)
6. Gui, C., Ni, W.N., Wang, X.: Further study on a nonlinear heat equation. J. Differ. Equ. **169**, 588–613 (2001)
7. Kalf, H., Schmincke, U.W., Walter, J., Wüst, R.: On the Spectral Theory of Schrödinger and Dirac Operators with Strongly Singular Potentials. Lecture Notes in Mathematics, vol. 448, pp. 182–226. Springer, Berlin (1975)
8. Mizoguchi, N.: Rate of type II blowup for a semilinear heat equation. Math. Ann. **339**, 839–877 (2007)
9. Polacik, P., Yanagida, E.: On bounded and unbounded global solutions of a supercritical semilinear heat equation. Math. Ann. **327**, 745–771 (2003)
10. Rellich, F.: Perturbation Theory of Eigenvalue Problems. New York University Lecture Notes, vol. 1954. Reprinted by Gordon and Breach, New York/London (1969)
11. Ruess, W.M.: Linearized stability for nonlinear evolution equation. J. Evol. Equ. **3**, 361–373 (2003)
12. Wang, X.: On the Cauchy problem for reaction-diffusion equations. Trans. Am. Math. Soc. **337**, 549–589 (1993)

Chapter 9
Abstract Elliptic Problems Depending on a Parameter and Parabolic Problems with Dynamic Boundary Conditions

Davide Guidetti

Dedicated to the memory of Alfredo Lorenzi

Abstract We study abstract elliptic problems depending on a complex parameter. Such parameter appears also in the boundary conditions. Next, we consider abstract parabolic systems with dynamic boundary conditions. Applications are given to parameter elliptic boundary value problems and to concrete parabolic problems.

9.1 Introduction

The aim of this paper is twofold. First, we want to study abstract elliptic problems of the form

$$\begin{cases} \lambda u(x) - u''(x) + Au(x) = f(x), \ x > 0, \\ \lambda u(0) - \gamma u'(0) + Bu(0) = h, \end{cases} \tag{9.1}$$

depending on the complex parameter λ. The assumptions that we shall make, concerning (9.1), are modeled on the concrete situation that A is a second order strongly elliptic operator in the variable $y \in \mathbb{R}^{n-1}$, so that $D_x^2 + A$ may be thought as a second order strongly elliptic operator in the variables (x, y), u and f are functions of $x \in \mathbb{R}^+$, with values in a space of mappings with domain $\mathbb{R}^{n-1}$, so that they depend, in fact, on $(x, y) \in \mathbb{R}^+ \times \mathbb{R}^{n-1}$. In this order of ideas, $u(0)$ and $u'(0)$ represent, respectively, the trace and the normal derivative of u on the subspace $x = 0$, while B is a (usually) unbounded operator defined in a space of functions with domain $\mathbb{R}^{n-1}$ and $\gamma \in \mathbb{R}^+$. In applications, B may be a first order differential operator in the variable y, so that (9.1) may be thought as a

D. Guidetti (✉)
Dipartimento di Matematica, Piazza di Porta S. Donato 5, 40126 Bologna, Italy
e-mail: davide.guidetti@unibo.it

© Springer International Publishing Switzerland 2014
A. Favini et al. (eds.), *New Prospects in Direct, Inverse and Control Problems for Evolution Equations*, Springer INdAM Series 10,
DOI 10.1007/978-3-319-11406-4_9

nonhomogeneous boundary value problem, with the complex parameter λ appearing also in the boundary condition.

The precise abstract assumptions concerning (9.1) are denominated $(B1) - (B2)$ and are explicitly written in Sect. 9.3. The concrete inspiring model, which we have briefly described, is treated in more detail in Sect. 9.5 and is used as a basis to study more complicated situations. Concrete elliptic boundary value problems with a complex parameter in the boundary conditions of the form

$$\begin{cases} \lambda u(\xi) - A(\xi, D_\xi)u(\xi) = f(\xi), & \xi \in \Omega, \\ \lambda u(\xi') + B(\xi', D_x)u(\xi')) = h(\xi'), & \xi' \in \partial\Omega \end{cases} \tag{9.2}$$

are studied in [16] and [10], with Ω domain in $\mathbb{R}^n$ with smooth boundary $\partial\Omega$, $A(\xi, D_\xi)$ strongly elliptic of the second order and $B(\xi', D_\xi)$ appropriate first order operator. From our point of view, the main result of these researches seems to be the following (see [10, Theorem 2] and [16]): let $p \in (1, \infty)$ and let us consider the operator

$$\begin{cases} \mathscr{A} & : \{(u, g) \in W^{2,p}(\Omega) \times W^{2-1/p,p}(\partial\Omega) : g = u_{|\partial\Omega}\} \\ & \rightarrow L^p(\Omega) \times W^{1-1/p,p}(\partial\Omega), \\ \mathscr{A}(u, g) & = (A(\cdot, D_\xi)u, -B(\cdot, D_\xi)u_{|\partial\Omega}). \end{cases} \tag{9.3}$$

Then, under suitable assumptions (which we are going to mention in the sequel) $\mathscr{A}$ is the infinitesimal generator of an analytic semigroup in $L^p(\Omega) \times W^{1-1/p,p}(\partial\Omega)$.

We pass to consider abstract elliptic systems with a structure which is close to that of (9.1) and compare it with the corresponding results in this paper. We quote [2, 3], where systems of the form

$$\begin{cases} \lambda u(x) - u''(x) + Au(x) = f(x), \, 0 < x < 1, \\ \lambda u(0) - \alpha u'(0) = f_1, \\ \lambda u(1) + \beta u'(1) = f_2 \end{cases} \tag{9.4}$$

are considered. Here A is a selfadjoint, positive operator in the Hilbert space H, α and β are complex numbers belonging to appropriate sectors, $f \in L^p((0, 1); H)$ $(1 < p < \infty)$, for each $k \in \{1, 2\}$ f_k belongs to the real interpolation space $(H, D(A))_{1-m_k/2-1/(2p),p}$, with $m_k = 0$ if $\alpha = 0$ (boundary condition of Dirichlet type), $m_k = 1$ if $\alpha \neq 0$ (boundary condition of Robin type). It is proved that, if λ belongs to an appropriate sector and $|\lambda|$ is sufficiently large, (9.4) has a unique solution u in $W^{2,p}((0, 1); H) \cap L^p((0, 1); D(A))$. Moreover, an estimate of the type

$$|\lambda| \|u\|_{L^p((0,1);H)} + \|u''\|_{L^p((0,1);H)} + \|Au\|_{L^p((0,1);H)}$$
$$\leq C[|\lambda|^{\max\{m_k\}}\|f\|_{L^p((0,1);H)} + \sum_{k=1}^{2} |\lambda|^{-1+m_k}(\|f_k\|_{H,D(A))_{1-m_k/2-1/(2p),p}}$$
$$+ |\lambda|^{1-m_k/2-1/(2p)}\|f_k\|_H)] \tag{9.5}$$

holds. Perturbations of the abstract boundary conditions are also treated.

Concerning system (9.1), we assume that A is an operator in a *UMD* Banach space E, such that, for some $\phi_A \in [0, \pi)$,

$$\Sigma_{\phi_A} := \{\lambda \in \mathbb{C} \setminus \{0\} : |Arg(\lambda)| \leq \phi_A\} \cup \{0\} \subseteq \rho(-A).$$

Moreover, $\{\lambda(\lambda + A)^{-1} : \lambda \in \Sigma_{\phi_A}\}$ is R-bounded in $\mathscr{L}(E)$, $\gamma \in \mathbb{R}^+$, $-B$ is the infinitesimal generator of a strongly continuous (not necessarily analytic) semigroup in E, commuting with A in the sense of the resolvent and such that its domain $D(B)$ contains $D(A^{1/2})$. Under these conditions, we are able to show that, if $p \in (1, \infty)$ and $\lambda \in \Sigma_{\min\{\phi_A, \pi/2\}}$, there exist ω, M in $\mathbb{R}^+$ such that, if $|\lambda| \geq \omega$, $f \in L^p(\mathbb{R}^+; E)$, $h \in (E, D(A))_{1/2-1/(2p), p}$, (9.1) has a unique solution u in $W^{2,p}(\mathbb{R}^+; E) \cap L^p(\mathbb{R}^+; D(A))$; moreover an estimate of the form

$$|\lambda| \|u\|_{L^p(\mathbb{R}^+; H)} + \|u''\|_{L^p(\mathbb{R}^+; H)} + \|Au\|_{L^p(\mathbb{R}^+; H)} \\ \leq M(\|f\|_{L^p(\mathbb{R}^+; E)} + \|h\|_{(E, D(A))_{1/2-1/(2p), p}}) \tag{9.6}$$

holds (Theorem 9.5). Differently from [2], we have chosen to work in $\mathbb{R}^+$ instead of $[0, 1]$. Employing our results, it is not difficult to obtain a theorem which is analogous to Theorem 9.5, in $[0, 1]$ (see, for this kind of arguments, [11]): for (9.4), in case $\alpha \neq 0$ and $\beta \neq 0$, we can obtain existence and uniqueness, together with an estimate in the form

$$|\lambda| \|u\|_{L^p((0,1); H)} + \|u''\|_{L^p((0,1); H)} + \|Au\|_{L^p((0,1); H)} \\ \leq M(\|f\|_{L^p((0,1); E)} + \|f_1\|_{(E, D(A))_{1/2-1/(2p), p}} + \|f_2\|_{(E, D(A))_{1/2-1/(2p), p}}),$$

which is better than (9.5).

So Theorem 9.5 seems to extend the quoted result of [2], in case $m_1 = m_2 = 1$, in many directions: we may replace a Hilbert space with a *UMD* space and we can afford an operator B in the second equation which is not covered by the perturbation result of [2]. Moreover estimate (9.6) seems better that (9.5) (in case $m_1 = m_2 = 1$), and it allows to prove the following fact, that, as we shall see in the applications, may be thought as an abstract generalization of the result already discussed, concerning $\mathscr{A}$ in (9.3): assume that $\phi_A \geq \frac{\pi}{2}$; let G be defined as follows:

$$\begin{cases} D(G) := \{(u, g) \in [W^{2,p}(\mathbb{R}^+; E) \cap L^p(\mathbb{R}^+; D(A))] \times (E, D(A))_{1-1/(2p), p} : \\ \qquad u(0) = g\}, \\ G(u, g) := (D_x^2 u - Au, \gamma D_x u(0) - Bu(0)) = (D_x^2 u - Au, \gamma D_x u(0) - Bg). \end{cases} \tag{9.7}$$

Then G is the infinitesimal generator of an analytic semigroup in $L^p(\mathbb{R}^+; E) \times (E, D(A))_{1/2-1/(2p), p}$ (Corollary 9.3).

The second main subject of the paper is the abstract parabolic problem

$$\begin{cases} D_t u(t, x) - D_x^2 u(t, x) + Au(t, x) = f(t, x), \, t \in (0, T), x \in \mathbb{R}^+, \\ D_t u(t, 0) - \gamma D_x u(t, 0) + Bu(t, 0) = h(t), \quad t \in (0, T), \\ u(0, x) = u_0(x), \qquad\qquad\qquad\qquad x \in \mathbb{R}^+. \end{cases} \tag{9.8}$$

The assumptions that we employ, concerning A, γ, B, are the same as in the elliptic case, with the supplementary condition $\phi_A \geq \frac{\pi}{2}$. Of course, we can already get some information using the fact that G is the infinitesimal generator of an analytic semigroup and, moreover, we are able to find necessary and sufficient conditions in order that (9.8) have solutions u in $L^p((0, T) \times \mathbb{R}^+; E)$, with $D_t u$, $D_x^2 u$, Au in the same class and $u_{|x=0}$ in $W^{1,p}((0, T); (E, D(A))_{1/2-1/(2p),p}) \cap L^p((0, T); (E, D(A))_{1-1/(2p),p}))$, for every $p \in (1, \infty) \setminus \{\frac{3}{2}\}$. We remark that, in case $p < \frac{3}{2}$, we have not uniqueness, because we are able to prescribe $u_{|x=0}$ at the initial time $t = 0$ (see Theorem 9.6).

This result has some connections with the work of J. Prüss [20]. In this paper the author considers, among other things, a problem which is similar to (9.8) (see Theorem 4.3). The main differences are the following: first of all, the main assumption (in Prüss' work), concerning A and B, is that they have equibounded purely imaginary powers, with power angles less than $\frac{\pi}{2}$. This implies, in particular (see [9, Corollary 2.9]), that, at least in case $]-\infty, 0] \subseteq \rho(B)$, the spectrum $\sigma(B)$ of B is contained in the set of complex numbers z such that $|Arg(z)| \leq \theta_0$, for some $\theta_0 < \frac{\pi}{2}$, which does not necessarily happen, in case if $-B$ is the infinitesimal generator of a strongly continuous semigroup. In particular, it does not seem to hold for the application in Sect. 9.5. On the other hand, in [20] it is not assumed that $D(A^{1/2}) \subseteq D(B)$ and the solution is searched in a more restricted class than ours.

Concerning work of other authors, we want to mention also the paper [18], where the author studies a system, in a Hilbert space setting, which has some connections with (9.8) and has applications to networks with coupled dynamic boundary conditions.

Now we describe, more in detail, how the paper is organized. It consists of five sections, with this introduction.

Section 9.2 treats preliminary technical material, concerning fractional powers of operators, interpolation theory, abstract Sobolev spaces, R-boundedness, which is employed throughout the paper.

Sections 9.3 and 9.4 are dedicated to the aforementioned abstract elliptic and parabolic systems (9.1) and (9.8).

Section 9.5 is dedicated to applications. We have avoided to apply (as it would be possible) the abstract results to elliptic and parabolic problems in cylindrical space domains (see, for example, [11, 26]). After some simple applications to elliptic and parabolic problems with constant coefficients in half-spaces, we employ them to give a different proof of the already mentioned generation result in [10].

In this paper it was treated the case of second order strongly elliptic systems in variational form, with conormal derivative as operator B. We treat the case of a scalar second order equation in non variational form, with B general first order operator such that the boundary is not characteristic (see Theorem 9.9 and Corollary 9.8). Moreover, we are able to precise necessary and sufficient conditions so that it has solutions in $W^{1,p}((0, T); L^p(\Omega)) \cap L^p((0, T); W^{2,p}(\Omega))$, with $u_{|\partial\Omega}$ in $W^{1,p}((0, T); W^{1-1/p,p}(\partial\Omega)) \cap L^p((0, T); W^{2-1/p,p}(\partial\Omega))$ (Theorem 9.10). The main tool is the abstract parabolic maximal regularity result in Theorem 9.6. This final achievement should be new.

The final section is an appendix, with some results, concerning traces, which are, in essence, known, but we were not able to find in this generality in mathematical literature.

9.2   Preliminaries

We introduce some notations and some preliminary results, which will be useful in the sequel.

We shall indicate with $\mathbb{N}$ and $\mathbb{N}_0$ the sets of positive and nonnegative integers, respectively, with $\mathbb{R}^+$ the set of positive real numbers. $*$ will be used to indicate the convolution in the one dimensional time variable t in $(0, T)$ $(0 < T \leq \infty)$.

If Ω is an open subset of $\mathbb{R}^n$, we shall indicate with $\mathscr{D}(\Omega)$ the space of elements of $C^\infty(\Omega)$, with compact support. $\mathscr{S}(\mathbb{R}^n)$ will indicate the space of elements of $C^\infty(\mathbb{R}^n)$ rapidly decaying, with all their derivatives. $\mathscr{F}$ will be used to indicate the Fourier transform.

The letter C will indicate a positive constant, which will be allowed to be different from time to time. In a sequence of inequalities, in order to stress the fact that the constants may change, we shall write $C_0, C_1, \ldots$. If we want to stress the dependence of C on α, β, we shall write $C(\alpha, \beta, \ldots)$.

Let E and F be complex Banach spaces. We shall indicate with $\mathscr{L}(E, F)$ the Banach space of bounded linear operators from E to F. In case $E = F$, we shall simply write $\mathscr{L}(E)$. The notation $E \hookrightarrow F$ will mean that the space E is continuously embedded into the space F.

Customarily we shall work in a fixed Banach space E, with a norm $\| \cdot \|$, without specifying this every time.

If $A : D(A) \to E$ is linear, with $D(A)$ linear subspace of E, we shall say that A is an operator in E. We shall indicate with $\rho(A)$ its resolvent set. If A is a closed linear operator in E, its domain $D(A)$ becomes a Banach space if we equip it with the norm

$$\|x\|_{D(A)} := \|x\|_E + \|Ax\|, \quad x \in D(A).$$

We introduce the following

Definition 9.1 Let A and B be linear operators in E. We shall say that they commute in the sense of the resolvent if, $\forall \lambda \in \rho(A)$, $\forall \mu \in \rho(B)$,

$$(\lambda - A)^{-1}(\mu - B)^{-1} = (\mu - B)^{-1}(\lambda - A)^{-1}.$$

Remark 9.1 It is well known that, in order that A and B commute in the sense of resolvents, it suffices that there exist $\lambda_0 \in \rho(A)$, $\mu_0 \in \rho(B)$, such that $(\lambda_0 - A)^{-1}$ and $(\mu_0 - B)^{-1}$ commute.

Lemma 9.1 *Let A and B be linear operators in E, commuting in the sense of the resolvent. Then:*

(I) if $x \in D(B)$ and $\lambda \in \rho(A)$, $(\lambda - A)^{-1}x \in D(B)$. Moreover, $B(\lambda - A)^{-1}x = (\lambda - A)^{-1}Bx$;

(II) if $-A$ and $-B$ are infinitesimal generators of strongly continuous semigroups $(e^{-tA})_{t \geq 0}$ and $(e^{-tB})_{t \geq 0}$, $\forall t, s \in [0, \infty)$,

$$e^{-tA}e^{-sB} = e^{-sB}e^{-tA}.$$

Moreover, $(e^{-tA}e^{-tB})_{t \geq 0}$ is a strongly continuous semigroup in E.

Proof It is elementary. We observe only that (II) follows from Hille's construction of e^{-tA} (see [23, Chapter 3]), showing that e^{-tA} coincides with $\lim_{n \to \infty} (1 + n^{-1}A)^{-n}$ in a strong sense. $\qquad\square$

Definition 9.2 Let E be a Banach space, let A be a densely defined linear operator in E and $\phi \in [0, \pi)$. We shall say that $A \in O(\phi)$ if

$$\Sigma_\phi := \{\lambda \in \mathbb{C} \setminus \{0\} : |Arg(\lambda)| \leq \phi\} \cup \{0\} \subseteq \rho(-A),$$

and $\{\lambda(\lambda + A)^{-1} : \lambda \in \Sigma_\phi\}$ is bounded in $\mathscr{L}(E)$.

Clearly, if $0 \leq \phi_1 \leq \phi_2 < \pi$, $O(\phi_2) \subseteq O(\phi_1)$. If $A \in O(0)$, we shall say that it is positive.

Remark 9.2 If $A \in O(\phi)$, with $\phi \in [0, \pi)$, then it is of type $(\pi - \phi, M)$, for some $M \geq 1$, in the sense of Definition 2.3.1 in [23]. We observe also that, as a consequence of a simple perturbation argument, if $A \in O(\phi)$, there exists $\epsilon \in R^+$, such that $A \in O(\phi + \epsilon)$. Finally, it is easily seen that, if $A \in O(\phi)$ for some $\phi \in [0, \pi)$, $\lambda \in \mathbb{C} \setminus \{0\}$ and $|Arg(\lambda)| \leq \phi$, $\lambda + A$ is positive.

If A is a positive operator in the Banach space E, its factional powers A^α ($\alpha \geq 0$) are well defined (see, for example, [23, Chapter 2]). If $0 \leq \alpha \leq \beta$, $D(A^\beta) \subseteq D(A^\alpha)$. Moreover, there exists $C(\alpha, \beta)$ in $\mathbb{R}^+$, such that, if $f \in D(A^\beta)$,

$$\|A^\alpha f\| \leq C(\alpha, \beta)\|f\|^{1-\alpha/\beta}\|A^\beta f\|^{\alpha/\beta}. \tag{9.9}$$

It is known that, if $A \in O(\phi)$ $(0 \le \phi < \pi)$ and $\alpha \in (0, 1)$, $A^\alpha \in O((1 - \alpha)\pi + \alpha\phi)$ (see [23, Proposition 2.3.2]). In particular, if $\alpha \le \frac{1}{2}$, $A^\alpha \in O(\psi)$ for some $\psi \ge \frac{\pi}{2}$, implying that $-A^\alpha$ is the infinitesimal generator of an exponentially decaying analytic semigroup $(e^{-tA^\alpha})_{t \ge 0}$.

We shall employ the following

Theorem 9.1 *Let $A \in O(\phi)$, for some $\phi \in [0, \pi)$, in the Banach space E and let $\alpha \in (0, 1]$. Then, $\forall \lambda \in \Sigma_\phi$, $D((\lambda + A)^\alpha) = D(A^\alpha)$. Moreover, $(\lambda + A)^\alpha - A^\alpha$ can be extended to an element of $\mathscr{L}(E)$, which we indicate with the same notation. Finally, there is $C \in \mathbb{R}^+$, such that $\forall \lambda \in \Sigma_\phi$*

$$\|(\lambda + A)^\alpha - A^\alpha\|_{\mathscr{L}(E)} \le C |\lambda|^\alpha.$$

For a proof, see [23, Lemma 2.3.5].

If E_0 and E_1 are compatible Banach spaces, $0 < \theta < 1$, $p \in [1, \infty]$, we shall indicate with $(E_0, E_1)_{\theta,p}$ the corresponding real interpolation space (see, for example, [5, 24]). We shall freely use the main properties of these interpolation functors, for which we refer to (for example) [5, 12, 24]. In some cases these interpolation spaces can be characterized quite precisely. We examine some results in this direction.

If E is a Banach space, we shall indicate with $L_*^p(\mathbb{R}^+; E)$ the standard vector valued L^p space, with respect to the measure μ such that $d\mu = \frac{1}{t} dt$.

Theorem 9.2 *Let E, F be Banach spaces, with $F \hookrightarrow E$ and let A be a positive operator in E, $\theta \in (0, 1]$, $p \in [1, \infty]$. Then*

$$(E, D(A))_{\theta,p} = \{x \in E : t \to t^\theta A(t + A)^{-1}x \in L_*^p(\mathbb{R}^+; E)\}.$$

Moreover, an equivalent norm in $(E, D(A))_{\theta,p}$ is

$$x \to \|x\| + \|t^\theta A(t + A)^{-1}x\|_{L_*^p(\mathbb{R}^+;E)};$$

(II) assume that $\forall t \in \mathbb{R}^+$ $(t + A)^{-1}(F) \subseteq F$ and there exists $C \in \mathbb{R}^+$, such that $\|(t + A)^{-1}_{|F}\|_{\mathscr{L}(F)} \le Ct^{-1}$. Then

$$(E, D(A) \cap F)_{\theta,p} = (E, D(A))_{\theta,p} \cap (E, F)_{\theta,p}.$$

For a proof, see [24, Subsection 1.14.2] for (I), [14, Theorem 5] for (II).

A simple consequence of Theorem 9.2 is the following

Proposition 9.1 *Let A be a positive operator in the complex Banach space E and let Ω be a measure space. Then, if $p \in (1, \infty)$, $\theta \in (0, 1)$,*

$$(L^p(\Omega; E); L^p(\Omega; D(A)))_{\theta,p} = L^p(\Omega; (E, D(A))_{\theta,p}),$$

with equivalent norms.

Proof It follows easily from Theorem 9.2 and the theorem of Fubini, thinking of $L^p(\Omega; D(A))$ as the domain of the positive operator in $L^p(\Omega; E)$ $\mathscr{A}(u)(\omega) := A[u(\omega)]$. □

We shall employ also the connection between real interpolation spaces and trace spaces:

Theorem 9.3 *Let E_0, E_1 be compatible Banach spaces, $1 < p < \infty$, $0 \le j < m$, with $j, m \in \mathbb{Z}$. Then*

$$\{u^{(j)}(0) : u \in L^p(\mathbb{R}^+; E_1), u^{(m)} \in L^p(\mathbb{R}^+; E_0)\} = (E_0, E_1)_{\frac{m-p^{-1}-j}{m},p}.$$

Proof See [14, Theorem 2] or [12, Chapter II]. □

In the following lemma we collect some well known facts, concerning fractional powers and interpolation theory:

Lemma 9.2 *Let A be a positive operator in the Banach space E. Then:*

(I) $\forall \alpha \in (0, 1)$,

$$(E, D(A))_{\alpha,1} \hookrightarrow D(A^\alpha) \hookrightarrow (E, D(A))_{\alpha,\infty};$$

(II) *if $0 \le \alpha < \beta \le 1$, $\theta \in (0, 1)$, $p \in [1, \infty]$,*

$$(D(A^\alpha), D(A^\beta))_{\theta,p} = (E, D(A))_{(1-\theta)\alpha+\theta\beta,p},$$

with equivalent norms;
(III) *if $\alpha, \beta \in \mathbb{R}^+$, A^α and A^β commute in the sense of resolvents.*
     Let $\alpha, \beta \in [0, 1)$, with $\alpha + \beta < 1$, $1 \le p \le \infty$. Then:
(IV) *A^α maps continuously $D(A^{\alpha+\beta})$ onto $D(A^\beta)$, and $(E, D(A))_{\alpha+\beta,p}$ onto $(E, D(A))_{\beta,p}$;*
(V) *in particular, if $\alpha \le \frac{1}{2}$ and $p < \infty$, the part of A^α in $(E, D(A))_{\beta,p}$, with domain $(E, D(A))_{\alpha+\beta,p}$, is the infinitesimal generator of an analytic exponentially decreasing semigroup in $(E, D(A))_{\beta,p}$.*

Proof (I) is proved in [24, Section 1.15.2];
 (II) follows from (I) and the reiteration theorem;
 (III) follows from the fact that

$$(A^\alpha)^{-1}(A^\beta)^{-1} = A^{-\alpha}A^{-\beta} = A^{-\alpha-\beta} = (A^\beta)^{-1}(A^\alpha)^{-1}$$

(see [23, Subsection 2.3.1]), and Remark 9.1;
 (IV) follows immediately from the definition of fractional powers (see [23, Subsection 2.3.1]), (II) and the interpolation property, employing also the fact that $A^{-\alpha}$ maps $D(A^\beta)$ into $D(A^{\alpha+\beta})$);

(V) first of all, we observe that, as $p < \infty$ and $D(A)$ is dense in E, $D(A)$ is dense in $(E, D(A))_{\beta,p}$ (see [5, Theorem 3.4.2]). We deduce that $(E, D(A))_{\alpha+\beta,p}$ is dense in $(E, D(A))_{\beta,p}$. Moreover, by (III), it is easily seen that e^{-sA^α} and $(A + s)^{-1}$ commute. So, if $f \in (E, D(A))_{\beta,p}$, in case $\|e^{-sA^\alpha}\|_{\mathscr{L}(E)} \le Me^{-\eta s}$ (for certain $M, \eta \in \mathbb{R}^+$, $\forall s \ge 0$),

$$\int_0^\infty \|t^\beta A(t + A)^{-1} e^{-sA^\alpha} f\|^p \frac{dt}{t} = \int_0^\infty \|e^{-sA^\alpha}[t^\beta A(t + A)^{-1} f]\|^p \frac{dt}{t}$$
$$\le (Me^{-\eta s})^p \int_0^\infty \|[t^\beta A(t + A)^{-1} f]\|^p \frac{dt}{t},$$

implying, as an application of Theorem 9.2, that, for each $s \ge 0$, e^{-sA^α} maps $(E, D(A))_{\beta,p}$ into itself and the norm of its part in this space is exponentially decreasing. The strong continuity is a consequence of the fact that

$$\lim_{s \to 0} \int_0^\infty \|t^\beta A(t + A)^{-1} (e^{-sA^\alpha} f - f)\|^p \frac{dt}{t} = 0,$$

by the dominated convergence theorem. □

Let I be an open interval in $\mathbb{R}$ and let E be a Banach space. If $m \in \mathbb{N}_0$, we shall indicate with $W^{m,p}(I; E)$ the space of elements of $L^p(I; E)$ whose derivatives $u^{(j)}$, with $j \le m$, in the sense of E−valued distributions, belong to $L^p(I; E)$, equipped with the norm

$$\|f\|_{W^{m,p}(I;E)} := \sum_{j=0}^m \|f^{(j)}\|_{L^p(I;E)}. \tag{9.10}$$

We refer, for these spaces, to [12] and [4].

We shall need the following version of Da Prato-Grisvard's theory:

Theorem 9.4 *Let B and G be linear operators in the Banach space E. Assume the following:*

(I) $B \in O(\phi_B)$, $G \in O(\phi_G)$, *for some* ϕ_B, ϕ_G *in* $[0, \pi)$, *such that* $\phi_B + \phi_G > \pi$;
(II) *B and G commute in the sense of the resolvent.*

Then:

I) $G + B$ *(with domain* $D(B) \cap D(G)$*) is closable;*
(II) *there exists a linear operator P, which is an extension of the closure of* $G + B$, *such that P is positive;*
(III) *if* $y \in (E, D(G))_{\beta,p}$, *for some* $\beta \in (0, 1)$ *and* $p \in [1, \infty]$ *and* $\lambda \in [0, \infty)$, $(\lambda + P)^{-1}y \in D(G) \cap D(B)$ *and* $G(\lambda + P)^{-1}y$ *and* $B(\lambda + P)^{-1}y$ *belong to* $(E, D(G))_{\beta,p}$;

(IV) there exists $M \in \mathbb{R}^+$, such that, if $\lambda \in [0, \infty)$, $y \in (E, D(G))_{\beta,p}$,

$$(1+\lambda)\|(\lambda + P)^{-1}y\|_{(E,D(G))_{\beta,p}} + \|G(\lambda + P)^{-1}y\|_{(E,D(G))_{\beta,p}}$$
$$+\|B(\lambda + P)^{-1}y\|_{(E,D(G))_{\beta,p}} \leq M\|y\|_{(E,D(G))_{\beta,p}}.$$

For a proof, see [6], and also [7, Theorem 2.2].

A well known application of Theorem 9.4 is the following

Proposition 9.2 *Let $-A$ be the infinitesimal generator of an analytic semigroup in the complex Banach space E. Let $T \in \mathbb{R}^+$, $p \in [1, \infty)$, $\theta \in (0, 1)$ and consider the mild solution*

$$u(t) := \int_0^t e^{-(t-s)A} f(s)ds, \quad t \in (0, T)$$

of the Cauchy problem

$$\begin{cases} u'(t) + Au(t) = f(t), & t \in (0, T), \\ u(0) = 0. \end{cases} \tag{9.11}$$

Then, if $f \in L^p((0,T); (E, D(A))_{\theta,p})$, $u \in W^{1,p}((0,T); (E, D(A))_{\theta,p})$, $u(t) \in D(A)$ for almost every t, $Au \in L^p((0,T); (E, D(A))_{\theta,p})$ and (9.11) holds almost everywhere.

Proof See [6, Theorem 4.7].   □

We shall employ complex valued Sobolev spaces $W^{s,p}(\Omega)$, Besov spaces $B_{p,p}^s(\Omega)$, Bessel potential spaces $H^{s,p}(\Omega)$ of several real variables with Ω open subset of $\mathbb{R}^n$. We recall only some basic facts referring to [1] and [24] for more detailed presentations.

Let Ω be an open subset of $\mathbb{R}^n$, with suitably smooth boundary $\partial\Omega$, or $\Omega = \mathbb{R}^n$, and let $p \in (1, \infty)$. If $m \in \mathbb{N}_0$, we indicate with $W^{m,p}(\Omega)$ the standard Sobolev space. If $\alpha \in (0, 1)$ and $(\cdot, \cdot)_\alpha$ stands for the complex interpolation functor, we have

$$(L^p(\Omega), W^{m,p}(\Omega))_\alpha = (L^p(\Omega), H^{m,p}(\Omega))_\alpha = H^{m\alpha,p}(\Omega), \tag{9.12}$$

while

$$(L^p(\Omega), W^{m,p}(\Omega))_{\alpha,p} = B_{p,p}^{m\alpha}(\Omega). \tag{9.13}$$

It is also known that $H^{m\alpha,p}(\Omega) = W^{m\alpha,p}(\Omega)$ if $m\alpha \in \mathbb{N}$, while $B_{p,p}^{m\alpha}(\Omega) = W^{m\alpha,p}(\Omega)$ if $2\alpha \notin \mathbb{N}$. In the particular case $p = 2$, we have

$$W^{m\alpha,2}(\Omega) = H^{m\alpha,2}(\Omega) = B_{2,2}^{m\alpha}(\Omega) \quad \forall \alpha \in (0, 1).$$

These facts can be extended to corresponding spaces in differential manifolds. It is
also of interest to treat vector valued spaces (see, in particular, [4]). In particular, if
I is an open interval in $\mathbb{R}$ and E is a Banach space, it is still true that, in case $\alpha \notin \mathbb{N}$,
$W^{\alpha,p}(I;E)$ coincides with the Besov space $B^{\alpha}_{p,p}(I;E)$ (see [12,22,24]).

If I is an interval in $\mathbb{R}$ and $m \in \mathbb{N}_0$ we shall indicate with $BUC^m(I;E)$ the
Banach space of functions $f : I \to E$, which are uniformly continuous and
bounded, with their derivatives or order less or equal to m, equipped with a natural
norm.

The following facts hold and will be used in the sequel:

Proposition 9.3 *(I) If $0 \le \alpha_0 < \alpha_1$, $\theta \in (0,1)$, $p \in [1,\infty]$,*

$$(W^{\alpha_0,p}(I;E), W^{\alpha_1,p}(I;E))_{\theta,p} = B^{(1-\theta)\alpha_0+\theta\alpha_1}_{p,p}(I;E),$$

with equivalent norms;
(II) if $\alpha > m + \frac{1}{p}$ with $m \in \mathbb{N}$, $B^{\alpha}_{p,p}(I;E) \hookrightarrow BUC^m(I;E)$.

Proof Concerning (I), see [13, Proposition 5.9]; for (II), see [22, Corollary 26]. □

In the following we shall employ the notions of *UMD* Banach space with
property (α) and of *R*-boundedness. We introduce the elements of this theory that
we shall need. The main reference is [17].

In the following we shall indicate with $(r_n)_{n\in\mathbb{N}}$ the Rademacher sequence in
$[0,1]$.

Definition 9.3 Let E_0, E_1 be Banach spaces, $\mathscr{B} \subseteq \mathscr{L}(E_0, E_1)$. We shall say
that $\mathscr{B}$ is *R*-bounded in $\mathscr{L}(E_0, E_1)$ if there exists $C \in \mathbb{R}^+$ such that, $\forall n \in \mathbb{N}$,
$\forall T_1,\ldots,T_n \in \mathscr{B}$, $\forall f_1,\ldots,f_n \in E_0$,

$$\left\| \sum_{k=1}^{n} r_k T_k f_k \right\|_{L^2((0,1);E_1)} \le C \left\| \sum_{k=1}^{n} r_k f_k \right\|_{L^2((0,1);E_0)}.$$

Remark 9.3 If $\mathscr{B}$ is *R*-bounded, it is bounded, but the converse does not in general
hold, except in the case that E_0 and E_1 are Hilbert spaces.

Definition 9.4 The complex Banach space E has property (α) if there exists $C \in$
$\mathbb{R}^+$, such that, $\forall n \in \mathbb{N}$, $\forall \alpha_{ij} \in \mathbb{C}$ within $|\alpha_{ij}| \le 1$, $\forall f_{ij} \in E$ $(1 \le i, j \le n)$,

$$\int_{[0,1]\times[0,1]} \left\| \sum_{i=1}^{n} \sum_{j=1}^{n} r_i(u) r_j(v) \alpha_{ij} f_{ij} \right\| du\, dv \le C \int_{[0,1]\times[0,1]} \left\| \sum_{i=1}^{n} \sum_{j=1}^{n} r_i(u) r_j(v) f_{ij} \right\| du\, dv.$$

Definition 9.5 A complex Banach space E is *UMD* if the Hilbert transform is
a linear bounded operator in $L^p(\mathbb{R}; E)$ for some (or, equivalently, for any) $p \in$
$(1,\infty)$.

Remark 9.4 Prototypes of *UMD* spaces with property (α) are spaces $L^p(\Omega, \mu)$, with $1 < p < \infty$ and μ σ-finite measure (see [17, Sections 3.14 and 4.10]).

The following fact holds (see [11, Lemma 1.4]):

Lemma 9.3 *Let E be a UMD Banach space and let A be a positive operator in E, such that $\{\lambda(\lambda + A)^{-1} : \lambda \geq 0\}$ is R-bounded in $\mathscr{L}(E)$. Then, if I is an interval in $\mathbb{R}$ with nonempty interior, $p \in (1, \infty)$, $m \in \mathbb{N}$, $u \in L^p(I; D(A))$ and $u^{(m)} \in L^p(I; E)$, for each $k \in \{1, \ldots, m-1\}$ $u^{(k)} \in L^p(I; D(A^{1-k/m}))$. Moreover, there exists C in $\mathbb{R}^+$, such that*

$$\|u^{(k)}\|_{L^p(I;D(A^{1-k/m}))} \leq C(\|u\|_{L^p(I;D(A))} + \|u^{(m)}\|_{L^p(I;E)}),$$

with C independent of u.

9.3 Abstract Elliptic Problem Depending on a Complex Parameter

In this section we study some abstract elliptic problems depending on the complex parameter λ. We start with the following:

$$\begin{cases} \lambda u(x) - u''(x) + Au(x) = f(x), \ x > 0, \\ u(0) = g \end{cases} \tag{9.14}$$

with $\lambda \in \mathbb{C}$ and the assumptions:

(A1) *E is a UMD Banach space with property (α);*
(A2) *A is a closed, densely defined operator in E. Moreover, for some $\phi_A \in [0, \pi)$, $A \in O(\phi_A)$ and $\{\lambda(\lambda + A)^{-1} : \lambda \in \Sigma_{\phi_A}\}$ is R– bounded in $\mathscr{L}(E)$.*

We start by considering the case $g = 0$. To this aim, we introduce the following operator $\mathscr{A}$, with $p \in (1, \infty)$:

$$\begin{cases} \mathscr{A} : \{u \in W^{2,p}(\mathbb{R}^+; E) \cap L^p(\mathbb{R}^+; D(A)) : u(0) = 0\} \to L^p(\mathbb{R}^+; E), \\ \mathscr{A}u = -u'' + Au. \end{cases} \tag{9.15}$$

Proposition 9.4 *Assume that (A1)–(A2) hold. Let $p \in (1, \infty)$. Then:*

(I) $W^{2,p}(\mathbb{R}^+; E) \cap L^p(\mathbb{R}^+; D(A)) \subseteq W^{1,p}(\mathbb{R}^+; D(A^{1/2}))$;
(II) $\mathscr{A}$, *as operator in $L^p(\mathbb{R}^+; E)$, belongs to $O(\phi_A)$. Moreover $\{\lambda(\lambda + \mathscr{A})^{-1} : \lambda \in \Sigma_{\phi_A}\}$ is R-bounded in $\mathscr{L}(L^p(\mathbb{R}^+; E))$; more precisely, if i, j, k are*

nonnegative integers, such that $i + j + k = 2$, $\{|\lambda|^{i/2} A^{j/2} D_x^k (\lambda + \mathscr{A})^{-1} :$
$\lambda \in \Sigma_{\phi_A}\}$ *is R-bounded in* $\mathscr{L}(L^p(\mathbb{R}^+; E))$;
(III) in case $\phi_A \geq \frac{\pi}{2}$, $-\mathscr{A}$ *is the infinitesimal generator of an analytic semigroup
in* $L^p(\mathbb{R}^+; E)$.

Proof (I) follows from Lemma 9.3.

Concerning (II), the case $\phi_A = \frac{\pi}{2}$ is treated in Step 2 of the proof of Theorem 2.3
in [11]. The general case can be obtained analogously, with simple modifications of
the previous Lemma 2.4 and Theorem 2 in [11] (replacing $\frac{\pi}{2}$ with ϕ_A). □

We consider (9.14), in case $f \equiv 0$, and observe that, by virtue of Theorem 9.3,
a necessary condition, in order that there exist a solution u in $W^{2,p}(\mathbb{R}^+; E) \cap$
$L^p(\mathbb{R}^+; D(A))$, is $g \in (E, D(A))_{1-1/(2p),p}$. The following result holds:

Proposition 9.5 *Let E be a Banach space and let A be an operator in E, $A \in$
$O(\phi_A)$, for some $\phi_A \in [0, \pi)$. Then, $\forall \lambda \in \Sigma_{\phi_A}$, $\forall g \in (E, D(A))_{1-1/(2p),p}$, the
system*

$$\begin{cases} \lambda u(x) - u''(x) + Au(x) = 0, \, x > 0, \\ u(0) = g \end{cases}$$

has a unique solution u in $W^{2,p}(\mathbb{R}^+; E) \cap L^p(\mathbb{R}^+; D(A))$. Precisely,

$$u(x) = e^{-x(\lambda+A)^{1/2}} g, \quad x \in \mathbb{R}^+.$$

*Moreover, there exists $C \in \mathbb{R}^+$, independent of λ and g, such that, if $i, j, k \in \mathbb{N}_0$
and $i + j + k = 2$,*

$$|\lambda|^{i/2} \|D_x^j A^{k/2} u\|_{L^p(\mathbb{R}^+;E)} \leq C(\|g\|_{(E,D(A))_{1-1/(2p),p}} + |\lambda|^{1-1/(2p)} \|g\|).$$

Proof The first statement follows from (for example) [11, Lemma 2.2]. Moreover,
by Theorem 5.4.2/1 in [26],

$$\|(\lambda + A)u\|_{L^p(\mathbb{R}^+;E)} \leq C(\|g\|_{(E,D(A))_{1-1/(2p),p}} + |\lambda|^{1-1/(2p)} \|g\|),$$

which implies the case $i = k = 0$, $j = 2$. From the inequality $|\lambda| \|u\| + \|Au\| \leq$
$C \|(\lambda + A)u\|$, we obtain the cases $i = 2, j = k = 0$ and $i = j = 0, k = 2$.
The other cases follow from the foregoing, the identity $u'(x) = -(\lambda + A)^{1/2} u(x)$,
Theorem 9.1 and (9.9). □

Corollary 9.1 *Assume that (A1)–(A2) hold. Consider system (9.14). Let $p \in (1, \infty)$. Then, if $\lambda \in \Sigma_{\phi_A}$, if $f \in L^p(\mathbb{R}^+; E)$ and $g \in (E, D(A))_{1-1/(2p), p}$, (9.14) has a unique solution u in $W^{2,p}(\mathbb{R}^+; E) \cap L^p(\mathbb{R}^+; D(A))$. Moreover, there exists $C \in \mathbb{R}^+$, such that, if $i, j, k \in \mathbb{N}_0$ and $i + j + k = 2$,*

$$\||\lambda|^{i/2} D_x^j A^{k/2} u\|_{L^p(\mathbb{R}^+; E)} \leq C(\|f\|_{L^p(\mathbb{R}^+; E)} + \|g\|_{(E, D(A))_{1-1/(2p), p}} + |\lambda|^{1-1/(2p)} \|g\|).$$

Proof It follows immediately from Propositions 9.4 and 9.5.   □

We introduce the following notations: if $\lambda \in \Sigma_{\phi_A}$, $f \in L^p(\mathbb{R}^+; E)$ and $g \in (E, D(A))_{1-1/(2p), p}$, we indicate with

$$S(\lambda, f, g) \tag{9.16}$$

the solution in $W^{2,p}(\mathbb{R}^+; E) \cap L^p(\mathbb{R}^+; D(A))$ of (9.14).

Now we want to study the abstract elliptic problem depending on the parameter λ (9.1), with the following assumptions:

(B1) (A1)–(A2) hold, $\gamma \in \mathbb{R}^+$;
(B2) $-B$ is the infinitesimal generator of a strongly continuous semigroup in E, A and B commute in the sense of the resolvent and $D(A^{1/2}) \subseteq D(B)$.

We start by considering the case $f \equiv 0$. Looking for solutions u in $W^{2,p}(\mathbb{R}^+; E) \cap L^p(\mathbb{R}^+; D(A))$, we are reduced, by Proposition 9.5, to search for g in the space $(E, D(A))_{1-1/(2p), p}$ such that

$$[\lambda + \gamma(\lambda + A)^{1/2} + B]g = h. \tag{9.17}$$

We shall solve (20.4) thinking of it as a perturbation to the following

$$(\lambda + \gamma A^{1/2} + B)g = h. \tag{9.18}$$

We recall that, as A is positive, $-\gamma A^{1/2}$ is the infinitesimal generator of an exponentially decreasing analytic semigroup. So we shall employ the following

Lemma 9.4 *Let B and G be closed densely defined operators in the Banach space E. Assume the following:*

(a) $-B$ *is the infinitesimal generator of a strongly continuous semigroup in E;*
(b) $-G$ *is the infinitesimal generator of an analytic semigroup in E;*
(c) B *and* G *commute in the sense of the resolvent;*
(d) $D(G) \subseteq D(B)$.

Let $\beta \in (0, 1)$ and $p \in [1, \infty)$. Then:

(I) B maps $(E, D(G))_{1+\beta,p} := \{x \in D(G) : Gx \in (E, D(G))_{\beta,p}\}$ into $(E, D(G))_{\beta,p}$; as a consequence, the following operator

$$\begin{cases} F : (E, D(G))_{1+\beta,p} \to (E, D(G))_{\beta,p}, \\ Fx := -Bx - Gx \end{cases}$$

is well defined;

(II) F is the infinitesimal generator of an analytic semigroup in $(E, D(G))_{\beta,p}$.

Proof (I) Let $t \in \rho(-G)$. Then $B(t + G)^{-1} \in \mathcal{L}(E)$. It maps also $D(G)$ into itself, as, in case $g \in D(G)$, $B(t + G)^{-1}g = (t + G)^{-1}Bg$ (see Lemma 9.1 (I)). So, by interpolation, it maps $(E, D(G))_{\beta,p}$ into itself. Assume that $g \in (E, D(G))_{1+\beta,p}$. Then $Bg = B(t + G)^{-1}(t + G)g \in (E, D(G))_{\beta,p}$, as $(t + G)g \in (E, D(G))_{\beta,p}$.

(II) We set

$$T_0(t) := e^{-tB}e^{-tG} = e^{-tG}e^{-tB}, \quad t \geq 0. \tag{9.19}$$

Then, by Lemma 9.1(II), $(T_0(t))_{t\geq 0}$ is a strongly continuous semigroup in E. Arguing as in the proof of Lemma 9.2(V) (replacing A with $G + t_0$, with $t_0 \in \mathbb{R}$ such that $t_0 + G$ is positive, and e^{-sA^α} with $T_0(s)$), we can show that $\forall t \geq 0$, $T_0(t)$ maps $(E, D(G))_{\beta,p}$ into itself. Moreover, if we indicate with $T(t)$ the restriction of $T_0(t)$ to $(E, D(G))_{\beta,p}$, we obtain that $T(t) \in \mathcal{L}((E, D(G))_{\beta,p})$ and $(T(t))_{t\geq 0}$ is a strongly continuous semigroup in $(E, D(G))_{\beta,p}$. We observe that the same happens also for $(e^{-tB})_{t\geq 0}$: the restrictions of these operators to $(E, D(G))_{\beta,p}$ give rise to a strongly continuous semigroup in the same space. We show that the infinitesimal generator of $(T(t))_{t\geq 0}$ is F.

Let $g \in (E, D(G))_{1+\beta,p}$. Then, if $s > 0$,

$$T(s)g - g = e^{-sB}(e^{-sG} - 1)g + (e^{-sB} - 1)g$$
$$= -e^{-sB}\int_0^s e^{-\sigma G}Gg d\sigma - \int_0^s e^{-\sigma B}Bg d\sigma.$$

As Gg and Bg belong to $(E, D(G))_{\beta,p}$, the functions $s \to e^{-\sigma G}Gg$ and $s \to e^{-\sigma B}Bg$ are continuous with values in $(E, D(G))_{\beta,p}$. We deduce that

$$\frac{T(s)g - g}{s} = -e^{-sB}\frac{1}{s}\int_0^s e^{-\sigma G}Gg d\sigma - \frac{1}{s}\int_0^s e^{-\sigma B}Bg d\sigma \to -Gg - Bg \quad (s \to 0)$$

in $(E, D(G))_{\beta,p}$. So, if we indicate with F' the infinitesimal generator of $(T(t))_{t\geq 0}$, we can say that F' is an extension of F. To show the inverse, it suffices to observe that, by Theorem 9.4, $\rho(F)$ contains $[t_0, \infty)$, for some $t_0 \in \mathbb{R}$.

It remains to show that $(T(t))_{t\geq 0}$ is analytic. To this aim, we can try to show that, if $t > 0$, $T(t)$ maps $(E, D(G))_{\beta,p}$ into $(E, D(G))_{1+\beta,p}$, and there exists $M \in \mathbb{R}^+$, such that

$$\|FT(t)\|_{\mathscr{L}((E,D(G))_{\beta,p})} \leq Mt^{-1}$$

for $t \in (0, 1]$ (see [19, Theorem 5.2 in Chapter 2]). We start by recalling that $(e^{-tG})_{t\geq 0}$ is analytic. It is clear that this holds even for its restriction to $(E, D(G))_{\beta,p}$. We deduce that, for $t > 0$, e^{-tG} maps $(E, D(G))_{\beta,p}$ into the domain of the part of G in $(E, D(G))_{\beta,p}$, which is $(E, D(G))_{1+\beta,p}$. Moreover, for some $M_0 \in \mathbb{R}^+$,

$$\|Ge^{-tG}\|_{\mathscr{L}((E,D(G))_{\beta,p})} \leq M_0 t^{-1}, \quad t \in (0, 1].$$

So, it is clear that, $\forall t \in \mathbb{R}^+$, $T(t)$ maps $(E, D(G))_{\beta,p}$ into $(E, D(G))_{1+\beta,p}$. Moreover, if $t \in (0, 1]$, and $f \in (E, D(G))_{\beta,p}$, for certain positive constants $C_0, C_1, C_2 \in \mathbb{R}^+$, as $G + B \in \mathscr{L}((E, D(G))_{1+\beta,p}, (E, D(G))_{\beta,p})$,

$$\|(G + B)T(t)f\|_{(E,D(G))_{\beta,p}} \leq C_0(\|T(t)f\|_{(E,D(G))_{\beta,p}} + \|Ge^{-tG}e^{-tB}f\|_{(E,D(G))_{\beta,p}})$$
$$\leq C_1(\|f\|_{(E,D(G))_{\beta,p}} + t^{-1}\|e^{-tB}f\|_{(E,D(G))_{\beta,p}}) \leq C_2 t^{-1}\|f\|_{(E,D(G))_{\beta,p}}.$$

The proof is complete.   □

Corollary 9.2 Let $\beta \in (0, \frac{1}{2})$, $p \in [1, \infty)$. Then:

(I) consider Eq. (9.18); there exists ω, C in $\mathbb{R}^+$ such that, if $\lambda \in \Sigma_{\pi/2}$, $|\lambda| \geq \omega$ and $h \in (E, D(A))_{\beta,p}$, (9.18) has a unique solution g in $(E, D(A))_{\beta+1/2,p}$. Moreover,

$$|\lambda|\|g\|_{(E,D(A))_{\beta,p}} + \|g\|_{(E,D(A))_{1/2+\beta,p}} \leq C\|h\|_{(E,D(A))_{\beta,p}}. \tag{9.20}$$

(II) Consider Eq. (9.17). Then there exist ω', C' in $\mathbb{R}^+$ such that, if $\lambda \in \Sigma_{\min\{\phi_A,\pi/2\}}$, $|\lambda| \geq \omega$ and $h \in (E, D(A))_{\beta,p}$, (9.18) has a unique solution g in $(E, D(A))_{\beta+1/2,p}$. Moreover an estimate like (9.20) holds.

Proof (I) We set $G := \gamma A^{1/2}$. From the explicit expression of $(\lambda - A^{1/2})^{-1}$ (see [23, Proposition 2.3.2]), we deduce that G and B commute in the sense of the resolvent. Moreover, $(E, D(A))_{\beta,p} = (E, D(G))_{2\beta,p}$ and, by Lemma 9.2 (IV), $(E, D(G))_{1+2\beta,p} = (E, D(A))_{\beta+1/2,p}$. So the conclusion follows from Lemma 9.4 and Theorem 5.2 in [19] (characterizing the resolvent of infinitesimal generators of analytic semigroups).

(II) We consider (9.17), with $\lambda \in \Sigma_{\min\{\phi_A,\pi/2\}}$, $|\lambda| \geq \omega$, and $h \in (E, D(A))_{\beta,p}$. We write it in the form

$$(\lambda + \gamma A^{1/2} + B)g = h + \gamma[A^{1/2} - (A + \lambda)^{1/2}]g.$$

Setting

$$v := (\lambda + \gamma A^{1/2} + B)g,$$

we obtain that v should solve

$$v = \gamma[A^{1/2} - (\lambda + A)^{1/2}](\lambda + \gamma A^{1/2} + B)^{-1}v + h. \tag{9.21}$$

Applying Theorem 9.1 to the part of A in $(E; D(A))_{\beta,p}$, we obtain, from (I),

$$\|\gamma[A^{1/2} - (\lambda + A)^{1/2}](\lambda + \gamma A^{1/2} + B)^{-1}\|_{\mathscr{L}((E;D(A))_{\beta,p})}$$
$$\leq C_0|\lambda|^{1/2}\|(\lambda + \gamma A^{1/2} + B)^{-1}\|_{\mathscr{L}((E;D(A))_{\beta,p})} \leq C_1|\lambda|^{-1/2}. \tag{9.22}$$

We deduce that, if $|\lambda|$ is sufficiently large, (9.21) has a unique solution v in $(E; D(A))_{\beta,p}$ and, for some C_2 independent of λ and h,

$$\|v\|_{(E;D(A))_{\beta,p}} \leq C_2\|h\|_{(E;D(A))_{\beta,p}}.$$

We conclude that (9.17) has the unique solution

$$g = (\lambda + \gamma A^{1/2} + B)^{-1}v,$$

so that, by (I),

$$|\lambda|\|g\|_{(E;D(A))_{\beta,p}} + \|g\|_{(E;D(A))_{1/2+\beta,p}} \leq C_3\|v\|_{(E;D(A))_{\beta,p}} \leq C_4\|h\|_{(E;D(A))_{\beta,p}}.$$

$\square$

Now we are able to study (9.1).

Theorem 9.5 *Assume that (B1)–(B2) are satisfied. Let* $p \in (1,\infty)$, $\lambda \in \Sigma_{\min\{\phi_A,\pi/2\}}$. *Then there exist* ω, M *in* $\mathbb{R}^+$ *such that:*

(I) if $|\lambda| \geq \omega$, $f \in L^p(\mathbb{R}^+; E)$, $h \in (E, D(A))_{1/2-1/(2p),p}$, *(9.1) has a unique solution* u *in* $W^{2,p}(\mathbb{R}^+; E) \cap L^p(\mathbb{R}^+; D(A))$;
(II) if i, j, k *are nonnegative integers, such that* $i + j + k = 2$,

$$|\lambda|^{i/2}\|A^{j/2}D_x^k u\|_{L^p(\mathbb{R}^+;E)} \leq M(\|f\|_{L^p(\mathbb{R}^+;E)} + \|h\|_{(E,D(A))_{1/2-1/(2p),p}}).$$

Proof As a first step, we consider the case $f \equiv 0$. Then, by Proposition 9.5 and Corollary 9.2 (taking $\beta = \frac{1}{2} - \frac{1}{2p}$), we have that, if $|\lambda| \geq \omega$, (9.1) has the unique solution

$$u(x) = e^{-x(\lambda+A)^{1/2}}[\lambda + \gamma(\lambda + A)^{1/2} + B]^{-1}h.$$

We deduce

$$
\begin{aligned}
|\lambda|^{i/2}\|A^{j/2}D_x^k u\|_{L^p(\mathbb{R}^+;E)} &\leq C(\|[\lambda + \gamma(\lambda + A)^{1/2} + B]^{-1}h\|_{(E,D(A))_{1-1/(2p),p}} \\
&\quad + |\lambda|^{1-1/(2p)}\|[\lambda + \gamma(\lambda + A)^{1/2} + B]^{-1}h\|) \\
&\leq C(\|[\lambda + \gamma(\lambda + A)^{1/2} + B]^{-1}h\|_{(E,D(A))_{1-1/(2p),p}} \\
&\quad + |\lambda|^{1-1/(2p)}\|[\lambda + \gamma(\lambda + A)^{1/2} + B]^{-1}h\|_{E,D(A))_{1/2-1/(2p),p}}) \\
&\leq C_1\|h\|_{(E,D(A))_{1/2-1/(2p),p}}.
\end{aligned}
$$

Now let us consider the general case. Let u_1 be the unique solution in $W^{2,p}(\mathbb{R}^+;E) \cap L^p(\mathbb{R}^+;D(A))$ of

$$
\begin{cases}
\lambda u_1(x) - u_1''(x) + A u_1(x) = f(x), \ x \in \mathbb{R}^+, \\
u_1(0) = 0,
\end{cases}
\tag{9.23}
$$

existing by Corollary 9.1. We consider the system

$$
\begin{cases}
\lambda u_2(x) - u_2''(x) + A u_2(x) = 0, \quad x > 0, \\
\lambda u_2(0) - \gamma u_2'(0) + B u_2(0) = h - [\lambda u_1(0) - \gamma u_1'(0) + B u_1(0)] = h + \gamma u_1'(0).
\end{cases}
\tag{9.24}
$$

We observe that, by Theorem 9.3, $u_1'(0) \in (E, D(A))_{1/2-1/(2p),p}$. Then, applying what we have seen in the first step, in case $|\lambda| \geq \omega$, (9.24) has a unique solution in $W^{2,p}(\mathbb{R}^+;E) \cap L^p(\mathbb{R}^+;D(A))$ and it is clear that (9.1) has the only solution $u := u_1 + u_2$. It remains to estimate u. By Corollary 9.1, if $i, j, k \in \mathbb{N}_0, i+j+k = 2$,

$$
|\lambda|^{i/2}\|A^{j/2}D_x^k u_1\|_{L^p(\mathbb{R}^+;E)} \leq C_1\|f\|_{L^p(\mathbb{R}^+;E)}.
\tag{9.25}
$$

Employing what we have seen in the first step, we deduce, owing to Theorem 9.3 and (again) Corollary 9.1,

$$
\begin{aligned}
|\lambda|^{i/2}\|A^{j/2}D_x^k u_2\|_{L^p(\mathbb{R}^+;E)} &\leq C_2(\|h\|_{(E,D(A))_{1/2-1/(2p),p}} + \|u_1'(0)\|_{(E,D(A))_{1/2-1/(2p),p}}) \\
&\leq C_3(\|h\|_{(E,D(A))_{1/2-1/(2p),p}} + \|u_1\|_{L^p(\mathbb{R}^+;D(A))} + \|D_x^2 u_1\|_{L^p(\mathbb{R}^+;E)}) \\
&\leq C_4(\|f\|_{L^p(\mathbb{R}^+;E)} + \|h\|_{(E,D(A))_{1/2-1/(2p),p}}).
\end{aligned}
$$

The proof is complete.   □

Remark 9.5 By inspection of the previous proof, recalling the notation (9.16), we can see that the solution u of (9.1) can be represented in the form

$$
u = S(\lambda, f, g),
$$

with

$$
g = [\lambda + \gamma(\lambda + A)^{1/2} + B]^{-1}[\gamma D_x S(\lambda, f, 0)(0) + h].
$$

Corollary 9.3 *Assume that (B1)–(B2) hold, with $\phi_A \geq \frac{\pi}{2}$. Let G be the operator defined in (9.7). Then G is the infinitesimal generator of an analytic semigroup in $L^p(\mathbb{R}^+; E) \times (E, D(A))_{1/2-1/(2p),p}$.*

Proof By Theorems 9.5 and 9.3, $\rho(G)$ contains $\{\lambda \in \mathbb{C} : |\lambda| \geq R, Re(\lambda) \geq 0\}$, for some R in $\mathbb{R}^+$. Observe that we have also, in case $(u, g) = (\lambda - G)^{-1}(f, h)$,

$$\|g\|_{(E,D(A))_{1-1/(2p),p}} = \|u(0)\|_{(E,D(A))_{1-1/(2p),p}}$$
$$\leq C_0(\|D_x^2 u\|_{L^p(\mathbb{R}^+;E)} + \|u\|_{L^p(\mathbb{R}^+;D(A))})$$
$$\leq C_1(\|f\|_{L^p(\mathbb{R}^+;E)} + \|h\|_{(E,D(A))_{1/2-1/(2p),p}}),$$

and

$$\|u\|_{L^p(\mathbb{R}^+;E)} + \|g\|_{(E,D(A))_{1/2-1/(2p),p}}$$
$$\leq |\lambda|^{-1}(\|D_x^2 u\|_{L^p(\mathbb{R}^+;E)} + \|Au\|_{L^p(\mathbb{R}^+;E)} + \|f\|_{L^p(\mathbb{R}^+;E)}$$
$$+\gamma\|D_x u(0)\|_{(E,D(A))_{1/2-1/(2p),p}} + \|Bu(0)\|_{E,D(A))_{1/2-1/(2p),p}} + \|h\|_{E,D(A))_{1/2-1/(2p),p}})$$
$$\leq C_0|\lambda|^{-1}(\|f\|_{L^p(\mathbb{R}^+;E)} + \|h\|_{E,D(A))_{1/2-1/(2p),p}}).$$

It remains to show that $D(G)$ is dense in $L^p(\mathbb{R}^+; E) \times (E, D(A))_{1/2-1/(2p),p}$.

To this aim, we start by observing that

$$\{u \in W^{2,p}(\mathbb{R}^+; E) \cap L^p(\mathbb{R}^+; D(A)) : u(0) = 0\}$$

is dense in $L^p(\mathbb{R}^+; E)$. In fact, as $D(A)$ is dense in E, it is easily seen that $L^p(\mathbb{R}^+; D(A))$ is dense in $L^p(\mathbb{R}^+; E)$. Given z in $L^p(\mathbb{R}^+; D(A))$, we indicate with $\tilde{z}$ its trivial extension to $\mathbb{R}$. Pick ω in $C_0^\infty(\mathbb{R}^+)$, such that $\int_\mathbb{R} \omega(y)dy = 1$ and set, for $k \in \mathbb{N}$, $\omega_k(y) := k\omega(ky)$. Then it is a standard fact that $\omega_k * \tilde{z}$ belongs to $W^{2,p}(\mathbb{R}; D(A))$ and converges to $\tilde{z}$ in $L^p(\mathbb{R}; D(A))$ as $k \to \infty$. We deduce that $W^{2,p}(\mathbb{R}^+; E) \cap L^p(\mathbb{R}^+; D(A))$ is dense in $L^p(\mathbb{R}^+; E)$. Let $z \in W^{2,p}(\mathbb{R}^+; E) \cap L^p(\mathbb{R}^+; D(A))$. We fix ϕ in $C^\infty([0, \infty))$, such that $\phi(y) = 1$ if $y \geq 2$, $\phi(y) = 0$ if $0 \leq y \leq 1$, Set, for $k \in \mathbb{N}$, $\phi_k(y) := \phi(ky)$. Then, $\forall k \in \mathbb{N}$, $\phi_k z \in W^{2,p}(\mathbb{R}^+; E) \cap L^p(\mathbb{R}^+; D(A)$, $\phi_k(0)z(0) = 0$ and $(\phi_k z)_{k\in\mathbb{N}}$ converges to z in $L^p(\mathbb{R}^+; E)$ as $k \to \infty$. We conclude that $\{u \in W^{2,p}(\mathbb{R}^+; E) \cap L^p(\mathbb{R}^+; D(A)) : u(0) = 0\}$ is dense in $L^p(\mathbb{R}^+; E)$.

Let now $(f, h) \in L^p(\mathbb{R}^+; E) \times (E, D(A))_{1/2-1/(2p),p}$. As $(E, D(A))_{1-1/(2p),p}$ is dense in $(E, D(A))_{\frac{1}{2}-\frac{1}{2p},p}$ (see [5, Theorem 3.4.2]), we take a sequence $(g_k)_{k\in\mathbb{N}}$ in $(E, D(A))_{1-1/(2p),p}$, such that

$$\|g_k - h\|_{(E,D(A))_{1/2-1/(2p),p}} \to 0 \quad (k \to \infty).$$

For each k, we take v_k in $W^{2,p}(\mathbb{R}^+; E) \cap L^p(\mathbb{R}^+; D(A))$, such that $v_k(0) = g_k$. As $\{u \in W^{2,p}(\mathbb{R}^+; E) \cap L^p(\mathbb{R}^+; D(A)) : u(0) = 0\}$ is dense in $L^p(\mathbb{R}^+; E)$, we may choose w_k in $W^{2,p}(\mathbb{R}^+; E) \cap L^p(\mathbb{R}^+; D(A))$, such that $w_k(0) = 0$ and

$\|f - v_k - w_k\|_{L^p(\mathbb{R}^+;E)} \to 0 \ (k \to \infty)$. We set $u_k := v_k + w_k$. Then $u_k \in W^{2,p}(\mathbb{R}^+; E) \cap L^p(\mathbb{R}^+; D(A))$ and

$$\|u_k - f\|_{L^p(\mathbb{R}^+;E)} + \|u_k(0) - h\|_{(E,D(A))_{1/2 - 1/(2p),p}} \to 0 \quad (k \to \infty). \qquad \square$$

9.4 Abstract Parabolic Problems

In this section we are going to study the abstract parabolic problem (9.8), with the following assumptions:

(C) E is a UMD Banach space with property (α). A is a closed, densely defined operator in E belonging to the class $O(\frac{\pi}{2})$ and $\{\lambda(\lambda + A)^{-1} : \lambda \in \Sigma_{\frac{\pi}{2}}\}$ is R-bounded in $\mathscr{L}(E)$. $-B$ is the infinitesimal generator of a strongly continuous semigroup in E, A and B commute in the sense of the resolvent and $D(A^{1/2}) \subseteq D(B)$. $\gamma \in \mathbb{R}^+$.

In order to study system (9.8), we shall need to study even the system

$$\begin{cases} D_t u(t, x) - D_x^2 u(t, x) + A u(t, x) = f(t, x), \ t \in (0, T), x \in \mathbb{R}^+, \\ u(t, 0) = g(t), & t \in (0, T), \\ u(0, x) = u_0(x), & x \in \mathbb{R}^+ \end{cases} \qquad (9.26)$$

considered also in [20], under slightly different assumptions. We shall look for solutions u in the class $W_p^{1,2,A}((0, T) \times \mathbb{R}^+; E)$, defined as follows: if $p \in [1, \infty]$,

$$W_p^{1,2,A}((0, T) \times \mathbb{R}^+; E)$$
$$:= \{u \in L^p((0, T) \times \mathbb{R}^+; D(A)) : D_t u, D_x u, D_x^2 u \in L^p((0, T) \times \mathbb{R}^+; E)\}. \qquad (9.27)$$

$W_p^{1,2,A}((0, T) \times \mathbb{R}^+; E)$ will be equipped with the natural norm

$$\|u\|_{W_p^{1,2,A}((0,T)\times\mathbb{R}^+;E)}$$
$$:= \|u\|_{L^p((0,T)\times\mathbb{R}^+;D(A))} + \|D_t u\|_{L^p((0,T)\times\mathbb{R}^+;E)} + \sum_{k=1}^2 \|D_x^k u\|_{L^p((0,T)\times\mathbb{R}^+;E)}. \qquad (9.28)$$

We observe that, as a consequence of Lemma 9.3, if $u \in W_p^{1,2,A}((0, T) \times \mathbb{R}^+; E)$, $D_x u \in L^p((0, T) \times \mathbb{R}^+; D(A^{1/2}))$.

If $u \in W_p^{1,2,A}((0, T) \times \mathbb{R}^+; E)$, we set

$$\tau u := u_{|x=0}. \qquad (9.29)$$

Now we are going to characterize the functions of the form $u(\cdot, 0)$, $u(0, \cdot)$, $D_x u(\cdot, 0)$, with $u \in W_p^{1,2,A}((0, T) \times \mathbb{R}^+; E)$.

Proposition 9.6 *Let $p \in (1, \infty)$ and let $A \in O(\phi_A)$, for some $\phi_A \in (0, \pi)$. Then:*

(I)

$$\{u(0, \cdot) : u \in W_p^{1,2,A}((0, T) \times \mathbb{R}^+; E)\}$$
$$= B_{p,p}^{2-2/p}(\mathbb{R}^+; E) \cap L^p(\mathbb{R}^+; (E, D(A))_{1-1/p,p});$$

(II) if $p \neq \frac{3}{2}$,

$$\{u(0, \cdot) : u \in W_p^{1,2,A}((0, T) \times \mathbb{R}^+; E), u(\cdot, 0) \equiv 0\}$$
$$= \begin{cases} B_{p,p}^{2-2/p}(\mathbb{R}^+; E) \cap L^p(\mathbb{R}^+; (E, D(A))_{1-1/p,p}) \\ \quad \text{if } 1 < p < \frac{3}{2}, \\ \{u_0 \in B_{p,p}^{2-2/p}(\mathbb{R}^+; E) \cap L^p(\mathbb{R}^+; (E, D(A))_{1-1/p,p}) \\ \quad : u_0(0) = 0\}, \text{ if } \frac{3}{2} < p < \infty; \end{cases}$$

(III)

$$\{u(\cdot, 0) : u \in W_p^{1,2,A}((0, T) \times \mathbb{R}^+; E)\}$$
$$= B_{p,p}^{1-1/(2p)}((0, T); E) \cap L^p((0, T); (E, D(A))_{1-1/(2p),p}).$$

Let $p \in (1, \infty) \setminus \{\frac{3}{2}\}$, $u_0 \in B_{p,p}^{2-2/p}(\mathbb{R}^+; E) \cap L^p(\mathbb{R}^+; (E, D(A))_{1-1/p,p})$, $g \in B_{p,p}^{1-1/(2p)}((0, T); E) \cap L^p((0, T); (E, D(A))_{1-1/(2p),p})$; then:
(IV) if $p < \frac{3}{2}$, there exists $u \in W_p^{1,2,A}((0, T) \times \mathbb{R}^+; E)$ such that $u(0, \cdot) = u_0$ and $u(\cdot, 0) = g$;
(V) if $p > \frac{3}{2}$, there exists $u \in W_p^{1,2,A}((0, T) \times \mathbb{R}^+; E)$ such that $u(0, \cdot) = u_0$ and $u(\cdot, 0) = g$ if and only if $u_0(0) = g(0)$;
(VI) $\{D_x u(\cdot, 0) : u \in W_p^{1,2,A}((0, T) \times \mathbb{R}^+; E)\}$

$$= W^{1/2-1/(2p),p}((0, T); E) \cap L^p((0, T); (E, D(A))_{1/2-1/(2p),p}).$$

Proof We postpone the proof to Sect. 9.6.   □

Remark 9.6 One should preliminarily observe (see the proof of Proposition 9.6 in Sect. 9.6) that, if $u \in W_p^{1,2,A}((0, T) \times \mathbb{R}^+; E)$ and $p > \frac{3}{2}$,

$$u \in C([0, T]; (L^p(\mathbb{R}^+; E), W^{2,p}(\mathbb{R}^+; E) \cap L^p(\mathbb{R}^+; D(A)))_{1-1/p,p})$$
$$\hookrightarrow C([0, T]; B^{2-2/p}(\mathbb{R}^+; E)).$$

So $u(0, \cdot) \in B^{2-2/p}(\mathbb{R}^+; E)) \hookrightarrow BC([0, \infty); E)$, while $u(\cdot, 0) \in B_{p,p}^{1-1/(2p)}((0, T); E) \hookrightarrow C([0, T]; E)$.

Corollary 9.4 *We assume that (A1)–(A2) hold, with $\phi_A \geq \frac{\pi}{2}$, and consider system (9.26). Let $p \in (1, \infty) \setminus \{\frac{3}{2}\}$. Then, the following conditions are necessary and sufficient, in order that (9.26) have a unique solution u in $W_p^{1,2,A}((0, T) \times \mathbb{R}^+; E)$:*

(I) $f \in L^p((0, T) \times \mathbb{R}^+; E)$;
(II) $g \in W^{1-1/(2p),p}((0, T); E) \cap L^p((0, T); (E, D(A))_{1-1/(2p),p})$;
(III) $u_0 \in B_{p,p}^{2-2/p}(\mathbb{R}^+; E) \cap L^p(\mathbb{R}^+; (E, D(A))_{1-1/p,p})$;
(IV) in case $p > \frac{3}{2}$, $g(0) = u_0(0)$.

Proof It follows from Proposition 9.6 that the conditions (I)–(IV) are necessary.

In order to show that they are sufficient, we start by observing that the case $g \equiv 0$ was essentially treated, in more general form, in [11, Theorem 3.2 and Proposition 3.1], replacing $\mathbb{R}^+$ with $(0, 1)$. So, we obtain that, in case $g \equiv 0$, we have a unique solution u in $W_p^{1,2,A}((0, T) \times \mathbb{R}^+; E))$ if and only if $f \in L^p((0, T) \times \mathbb{R}^+; E)$, $u_0 \in B_{p,p}^{2-2/p}(\mathbb{R}^+; E) \cap L^p(\mathbb{R}^+; (E, D(A))_{1-1/p,p})$ and $u_0(0) = 0$ if $p > \frac{3}{2}$.

In general, applying again Proposition 9.6, we may pick $v \in W_p^{1,2,A}((0, T) \times \mathbb{R}^+; E)$, such that $v(0, \cdot) = u_0$ and $v(\cdot, 0) = g$. Taking as new unknown $u - v$, we are reduced to the case $u_0 = 0$ and $g = 0$, replacing f with $f - D_t v + D_x^2 v - Av \in L^p((0, T) \times \mathbb{R}^+; E)$. □

Remark 9.7 A result quite similar to Corollary 9.4 (with slightly different assumptions) was proved in [20, Theorem 4.1].

Now we study (9.8). We shall employ the following simple general fact:

Lemma 9.5 *Let G be the infinitesimal generator of the strongly continuous semigroup $(e^{tG})_{t \geq 0}$ in the Banach space E and let $u \in W^{1,1}((0, T); E) \cap L^1((0, T); D(G))$. Then:*

$$u(t) = e^{tG} u(0) + \int_0^t e^{(t-s)G}(u'(s) - Gu(s)) ds$$

almost everywhere in $(0, T)$.

Proof See [19, Section 4.2]. □

Now we consider the operator G defined in (9.7). By Remark 9.5, if $h \in (E, D(A))_{1/2-1/(2p),p}$, the second component g of $(\lambda - G)^{-1}(0, h)$ coincides with $[\lambda + \gamma(\lambda + A)^{1/2} + B]^{-1}h$. So the second component of $e^{tG}(0, g)$ coincides with $K(t)g$, with

$$K(t) = \frac{1}{2\pi i} \int_\Gamma e^{\lambda t} [\lambda + \gamma(\lambda + A)^{1/2} + B]^{-1} d\lambda, \qquad (9.30)$$

with Γ piecewise regular path, contained in the resolvent of G, connecting $\infty e^{-i\theta_0}$ to $\infty e^{i\theta_0}$, for some $\theta_0 \in (\frac{\pi}{2}, \pi)$. K is analytic from $\mathbb{R}^+$ to $\mathscr{L}((E, D(A))_{1/2-1/(2p),p}; (E, D(A))_{1-1/(2p),p})$, and, if $0 < t \leq T$,

$$t\|K(t)\|_{\mathscr{L}((E,D(A))_{1/2-1/(2p),p},(E,D(A))_{1-1/(2p),p})} + \|K(t)\|_{\mathscr{L}((E,D(A))_{1/2-1/(2p),p}} \leq C(T).$$

Corollary 9.5 *Assume that (C) holds.*

(I) Let $u \in W_p^{1,2,A}((0,T) \times \mathbb{R}^+; E)$ be such that $\tau u = u(\cdot, 0) \in W^{1,p}((0,T);$ $(E, D(A))_{1/2-1/(2p),p}) \cap L^p((0,T); (E, D(A))_{1-1/(2p),p})$, for some $p \in (1, \infty)$. We set

$$h(t) := D_t u(t, 0) - \gamma D_x u(t, 0) + Bu(t, 0), \quad t \in (0, T).$$

Then, if $u(0, \cdot) = 0$, $\tau u(0) = 0$ and $D_t u - D_x^2 u + Au \equiv 0$ in $(0, T) \times \mathbb{R}^+$, u coincides with the solution in $W_p^{1,2,A}((0,T); E)$ of (9.26) with $f \equiv 0$, $u_0 = 0$ and

$$g(t) = \int_0^t K(t - s)h(s)ds. \tag{9.31}$$

(II) If $h \in L^p((0,T); (E, D(A))_{1/2-1/(2p),p})$ is such that g (defined in (9.31)), belongs to $W^{1,p}((0,T); (E, D(A))_{1/2-1/(2p),p}) \cap L^p((0,T); (E, D(A))_{1-1/(2p),p})$, and u is the solution in $W_p^{1,2,A}((0,T); E)$ of (9.26), with $f \equiv 0$ and $u_0 = 0$, then u solves also (9.8), again with $f \equiv 0$ and $u_0 = 0$.

Proof (I) This follows from the fact that, by Lemma 9.5, $u(t, 0)$ coincides with (9.31).

(II) We fix a sequence $(\omega_k)_{k \in \mathbb{N}}$ in $\mathscr{D}((0, T))$, such that $\int_0^T \omega_k(s)ds = 1 \,\forall k \in \mathbb{N}$ and $\omega_k(t) = 0$ if $t \geq k^{-1}$, and set $h_k := \omega_k * h$. Then

$$h_k \in C^1([0, T]; (E, D(A))_{1/2-1/(2p),p}).$$

We consider the system (9.8) taking $f \equiv 0$, $u_0 = 0$ and replacing h with h_k. Then, by Corollary 9.3 and well known properties of analytic semigroups, there is a unique solution u_k in $C^1([0, T]; L^p(\mathbb{R}^+; E)) \cap C([0, T]; W^{2,p}(\mathbb{R}^+; E) \cap L^p(\mathbb{R}^+; D(A)))$, with $\tau u_k \in C^1([0, T]; (E, D(A))_{1/2-1/(2p),p}) \cap C([0, T]; (E, D(A))_{1-1/(2p),p})$. Moreover,

$$g_k(t) := u_k(t, 0) = \int_0^t K(t - s)h_k(s)ds, \quad t \in [0, T].$$

Obviously, u_k is also the solution of (9.26), taking $f \equiv 0$, $u_0 = 0$ and replacing g with g_k. Now, we observe that $g_k = K * h_k = K * (\omega_k * h) = \omega_k * g$, and

$D_t g_k = \omega_k * D_t g$, because $g(0) = 0$. We deduce that $(g_k)_{k \in \mathbb{N}}$ converges to g in $W^{1,p}((0,T); (E, D(A))_{1/2-1/(2p),p}) \cap L^p((0,T); (E, D(A))_{1-1/(2p),p})$, so that, by Corollary 9.4, the sequence $(u_k)_{k \in \mathbb{N}}$ converges to u in $W_p^{1,2,A}((0,T); E)$. So, employing Proposition 9.6(VI), we obtain

$$
\begin{aligned}
h_k &= D_t u_k(\cdot, 0) - \gamma D_x u_k(\cdot, 0) + B u_k(\cdot, 0) \\
&= D_t g_k - \gamma D_x u_k(\cdot, 0) + B g_k \to D_t g - \gamma D_x u(\cdot, 0) + B g \\
&= D_t u(\cdot, 0) - \gamma D_x u(\cdot, 0) + B u(\cdot, 0) \quad (k \to \infty)
\end{aligned}
$$

in $L^p((0,T); (E, D(A))_{1/2-1/(2p),p})$. But, as $h_k \to h$ $(k \to \infty)$ in the same space, we deduce

$$
D_t u(\cdot, 0) - \gamma D_x u(\cdot, 0) + B u(\cdot, 0) \equiv h. \qquad \square
$$

Lemma 9.6 *Assume that (C) is satisfied. Let* $p \in [1, \infty)$, $\theta \in (0, \frac{1}{2})$ *and let* g *be defined as in (9.31). Then, if* $h \in L^p((0,T); (E, D(A))_{\theta,p})$, $g \in W^{1,p}((0,T); (E, D(A))_{\theta,p}) \cap L^p((0,T); (E, D(A))_{\theta+1/2,p})$,

Proof We start by considering the problem

$$
\begin{cases}
D_t w(t) + (A^{1/2} + B) w(t) = h(t), \ t \in (0,T), \\
w(0) = 0.
\end{cases}
\tag{9.32}
$$

By Corollary 9.2, for every $\beta \in (0, \frac{1}{2})$, the part of $-(\gamma A^{1/2} + B)$ in $(E, D(A))_{\beta,p}$, with domain $(E, D(A))_{1/2+\beta,p}$, is the infinitesimal generator of an analytic semigroup in $(E, D(A))_{\beta,p}$. By the reiteration theorem, if $0 < \beta < \theta < \frac{1}{2}$,

$$
(E, D(A))_{\theta,p} = ((E, D(A))_{\beta,p}, (E, D(A))_{\beta+1/2,p})_{2(\theta-\beta),p}.
$$

So, by Proposition 9.2, if $h \in L^p((0,T); (E, D(A))_{\theta,p})$, (9.32) has a unique solution w in $W^{1,p}((0,T); (E, D(A))_{\theta,p}) \cap L^p((0,T); (E, D(A))_{1/2+\theta,p})$, which can be represented in the form

$$
w(t) = \int_0^t K_2(t-s) h(s) ds,
$$

with

$$
K_2(t) = \frac{1}{2\pi i} \int_\Gamma e^{\lambda t} (\lambda + \gamma A^{1/2} + B)^{-1} d\lambda.
$$

Now we observe that

$$
\begin{aligned}
& [\lambda + \gamma(\lambda + A)^{1/2} + B]^{-1} \\
= {} & (\lambda + \gamma A^{1/2} + B)^{-1}(\lambda + \gamma A^{1/2} + B)[\lambda + \gamma(\lambda + A)^{1/2} + B]^{-1} \\
= {} & (\lambda + \gamma A^{1/2} + B)^{-1} + (\lambda + \gamma A^{1/2} + B)^{-1} \\
& \times \gamma[A^{1/2} - (\lambda + A)^{1/2}][\lambda + \gamma(\lambda + A)^{1/2} + B]^{-1}.
\end{aligned}
\tag{9.33}
$$

We set

$$
K_3(t) := \frac{1}{2\pi i} \int_\Gamma e^{\lambda t} \gamma[A^{1/2} - (\lambda + A)^{1/2}][\lambda + \gamma(\lambda + A)^{1/2} + B]^{-1} d\lambda.
$$

From

$$
\|\gamma[A^{1/2} - (\lambda + A)^{1/2}][\lambda + \gamma(\lambda + A)^{1/2} + B]^{-1}\|_{\mathscr{L}((E,D(A))_{\theta,p}} \le C|\lambda|^{-1/2}
$$

we obtain, for $t \in (0, T)$,

$$
\|K_3(t)\|_{\mathscr{L}((E,D(A))_{\theta,p})} \le C t^{-1/2},
$$

and $K_3 \in L^1((0, T); \mathscr{L}((E, D(A))_{\theta,p}))$. We immediately deduce that the convolution operator $h \to K_3 * h$ maps $L^p((0, T); (E, D(A))_{\theta,p})$ into itself. By (9.33), we have

$$
g(t) = \int_0^t K_2(t - s)h(s)ds + \int_0^t K_2(t - s)(K_3 * h)(s)ds,
$$

which implies the statement.   □

Now we are able to prove the following

Theorem 9.6 *We consider the Cauchy problem (9.8), with assumption (C). Let $p \in (1, \infty) \setminus \{\frac{3}{2}\}$. Then:*

(I) the following conditions are necessary and sufficient in order that (9.8) have a solution u in $W_p^{1,2,A}((0, T) \times \mathbb{R}^+; E)$, with

$$
u(\cdot, 0) \in W^{1,p}((0, T); (E, D(A))_{1/2-1/(2p),p}) \cap L^p((0, T); (E, D(A))_{1-1/(2p),p}):
$$

(a) $f \in L^p((0, T) \times \mathbb{R}^+; E)$;

(b) $h \in L^p((0, T); (E, D(A))_{1/2-1/(2p),p})$;

(c) $u_0 \in B_{p,p}^{2-2/p}(\mathbb{R}^+; E) \cap L^p(\mathbb{R}^+; (E, D(A))_{1-1/p,p})$, and, in case $p > \frac{3}{2}$, $u_0(0) \in (E, D(A))_{1-1/p,p}$.

(II) If $p > \frac{3}{2}$ the solution is unique;

(III) in case $1 < p < \frac{3}{2}$, the solution is not unique: more precisely, for each g_0 in $(E, D(A))_{1-1/p,p}$, (9.8) has a unique solution u such that $\tau u(0) = g_0$.

Proof We begin with uniqueness. We start by observing that, owing to Corollary 9.3 and Lemma 9.5, if $f \equiv 0$, $g \equiv 0$, $u_0 \equiv 0$, the only solution u with the declared regularity such that $u(\cdot, 0)(0) = 0$ is the trivial one. In case $p > \frac{3}{2}$, by Proposition 9.6 (V), if $u_0 \equiv 0$, $u(\cdot, 0)(0) = 0$, so that we have uniqueness in a full sense.

Concerning the existence, the necessity of conditions (a)-(b) and of the belonging of u_0 to $B^{2-2/p,p}(\mathbb{R}^+; E) \cap L^p(\mathbb{R}^+; (E, D(A))_{1-1/p,p})$ follow from Proposition 9.6. Assume $p > \frac{3}{2}$. If a solution u with the prescribed regularity exists, we set $g := \tau u$. As $g \in W^{1,p}((0, T); (E, D(A))_{1/2-1/(2p),p}) \cap L^p((0, T); (E, D(A))_{1-1/(2p),p})$, by Theorem 9.2 $g(0)$ belongs to

$$((E, D(A))_{1/2-1/(2p),p}, (E, D(A))_{1-1/(2p),p})_{1-1/p,p} = (E, D(A))_{1-1/p,p}.$$

So the belonging of $u_0(0)$ to $(E, D(A))_{1-1/p,p}$ follows from Proposition 9.6 (V).

Now we assume that (a)–(c) are satisfied. In case $p > \frac{3}{2}$, we set $g_0 := u_0(0)$, while, if $1 < p < \frac{3}{2}$, we fix it arbitrarily in $(E, D(A))_{1-1/p,p}$. We shall show there there exists a solution of (9.8), such that $u(\cdot, 0)(0) = g_0$. We start by fixing v in $W^{1,p}((0, T); (E, D(A))_{1/2-1/(2p),p}) \cap L^p((0, T); (E, D(A))_{1-1/(2p),p})$, such that $v(0) = g_0$. As

$$W^{1,p}((0, T); (E, D(A))_{1/2-1/(2p),p}) \cap L^p((0, T); (E, D(A))_{1-1/(2p),p})$$
$$\hookrightarrow W^{1-1/(2p),p}((0, T); E) \cap L^p((0, T); (E, D(A))_{1-1/(2p),p}),$$

we can say that, in force of Corollary 9.4, the problem

$$\begin{cases} D_t U(t, x) - D_x^2 U(t, x) + A U(t, x) = f(t, x), \ t \in (0, T), x \in \mathbb{R}^+, \\ U(t, 0) = v(t), & t \in (0, T), \\ U(0, x) = u_0(x), & x \in \mathbb{R}^+ \end{cases} \quad (9.34)$$

has a unique solution U in $W_p^{1,2,A}((0, T) \times \mathbb{R}^+; E)$. Taking $z := u - U$ as new unknown, we are reduced to the problem

$$\begin{cases} D_t z(t, x) - D_x^2 z(t, x) + A z(t, x) = 0, & t \in (0, T), x \in \mathbb{R}^+, \\ D_t z(t, 0) - \gamma D_x z(t, 0) + B z(t, 0) = h_0(t), \ t \in (0, T), \\ z(0, x) = 0, & x \in \mathbb{R}^+, \end{cases} \quad (9.35)$$

with

$$h_0(t) := h(t) - D_t v(t) + \gamma D_x U(t, 0) - B v(t).$$

We set

$$g(t) := \int_0^t K(t-s)h_0(s)ds.$$

Then, by Lemma 9.6,

$$g \in W^{1,p}((0,T);(E,D(A))_{1/2-1/(2p),p}) \cap L^p((0,T);(E,D(A))_{1-1/(2p),p}).$$

So, by Corollary 9.5(II), if we indicate with z the solution in $W_p^{1,2,A}((0,T);E)$ of (9.26), with $f \equiv 0$ and $u_0 = 0$, we can say that z is a solution to (9.35). □

Remark 9.8 In case $p < \frac{3}{2}$, if $u \in W_p^{1,2,A}((0,T) \times \mathbb{R}^+; E)$, $u(0,0)$ is usually not defined. However, if we assume that $u(\cdot,0) \in W^{1,p}((0,T);(E,D(A))_{1/2-1/(2p),p}) \cap L^p((0,T);(E,D(A))_{1-1/(2p),p})$, $u(\cdot,0)(0)$ is certainly defined. So in this case of relatively low regularity, we are able to prescribe the initial values of the solution and of its trace in independent way.

9.5 Examples and Applications

We begin by considering simple problems of the form

$$\begin{cases} \lambda u(x,y) - D_x^2 u(x,y) + A(D_y)u(x,y) = f(x,y), \\ (x,y) \in \mathbb{R}^+ \times \mathbb{R}^{n-1}, \\ \lambda u(0,y) - \gamma D_x u(0,y) + B(D_y)u(0,y) = h(y), \quad y \in \mathbb{R}^{n-1}, \end{cases} \tag{9.36}$$

and

$$\begin{cases} D_t u(t,x,y) - D_x^2 u(t,x,y) + A(D_y)u(t,x,y) = f(t,x,y), \\ (t,x,y) \in (0,T) \times \mathbb{R}^+ \times \mathbb{R}^{n-1}, \\ D_t u(t,0,y) - \gamma D_x u(t,0,y) + B(D_y)u(t,0,y) = h(t,y), \quad t \in (0,T), y \in \mathbb{R}^{n-1}, \\ u(0,x,y) = u_0(x,y), \quad (x,y) \in \mathbb{R}^+ \times \mathbb{R}^{n-1}. \end{cases} \tag{9.37}$$

We assume that the following conditions hold:

(E1) $m \in \mathbb{N}$; $A(D_y) = \sum_{|\alpha| \le 2m} a_\alpha D_y^\alpha$; $B(D_y) = \sum_{|\alpha| \le m} b_\alpha D_y^\alpha$;

(E2) $Re\{\sum_{|\alpha|=2m} a_\alpha(i\xi)^\alpha\} > 0$, $\forall \xi \in \mathbb{R}^n \setminus \{0\}$;

(E3) *for some* $p \in (1,\infty)$, *we set* $D(B_0) := \{g \in L^p(\mathbb{R}^{n-1}) : B(D_y)g \in L^p(\mathbb{R}^{n-1})\}$, $B_0 g := B(D_y)g$ *(in the sense of distributions). Then* $-B_0$ *is the infinitesimal generator of a strongly continuous semigroup in* $L^p(\mathbb{R}^{n-1})$.

Remark 9.9 Examples of differential operators satisfying (E3) are first order differential operator $B(D_y) = v \cdot \nabla_y$, with $v \in \mathbb{R}^{n-1}$. In this case $-B$ generates the group of translations $(e^{-tB})_{t \geq 0}$ with $e^{-tB} g = g(\cdot - tv)$.

Other examples are strongly elliptic operators in the form $B(D_y) = \sum_{|\beta| \leq k} b_\beta D_y^\beta$, with $k \leq m$, $Re\{\sum_{|\beta|=k} b_\beta (i\xi)^\beta\} > 0$, $\forall \xi \in \mathbb{R}^{n-1} \setminus \{0\}$. In this case, $D(B) = W^{k,p}(\mathbb{R}^{n-1})$ and the semigroup is analytic.

Finally, it is not difficult to see that, if $B_1(D_y)$ and $B_2(D_y)$ satisfies (E3), the same holds for $B(D_y) = B_1(D_y) + B_2(D_y)$ and $e^{-tB} = e^{-tB_1} e^{-tB_2}$, $\forall t \geq 0$.

We set

$$E := L^p(\mathbb{R}^{n-1}), \tag{9.38}$$

and introduce the following operator A_0:

$$\begin{cases} \quad D(A_0) := W^{2m,p}(\mathbb{R}^{n-1}), \\ A_0 u(y) := A(D_y)u(y), \quad y \in \mathbb{R}^{n-1}. \end{cases} \tag{9.39}$$

It is known (see, for example, [23, Chapter 3.7]) that there exists $\lambda_0 \geq 0$, such that, if we set

$$A := A_0 + \lambda_0, \tag{9.40}$$

$\Sigma_{\pi/2} \subseteq \rho(-A)$; moreover, $\{\lambda(\lambda + A)^{-1} : \lambda \in \Sigma_{\pi/2}\}$ is R-bounded in $\mathcal{L}(E)$ (see [8, Chapter II]). Moreover, by Proposition 13.11 in [17], A has equibounded purely imaginary powers. This implies (see [24, Section 1.15.3]) that, for each $\alpha \in (0, 1)$, $D(A^\alpha)$ coincides with the complex interpolation space $[E, D(A_0)]_\alpha$. So

$$D(A^{1/2}) = H^{m,p}(\mathbb{R}^{n-1}) = W^{m,p}(\mathbb{R}^{n-1}). \tag{9.41}$$

Therefore, replacing λ with $\lambda - \lambda_0$, $A(D_y)$ with $A(D_y) + \lambda_0$, $B(D_y)$ with $B(D_y) + \lambda_0$ and applying Theorems 9.5 and 9.6, we are able to prove the following Theorems 9.7 and 9.8:

Theorem 9.7 *Consider problem (9.36), with the assumptions (E1)-(E3); let* $p \in (1, \infty)$. *Then, there exist* M, ω *in* $\mathbb{R}^+$, *such that, if* $Re(\lambda) \geq 0$ *and* $|\lambda| \geq \omega$, $f \in L^p(\mathbb{R}^+ \times \mathbb{R}^{n-1})$, $h \in B_{p,p}^{m(1-\frac{1}{p})}(\mathbb{R}^{n-1})$, *(9.36) has a unique solution* u *in* $W^{2,p}(\mathbb{R}^+;$ $L^p(\mathbb{R}^{n-1})) \cap L^p(\mathbb{R}^+; W^{2m,p}(\mathbb{R}^{n-1}))$.

u belongs also to $W^{1,p}(\mathbb{R}^+; W^{m,p}(\mathbb{R}^{n-1}))$; *moreover, if* i, j, k *are nonnegative integers, such that* $i + j + k = 2$,

$$|\lambda|^{i/2}\|D_x^k u\|_{L^p(\mathbb{R}^+; W^{jm,p}(\mathbb{R}^{n-1}))} \leq M(\|f\|_{L^p(\mathbb{R}^+\times\mathbb{R}^{n-1})} + \|h\|_{B_{p,p}^{m(1-\frac{1}{p})}(\mathbb{R}^{n-1})}).$$

Theorem 9.8 *Consider problem (9.37), with the assumptions (E1)–(E3), and* $p \in (1, \infty) \setminus \{\frac{3}{2}\}$. *Then*

(I) *the following conditions are necessary and sufficient, in order that (9.37) have a solution* u *in* $W^{1,p}((0,T); L^p(\mathbb{R}^+\times\mathbb{R}^{n-1})) \cap L^p((0,T); W^{2,p}(\mathbb{R}^+; L^p(\mathbb{R}^{n-1})))$
$\cap L^p((0,T)\times\mathbb{R}^+; W^{2m,p}(\mathbb{R}^{n-1}))$, *with* $u_{|x=0} = \tau u \in W^{1,p}((0,T); B_{p,p}^{m(1-1/p)}(\mathbb{R}^{n-1})) \cap L^p((0,T); B_{p,p}^{m(2-1/p)}(\mathbb{R}^{n-1}))$:

    (a) $f \in L^p((0,T)\times\mathbb{R}^+\times\mathbb{R}^{n-1})$;

    (b) $h \in L^p((0,T); B_{p,p}^{m(1-1/p)}(\mathbb{R}^{n-1}))$;

    (c) $u_0 \in B_{p,p}^{2-2/p}(\mathbb{R}^+; L^p(\mathbb{R}^{n-1})) \cap L^p(\mathbb{R}^+; B_{p,p}^{2m(1-1/p)}(\mathbb{R}^{n-1}))$ *and, in case* $p > \frac{3}{2}$, $u_{0|x=0} \in B_{p,p}^{2m(1-1/p)}(\mathbb{R}^{n-1})$.

(II) *If* $p > \frac{3}{2}$, *the solution with the declared regularity is unique;*

(II) *in case* $p < \frac{3}{2}$, *the solution is not unique:* $\forall g_0 \in B_{p,p}^{2m(1-1/p)}(\mathbb{R}^{n-1})$ *there exists a unique solution such that* $\tau u(0) = g_0$.

Remark 9.10 The belonging of u to $W^{1,p}((0,T); L^p(\mathbb{R}^+\times\mathbb{R}^{n-1})) \cap L^p((0,T); W^{2,p}(\mathbb{R}^+; L^p(\mathbb{R}^{n-1}))) \cap L^p((0,T)\times\mathbb{R}^+; W^{2m,p}(\mathbb{R}^{n-1}))$ is equivalent to the fact that u, together with $D_t u$, $D_x^\alpha u$ ($\alpha \leq 2$), $D_y^\beta u$ ($|\beta| \leq 2m$), belongs to $L^p((0,T)\times\mathbb{R}^+\times\mathbb{R}^{n-1})$.

In particular, we get the following

Corollary 9.6 *Consider the problem*

$$\begin{cases} \lambda u(x,y) - D_x^2 u(x,y) - \Delta_y u(x,y) + A_1(D_x, D_y)u(x,y) = f(x,y), \\ (x,y) \in \mathbb{R}^+\times\mathbb{R}^{n-1}, \\ \lambda u(0,y) - \gamma D_x u(0,y) + v \cdot \nabla_y u(0,y) + c_0 u(0,y) = g(y), \qquad y \in \mathbb{R}^{n-1} \end{cases}$$
$$\tag{9.42}$$

with $\gamma \in \mathbb{R}^+$, $v \in \mathbb{R}^{n-1}$, $A_1(D_x, D_y)$ *first order differential operator with constant coefficients,* $c_0 \in \mathbb{C}$; *let* $p \in (1, \infty)$. *Then, there exist* M, ω *in* $\mathbb{R}^+$, *such that, if* $Re(\lambda) \geq 0$ *and* $|\lambda| \geq \omega$, $f \in L^p(\mathbb{R}^+\times\mathbb{R}^{n-1})$, $g \in B_{p,p}^{1-\frac{1}{p}}(\mathbb{R}^{n-1})$, *(9.42) has a unique solution* u *in* $W^{2,p}(\mathbb{R}^+\times\mathbb{R}^{n-1})$. *Moreover, if* i, j, k *are nonnegative integers, such that* $i + j + k = 2$,

$$|\lambda|^{i/2}\|D_x^k u\|_{W^{j,p}(\mathbb{R}^+\times\mathbb{R}^{n-1})} \leq M(\|f\|_{L^p(\mathbb{R}^+\times\mathbb{R}^{n-1})} + \|h\|_{B_{p,p}^{1-\frac{1}{p}}(\mathbb{R}^{n-1})}).$$

This is a particular case of the estimates in a half-space contained in [16]. Moreover, we have:

Corollary 9.7 *Consider the system*

$$\begin{cases} D_t u(t,x,y) - D_x^2 u(t,x,y) - \Delta_y u(t,x,y) + A_1(D_x, D_y)u(t,x,y) = f(t,x,y), \\ (t,x,y) \in (0,T) \times \mathbb{R}^+ \times \mathbb{R}^{n-1}, \\ D_t u(t,0,y) - \gamma D_x u(t,0,y) + v \cdot \nabla_y u(t,0,y) = g(t,y), \quad (t,y) \in (0,T) \times \mathbb{R}^{n-1}, \\ u(0,x,y) = u_0(x,y), \quad (x,y) \in \mathbb{R}^+ \times \mathbb{R}^{n-1}. \end{cases}$$

$$(9.43)$$

Let $p \in (1,\infty) \setminus \{\frac{3}{2}\}$. Then

(I) the following conditions are necessary and sufficient, in order that (9.43) have a solution u in

$$W^{1,p}((0,T); L^p(\mathbb{R}^+ \times \mathbb{R}^{n-1})) \cap L^p((0,T); W^{2,p}(\mathbb{R}^+; L^p(\mathbb{R}^+ \times \mathbb{R}^{n-1})),$$

with $u|_{x=0} \in W^{1,p}((0,T); B_{p,p}^{1-1/p}(\mathbb{R}^{n-1})) \cap L^p((0,T); B_{p,p}^{2-1/p}(\mathbb{R}^{n-1}))$:

(a) $f \in L^p((0,T) \times \mathbb{R}^+ \times \mathbb{R}^{n-1})$;

(b) $h \in L^p((0,T); W^{1-1/p,p}(\mathbb{R}^{n-1}))$;

(c) $u_0 \in B_{p,p}^{2-2/p}(\mathbb{R}^+ \times \mathbb{R}^{n-1}))$ and, in case $p > \frac{3}{2}$, $u_{0|x=0} \in B_{p,p}^{2-2/p}(\mathbb{R}^{n-1})$.

(II) If $p > \frac{3}{2}$, the solution with the declared regularity is unique;

(II) in case $p < \frac{3}{2}$, the solution is not unique: $\forall g_0 \in B_{p,p}^{2-2/p}(\mathbb{R}^{n-1})$ there exists a unique solution such that $\tau u(0,\cdot) = g_0$.

Proof Applying Theorem 9.8, we have only to show that

$$B_{p,p}^{2-2/p}(\mathbb{R}^+; L^p(\mathbb{R}^{n-1})) \cap L^p(\mathbb{R}^+, B_{p,p}^{2-2/p}(\mathbb{R}^{n-1})) = B_{p,p}^{2-2/p}(\mathbb{R}^+ \times \mathbb{R}^{n-1}).$$

In fact, employing Proposition 9.1, Theorem 9.2, Proposition 9.3 and the following (9.59),

$$B_{p,p}^{2-2/p}(\mathbb{R}^+; L^p(\mathbb{R}^{n-1})) \cap L^p(\mathbb{R}^+, B_{p,p}^{2-2/p}(\mathbb{R}^{n-1}))$$
$$= (L^p(\mathbb{R}^+; E), W^{2,p}(\mathbb{R}^+; E))_{1-1/p,p} \cap (L^p(\mathbb{R}^+; E), L^p(\mathbb{R}^+, D(A)))_{1-1/p,p}$$
$$= (L^p(\mathbb{R}^+; E), W^{2,p}(\mathbb{R}^+; E) \cap L^p(\mathbb{R}^+, D(A)))_{1-1/p,p}.$$

As a consequence of Lemma 9.3, we have $W^{2,p}(\mathbb{R}^+, L^p(\mathbb{R}^{n-1})) \cap L^p(\mathbb{R}^+; D(A)) = W^{2,p}(\mathbb{R}^+ \times \mathbb{R}^{n-1})$. So

$$(L^p(\mathbb{R}^+; E), W^{2,p}(\mathbb{R}^+; E) \cap L^p(\mathbb{R}^+, D(A)))_{1-1/p,p}$$
$$= (L^p(\mathbb{R}^+ \times \mathbb{R}^{n-1}), W^{2,p}(\mathbb{R}^+ \times \mathbb{R}^{n-1}))_{1-1/p,p}$$
$$= B_{p,p}^{2-2/p}(\mathbb{R}^+ \times \mathbb{R}^{n-1}).$$

$\square$

Corollaries 9.6 and 9.7 admit the following generalization:

Proposition 9.7 *Let* $A(D_x, D_y) = \sum_{|\alpha| \leq 2} a_\alpha D_x^{\alpha_1} D_y^{\alpha_2}$, *with* $a_\alpha \in \mathbb{R}$ *for every* $\alpha = (\alpha_1, \alpha_2) \in \mathbb{N}_0 \times \mathbb{N}_0^{n-1}$ *with* $|\alpha| = 2$ *and* $\sum_{|\alpha|=2} a_\alpha \xi^\alpha > 0 \ \forall \xi \in \mathbb{R}^n \setminus \{0\}$. *Let* $B(D_x, D_y) := -\gamma D_x + v \cdot \nabla_y + c_0$, *with* $\gamma \in \mathbb{R}^+$, $v \in \mathbb{R}^{n-1}$, $c_0 \in \mathbb{C}$.

(I) Consider the system

$$\begin{cases} \lambda u(x, y) - A(D_x, D_y)u(x, y) = f(x, y), & (x, y) \in \mathbb{R}^+ \times \mathbb{R}^{n-1}, \\ \lambda u(0, y) + B(D_x)u(0, y) = g(y), & y \in \mathbb{R}^{n-1}. \end{cases} \tag{9.44}$$

Let $p \in (1, \infty)$. *Then, there exist* M, ω *in* $\mathbb{R}^+$, *such that, if* $Re(\lambda) \geq 0$ *and* $|\lambda| \geq \omega$, $f \in L^p(\mathbb{R}^+ \times \mathbb{R}^{n-1})$, $g \in B_{p,p}^{1-\frac{1}{p}}(\mathbb{R}^{n-1})$, *(9.44) has a unique solution* u *in* $W^{2,p}(\mathbb{R}^+ \times \mathbb{R}^{n-1}))$. *Moreover, if* i, j, k *are nonnegative integers, such that* $i + j + k = 2$,

$$|\lambda|^{i/2} \|D_x^k u\|_{W^{j,p}(\mathbb{R}^+ \times \mathbb{R}^{n-1})} \leq M(\|f\|_{L^p(\mathbb{R}^+ \times \mathbb{R}^{n-1})} + \|h\|_{B_{p,p}^{1-\frac{1}{p}}(\mathbb{R}^{n-1})}).$$

(II) Consider the system

$$\begin{cases} D_t u(t, x, y) - A(D_x, D_y)u(t, x, y) = f(t, x, y), & (t, x, y) \in (0, T) \times \mathbb{R}^+ \times \mathbb{R}^{n-1}, \\ D_t u(t, 0, y) + B(D_x, D_y)u(t, 0, y) = g(t, y), & (t, y) \in (0, T) \times \mathbb{R}^{n-1}, \\ u(0, x, y) = u_0(x, y), & (x, y) \in \mathbb{R}^+ \times \mathbb{R}^{n-1}. \end{cases} \tag{9.45}$$

Let $p \in (1, \infty) \setminus \{\frac{3}{2}\}$. *Then the following conditions are necessary and sufficient, in order that (9.45) have a solution* u *in*

$$W^{1,p}((0, T); L^p(\mathbb{R}^+ \times \mathbb{R}^{n-1})) \cap L^p((0, T); W^{2,p}(\mathbb{R}^+; L^p(\mathbb{R}^+ \times \mathbb{R}^{n-1})),$$

with $u_{|x=0} \in W^{1,p}((0, T); B_{p,p}^{1-1/p}(\mathbb{R}^{n-1})) \cap L^p((0, T); B_{p,p}^{2-1/p}(\mathbb{R}^{n-1}))$:

(a) $f \in L^p((0, T) \times \mathbb{R}^+ \times \mathbb{R}^{n-1})$;
(b) $h \in L^p((0, T); W^{1-1/p,p}(\mathbb{R}^{n-1}))$;
(c) $u_0 \in B_{p,p}^{2-2/p}(\mathbb{R}^+ \times \mathbb{R}^{n-1}))$ *and, in case* $p > \frac{3}{2}$, $u_{0|x=0} \in B_{p,p}^{2-2/p}(\mathbb{R}^{n-1})$.

If $p > \frac{3}{2}$, *the solution with the declared regularity is unique; in case* $p < \frac{3}{2}$, *the solution is not unique:* $\forall g_0 \in B_{p,p}^{2-2/p}(\mathbb{R}^{n-1})$ *there exists a unique solution such that, if* $\tau u = u_{|x=0}$, $\tau u(0, \cdot) = g_0$.

Proof By [15, Section 4], there exists a linear automorphism H of $\mathbb{R}^n$ mapping $\mathbb{R}_+ \times \mathbb{R}^{n-1}$ into itself and such that $[A(D_x, D_y)u][H(\xi, \eta)] = \Delta_{\xi, \eta}(u \circ H)(\xi, \eta)$, $\forall (\xi, \eta) \in \mathbb{R}^+ \times \mathbb{R}^{n-1}$. Employing this change of variables, we are reduced to problems in the forms (9.42) and (9.43) and we can prove (I)–(II).   □

We pass to consider the elliptic problem depending on the complex parameter λ

$$\begin{cases} \lambda u(\xi) - A(\xi, D_\xi)u(\xi) = f(\xi), & \xi \in \Omega, \\ \lambda u(\xi') + B(\xi', D_\xi)u(\xi') = g(\xi'), & \xi' \in \partial\Omega, \end{cases} \tag{9.46}$$

with the following conditions:

(D1) Ω is an open bounded subset of $\mathbb{R}^n$, lying on one side of its boundary $\partial\Omega$, which is a submanifold of class C^2 of $\mathbb{R}^n$;

(D2) $A(\xi, D_\xi) = \sum_{|\alpha| \leq 2} a_\alpha(\xi) D_\xi^\alpha$, $a_\alpha \in C(\overline{\Omega})$ $\forall \alpha$ with $|\alpha| \leq 2$; if $|\alpha| = 2$, a_α is real valued and $\sum_{|\alpha|=2} a_\alpha(\xi)\eta^\alpha \geq N|\eta|^2$ for some $N \in \mathbb{R}^+$, $\forall \xi \in \overline{\Omega}$, $\forall \eta \in \mathbb{R}^n$;

(D3) $B(\xi', D_\xi) = \sum_{|\alpha| \leq 1} b_\alpha(\xi') D_\xi^\alpha$, $b_\alpha \in C^1(\partial\Omega)$ $\forall \alpha$ with $|\alpha| \leq 1$; if $|\alpha| = 1$, b_α is real valued and $\sum_{|\alpha|=1} b_\alpha(\xi')v(\xi')^\alpha < 0$ $\forall \xi' \in \partial\Omega$, where we have indicated with $v(\xi')$ the unit normal vector to $\partial\Omega$ in ξ' pointing inside Ω.

The following result holds:

Theorem 9.9 *Assume that (D1)–(D3) hold. Let $p \in (1, \infty)$. Then there exists $R \in \mathbb{R}^+$, such that, $\forall \lambda \in \mathbb{C}$ with $|\lambda| \geq R$ and $|Arg(\lambda)| \leq \frac{\pi}{2}$, $\forall f \in L^p(\Omega)$, $\forall g \in W^{1-1/p,p}(\partial\Omega)$, (9.46) has a unique solution in $W^{2,p}(\Omega)$. Moreover, there exists $C \in \mathbb{R}^+$, such that*

$$|\lambda| \|u\|_{L^p(\Omega)} + \|u\|_{W^{2,p}(\Omega)} \leq C(\|f\|_{L^p(\Omega)} + \|g\|_{W^{1-1/p,p}(\partial\Omega)}). \tag{9.47}$$

Proof We start with an a priori estimate. Assume that u solves (9.46) and vanishes outsides the neighborhood U of the point $\xi^0 \in \overline{\Omega}$. Employing a suitable change of variable χ and setting $v := u \circ \chi$, we are locally reduced to estimate v in $W^{2,p}(\mathbb{R}^+ \times \mathbb{R}^{n-1})$ such that

$$\begin{cases} \lambda v(x, y) - A(D_x, D_y)v(x, y) = A^\sharp(x, y, D_{x,y})v(x, y) + f^\sharp(x, y), \\ (x, y) \in \mathbb{R}^+ \times \mathbb{R}^{n-1}, \\ \lambda v(0, y) - \gamma D_x v(0, y) + v \cdot \nabla_y v(0, y) = B^\sharp(0, y, D_x, D_y)v(x, 0) + g^\sharp(y), \ y \in \mathbb{R}^{n-1}, \end{cases}$$

with $A(D_x, D_y)$ as in the statement of $\gamma \in \mathbb{R}^+$, $A^\sharp(x, y, D_{x,y}) = \sum_{|\alpha| \leq 2} a_\alpha^\sharp(x, y) D_{x,y}^\alpha$, $B^\sharp(x, y, D_{x,y}) = \sum_{|\alpha| \leq 1} b_\alpha^\sharp(x, y) D_{x,y}^\alpha$, $f^\sharp = f \circ \chi$, $g^\sharp = g \circ \chi(0, \cdot)$,

$$\sum_{|\alpha|=2} \|a_\alpha^\sharp\|_{L^\infty(\mathbb{R}^+ \times \mathbb{R}^{n-1})} + \sum_{|\alpha|=1} \|b_\alpha^\sharp\|_{L^\infty(\mathbb{R}^{n-1})} \leq \epsilon,$$

with $\epsilon \in \mathbb{R}^+$. We shall use the following simple inequality: if $f \in W^{\alpha,p}(\mathbb{R}^{n-1})$, with $0 < \alpha < 1$,

$$\begin{aligned} \|af\|_{W^{\alpha,p}(\mathbb{R}^{n-1})} &\leq \|a\|_{L^\infty(\mathbb{R}^{n-1})} \|f\|_{W^{\alpha,p}(\mathbb{R}^{n-1})} \\ &+ C(\|a\|_{L^\infty(\mathbb{R}^{n-1})} + \|\nabla a\|_{L^\infty(\mathbb{R}^{n-1})}) \|f\|_{L^p(\mathbb{R}^{n-1})}, \end{aligned}$$

for some $C \in \mathbb{R}^+$, independent of a and f (for inequalities of this type, see [21, Chapter 5.3.7]).

Employing Proposition 9.7 (I) and standard interpolation inequalities and trace theorems, if $|\lambda|$ is sufficiently large, we obtain, $\forall \sigma \in (0, 2)$,

$$
\begin{aligned}
&\sum_{j=0}^{2} |\lambda|^{\frac{2-j}{2}} \|v\|_{W^{j,p}(\mathbb{R}^+ \times \mathbb{R}^{n-1})} + |\lambda|^{\frac{2-\sigma}{2}} \|v\|_{W^{\sigma,p}(\mathbb{R}^+ \times \mathbb{R}^{n-1})} \\
&\leq C_0[(\|f^\sharp\|_{L^p(\mathbb{R}^+ \times \mathbb{R}^{n-1})} + \|g^\sharp\|_{W^{1-1/p,p}(\mathbb{R}^{n-1})} \\
&+ \|A^\sharp(\cdot, \cdot, D_{x,y})v\|_{L^p(\mathbb{R}^+ \times \mathbb{R}^{n-1})} + \|B^\sharp(0, \cdot, D_{x,y})v(0, \cdot)\|_{W^{1-1/p,p}(\mathbb{R}^{n-1})}) \\
&\leq C_1[\|f^\sharp\|_{L^p(\mathbb{R}^+ \times \mathbb{R}^{n-1})} + \|g^\sharp\|_{W^{1-1/p,p}(\mathbb{R}^{n-1})} \\
&+ \epsilon(\|v\|_{W^{2,p}(\mathbb{R}^+ \times \mathbb{R}^{n-1})} + \|D_x v(0, \cdot)\|_{W^{1-1/p,p}(\mathbb{R}^{n-1})} \\
&+ \sum_{j=1}^{n-1} \|D_{y_j} v(0, \cdot)\|_{W^{1-1/p,p}(\mathbb{R}^{n-1})}) + \|D_x v(0, \cdot)\|_{L^p(\mathbb{R}^{n-1})} \\
&+ \sum_{j=1}^{n-1} \|D_{y_j} v(0, \cdot)\|_{L^p(\mathbb{R}^{n-1})} + \|v\|_{W^{1,p}(\mathbb{R}^+ \times \mathbb{R}^{n-1})}] \\
&\leq C_2[\|f^\sharp\|_{L^p(\mathbb{R}^+ \times \mathbb{R}^{n-1})} + \|g^\sharp\|_{W^{1-1/p,p}(\mathbb{R}^{n-1})} \\
&+ \epsilon\|v\|_{W^{2,p}(\mathbb{R}^+ \times \mathbb{R}^{n-1})} + \|v\|_{W^{\sigma,p}(\mathbb{R}^+ \times \mathbb{R}^{n-1})}]
\end{aligned}
\tag{9.48}
$$

for some $\sigma < 2$. If we take U so small that $C_2\epsilon \leq \frac{1}{2}$, and $|\lambda|$ sufficiently large, from (9.48) we deduce

$$
\sum_{j=0}^{2} |\lambda|^{\frac{2-j}{2}} \|v\|_{W^{j,p}(\mathbb{R}^+ \times \mathbb{R}^{n-1})} \leq C_1[(\|f^\sharp\|_{L^p(\mathbb{R}^+ \times \mathbb{R}^{n-1})} + \|g^\sharp\|_{W^{1-1/p,p}(\mathbb{R}^{n-1})})
$$

and so

$$
\sum_{j=0}^{2} |\lambda|^{\frac{2-j}{2}} \|u\|_{W^{j,p}(\Omega)} \leq C_1[(\|f\|_{L^p(\Omega)} + \|g\|_{W^{1-1/p,p}(\partial\Omega)}).
\tag{9.49}
$$

More generally, we fix a suitable partition of unity $\{\phi_k : k \in \{1, \ldots, N\}\}$ in $\overline{\Omega}$, in such a way that (9.49) is applicable to $\phi_k u$ for each $k \in \{1, \ldots, N\}$. $\phi_k u$ is such that

$$
\begin{cases}
\lambda(\phi_k u)(\xi) - A(\xi, D_\xi)(\phi_k u)(\xi) = \phi_k(\xi)f(\xi) + A_k(\xi, D_\xi)u(\xi), & \xi \in \Omega, \\
\lambda(\phi_k u)(\xi') + B(\xi', D_\xi)(\phi_k u)(\xi') = \phi_k(\xi')g(\xi') + c_k(\xi')u(\xi'), & \xi' \in \partial\Omega,
\end{cases}
$$

with $A_k(\xi, D_\xi)$ differential operator of order one. From (9.49) we deduce

$$
\begin{aligned}
\sum_{j=0}^{2} |\lambda|^{\frac{2-j}{2}} \|u\|_{W^{j,p}(\Omega)} &\leq \sum_{j=0}^{2} \sum_{k=1}^{N} |\lambda|^{\frac{2-j}{2}} \|\phi_k u\|_{W^{j,p}(\Omega)} \\
&\leq C_0(\|f\|_{L^p(\Omega)} + \|g\|_{W^{1-1/p,p}(\partial\Omega)} + \|u\|_{W^{1,p}(\Omega)}),
\end{aligned}
$$

which implies that (9.49) is valid without restrictions on the support, if $|\lambda|$ is sufficiently large.

It remains to show the existence of a solution to (9.46). First of all, it is well known that there exists $\lambda_0 \in \mathbb{C}$, such that the problem

$$\begin{cases} \lambda_0 u(\xi) - A(\xi, D_\xi)u(\xi) = f(\xi), & \xi \in \Omega, \\ B(\xi', D_\xi)u(\xi') = g(\xi'), & \xi' \in \partial\Omega \end{cases} \tag{9.50}$$

has a unique solution u in $W^{2,p}(\Omega)$ for every $(f, g) \in L^p(\Omega) \times W^{1-1/p,p}(\partial\Omega)$ (see [23, Chapter 3.7]). We set $K(f, g) := (u, u_{|\partial\Omega})$. We think of K as a linear bounded operator from $L^p(\Omega) \times W^{1-1/p,p}(\partial\Omega)$ into itself. As it is an element of $\mathcal{L}(L^p(\Omega) \times W^{1-1/p,p}(\partial\Omega), W^{2,p}(\Omega) \times W^{2-1/p,p}(\partial\Omega))$ and Ω is bounded, K is a compact operator. Consider the following operator K_λ in $\mathcal{L}(L^p(\Omega) \times W^{1-1/p,p}(\partial\Omega))$:

$$K_\lambda(\phi, \psi) := (\phi, \psi) - K((\lambda_0 - \lambda)\phi, -\lambda\psi).$$

Then, if λ is such that (9.47) holds, K_λ is a linear and topological isomorphism of $L^p(\Omega) \times W^{1-1/p,p}(\partial\Omega)$ into itself: to show this, it suffices to prove, thanks to the compactness, that, $Ker(K_\lambda) = \{(0,0)\}$. In fact, if $K_\lambda(\phi, \psi) = (0,0)$, $(\phi, \psi) \in W^{2,p}(\Omega) \times W^{2-1/p,p}(\partial\Omega)$ and $\psi = \phi_{|\partial\Omega}$. Moreover, ϕ solves (9.46) with $f = 0$ and $g = 0$. So $\phi = 0$ and $\psi = 0$ by the a priori-estimate. We deduce that the equation

$$K_\lambda(\phi, \psi) = K(f, g)$$

has a unique solution (ϕ, ψ) in $L^p(\Omega) \times W^{1-1/p,p}(\partial\Omega)$ and it is easy to show that $\phi \in W^{2,p}(\Omega)$, $\psi = \phi_{|\partial\Omega}$ and ϕ solves (9.46). □

As a consequence, we obtain the following variation of Theorem 2 in [10]:

Corollary 9.8 *Assume that (D1)–(D3) hold. Let*

$$\begin{cases} D(G) := \{(u, g) \in W^{2,p}(\Omega) \times W^{2-1/p,p}(\partial\Omega)u_{|\partial\Omega} = g\} \\ \qquad\quad \to L^p(\Omega) \times W^{1-1/p,p}(\partial\Omega), \\ G(u, g) \qquad := (A(\cdot, D_\xi)u, -B(\cdot, D_\xi)u_{|\partial\Omega}). \end{cases} \tag{9.51}$$

Then G is the infinitesimal generator of an analytic semigroup in $L^p(\Omega) \times W^{1-1/p,p}(\partial\Omega)$.

Proof This follows from Theorem 9.9. The density of $D(G)$ in $L^p(\Omega) \times W^{1-1/p,p}(\partial\Omega)$ can be shown slightly modifying the argument in the proof of Corollary 9.3.

□

We conclude with the following maximal regularity result, which seems to be new:

Theorem 9.10 *Consider the system*

$$\begin{cases} D_t u(t,\xi) - A(\xi, D_\xi) u(t,\xi) = f(t,\xi), & t \in (0,T), \xi \in \Omega, \\ D_t u(t,\xi') + B(\xi', D_\xi) u(t,\xi') = h(t,\xi'), & t \in (0,T), \xi' \in \partial\Omega, \\ u(0,\xi) = u_0(\xi), & \xi \in \Omega \end{cases} \quad (9.52)$$

with the assumptions (D1)–(D3). Let $p \in (1,\infty) \setminus \{\frac{3}{2}\}$. Then the following conditions are necessary and sufficient, in order that there exists a solution u in $W^{1,p}((0,T); L^p(\Omega)) \cap L^p((0,T); W^{2,p}(\Omega))$, such that

$$u_{|(0,T)\times\partial\Omega} \in W^{1,p}((0,T); W^{1-1/p,p}(\partial\Omega)) \cap L^p((0,T); W^{2-1/p,p}(\partial\Omega)) :$$

(a) $f \in L^p((0,T) \times \Omega)$;
(b) $h \in L^p((0,T); W^{1-1/p,p}(\partial\Omega))$;
(c) $u_0 \in B_{p,p}^{2-2/p}(\Omega)$ and, in case $p > \frac{3}{2}$, $u_{0|\partial\Omega} \in B_{p,p}^{2-2/p}(\Omega)$.

In case $p > \frac{3}{2}$, the solution is unique. If $p < \frac{3}{2}$, the solution is not unique: for every $g_0 \in B_{p,p}^{2-2/p}(\partial\Omega)$, there exists a unique solution u such that, if $\tau u = u_{|(0,T)\times\partial\Omega}$, $\tau u(0, \cdot) = g_0$. u and $u_{|(0,T)\times\partial\Omega}$ can be represented in the form

$$(u(t,\cdot), u(t,\cdot)_{|\partial\Omega}) = e^{tG}(u_0, g_0) + \int_0^t e^{(t-s)G}(f(s,\cdot), h(s,\cdot))ds, \quad (9.53)$$

with G as in (9.51).

Proof We begin with the necessity of (a)–(c). (a) and (b) are clear, by classical trace results. Moreover, by Theorem 9.3, u_0 should belong to $(L^p(\Omega), W^{2,p}(\Omega))_{1-1/p,p} = B_{p,p}^{2-2/p}(\Omega)$ and $g_0 := \tau u(0, \cdot)$ should be in

$$(W^{1-1/p,p}(\partial\Omega), W^{2-1/p,p}(\partial\Omega))_{1-1/p,p} = B_{p,p}^{2-2/p}(\partial\Omega).$$

In case $p > \frac{3}{2}$, u_0 admits a trace on $\partial\Omega$, which is necessarily g_0.

We prove the sufficiency in some steps. We assume that $g_0 \in B_{p,p}^{2-2/p}(\partial\Omega)$ and, in case $p > \frac{3}{2}$, $g_0 = u_{0|\partial\Omega}$.

Step 1 We have already seen (Lemma 9.5) that, if a solution with the declared properties exists, it can be represented in the form (9.53). As a consequence we immediately obtain the following a priori-estimate:

$$\|u\|_{L^p((0,T)\times\Omega)} + \|u_{|(0,T)\times\partial\Omega}\|_{L^p((0,T);W^{1-1/p,p}(\partial\Omega)}$$
$$\leq C(\|f\|_{L^p((0,T)\times\Omega)} + \|h\|_{L^p((0,T);W^{1-1/p,p}(\partial\Omega))} + \|u_0\|_{L^p(\Omega)} + \|g_0\|_{W^{1-1/p,p}(\partial\Omega)}).$$
$$(9.54)$$

Step 2 Now we prove the main a priori estimate: assume that u and $u_{|(0,T)\times\partial\Omega}$ have the declared regularity. We suppose also that u is a solution to (9.52), and f, h, u_0, g_0 satisfy (a)–(d). Moreover, we suppose that u vanishes outsides $[0,T]\times U$, with U as in the proof of Theorem 9.9. Employing a suitable change of variable χ and setting $v(t,x,y) = u(t,\chi(x,y))$, we are locally reduced to estimate v in $L^p((0,T);W^{2,p}(\mathbb{R}^+\times\mathbb{R}^{n-1}))\cap W^{1,p}((0,T);L^p(\mathbb{R}^+\times\mathbb{R}^{n-1}))$ and $v(\cdot,0,\cdot)$ in $W^{1,p}((0,T);W^{1-1/p,p}(\mathbb{R}^{n-1}))\cap L^p((0,T);W^{2-1/p,p}(\mathbb{R}^{n-1}))$. v satisfies the system

$$\begin{cases} D_t v(t,x,y) - A(D_x,D_y)v(t,x,y) = A^\sharp(x,y,D_{x,y})v(t,x,y) + f^\sharp(t,x,y), \\ (t,x,y)\in(0,T)\times\mathbb{R}^+\times\mathbb{R}^{n-1}, \\ D_t v(t,0,y) - \gamma D_x v(t,0,y) + v\cdot\nabla_y v(t,0,y) = B^\sharp(y,D_{x,y})v(t,x,0) + h^\sharp(t,y), \\ y\in\mathbb{R}^{n-1}, \\ v(0,x,y) = u_0^\sharp(x,y), \quad (x,y)\in\mathbb{R}^+\times\mathbb{R}^{n-1}, \\ v(\cdot,0,\cdot)_{|t=0} = g_0^\sharp, \end{cases}$$

with $\gamma\in\mathbb{R}^+$, $A^\sharp(x,y,D_{x,y}) = \sum_{|\alpha|\leq 2} a_\alpha^\sharp(x,y)D_{x,y}^\alpha$, $B^\sharp(y,D_{x,y}) = \sum_{|\alpha|\leq 1} b_\alpha^\sharp(y)D_{x,y}^\alpha$, $f^\sharp(t,x,y) = f(t,\chi(x,y))$, $h^\sharp(t,y) = h(t,\chi(0,y))$, $u_0^\sharp(x,y) = u_0(\chi(x,y))$, $g_0^\sharp(y) = g_0(\chi(0,y))$,

$$\sum_{|\alpha|=2}\|a_\alpha^\sharp\|_{L^\infty(\mathbb{R}^+\times\mathbb{R}^{n-1})} + \sum_{|\alpha|=1}\|b_\alpha^\sharp\|_{L^\infty(\mathbb{R}^{n-1})}\leq\epsilon,$$

with $\epsilon\in\mathbb{R}^+$. We observe that, in case $p > \frac{3}{2}$, $u_0^\sharp(0,\cdot) = g_0^\sharp$. Then, from Proposition 9.7 we deduce the estimate

$$\|v\|_{W^{1,p}((0,T);L^p(\mathbb{R}^+\times\mathbb{R}^{n-1}))} + \|v\|_{L^p((0,T);W^{2,p}(\mathbb{R}^+\times\mathbb{R}^{n-1}))}$$
$$+\|v(\cdot,0,\cdot)\|_{W^{1,p}((0,T);W^{1-1/p,p}(\mathbb{R}^{n-1}))} + \|v(\cdot,0,\cdot)\|_{L^p((0,T);W^{2-1/p,p}(\mathbb{R}^{n-1}))}$$
$$\leq C_0(\|f^\sharp\|_{L^p((0,T)\times\mathbb{R}^+\times\mathbb{R}^{n-1})} + \|h^\sharp\|_{L^p((0,T);W^{1-1/p,p}(\partial\Omega))} + \|u_0^\sharp\|_{B_{p,p}^{2-2/p}(\mathbb{R}^+\times\mathbb{R}^{n-1})}$$
$$+\|g_0^\sharp\|_{B_{p,p}^{2-2/p}(\mathbb{R}^{n-1})} + \|A^\sharp(x,y,D_{x,y})v\|_{L^p((0,T)\times\mathbb{R}^+\times\mathbb{R}^{n-1})}$$
$$+\|B^\sharp(y,D_{x,y})v\|_{L^p((0,T);W^{1-1/p,p}(\mathbb{R}^{n-1}))})$$
$$\leq C_1(\|f^\sharp\|_{L^p((0,T)\times\mathbb{R}^+\times\mathbb{R}^{n-1})} + \|h^\sharp\|_{L^p((0,T);W^{1-1/p,p}(\mathbb{R}^{n-1}))} + \|u_0^\sharp\|_{B_{p,p}^{2-2/p}(\mathbb{R}^+\times\mathbb{R}^{n-1})}$$
$$+\|g_0^\sharp\|_{B_{p,p}^{2-2/p}(\mathbb{R}^{n-1})} + \epsilon\|v\|_{L^p((0,T);W^{2,p}(\mathbb{R}^+\times\mathbb{R}^{n-1}))} + \|v\|_{L^p((0,T);W^{\sigma,p}(\mathbb{R}^+\times\mathbb{R}^{n-1}))}),$$
$$(9.55)$$

for some $\sigma < 2$. If U is suitably small, we may assume that $C_1 \epsilon \leq \frac{1}{2}$, so that, from (9.55) we deduce

$$
\begin{aligned}
\|v\|_{W^{1,p}((0,T);L^p(\mathbb{R}^+\times\mathbb{R}^{n-1}))} &+ \|v\|_{L^p((0,T);W^{2,p}(\mathbb{R}^+\times\mathbb{R}^{n-1}))} \\
+\|v(\cdot,0,\cdot)\|_{W^{1,p}((0,T);W^{1-1/p,p}(\mathbb{R}^{n-1}))} &+ \|v(\cdot,0,\cdot)\|_{L^p((0,T);W^{2-1/p,p}(\mathbb{R}^{n-1}))} \\
\leq C_2(\|f^{\sharp}\|_{L^p((0,T)\times\mathbb{R}^+\times\mathbb{R}^{n-1})} &+ \|h^{\sharp}\|_{L^p((0,T);W^{1-1/p,p}(\mathbb{R}^{n-1}))} \\
+\|u_0^{\sharp}\|_{B_{p,p}^{2-2/p}(\mathbb{R}^+\times\mathbb{R}^{n-1})} &+ \|g_0^{\sharp}\|_{B_{p,p}^{2-2/p}(\mathbb{R}^{n-1})} \\
+\|v\|_{L^p((0,T);W^{\sigma,p}(\mathbb{R}^+\times\mathbb{R}^{n-1}))}&).
\end{aligned}
$$

Applying a suitable partition of unity, as in the proof of Theorem 9.9, we deduce the a priori estimate:

$$
\begin{aligned}
\|u\|_{W^{1,p}((0,T);L^p(\Omega))} &+ \|u\|_{L^p((0,T);W^{2,p}(\Omega))} \\
+\|u_{|(0,T)\times\partial\Omega}\|_{W^{1,p}((0,T);W^{1-1/p,p}(\partial\Omega))} &+ \|u_{|(0,T)\times\partial\Omega}\|_{L^p((0,T);W^{2-1/p,p}(\partial\Omega))} \\
\leq C(\|f\|_{L^p((0,T)\times\Omega)} &+ \|h\|_{L^p((0,T);W^{1-1/p,p}(\partial\Omega))} + \|u_0\|_{B_{p,p}^{2-2/p}(\Omega)} \\
+\|g_0\|_{B_{p,p}^{2-2/p}(\partial\Omega)} &+ \|u\|_{L^p((0,T);W^{\sigma,p}(\Omega))}),
\end{aligned}
\tag{9.56}
$$

for some $\sigma < 2$. Now, for every $\epsilon \in \mathbb{R}^+$ there exists $C(\epsilon) \in \mathbb{R}^+$, such that

$$
\|u\|_{L^p((0,T);W^{\sigma,p}(\Omega))} \leq \epsilon \|u\|_{L^p((0,T);W^{2,p}(\Omega))} + C(\epsilon)\|u\|_{L^p((0,T)\times\Omega)}.
$$

So (9.56) and (9.54) imply the a priori estimate

$$
\begin{aligned}
\|u\|_{W^{1,p}((0,T);L^p(\Omega))} &+ \|u\|_{L^p((0,T);W^{2,p}(\Omega))} \\
+\|u_{|(0,T)\times\partial\Omega}\|_{W^{1,p}((0,T);W^{1-1/p,p}(\partial\Omega))} &+ \|u_{|(0,T)\times\partial\Omega}\|_{L^p((0,T);W^{2-1/p,p}(\partial\Omega))} \\
\leq C(\|f\|_{L^p((0,T)\times\Omega)} &+ \|h\|_{L^p((0,T);W^{1-1/p,p}(\partial\Omega))} \\
+\|u_0\|_{B_{p,p}^{2-2/p}(\Omega)} &+ \|g_0\|_{B_{p,p}^{2-2/p}(\partial\Omega)}).
\end{aligned}
\tag{9.57}
$$

Step 3 Now we show that, if (a)–(d) hold, a solution with the declared regularity really exists. We fix $(f_k)_{k\in\mathbb{N}}$ in $C^1([0,T];L^p(\Omega))$, such that $\|f_k - f\|_{L^p((0,T)\times\Omega)} \to 0$, $(h_k)_{k\in\mathbb{N}}$ in $C^1([0,T];W^{1-1/p,p}(\partial\Omega))$, such that $\|h_k - h\|_{L^p((0,T);W^{1-1/p,p}(\partial\Omega))} \to 0$, $(v_{0k})_{k\in\mathbb{N}}$ in $W^{2,p}(\Omega))$, such that $\|v_{0k} - u_0\|_{B_{p,p}^{2-2/p}(\Omega)} \to 0$, $(g_{0k})_{k\in\mathbb{N}}$ in $W^{2-1/p,p}(\partial\Omega))$, such that $\|g_{0k} - g_0\|_{B_{p,p}^{2-2/p}(\partial\Omega)} \to 0$.

We assume first that $p < \frac{3}{2}$. Then, as $2 - \frac{2}{p} < \frac{1}{p}$, $\mathscr{D}(\Omega)$ is dense in $B_{p,p}^{2-2/p}(\Omega)$ (see [24, Section 4.3.2]). So $\{v \in W^{2,p}(\Omega) : v_{|\partial\Omega} = 0\}$ is dense in $B_{p,p}^{2-2/p}(\Omega)$. By translation, $\forall z \in W^{2-1/p,p}(\partial\Omega)$ $\{v \in W^{2,p}(\Omega) : v_{|\partial\Omega} = z\}$ is dense in $B_{p,p}^{2-2/p}(\Omega)$. As a consequence, we can construct a sequence $(z_{0k})_{k\in\mathbb{N}}$ in $W^{2,p}(\Omega)$ such that $z_{0k|\partial\Omega} = g_{0,k} - v_{0k|\partial\Omega}$ and $\|z_{0k}\|_{B_{p,p}^{2-2/p}(\Omega)} \to 0$ $(k \to \infty)$. So, if we set $u_{0k} := v_{0k} + z_{0k}$, we have that $\|u_{0k} - u_0\|_{B_{p,p}^{2-2/p}(\Omega)} \to 0$ and $\|u_{0k|\partial\Omega} - g_0\|_{B_{p,p}^{2-2/p}(\partial\Omega)} \to 0$ $(k \to \infty)$.

We consider now the case $p > \frac{3}{2}$. In this case we take a linear bounded operator $P : B_{p,p}^{2-3/p}(\partial\Omega) \to B_{p,p}^{2-2/p}(\Omega)$ such that $(Pz)_{|\partial\Omega} = z \; \forall z \in B_{p,p}^{2-3/p}(\partial\Omega)$, whose restriction to $W^{2-1/p,p}(\partial\Omega)$ belongs to $\mathscr{L}(W^{2-1/p,p}(\partial\Omega), W^{2,p}(\Omega))$ (see [25, Section I.2.7.2]). Then, we set $z_{0k} := P(g_{0k} - v_{0k|\partial\Omega})$. As $\|g_{0k} - v_{0k|\partial\Omega}\|_{B_{p,p}^{2-3/p}(\partial\Omega)} \to 0$ $(k \to \infty)$ (because $u_{|\partial\Omega} = g_0$), we have that $z_{0k} \in W^{2,p}(\Omega)$ and $\|z_{0k}\|_{B_{p,p}^{2-2/p}(\Omega)} \to 0$ $(k \to \infty)$. So, if we set $u_{0k} := v_{0k} + z_{0k}$, $u_{0k} \in W^{2,p}(\Omega)$, $u_{0k|\partial\Omega} = g_{0k}$ and $\|u_{0k} - u_0\|_{B_{p,p}^{2-2/p}(\Omega)} \to 0$ $(k \to \infty)$.

Now, for every $k \in \mathbb{N}$, we consider the solution u_k of (9.52), replacing f with f_k, h with h_k, u_0 with u_{0k}. By Corollary 9.8 and well known properties of regularity of solutions in case of analytic semigroups, $u_k \in C^1([0,T]; L^p(\Omega)) \cap C([0,T]; W^{2,p}(\Omega)))$, $u_{k|[0,T]\times\partial\Omega} \in C^1([0,T]; W^{1-1/p,p}(\partial\Omega)) \cap C([0,T]; W^{1-1/p,p}(\partial\Omega))$. Moreover, by the a priori estimate (9.57), $(u_k)_{k\in\mathbb{N}}$ is a Cauchy sequence in

$$W^{1,p}((0,T); L^p(\Omega)) \cap L^p((0,T); W^{2,p}(\Omega)),$$

$(u_{k|(0,T)\times\partial\Omega})_{k\in\mathbb{N}}$ is a Cauchy sequence in

$$W^{1,p}((0,T); W^{1-1/p,p}(\partial\Omega)) \cap L^p((0,T); W^{2-1/p,p}(\Omega)).$$

It is clear that the limit u of $(u_k)_{k\in\mathbb{N}}$ in $W^{1,p}((0,T); L^p(\Omega)) \cap L^p((0,T); W^{2,p}(\Omega))$ is the solution of (9.52). $\square$

9.6 Appendix

This appendix is dedicated to the proof of Proposition 9.6. We shall employ the following result, which, in case, $E = \mathbb{C}$, is well known. In case E is a UMD space, with property (α), it can be deduced from [4, Theorem 4.9.1].

Lemma 9.7 (I) Let E be a Banach space, $p \in (1,\infty)$ and let $\theta \in (0,1) \setminus \{\frac{1}{2p}\}$. We set

$$S := \{u \in W^{2,p}(\mathbb{R}^+; E) : u(0) = 0\}. \tag{9.58}$$

Then

$$(L^p(\mathbb{R}^+; E), S)_{\theta,p} = \begin{cases} B_{p,p}^{2\theta}(\mathbb{R}^+; E) & \text{if } \theta < \frac{1}{2p}, \\ \{v \in B_{p,p}^{2\theta}(\mathbb{R}^+; E) : v(0) = 0\} & \text{if } \theta > \frac{1}{2p}. \end{cases}$$

Proof By Proposition 9.3 (I), for every θ in $(0, 1)$, $(L^p(\mathbb{R}^+; E), S)_{\theta, p}$ is continuously embedded into $B^{2\theta}_{p,p}(\mathbb{R}^+; E)$. On the other hand, if we set

$$S_0 := \{u \in W^{2,p}(\mathbb{R}^+; E) : u(0) = u'(0) = 0\},$$

we have, following the argument in the proof of Theorem 8 in [14], which is valid even for vector valued functions,

$$(L^p(\mathbb{R}^+; E), S_0)_{\theta, p} = \begin{cases} B^{2\theta}_{p,p}(\mathbb{R}^+; E) \text{ if } \theta < \frac{1}{2p}, \\ \{v \in B^{2\theta}_{p,p}(\mathbb{R}^+; E) : v(0) = 0\} \text{ if } \frac{1}{2p} < \theta < \frac{1}{2} + \frac{1}{2p}, \\ \{v \in B^{2\theta}_{p,p}(\mathbb{R}^+; E) : v(0) = v'(0) = 0\} \text{ if } \frac{1}{2} + \frac{1}{2p} < \theta < 1. \end{cases}$$

As $S_0 \hookrightarrow S$, we deduce that the conclusion holds if $\theta < \frac{1}{2p}$, and in case $\frac{1}{2p} < \theta < \frac{1}{2} + \frac{1}{2p}$,

$$\{v \in B^{2\theta}_{p,p}(\mathbb{R}^+; E) : v(0) = 0\} \subseteq (L^p(\mathbb{R}^+; E), S)_{\theta, p}.$$

We show that, if $\frac{1}{2p} < \theta$, $(L^p(\mathbb{R}^+; E), S)_{\theta, p} \subseteq \{v \in B^{2\theta}_{p,p}(\mathbb{R}^+; E) : v(0) = 0\}$. In fact, as $p < \infty$, S is dense in $(L^p(\mathbb{R}^+; E), S)_{\theta, p}$, which is continuously embedded into $B^{2\theta}_{p,p}(\mathbb{R}^+; E)$. We deduce that, if $v \in (L^p(\mathbb{R}^+; E), S)_{\theta, p}$, there is a sequence $(v_k)_{k \in \mathbb{N}}$ in S converging to v in $B^{2\theta}_{p,p}(\mathbb{R}^+; E)$. As $2\theta > \frac{1}{p}$, the sequence $(v_k(0))_{k \in \mathbb{N}}$ converges to $v(0)$ in E. We deduce that $v(0) = 0$.

So the conclusion holds in case $\theta < \frac{1}{2} + \frac{1}{2p}$.

Finally, let $\theta \geq \frac{1}{2} + \frac{1}{2p}$. We fix θ_0 in $(\frac{1}{2p}, \frac{1}{2} + \frac{1}{2p})$, and set

$$S_{\theta_0} := \{v \in B^{2\theta_0}_{p,p}(\mathbb{R}^+; E) : v(0) = 0\} = (L^p(\mathbb{R}^+; E), S)_{\theta_0, p}.$$

By the reiteration property,

$$(L^p(\mathbb{R}^+; E), S)_{\theta, p} = ((L^p(\mathbb{R}^+; E), S)_{\theta_0, p}, S)_{\frac{\theta - \theta_0}{1 - \theta_0}, p}.$$

Let $v \in B^{2\theta}_{p,p}(\mathbb{R}^+; E)$, with $v(0) = 0$. As

$$B^{2\theta}_{p,p}(\mathbb{R}^+; E) = (B^{2\theta_0}_{p,p}(\mathbb{R}^+; E), W^{2,p}(\mathbb{R}^+; E))_{\frac{\theta - \theta_0}{1 - \theta_0}, p},$$

there exist v_0, v_1, such that $t^{-\frac{\theta - \theta_0}{1 - \theta_0}} v_0 \in L^p_*(\mathbb{R}^+; B^{2\theta_0}_{p,p}(\mathbb{R}^+; E))$, $t^{-\frac{1 - \theta}{1 - \theta_0}} v_1 \in L^p_*(\mathbb{R}^+; W^{2,p}(\mathbb{R}^+; E))$, and $v = v_0(t) + v_1(t)$ $\forall t \in \mathbb{R}^+$. We fix $\phi \in C^\infty([0, \infty))$, such that $\phi(s) = 0$ if $s \geq 1$ and $\phi(0) = 1$. We set

$$u_j(t)(s) := v_j(t)(s) - \phi(s)v_j(t)(0), \quad j \in \{0, 1\}, t \in \mathbb{R}^+, s \in \mathbb{R}^+.$$

As $z \to z(0)$ is continuous and bounded from $B_{p,p}^{2\theta_0}(\mathbb{R}^+; E)$ to E, we deduce that, for some $C \in \mathbb{R}^+$ independent of t,

$$\|u_0(t)\|_{B_{p,p}^{2\theta_0}(\mathbb{R}^+;E))} \le C \|v_0(t)\|_{B_{p,p}^{2\theta_0}(\mathbb{R}^+;E))},$$
$$\|u_1(t)\|_{W^{2,p}(\mathbb{R}^+;E))} \le C \|v_1(t)\|_{W^{2,p}(\mathbb{R}^+;E))},$$

so that $t^{-\frac{\theta-\theta_0}{1-\theta_0}} u_0 \in L_*^p(\mathbb{R}^+; S_{\theta_0})$, $t^{-\frac{1-\theta}{1-\theta_0}} u_1 \in L_*^p(\mathbb{R}^+; S)$. Moreover, as $v_0(t)(0) + v_1(t)(0) = v(0) = 0$, $\forall t \in \mathbb{R}^+$, $u_0(t) + u_1(t) = v$ $\forall t \in \mathbb{R}^+$. We conclude that $v \in ((L^p(\mathbb{R}^+; E), S)_{\theta_0, p}, S)_{\frac{\theta-\theta_0}{1-\theta_0}, p}$. $\square$

Proof (of Proposition 9.6)

(I) We may identify $W_p^{1,2,A}((0,T) \times \mathbb{R}^+; E)$ with $\{u \in L^p((0,T); E_1) : D_t u \in L^p((0,T); E_0)\}$, if

$$E_1 = W^{2,p}(\mathbb{R}^+; E) \cap L^p(\mathbb{R}^+; D(A)), \quad E_0 = L^p(\mathbb{R}^+; E).$$

So, by Theorem 9.3, we have

$$\{u(0,\cdot) : u \in W_p^{1,2,A}((0,T) \times \mathbb{R}^+; E)\} = (E_0, E_1)_{1-1/p,p}.$$

We may identify $L^p(\mathbb{R}^+; D(A))$ with $D(A_0)$, setting

$$\begin{cases} A_0 : L^p(\mathbb{R}^+; D(A)) \to L^p(\mathbb{R}^+; E) \\ A_0 v(x) := Av(x). \end{cases}$$

Then, if $h \in W^{2,p}(\mathbb{R}^+; E)$ and $t \in \mathbb{R}^+$, $(t + A_0)^{-1} h \in W^{2,p}(\mathbb{R}^+; E)$ and

$$\|(t + A_0)^{-1} h\|_{W_p^2(\mathbb{R}^+;E)} \le Ct^{-1} \|h\|_{W_p^2(\mathbb{R}^+;E)}.$$

We deduce from Theorem 9.2 (II) that

$$(E_0, E_1)_{1-1/p,p}$$
$$= (L^p(\mathbb{R}^+; E), W^{2,p}(\mathbb{R}^+; E))_{1-1/p,p} \cap (L^p(\mathbb{R}^+; E), L^p(\mathbb{R}^+; D(A)))_{1-1/p,p}$$
$$= B_{p,p}^{2-2/p}(\mathbb{R}^+; E) \cap L^p(\mathbb{R}^+; (E, D(A))_{1-1/p,p}),$$
$$\tag{9.59}$$

by Propositions 9.3 and 9.1;

(II) a similar argument holds for (II), replacing $W^{2,p}(\mathbb{R}^+; E)$ with

$$\{u \in W^{2,p}(\mathbb{R}^+; E) : u(0) = 0\}$$

and employing Lemma 9.7. By the way, in case $p > \frac{3}{2}$, $u_0(0)$ is well defined, because $2 - \frac{2}{p} > \frac{1}{p}$ (see Proposition 9.3 (II));

(III) we may identify $W_p^{1,2,A}((0,T) \times \mathbb{R}^+; E)$ with

$$\{u \in L^p(\mathbb{R}^+; W^{1,p}((0,T); E) \cap L^p((0,T); D(A)))) : D_x^2 u \in L^p(\mathbb{R}^+; L^p((0,T); E))\}.$$

So, by Theorem 9.3, we have

$$\{u(\cdot,0) : u \in W_p^{1,2,A}((0,T) \times \mathbb{R}^+; E)\}$$
$$= (L^p((0,T); E), W^{1,p}((0,T); E) \cap L^p((0,T); D(A)))_{1-1/(2p),p}.$$

Arguing as in the proof of (I), we have

$$(L^p((0,T); E), W^{1,p}((0,T); E) \cap L^p((0,T); D(A)))_{1-1/(2p),p}$$
$$= (L^p((0,T); E), W^{1,p}((0,T); E))_{1-1/(2p),p}$$
$$\cap (L^p((0,T); E), L^p((0,T); D(A)))_{1-1/(2p),p}$$
$$= B_{p,p}^{1-1/(2p)}((0,T); E) \cap L^p((0,T); (E, D(A))_{1-1/(2p),p});$$

(IV) by (III), there exists $v \in W_p^{1,2,A}((0,T) \times \mathbb{R}^+; E)$, such that $v(\cdot,0) = g$. Then $v(0,\cdot) \in B_{p,p}^{2-2/p}((0,T); E) \cap L^p((0,T); (E, D(A))_{1-1/p,p})$. Subtracting v, we are reduced to show that there exists $w \in W_p^{1,p,A}((0,T) \times \mathbb{R}^+; E)$, such that $w(0,\cdot) = u_0 - v(0,\cdot)$, $w(\cdot,0) \equiv 0$. This follows from (II);

(V) can be proved with the same argument of (IV);

(VI) it can be obtained following the same argument in (III). □

References

1. Adams, R.A.: Sobolev Spaces. Academic, New York (1975)
2. Aliev, B.A., Yakubov, Y.Y.: Elliptic differential-operator problems with a spectral parameter in both the equation and boundary-operator conditions. Adv. Differ. Equ. **11**, 1081–1110 (2006)
3. Aliev, B.A., Yakubov, Y.Y.: Erratum to the paper "Elliptic differential-operator problems with a spectral parameter in both the equation and boundary-operator conditions". Adv. Differ. Equ. **12**, 1079 (2006)
4. Amann, H.: Anisotropic Function Spaces and Maximal Regularity for Parabolic Problems. Part 1. Function Spaces. Jindrich Necas Center for Mathematical Modeling Lecture Notes, vol. 23. Matfyzpress, Prague (2009)
5. Bergh, J., Löfström, J.: Interpolation Spaces. An Introduction. Grundlehren der Mathematischen Wissenschaften, vol. 223. Springer, Berlin (1976)
6. Da Prato, G., Grisvard, P., Sommes d'opérateurs linéaires et équations différentielles opérationelles. J. Math. Pures Appl. **54**, 301–387 (1975)
7. Di Cristo, M., Guidetti, D., A. Lorenzi, A.: Abstract parabolic equations with applications to problems in cylindrical space domains. Adv. Differ. Equ. **15**, 1–42 (2010)
8. Denk, R., Hieber, M., Prüss, J.: R-boundedness, Fourier Multipliers and Problems of Elliptic and Parabolic Type. Memoirs of the American Mathematical Society, vol. 788. American Mathematical Society, Providence (2003)
9. Dore, G., Venni, A.: On the closeness of the sum of two closed operators. Math. Z. **196**, 189–201 (1987)

10. Escher, J.: Quasilinear parabolic systems with dynamical boundary conditions. Commun. Partial Differ. Equ. **18**, 1309–1364 (1993)
11. Favini, A., Guidetti, D., Yakubov, Y.: Abstract elliptic and parabolic systems with applications to problems in cylindrical domains. Adv. Differ. Equ. **16**, 1139–1196 (2011)
12. Grisvard, P.: Commutativité de deux foncteurs d'interpolation et applications I, II. J. Math. Pures Appl. **45**, 207–290 (1966)
13. Grisvard, P.: Equations difféntielles abstraites. Ann. Sc. Ec. Norm. Sup. **2**, 311–395 (1969)
14. Grisvard, P.: Spazi di tracce e applicazioni. Rend. Mat. **6**, 657–729 (1972)
15. Guidetti, D.: Generation of analytic semigroups by elliptic operators with dirichlet boundary conditions in a cylindrical domain. Semigroup Forum **68**, 108–136 (2004)
16. Hintermann, T.: Evolution equations with dynamic boundary conditions. Proc. R. Soc. Edinb. **113**, 43–60 (1989)
17. Kunstmann, P.C., Weis, L.: Maximal L^p-regularity for parabolic equations, fourier multiplier theorems and H^∞-functional calculus. In: Functional Analytic Methods for Evolution Equations. Lecture Notes in Mathematics, vol. 1855. Springer, Berlin (2004)
18. Mugnolo, D.: Vector-valued heat equations and networks with coupled dynamic boundary conditions. Adv. Differ. Equ. **15**, 1125–1160 (2010)
19. Pazy, A.: Semigroups of Linear Operators and Applications to Partial Differential Equations. Springer, Berlin (1983)
20. Prüss, J.: Maximal regularity for abstract parabolic problems with inhomogeneous boundary data in L_p spaces. Math. Bohem. **127**, 311–327 (2002)
21. Runst, T., Sickel, W.: Sobolev Spaces of Fractional Order, Nemytskij Operators, and Nonlinear Partial Differential Equations. De Gruyter Series in Nonlinear Analysis and Applications. De Gruyter, Berlin (1996)
22. Simon, J.: Sobolev, Besov and Nikolskii fractional spaces: imbeddings an comparisons for vector valued spaces on an interval. Ann. Mat. Pura Appl. **157**, 117–148 (1990)
23. Tanabe, H.: Equations of Evolution. Monographs and Studies in Mathematics, 6. Pitman, Boston, Mass.-London (1979)
24. Triebel, H.: Interpolation Theory, Function Spaces, Differential Operators. North-Holland, Amsterdam (1978). 2nd revised edition Leipzig, Barth (1995)
25. Triebel, H.: Theory of Function Spaces. Monographs in Mathematics. Birkhäuser, Basel (1983)
26. Yakubov, S., Yakubov, S.: Differential-Operator Equations. Ordinary and Partial Differential Equations. Chapman and Hall, London (2000)

Chapter 10
Increasing Stability of the Continuation for General Elliptic Equations of Second Order

Victor Isakov

To the memory of a good colleague and dear friend Alfredo Lorenzi

Abstract We consider the Cauchy problem for general second partial differential equations of elliptic type containing large parameter k (like in the Helmholtz equation). By using energy estimates and splitting solution in "low" and "high" frequency parts we obtain bounds of a solution in Sobolev spaces indicating increasing stability in the Cauchy problem with growing k. These bounds show Lipschitz stability of the "low" frequency part with decaying contribution of the a priori bound on the "high" frequency part. Increasing stability is of importance in (boundary) control theory and inverse problems (remote sensing).

10.1 Introduction

The Cauchy Problem (or equivalently the continuation of solutions) for partial differential equations has a long and rich history, starting with the Holmgren-John theorem on uniqueness for equations with analytic coefficients. It is of great importance in the theories of boundary control and of inverse problems. In 1938 T. Carleman introduced a special exponentially weighted energy (Carleman type) estimates to handle non analytic coefficients. These estimates imply in addition some conditional Hölder type stability estimates for solutions of this problem. In 1960 [11] F. John showed that for the continuation for the Helmholtz equation from inside of the unit disk onto any larger disk the stability estimate which is uniform with respect to the wave numbers is still of logarithmic type. Logarithmic stability is quite damaging for numerical solution of many inverse problems. In recent papers

V. Isakov (✉)
Dept. Mathem., Statis., and Physics, Wichita State University, Wichita,
KS 67260-0033, USA
e-mail: victor.isakov@wichita.edu

© Springer International Publishing Switzerland 2014
A. Favini et al. (eds.), *New Prospects in Direct, Inverse and Control Problems for Evolution Equations*, Springer INdAM Series 10,
DOI 10.1007/978-3-319-11406-4_10

[1, 2, 4, 6, 7, 9] it was shown that in a certain sense stability is always improving for larger k under (pseudo) convexity conditions on the geometry of the domain and of the coefficients of the elliptic equation.

In this paper we attempt to eliminate any convexity type condition on the elliptic operator or the domain. Due to John's counterexample, one can not expect increasing stability for all solutions. We show that (near Lipschitz) stability holds on a subspace of ("low frequency") solutions which is growing with the wave number k under some mild boundedness constraints on complementary "high frequency" part.

We will consider the Cauchy problem

$$(A + ck + k^2)u = f \text{ in } \Omega, \tag{10.1}$$

with the Cauchy data

$$u = u_0, \partial_\nu u = u_1 \text{ on } \Gamma_0 \subset \partial\Omega, \tag{10.2}$$

where ν is the outer unit normal to $\partial\Omega$ and

$$Au = \sum_{j,m=1}^{n} a_{jm} \partial_j \partial_m u + \sum_{j=1}^{n} a_j \partial_j u + au$$

is the general partial differential operator of second order satisfying the ellipticity condition

$$\varepsilon_0 |\xi|^2 \leq \sum_{j,l=1}^{n} a_{jl}(x) \xi_j \xi_l$$

for some positive number ε_0 and all $x \in \Omega$ and $\xi \in \mathbf{R}^n$. We assume that

$$a_{jm}, \partial_p a_{jm}, a_j, a, c \in L^\infty(\Omega).$$

We consider bounded open $\Omega \subset \mathbf{R}^{n-1} \times (0, 1)$, $\Gamma_0 = \partial\Omega \cap \{x_n = 0\}$, $\Gamma_1 = \partial\Omega \cap \{x_n = 1\}$, and $\Gamma = \partial\Omega \cap (\mathbf{R}^{n-1} \times (0, 1))$. Let V be a neighbourhood of $\bar{\Gamma}$ and $\omega = \Omega \cap V$.

We use the classical Sobolev spaces $H^{(p)}(\Omega)$ with the standard norm $\| \cdot \|_{(p)}(\Omega)$.

In what follows C denote generic constants depending only on $\Omega, \Gamma_0, \omega, \Gamma_1, A, c$, and ε.

Under additional a priori constraints on u near Γ and on a "high frequency" part of u we can claim

Theorem 10.1 *There are a monotone family of closed subspaces $H_{(1)}(\Omega; k)$ of $H_{(1)}(\Omega)$ with $\cup_k H_{(1)}(\Omega; k) = H_{(1)}(\Omega)$, linear continuous operators P_k from $H_{(1)}(\Omega)$ onto $H_{(1)}(\Omega; k)$ with $P_k u_l = u_l$ for $u_l \in H_{(1)}(\Omega; k)$, a semi norm $||| \cdot |||_{(1;k)}(\Omega)$ on $H_{(1)}(\Omega)$, which is zero on $H_{(1)}(\Omega; k)$ and decreasing with respect to k, and a constant C such that*

$$\|u_l\|_{(0)}(\Omega \setminus) \leq$$

$$C(\|u\|_{(0)}(\Gamma_0) + k^{-1}(\|f\|_{(0)}(\Omega) + \|u_1\|_{(0)}(\Gamma_0) + \|u\|_{(1)}(\omega) + \||u\||_{(1,k)}(\Omega))),$$
(10.3)

where $u_l = P_k u$, and

$$\|u\|_{(0)}(\Omega \setminus \bar{V}) \leq$$

$$C(\|u\|_{(0)}(\Gamma_0) + k^{-1}(\|f\|_{(0)}(\Omega) + \|u_1\|_{(0)}(\Gamma_0) + \|u\|_{(1)}(\omega) + \||u\||_{(1,k)}(\Omega)))$$
(10.4)

for all $u \in H_{(2)}(\Omega)$ solving (10.1), (10.2).

In the next result we will partially replace the Cauchy data on Γ_0 by a function u in $\omega = \Omega \cap V$.

Theorem 10.2 *Let $\theta > 0$.*

There are a monotone family of closed subspaces $H_{(2)}(\Omega; k)$ of $H_{(2)}(\Omega)$ with $\cup_k H_{(2)}(\Omega; k) = H_{(2)}(\Omega)$, linear continuous operators P_k from $H_{(2)}(\Omega)$ onto $H_{(2)}(\Omega; k)$ with $P_k u_l = u_l$ for $u_l \in H_{(2)}(\Omega; k)$, a semi norm $\|| \cdot \||_{(2;k)}(\Omega)$ on $H_{(2)}(\Omega)$, which is zero on $H_{(2)}(\Omega; k)$ and decreasing with respect to k, and constants $C, C(\theta)$ (depending on $\theta > 0$) such that

$$\|u_l\|_{(1)}(\Gamma_1 \setminus \bar{V}) + \|\nabla u_l\|_{(0)}(\Gamma_1 \setminus \bar{V}) + \|u_l\|_{(0)}(\Omega) \leq$$

$$CF + C(\theta)k^{-\frac{1}{2}+\theta}\||u\||_{(2,k)}(\Omega)),$$
(10.5)

where $u_l = P_k u$, and

$$\|u\|_{(1)}(\Gamma_1 \setminus \bar{V}) + \|\nabla u\|_{(0)}(\Gamma_1 \setminus \bar{V}) + \|u\|_{(0)}(\Omega) \leq CF + C(\theta)k^{-\frac{1}{2}+\theta}\||u\||_{(2,k)}(\Omega))$$
(10.6)

for all $u \in H_{(2)}(\Omega)$ solving (10.1), (10.2),
where $F = \|f\|_{(0)}(\Omega) + \|u_0\|_{(1)}(\Gamma_0) + \|u_1\|_{(0)}(\Gamma_0) + \|u\|_{(1)}(\omega)$.

Let χ be C^∞ function, $\chi = 1$ on $\Omega \setminus V$, $\chi = 0$ near Γ. We let $v = \chi u$ in Ω and $v = 0$ on $\mathbf{R}^{n-1} \times (0, 1)$. Obviously, $\sum_{j,m=1}^{n-1} a_{jm}(x)\xi_j\xi_m \leq E^2|\xi|^2$ for some number $E > 0$ and all $x \in \mathbf{R}^{n-1} \times (0, 1), \xi \in \mathbf{R}^{n-1}$. We introduce low and high frequency projectors

$$v_l(x) = \mathscr{F}^{-1}\chi_k \mathscr{F} v(x), \quad v_h = v - v_l,$$
(10.7)

where $\mathscr{F}$ is the (partial) Fourier transformation with respect to $x' = (x_1, \ldots, x_{n-1}, 0)$, $\chi_k(\xi') = 1$ when $|\xi'|^2 < (1 - \varepsilon)\frac{k^2}{E^2}$ and $\chi_k(\xi') = 0$ otherwise. We define $u_l = v_l$ on $\Omega \setminus \bar{V}$.

As can be seen from the proofs of Theorems 10.1, 10.2,

$$|||u|||_{(m,k)}(\Omega) = (\|v_h\|_{(m-1)}^2(\Omega) + \sum_{j=1}^{n-1}\|\partial_j v_h\|_{(m-1)}^2(\Omega))^{\frac{1}{2}}. \tag{10.8}$$

Corollary 10.1 *Let Ω be a C^2-diffeomorphic image of the unit ball, V be a neighbourhood of a boundary point of Ω, $\Gamma_1 = \partial\Omega \setminus V$, and $\omega = \Omega \cap V$. Let $\theta > 0$.*

There are a monotone family of closed subspaces $H_{(2)}(\Omega;k)$ of $H_{(2)}(\Omega)$ with $\cup_k H_{(2)}(\Omega;k) = H_{(2)}(\Omega)$, linear continuous operators P_k from $H_{(2)}(\Omega)$ onto $H_{(2)}(\Omega;k)$ with $P_k u_l = u_l$ for $u_l \in H_{(2)}(\Omega;k)$, a semi norm $||| \cdot |||_{(2;k)}(\Omega)$ on $H_{(2)}(\Omega)$ which is zero on $H_{(2)}(\Omega;k)$ and decreasing with respect to k, and constants $C, C(\theta)$ such that

$$\|u_l\|_{(1)}(\Gamma_1 \setminus \bar{V}) + \|\nabla u_l\|_{(0)}(\Gamma_1 \setminus \bar{V}) + \|u_l\|_{(0)}(\Omega) \leq$$

$$CF + C(\theta)k^{-\frac{1}{2}+\theta}|||u|||_{(2;k)}(\Omega), \tag{10.9}$$

where $u_l = P_k u$, and

$$\|u\|_{(1)}(\Gamma_1 \setminus \bar{V}) + \|\nabla u\|_{(0)}(\Gamma_1 \setminus \bar{V}) + \|u\|_{(0)}(\Omega) \leq$$

$$CF + C(\theta)k^{-\frac{1}{2}+\theta}|||u|||_{(2;k)}(\Omega) \tag{10.10}$$

for all $u \in H_{(2)}(\Omega)$ solving (10.1), where $F = \|f\|_{(0)}(\Omega) + \|u\|_{(1)}(\omega)$.

This corollary shows that the geometrical condition of Theorem 10.2 can be substantially relaxed. We will show that it directly follows from Theorem 10.2.

Since the operator A preserves ellipticity, we can use C^2 diffeomorphic substitution and hence we can assume that $\Omega = \{x : |x - e(n)| < 1\}$, where $e(n) = (0, \ldots, 0, 1)$, and the origin is contained in V. To make use of Theorem 10.2 we will use the inversion $y = -(2|x|)^{-2}x$. In y coordinates Γ will be a (bounded) part of the (hyper)plane $\{y_n = -1\}$, Ω will be the lower half-space $\{y_n < -1\}$, and ω will contain $\{y : y_n < -1, |y| > R\}$. After a scaling, a translation, and possible shrinking ω we can assume that $\Omega = \{y : 0 < y_n < 1, |y - e(n)| < 0.5\}$ and $\omega = \{y : 0 < y_n < 1, 0.4 < |y - e(n)| < 0.5\}$. Now we can apply Theorem 10.2 (with void Γ_0), and complete the derivation of Corollary. $\square$

Now, also for a particular Ω in $\mathbf{R}^2$ we will eliminate the constraint on u in ω. Let $\Omega = \{x : 1 < |x| < R\}$, $\Gamma_0 = \{x : |x| = 1\}$, and $\Gamma_1 = \{x : |x| = R\}$. The principal part of the operator A in the polar coordinates (φ, r) is $a^{22}\partial_r^2 + 2a^{12}\partial_r\partial_\varphi + a^{11}\partial_\varphi^2$. Let $E = sup(a^{11})^{\frac{1}{2}}$ over Ω and $\varepsilon > 0$. We will write the (angular) Fourier series

$$u(, \varphi) = \sum_m u(m)e(\varphi; m),$$

where $e(\varphi; m) = \frac{1}{\sqrt{2\pi}} e^{im\varphi}$, and introduce the low frequency part of u

$$u_l(, \varphi) = \sum_{E^2|m|^2 < (1-\varepsilon)k^2} u(m)e(\varphi; m). \qquad (10.11)$$

Under a constraint on the high frequency component of u we have

Theorem 10.3 *Let $\theta > 0$.*
There are $C, C(\theta)$ such that

$$k\|u\|_{(0)}(\Gamma_1) + \|\nabla u\|_{(0)}(\Gamma_1) + \|u\|_{(1)}(\Omega) \leq$$

$$C(k\|u_0\|_{(0)}(\Gamma_0) + \|u_1\|_{(0)}(\Gamma_0) + \|f\|_{(0)}(\Omega)) + C(\theta)k^{-\frac{1}{2}+\theta}\|u - u_l\|_{(2)}(\Omega) \qquad (10.12)$$

and

$$k\|u_l\|_{(0)}(\Gamma_1) + \|\partial_r u_l\|_{(0)}(\Gamma_1) + \|u_l\|_{(1)}(\Omega) \leq$$

$$C(k\|u_0\|_{(0)}(\Gamma_0) + \|u_1\|_{(0)}(\Gamma_0) + \|f\|_{(0)}(\Omega) + C(\theta)k^{-\frac{1}{2}+\theta}\|u - u_l\|_{(2)}(\Omega) \qquad (10.13)$$

for all $u \in H^{(2)}(\Omega)$ solving (10.1), (10.2).

10.2   Proof Under "High Frequency" and Local Energy A Priori Constraints

In this section we will prove Theorem 10.1.
  Since $v = \chi u$, from (10.1) by using the Leibniz formula we yield

$$\left(\sum_{j,m=1}^{n} a_{jm}\partial_j \partial_m v + \sum_{j=1}^{n} a_j \partial_j v + a + kc + k^2\right)v = \chi f + A_1 u, \qquad (10.14)$$

where $A_1 u = 2\sum_{j,m=1}^{n} a_{jm}\partial_j \chi \partial_m u + \sum_{j=1}^{n} a_j \partial_j \chi u$.
  Observe that

$$a_{nn}\partial_n^2 v \partial_n v e^{-\tau x_n} = \frac{1}{2}\partial_n(a_{nn}(\partial_n v)^2 e^{-\tau x_n}) + \tau \frac{1}{2}a_{nn}(\partial_n v)^2 e^{-\tau x_n} - \frac{1}{2}(\partial_n a_{nn})(\partial_n v)^2 e^{-\tau x_n},$$

$$2a_{jn}\partial_j \partial_n v \partial_n v e^{-\tau x_n} = \partial_j(a_{jn}(\partial_n v)^2 e^{-\tau x_n}) - (\partial_j a_{jn})(\partial_n v)^2 e^{-\tau x_n}, \qquad (10.15)$$

$j = 1, \ldots, n-1$ and τ is a (large) positive number to be chosen later on. Integrating by parts with respect to x_j,

$$\int_{\mathbf{R}^{n-1}\times(0,1)} \sum_{j=1}^{n-1} a_{jm}\partial_j\partial_m v\partial_n v e^{-\tau x_n} =$$

$$-\int_{\mathbf{R}^{n-1}\times(0,1)} \sum_{j=1}^{n-1} a_{jm}\partial_m v\partial_j\partial_n v e^{-\tau x_n} - \int_{\mathbf{R}^{n-1}\times(0,1)} \sum_{j=1}^{n-1} (\partial_j a_{jm})\partial_m v\partial_n v e^{-\tau x_n}.$$

$$(10.16)$$

We have

$$\sum_{j,m=1}^{n-1} a_{jm}\partial_m v\partial_j\partial_n v e^{-\tau x_n} = \frac{1}{2}\sum_{j,m=1}^{n-1} \partial_n(a_{jm}\partial_m v\partial_j v e^{-\tau x_n})+$$

$$\frac{\tau}{2}\sum_{j,m=1}^{n-1} a_{jm}\partial_m v\partial_j v e^{-\tau x_n} - \frac{1}{2}\sum_{j,m=1}^{n-1} (\partial_n a_{jm})\partial_m v\partial_j v e^{-\tau x_n}, \qquad (10.17)$$

due to symmetry of a_{jm}.

To form an energy integral we multiply the both sides of (10.14) by $\partial_n v e^{-\tau x_n}$ and integrate by parts over $\mathbf{R}^{n-1}\times(0,\theta), 0 < \theta \le 1$, with using (10.15), (10.16), and (10.17) to yield

$$\frac{1}{2}\int_{\mathbf{R}^{n-1}} a_{nn}(\partial_n v)^2(,\theta)e^{-\tau\theta} - \frac{1}{2}\int_{\mathbf{R}^{n-1}} a_{nn}(\partial_n v)^2(,0) + \frac{\tau}{2}\int_{\mathbf{R}^{n-1}\times(0,\theta)} a_{nn}(\partial_n v)^2 e^{-\tau x_n} -$$

$$\frac{1}{2}\int_{\mathbf{R}^{n-1}} \sum_{j,m=1}^{n-1} a_{jm}\partial_j v\partial_m v(,\theta)e^{-\tau\theta} + \frac{1}{2}\int_{\mathbf{R}^{n-1}} \sum_{j,m=1}^{n-1} a_{jm}\partial_j v\partial_m v(,0) -$$

$$\frac{\tau}{2}\int_{\mathbf{R}^{n-1}\times(0,\theta)} \sum_{j,m=1}^{n-1} a_{jm}(,x_n)\partial_j v\partial_m v(,x_n)e^{-\tau x_n} + \int_{\mathbf{R}^{n-1}\times(0,\theta)} ckv\partial_n v(,x_n)e^{-\tau x_n} +$$

$$\frac{k^2}{2}\int_{\mathbf{R}^{n-1}} v^2(,\theta)e^{-\tau\theta} - \frac{k^2}{2}\int_{\mathbf{R}^{n-1}} v^2(,0) + \frac{\tau k^2}{2}\int_{\mathbf{R}^{n-1}\times(0,\theta)} v^2 e^{-\tau x_n} + \ldots =$$

$$\int_{\mathbf{R}^{n-1}\times(0,\theta)} \partial_n v\chi f e^{-\tau x_n} + \int_{\mathbf{R}^{n-1}\times(0,\theta)} \partial_n v A_1 u e^{-\tau x_n}, \qquad (10.18)$$

where ... denotes the sum of terms bounded by

$$C\int_{\mathbf{R}^{n-1}\times(0,1)} (\sum_{j=1}^n (\partial_j v)^2 + k^2 v^2)e^{-\tau x_n}.$$

We have

$$\sum_{j,m=1}^{n-1} a_{jm}(,x_n)\partial_j v \partial_m v(,x_n) = \sum_{j,m=1}^{n-1} a_{jm}(,x_n)\partial_j (v_l + v_h)\partial_m (v_l + v_h)(,x_n) =$$

$$\sum_{j,m=1}^{n-1} (a_{jm}(,x_n)\partial_j v_l \partial_m v_l(,x_n) + 2a_{jm}(,x_n)\partial_j v_l \partial_m v_h(,x_n) + a_{jm}(,x_n)\partial_j v_h \partial_m v_h(,x_n)) \le$$

$$\sum_{j,m=1}^{n-1} a_{jm}(,x_n)\partial_j v \partial_m v(,x_n) + C\delta \sum_{j=1}^{n-1} (\partial_j v_l)^2(,x_n) + C\delta^{-1} \sum_{j=1}^{n-1} (\partial_j v_h)^2(,x_n), \quad (10.19)$$

where we used the elementary inequality $AB \le \frac{\delta}{2}A^2 + \frac{1}{2\delta}B^2$ with $A = \partial_j v_l$, $B = \partial_m v_h$ and assumed that $0 < \delta < 1$.

According to the definition of E,

$$-\int_{\mathbf{R}^{n-1}} \sum_{j,m=1}^{n-1} a_{jm}(,x_n)\partial_j v_l \partial_m v_l(,x_n) \ge -\int_{\mathbf{R}^{n-1}} E^2 \sum_{j=1}^{n-1} (\partial_j v_l)^2(,x_n) =$$

$$-\int_{\mathbf{R}^{n-1}} E^2 \sum_{j=1}^{n-1} \xi_j^2 |\mathscr{F} v_l|^2(,x_n) \ge -\int_{\mathbf{R}^{n-1}} k^2(1-\varepsilon)|\mathscr{F} v_l|^2(,x_n) =$$

$$-(1-\varepsilon)k^2 \int_{\mathbf{R}^{n-1}} v_l^2(,x_n), \quad (10.20)$$

where we used that $\mathscr{F} v_l(,\xi',x_n) = 0$ when $-E^2|\xi'|^2 < -(1-\varepsilon)k^2$, due to (10.7), and utilized the Parseval's identity. Similarly,

$$\int_{\mathbf{R}^{n-1}} (\sum_{j=1}^{n-1} \partial_j v_l)^2 \le Ck^2 \int_{\mathbf{R}^{n-1}} v_l^2. \quad (10.21)$$

Therefore, using (10.19) and (10.20) we obtain

$$-\int_{\mathbf{R}^{n-1}} \sum_{j,m=1}^{n-1} a_{jm}(,x_n)\partial_j v_l \partial_m v_l(,x_n) \ge$$

$$-(1-\varepsilon)k^2 \int_{\mathbf{R}^{n-1}} v_l^2(,x_n) - C\delta k^2 \int_{\mathbf{R}^{n-1}} v_l^2(,x_n) - \frac{C}{\delta} \int_{\mathbf{R}^{n-1}} \sum_{j=1}^{n-1} (\partial_j v_h)^2(,x_n).$$

$$(10.22)$$

Since

$$\int_{\mathbf{R}^{n-1}} v_l v_h(,x_n) = 0, \quad \int_{\mathbf{R}^{n-1}} \partial_j v_l \partial_j v_h(,x_n) = 0, \quad j = 1, \ldots, n,$$

we have

$$\int_{\mathbf{R}^{n-1}} v^2(,x_n) = \int_{\mathbf{R}^{n-1}} (v_l + v_h)^2(,x_n) = \int_{\mathbf{R}^{n-1}} v_l^2(,x_n) + \int_{\mathbf{R}^{n-1}} v_h^2(,x_n),$$

$$\int_{\mathbf{R}^{n-1}} (\partial_j v)^2(,x_n) = \int_{\mathbf{R}^{n-1}} (\partial_j v_l)^2(,x_n) + \int_{\mathbf{R}^{n-1}} (\partial_j v_h)^2(,x_n). \tag{10.23}$$

Hence from (10.18) and (10.22) by using the inequalities $2AB \le A^2 + B^2$ and $\frac{1}{C} < a_{nn}$ (due to the ellipticity of A) we conclude that

$$\frac{1}{C} \int_{\mathbf{R}^{n-1}} (\partial_n v)^2(,\theta) e^{-\tau\theta} + \frac{\tau}{C} \int_{\mathbf{R}^{n-1} \times (0,\theta)} (\partial_n v)^2 e^{-\tau x_n} +$$

$$(\varepsilon - C\delta) \frac{k^2}{2} \int_{\mathbf{R}^{n-1}} v_l^2(,\theta) e^{-\tau\theta} + \frac{\tau(\varepsilon - C\delta)k^2}{2} \int_{\mathbf{R}^{n-1} \times (0,\theta)} v_l^2 e^{-\tau x_n} +$$

$$\frac{k^2}{2} \int_{\mathbf{R}^{n-1}} v_h^2(,\theta) e^{-\tau\theta} + \frac{\tau}{2} \int_{\mathbf{R}^{n-1} \times (0,\theta)} v_l^2 e^{-\tau x_n} \le$$

$$C(\int_{\mathbf{R}^{n-1}} ((\partial_n v)^2(,0) + k^2 v^2(,0)) + \int_\Omega f^2 e^{-\tau x_n} + \int_\omega (A_1 u)^2 e^{-\tau x_n}) +$$

$$C(\int_{\mathbf{R}^{n-1}} (\frac{1}{\delta} \sum_{j=1}^{n-1} (\partial_j v_h)^2(,\theta) e^{-\tau} + \int_{\mathbf{R}^{n-1} \times (0,\theta)} (\frac{1}{\delta} \sum_{j=1}^{n-1} (\partial_j v)^2 + (\partial_n v)^2 + k^2 v^2) e^{-\tau x_n}).$$

Let $\delta = \frac{\varepsilon}{2C}$ and use (10.23) again, then we yield the inequality

$$\int_{\mathbf{R}^{n-1}} (\partial_n v)^2(,\theta) e^{-\tau\theta} + \tau \int_{\mathbf{R}^{n-1} \times (0,\theta)} (\partial_n v)^2 e^{-\tau x_n} +$$

$$k^2 \int_{\mathbf{R}^{n-1}} v^2(,\theta) e^{-\tau\theta} + \tau k^2 \int_{\mathbf{R}^{n-1} \times (0,1)} v^2 e^{-\tau x_n} \le$$

$$C(\int_{\mathbf{R}^{n-1}} ((\partial_n v)^2(,0) + k^2 v^2(,0)) + \int_\Omega f^2 e^{-\tau x_n} + \int_\omega (A_1 u)^2 e^{-\tau x_n}) +$$

$$C(\int_{\mathbf{R}^{n-1}} \sum_{j=1}^{n-1} (\partial_j v_h)^2(,\theta) e^{-\tau} +$$

$$\int_{\mathbf{R}^{n-1} \times (0,1)} (\sum_{j=1}^{n-1} (\partial_j v_h)^2 + \sum_{j=1}^{n-1} (\partial_j v_l)^2 + (\partial_n v)^2 + k^2 v^2) e^{-\tau x_n}). \tag{10.24}$$

Choosing and fixing sufficiently large τ (depending on the same parameters as C) to absorb the three last terms on the right side in (10.24) by the left side we obtain

$$
\int_{\mathbf{R}^{n-1}} (\partial_n v)^2(,\theta) + \int_{\mathbf{R}^{n-1}\times(0,\theta)} (\partial_n v)^2 + k^2 \int_{\mathbf{R}^{n-1}} v^2(,\theta) + k^2 \int_{\mathbf{R}^{n-1}\times(0,\theta)} v^2 \leq
$$

$$
C(\int_{\mathbf{R}^{n-1}} ((\partial_n v)^2(,0) + k^2 v^2(,0)) + \int_{\Omega} f^2 + \int_{\omega} (A_1 u)^2) +
$$

$$
\int_{\mathbf{R}^{n-1}} \sum_{j=1}^{n-1} (\partial_j v_h)^2(,\theta)) + \int_{\mathbf{R}^{n-1}\times(0,1)} \sum_{j=1}^{n-1} (\partial_j v_h)^2). \tag{10.25}
$$

Integrating the inequality (10.25) with respect θ over $(0, 1)$, dropping the first two terms on the left side, and recalling that $v = \chi u$ we yield

$$
k^2 \|v\|_{(0)}^2(\Omega) \leq C(\|u_1\|_{(0)}^2(\Gamma_0) + k^2 \|u_0\|_{(0)}^2(\Gamma_0)
$$

$$
+ \|f\|_{(0)}^2(\Omega) + \|u\|_{(1)}^2(\omega)
$$

$$
+ \sum_{j=1}^{n-1} \|\partial_j v_h\|_{(0)}^2(\mathbf{R}^{n-1}\times(0,1)).
$$

Recalling the definition of a high frequency norm (10.8), using that $u = v$ on $\Omega \setminus V$, and dividing by k^2 we obtain (10.4).

Due to (10.23), (10.3) follows from (10.4). $\qquad\square$

10.3 Proof Under High Frequency Constraints

In this section we will prove Theorem 10.2.

From the proof of Theorem 10.1, (10.25), we have

$$
\int_{\mathbf{R}^{n-1}} (\partial_n v)^2(,1) + \int_{\mathbf{R}^{n-1}\times(0,1)} (\partial_n v)^2 + k^2 \int_{\mathbf{R}^{n-1}} v^2(,1) + k^2 \int_{\mathbf{R}^{n-1}\times(0,1)} v^2 \leq
$$

$$
C(\int_{\mathbf{R}^{n-1}} ((\partial_n v)^2(,0) + k^2 v^2(,0)) + \int_{\Omega} f^2 + \int_{\omega} (A_1 u)^2) +
$$

$$
\int_{\mathbf{R}^{n-1}} \sum_{j=1}^{n-1} (\partial_j v_h)^2(,1) + \int_{\mathbf{R}^{n-1}\times(0,1)} \sum_{j=1}^{n-1} (\partial_j v_h)^2).
$$

Using (10.21) and (10.23) we obtain

$$\int_{\mathbf{R}^{n-1}} (\partial_n v)^2(,1) + \int_{\mathbf{R}^{n-1}} \sum_{j=1}^{n-1} (\partial_j v_l)^2(,1) + \int_{\mathbf{R}^{n-1} \times (0,1)} (\partial_n v)^2 + \int_{\mathbf{R}^{n-1} \times (0,1)} \sum_{j=1}^{n-1} (\partial_j v_l)^2 +$$

$$k^2 \int_{\mathbf{R}^{n-1}} v^2(,1) + k^2 \int_{\mathbf{R}^{n-1} \times (0,1)} v^2 \le$$

$$C \Big(\int_{\mathbf{R}^{n-1}} ((\partial_n v)^2(,0) + k^2 v^2(,0) + \int_\Omega f^2 + \int_\omega (A_1 u)^2) +$$

$$\int_{\mathbf{R}^{n-1}} \sum_{j=1}^{n-1} (\partial_j v_h)^2(,1)) + \int_{\mathbf{R}^{n-1} \times (0,1)} \sum_{j=1}^{n-1} (\partial_j v_h)^2).$$

By increasing C and using (10.23) again, it gives

$$\int_{\mathbf{R}^{n-1}} \sum_{j=1}^{n} (\partial_j v)^2(,1) + \int_{\mathbf{R}^{n-1} \times (0,1)} \sum_{j=1}^{n} (\partial_j v)^2 + k^2 \int_{\mathbf{R}^{n-1}} v^2(,1) + k^2 \int_{\mathbf{R}^{n-1} \times (0,1)} v^2 \le$$

$$C \Big(\int_{\mathbf{R}^{n-1}} ((\partial_n v)^2(,0) + k^2 v^2(,0)) + \int_\Omega f^2 + \int_\omega (A_1 u)^2) +$$

$$\int_{\mathbf{R}^{n-1}} \sum_{j=1}^{n-1} (\partial_j v_h)^2(,1) + \int_{\mathbf{R}^{n-1} \times (0,1)} \sum_{j=1}^{n-1} (\partial_j v_h)^2).$$

Therefore,

$$\|\partial_n v(,1)\|_{(0)}^2 (\mathbf{R}^{n-1}) + \|v(,1)\|_{(1)}^2 (\mathbf{R}^{n-1}) + \|v\|_{(1)}^2 (\mathbf{R}^{n-1} \times (0,1)) +$$

$$k^2 \|v(,1)\|_{(0)} (\mathbf{R}^{n-1}) + k^2 \|v(,1)\|_{(0)} (\mathbf{R}^{n-1} \times (0,1)) \le$$

$$C (\|u_1\|_{(0)}^2 (\Gamma_0) + \|u_0\|_{(1)}^2 (\Gamma_0) + \|f\|_{(0)}^2 (\Omega) + \|u\|_{(1)}^2 (\omega) +$$

$$\|v_h(,1)\|_{(1)}^2 (\mathbf{R}^{n-1}) + \|v_h\|_{(1)}^2 (\mathbf{R}^{n-1} \times (0,1)). \tag{10.26}$$

Let v_h^* be the extension of v_h onto $\mathbf{R}^n$ constructed in [12, p. 14]. As follows from [12], with natural choice of functional spaces,

$$\|v_h^*\|_{(2)} (\mathbf{R}^n) \le C \|v_h\|_{(2)} (\mathbf{R}^{n-1} \times (0,1)). \tag{10.27}$$

From the construction in [12] it follows that $(v_h)^* = (v^*)_h$, i.e. that the Fourier transform of $(v_h)^*$ with respect to $(x_1, \ldots, x_{n-1})$ is zero when $|\xi'| \le \sqrt{1 - \varepsilon \frac{k}{E}}$. By known trace theorems for Sobolev spaces [12, p. 42],

$$\|v_h(,1)\|_{(1)}^2 (\mathbf{R}^{n-1}) \le C(\theta) \|v_h^*\|_{(\frac{3}{2}+\theta)} (\mathbf{R}^n). \tag{10.28}$$

Let $V_h^*(\xi)$ be the Fourier transform of $v_h^*(x)$. As known [12, p. 30], for Sobolev norms,

$$\|v_h^*\|_{(\frac{3}{2}+\theta)}^2(\mathbf{R}^n) = \int (1 + |\xi'|^2 + \xi_n^2)^{\frac{3}{2}+\theta}|V_h^*(\xi)|^2 d\xi \leq$$

$$k^{-1+2\theta} \int k^{1-2\theta}(1 + |\xi'|^2 + \xi_n^2)^{-\frac{1}{2}+\theta}(1 + |\xi'|^2 + \xi_n^2)^2|V_h^*(\xi)|^2 d\xi \leq$$

$$Ck^{-1+2\theta} \int (1 + |\xi'|^2 + \xi_n^2)^2|V_h^*(\xi)|^2 d\xi =$$

$$Ck^{-1+2\theta}\|v_h^*\|_{(2)}^2(\mathbf{R}^n) \leq Ck^{-1+2\theta}\|v_h\|_{(2)}^2(\mathbf{R}^{n-1} \times (0,1)),$$

due to (10.27). Here we used that $k^{1-2\theta}(1 + |\xi'|^2 + \xi_n^2)^{-\frac{1}{2}+\theta} \leq C$ on the actual integration domain (where $|V_h^*(\xi)| > 0$ and hence $\frac{k}{C} < |\xi'|$). So using (10.28) we yield

$$\|v_h(,1)\|_{(1)}^2(\mathbf{R}^{n-1}) \leq Ck^{-1+2\theta}\|v_h\|_{(2)}^2(\mathbf{R}^{n-1} \times (0,1)). \tag{10.29}$$

Similarly,

$$\|v_h\|_{(1)}^2(\mathbf{R}^{n-1} \times (0,1)) \leq \|v_h^*\|_{(1)}^2(\mathbf{R}^n) = \int (1 + |\xi'|^2 + \xi_n^2)|V_h^*(\xi)|^2 d\xi =$$

$$k^{-2} \int (k^2(1 + |\xi'|^2 + \xi_n^2)^{-1})(1 + |\xi'|^2 + \xi_n^2)^2|V_h^*(\xi)|^2 d\xi \leq$$

$$Ck^{-2} \int (1 + |\xi'|^2 + \xi_n^2)^2|V_h^*(\xi)|^2 d\xi =$$

$$Ck^{-2}\|v_h^*\|_{(2)}^2(\mathbf{R}^n) \leq Ck^{-2}\|v_h\|_{(2)}^2(\mathbf{R}^{n-1} \times (0,1)),$$

due to (10.27). So

$$\|v_h\|_{(1)}^2(\mathbf{R}^{n-1} \times (0,1)) \leq Ck^{-2}\|v_h\|_{(2)}^2(\mathbf{R}^{n-1} \times (0,1)). \tag{10.30}$$

Using that $v = \chi u$, from (10.26), (10.29), and (10.30), we obtain (10.6). As in the proof of Theorem 10.1, (10.5) follows from (10.6) because of (10.23). The proof is complete. $\qquad\square$

10.4 Proof for Annular Domains

In this section we will prove Theorem 10.3. We will use polar coordinates and the operator A in these coordinates.

From (10.1) we yield

$$a^{22}\partial_r^2 u + 2a^{12}\partial_\varphi\partial_r u + a^{11}\partial_\varphi^2 u + a^1\partial_\varphi u + a^2\partial_r u + au + kcu + k^2 u = f \quad (10.31)$$

in $[0, 2\pi] \times (1, R)$. Repeating the argument from the proof of Theorem 10.1 (multiplying the both parts of (10.31) by $\partial_r u e^{-\tau r}$ and integrating by parts over Ω with using angular periodicity) we will have

$$\frac{1}{2}\int_{[0,2\pi]} a^{22}(\partial_r u)^2(, R)e^{-\tau R}R - \frac{1}{2}\int_{[0,2\pi]} a^{22}(\partial_r u)^2(, 1)e^{-\tau} + \frac{\tau}{2}\int_{\Omega} a^{22}(\partial_r u)^2 e^{-\tau r}r -$$

$$\frac{1}{2}\int_{[0,2\pi]} a^{11}(\partial_\varphi u)^2(, R)e^{-\tau R}R + \frac{1}{2}\int_{[0,2\pi]} a^{11}(\partial_\varphi u)^2(, 1)e^{-\tau} - \frac{\tau}{2}\int_{\Omega} a^{11}(\partial_\varphi u)^2 e^{-\tau r}r +$$

$$\frac{k^2}{2}\int_{[0,2\pi]} u^2(, R)e^{-\tau R}R - \frac{k^2}{2}\int_{[0,2\pi]} u^2(, 1) + \frac{\tau k^2}{2}\int_{\Omega} u^2 e^{-\tau r}r + \ldots =$$

$$\int_{\Omega} \partial_r u f e^{-\tau r}, \quad (10.32)$$

where ... denotes the sum of terms bounded by

$$C \int_{\Omega} ((\partial_\varphi u)^2 + (\partial_r u)^2 + k^2 u^2)e^{-\tau r}.$$

To handle the negative terms on the left side of (10.32) we use that

$$-\int_{[0,2\pi]} a^{11}(\partial_\varphi u)^2(, r) = -\int_{[0,2\pi]} a^{11}(\partial_\varphi u_l + \partial_\varphi u_h)^2(, r) \geq$$

$$-\int_{[0,2\pi]} a^{11}(\partial_\varphi u_l)^2(, r) - \delta\int_{[0,2\pi]} (\partial_\varphi u_l)^2(, r) - \frac{C}{\delta}\int_{[0,2\pi]} (\partial_\varphi u_h)^2(, r).$$

As in the proof of Theorem 10.1, using (10.11), from the Parseval's identity for the Fourier series, we have

$$-\int_{[0,2\pi]} a^{11}(\partial_\varphi u_l)^2(, r) \geq -\int_{[0,2\pi]} E^2(\partial_\varphi u_l)^2(, r) \geq$$

$$-(1-\varepsilon)k^2\int_{[0,2\pi]} u_l^2(, r) \geq -(1-\varepsilon)k^2\int_{[0,2\pi]} u^2(, r)$$

and

$$-\int_{[0,2\pi]} (\partial_\varphi u_l)^2(, r) \geq -Ck^2\int_{[0,2\pi]} (u_l)^2(, r) \geq -Ck^2\int_{[0,2\pi]} u^2(, r). \quad (10.33)$$

Hence from (10.32) we conclude that

$$\frac{1}{2}\int_{(0,2\pi)} a^{22}(\partial_r u)^2(, R)e^{-\tau R}R + \frac{\tau}{2}\int_{(0,2\pi)\times(1,R)} a^{22}(\partial_r u)^2 e^{-\tau r}r +$$

$$(\varepsilon - C\delta)\frac{k^2}{2}\int_{(0,2\pi)} u^2(, R)e^{-\tau R}R + \frac{\tau(\varepsilon - C\delta)k^2}{2}\int_{(0,2\pi)\times(1,R)} u^2 e^{-\tau r}r \le$$

$$C(\frac{1}{2}\int_{(0,2\pi)}((\partial_r u)^2(, 1) + k^2 u^2(, 1)) + \int_{(0,2\pi)\times(1,R)} f^2 e^{-\tau r}r +$$

$$\frac{C}{\delta}(\int_{(0,2\pi)}((\partial_r u_h)^2 + (\partial_\varphi u_h)^2)(, R)e^{-\tau R}R + \tau\int_{(0,2\pi)\times(1,R)}((\partial_r u_h)^2 + (\partial_\varphi u_h)^2)e^{-\tau r}r) +$$

$$\int_{(0,2\pi)\times(1,R)}((\partial_r u)^2 + (\partial_\varphi u)^2 + k^2 u^2)e^{-\tau r}r).$$

Choosing $\delta = \frac{\varepsilon}{2C}$ and using that $\frac{1}{C} < a^{22}$ we yield

$$\int_{(0,2\pi)}((\partial_r u)^2(, R) + k^2 u^2(, R))e^{-\tau R}R + \tau\int_{(0,2\pi)\times(1,R)}((\partial_r u)^2 + k^2 u^2)e^{-\tau r}r \le$$

$$C(\int_{(0,2\pi)}((\partial_r u)^2(, 1) + k^2 u^2(, 1)) + \int_{(0,2\pi)\times(1,R)} f^2 e^{-\tau r}r +$$

$$\int_{(0,2\pi)}(\partial_r u_h)^2(, R)e^{-\tau R}R + \tau\int_{(0,2\pi)\times(1,R)}(\partial_r u_h)^2 e^{-\tau r}r +$$

$$\int_{(0,2\pi)\times(1,R)}((\partial_r u)^2 + (\partial_\varphi u)^2 + k^2 u^2)e^{-\tau r}r). \tag{10.34}$$

From the definition of u_l we have

$$\int_{(0,2\pi)\times(1,R)}(\partial_\varphi u)^2 e^{-\tau r}r = \int_{(0,2\pi)\times(1,R)}((\partial_\varphi u_l)^2 + (\partial_\varphi u_h)^2)e^{-\tau r}r \le$$

$$Ck^2\int_{(0,2\pi)\times(1,R)} u^2 + \int_{(0,2\pi)\times(1,R)}(\partial_\varphi u_h)^2 e^{-\tau r}r,$$

when we apply (10.33). So from (10.34) we obtain

$$\int_{(0,2\pi)}((\partial_r u)^2(, R) + k^2 u^2(, R))e^{-\tau R}R + \tau\int_{(0,2\pi)\times(1,R)}((\partial_r u)^2 + k^2 u^2)e^{-\tau r}r \le$$

$$C(\int_{(0,2\pi)}((\partial_r u)^2(, 1) + k^2 u^2(, 1)) + \int_{(0,2\pi)\times(1,R)} f^2 e^{-\tau r}r +$$

$$\int_{(0,2\pi)} ((\partial_r u_h)^2 + (\partial_\varphi u_h)^2)(, R)e^{-\tau R}R + \tau \int_{(0,2\pi)\times(1,R)} ((\partial_r u_h)^2 + (\partial_\varphi u_h)^2)e^{-\tau r}r +$$

$$\int_{(0,2\pi)\times(1,R)} ((\partial_r u)^2 + k^2 u^2)e^{-\tau r}r).$$

Now, choosing and fixing τ sufficiently large (but depending on the same quantities as C) to absorb the last term on the right side by the left side we yield

$$\int_{(0,2\pi)} ((\partial_r u)^2(, R) + k^2 u^2(, R)) + \tau \int_{(0,2\pi)\times(1,R)} ((\partial_r u)^2 + k^2 u^2) \le$$

$$C(\int_{(0,2\pi)} ((\partial_r u)^2(, 1) + k^2 u^2(, 1)) + \int_{(0,2\pi)\times(1,R)} f^2) +$$

$$\int_{(0,2\pi)} ((\partial_r u_h)^2 + (\partial_\varphi u_h)^2)(, R) + \int_{(0,2\pi)\times(1,R)} ((\partial_r u_h)^2 + (\partial_\varphi u_h)^2)). \tag{10.35}$$

By trace theorems for Sobolev spaces

$$\|u_h(, R)\|_{(1)}(\Gamma_1) + \|\partial_r u_h(, R)\|_{(0)}(\Gamma_1) \le C(\theta)\|u_h\|_{(\frac{3}{2}+\theta))}(\Omega).$$

For the high frequency part

$$\|u_h\|^2_{(\frac{3}{2}+\theta)}(\Omega)) \le C k^{2\theta-1}\|u_h\|^2_{(2)}(\Omega)), \tag{10.36}$$

so from (10.35) we obtain (10.12).

Since the Sobolev norms of u_l are bounded by Sobolev norms of u, (10.13) follows from (10.12).

The proof is complete. $\qquad\square$

Conclusion

We think that the results of this paper can extended onto higher order elliptic equations and systems. An important question is about minimal a priori constraints on the high frequency part of a solution. It is feasible that semi norms $\||\cdot\||_{(m;k)}(\Omega), m = 2$, in Theorem 10.2 can be replaced by a similar semi norm with $m = 1$, imposing only natural energy constraints on the high frequency part of u. Moreover, a complete justification of increasing stability can be obtained by proving that there are growing invariant subspaces where the solution of the Cauchy problem (10.1), (10.2) is Lipschitz stable. We will give one of related conjectures.

(continued)

Let Ω be a Lipschitz bounded domain introduced in Theorem 10.1. Let us assume that there is an unique solution $u \in H^{(1)}(\Omega)$ of the following Neumann problem

$$Au + cku + k^2 u = 0 \text{ in } \Omega,$$

$$\partial_\nu u = 0 \text{ on } \partial\Omega \setminus \Gamma_1, \ \partial_\nu u = g \in H^{(-\frac{1}{2})}(\Gamma_1) \text{ on } \Gamma_1.$$

The operator B mapping g into $u_0 = u$ on Γ_0 is compact from $L^2(\Gamma_1)$ into $L^2(\Gamma_0)$. Hence it admits the singular value decomposition consisting of complete orthonormal system of functions $g_m, m = 1, 2, \ldots$ in $L^2(\Gamma_1)$ and corresponding singular values $\sigma_m \geq \sigma_{m+1} > 0$ (eigenfunction and square roots of eigenvalues of $B^* B$). The conjecture is that there are positive numbers δ_1, δ_2 depending only on A, c and Ω (but not on k) such that $\sigma_m > \delta_1$ when $m < \delta_2 k$. This conjecture for some interesting plane Ω was numerically confirmed in [9].

Use of low frequency zone does not need any convexity type assumptions and for this reason is very promising for applications. In the recent paper [9] we studied this phenomenon on more detail and gave regularization schemes for numerical solution incorporating the increasing stability. We gave several numerical examples of increasing stability for the Helmholtz equation in some interesting plane domains, admitting or not admitting explicit analytical solution and complete analytic justification. It is important to develop theory and collect numerical evidence of the increased stability for more complicated geometries and for systems by using results in [3,5].

The increasing stability is expected in the inverse source problem, where one looks for f in the Helmholtz equation $(\Delta + k^2)u = f$ (not depending on k) in $\Omega = \{x : 1 < |x| < R\}$ from the Cauchy data $u, \partial_\nu u$ on $\Gamma = \{x : |x| = 1\}$, $k_* < k < k^*$. One needs to obtain stability estimates improving with growing k^* and to give a numerical evidence of better resolution for larger k^*.

It was (numerically) observed, that use of only low frequency zone can produce a stable solution of the inverse problem, where one looks for a speed of the propagation from all possible boundary measurements. One can look at the linearized problem: find f (supported in $\Omega \subset \mathbf{R}^3$) from

$$\int_\Omega f(y) \frac{e^{ki|x-y|}}{|x-y|} \frac{e^{ki|z-y|}}{|z-y|} dy$$

given for $x, z \in \Gamma \subset \partial\Omega$. The closest analytic results on improving stability are obtained in [8, 10] for the Schrödinger potential.

Acknowledgements This research is supported in part by the Emylou Keith and Betty Dutcher Distinguished Professorship and the NSF grant DMS 10-08902.

References

1. Aralumallige Subbarayappa, D., Isakov, V.: On increased stability in the continuation of the Helmholtz equation. Inverse Probl. **23**, 1689–1697 (2007)
2. Aralumallige Subbarayappa, D., Isakov, V.: Increasing stability of the continuation for the Maxwell system. Inverse Probl. **26**, 074005 (2010)
3. Eller, M., Isakov, V., Nakamura, G., Tataru, D.: Uniqueness and stability in the cauchy problem for Maxwell' and elasticity systems. In: Cioranescu, D., Lions, J.-L. (eds.) Nonlinear Partial Differential Equations and Their Applications, College de France Seminar, vol. 14. Studies in Mathematics and its Applications, vol. 31, pp. 329–351. Elsevier Science, North-Holland (2002)
4. Hrycak, T., Isakov, V.: Increased stability in the continuation of solutions to the Helmholtz equation. Inverse Probl. **20**, 697–712 (2004)
5. Isakov, V.: Inverse Problems for Partial Differential Equations. Springer, New York (2006)
6. Isakov, V.: Increased stability in the continuation for the Helmholtz equation with variable coefficient. Contemp. Math. **426**, 255–269 (2007)
7. Isakov, V.: Increased stability in the Cauchy problem for some elliptic equations. In: Bardos, C., Fursikov, A. (eds.) Instability in Models Connected with Fliud Flow. International Mathematical Series, vols. 6–7, pp. 337–360. Springer, New York (2007)
8. Isakov, V.: Increasing stability for the Schrödinger potential from the Dirichlet-to Neumann map. Discrete Continuous Dyn. Syst. Ser. **4**, 631–641 (2011)
9. Isakov, V., Kindermann, S.: Regions of stability in the Cauchy problem for the Helmholtz equation. Methods Appl. Anal. **18**, 1–30 (2011)
10. Isakov, V., Wang, J.-N.: Increasing stability for determining the potential in the Schrödinger equation with attenuation from the Dirichlet-to Neumann map. Inverse Probl. Imag. **8** (2014). arXive:1309.2840 (to appear)
11. John, F.: Continuous dependence on data for solutions of partial differential equations with a prescribed bound. Commun. Pure Appl. Math. **13**, 551–587 (1960)
12. Lions, J.-L., Magenes, E.: Non-Homogeneous Boundary Value Problems and Applications, I. Springer, New York (1972)

Chapter 11
Simultaneous Observability of Plates

Vilmos Komornik and Paola Loreti

Abstract We consider a finite number of independently vibrating rectangular plates, having a common side. We investigate whether the observation of the total force exerted by the plates at this common side is sufficient for the determination of the complete behaviour of each plate.

11.1 Introduction

The boundary observability of vibrating strings, membranes, beams and plates has been investigated intensively during the last 30 years by several different methods: by using multipliers, Fourier series or microlocal analysis the most frequently, see, e.g., [7, 10] and their references. In [1–4] a system of vibrating beams or strings was considered, having a common endpoint. By using a generalization of a classical inequality of Ingham, it was established that by measuring the total force exerted by the strings at this point, it is possible to determine the separate behaviour of each string, under some natural assumptions on the lengths of the strings. The purpose of this note is to show how to generalize these results to vibrating plates.

V. Komornik (✉)
Département de Mathématique, Université de Strasbourg, 7 rue René Descartes, 67084 Strasbourg Cedex, France
e-mail: vilmos.komornik@math.unistra.fr

P. Loreti
Sapienza Università di Roma, Dipartimento di Scienze di Base e Applicate per l'Ingegneria, via A. Scarpa n. 16, 00161 Roma, Italy
e-mail: paola.loreti@sbai.uniroma1.it

© Springer International Publishing Switzerland 2014   219
A. Favini et al. (eds.), *New Prospects in Direct, Inverse and Control Problems for Evolution Equations*, Springer INdAM Series 10,
DOI 10.1007/978-3-319-11406-4_11

Given finitely many positive numbers $\ell, \ell_1, \ldots, \ell_N$, we introduce the rectangular domains $\Omega_j := (0, \ell_j) \times (0, \ell)$ with boundaries $\Gamma_j := \partial \Omega_j$, $j = 1, \ldots, N$, and we consider the following system of independently vibrating hinged plates:

$$
\begin{cases}
u_j'' + \Delta^2 u_j = 0 & \text{in} \quad \mathbb{R} \times \Omega_j, \\
u_j = \Delta u_j = 0 & \text{on} \quad \mathbb{R} \times \Gamma_j, \\
u_j(0) = u_{j0} \quad \text{and} \quad u_j'(0) = u_{j1} & \text{in} \quad \Omega_j, \\
j = 1, \ldots, N.
\end{cases}
\tag{11.1}
$$

We investigate whether the observation of the total force exerted by these plates at this common side $\Gamma := \{0\} \times (0, \ell)$ is sufficient for the determination of the complete behaviour of each plate. More precisely, introducing for some given positive number T the function

$$
f(t, y) := \sum_{j=1}^{N} \frac{\partial u_j}{\partial \nu}(t, 0, y), \quad (t, y) \in (0, T) \times (0, \ell),
$$

we ask whether the linear map

$$
(u_{10}, \ldots, u_{N0}, u_{11}, \ldots, u_{N1}) \mapsto f
$$

is one-to-one for some natural Hilbert space of the initial data.

In order to formulate our theorem we recall some earlier results. We need some notations.

For each $j = 1, \ldots, N$ we consider the vector space Z_j spanned by the eigenfunctions of $-\Delta$ in $H_0^1(\Omega_j)$, i.e. the vector space spanned by the functions

$$
e_{jkn}(x, y) := \sin \frac{k\pi x}{\ell_j} \sin \frac{n\pi y}{\ell}, \quad k = 1, 2, \ldots, \quad n = 1, 2, \ldots.
$$

For each real number s we denote by D_j^s the Hilbert space obtained by completion from Z_j endowed with the Euclidean norm

$$
\left\| \sum_{k,n} c_{kn} e_{jkn} \right\|_s := \left(\sum_{k,n} (k^2 + n^2)^s |c_{kn}|^2 \right)^{1/2}.
$$

We observe that

$$
D_j^0 = L^2(\Omega_j), \quad D_j^1 = H_0^1(\Omega_j) \quad \text{and} \quad D_j^2 = H^2(\Omega_j) \cap H_0^1(\Omega_j)
$$

with equivalent norms.

Let us also set $D^s := D_1^s \times \cdots \times D_N^s$.

Since the system (11.1) is uncoupled, it follows from classical results (see, e.g., [10] or [7]) that for any given

$$(u_{10}, \ldots, u_{N0}, u_{11}, \ldots, u_{N1}) \in D^s \times D^{s-2}$$

the system (11.1) has a unique solution

$$u \in C(\mathbb{R}; D^s) \cap C^1(\mathbb{R}; D^{s-2}).$$

If $s \geq 2$, then it follows from the "hidden regularity" results of Lions [8–10] that[1]

$$f_y \in L^2((0, T) \times (0, \ell)),$$

and

$$\int_0^T \int_0^\ell |f_y(t, y)|^2 \, dy \, dt \leq c_2 \left(\|u_0\|_2^2 + \|u_1\|_0^2 \right) \tag{11.2}$$

for all $(u_0, u_1) \in D^2 \times D^0$, with a constant depending only on T.

In what follows we consider only solutions of (11.1) corresponding to initial data $(u_0, u_1) \in D^2 \times D^0$. Our main result is the following inverse inequality:

Theorem 11.1 *For almost all choices of* $(\ell_1, \ldots, \ell_N) \in (0, \infty)^N$, *the solutions of* (11.1) *satisfy the estimates*

$$\|u_0\|_s^2 + \|u_1\|_{s-2}^2 \leq C_s \int_0^T \int_0^\ell |f_y(t, y)|^2 \, dy \, dt \tag{11.3}$$

for every $s < 1$, *with a suitable constant* C_s, *independent of the particular choice of the initial data.*

11.2 Review of the Simultaneous Observability of Beams

Let us first consider the analogous system of independently vibrating hinged beams:

$$\begin{cases} u_{j,tt} + u_{j,xxxx} = 0 & \text{in} \quad \mathbb{R} \times (0, \ell_j), \\ u_j = u_{j,xx} = 0 & \text{on} \quad \mathbb{R} \times \{0, \ell_j\}, \\ u_j(0) = u_{j0} \quad \text{and} \quad u'_j(0) = u_{j1} & \text{in} \quad (0, \ell_j), \\ j = 1, \ldots, N \end{cases} \tag{11.4}$$

---

[1]The subscript y stands for the corresponding partial derivative.

with the observation

$$f(t) := \sum_{j=1}^{N} u_{j,x}(t,0), \quad t \in (0,T).$$

Now the vector spaces Z_j are spanned by the functions

$$e_{jk}(x) := \sin \frac{k\pi x}{\ell_j}, \quad k = 1, 2, \ldots,$$

and the Hilbert spaces D_j^s are obtained by completing Z_j with respect to the Euclidean norms

$$\left\| \sum_k c_k e_{jk} \right\|_s := \left(\sum_k k^{2s} |c_k|^2 \right)^{1/2}.$$

We have

$$D_j^0 = L^2(0,\ell_j), \quad D_j^1 = H_0^1(0,\ell_j) \quad \text{and} \quad D_j^2 = H^2(0,\ell_j) \cap H_0^1(0,\ell_j)$$

with equivalent norms.

Setting $D^s := D_1^s \times \cdots \times D_N^s$ again, it follows from classical results that for any given

$$(u_{10}, \ldots, u_{N0}, u_{11}, \ldots, u_{N1}) \in D^s \times D^{s-2}$$

the system (13.19) has a unique solution

$$u \in C(\mathbb{R}; D^s) \cap C^1(\mathbb{R}; D^{s-2}).$$

The results of Lions [8] yield again the estimate

$$\int_0^T |f(t)|^2 \, dt \le c_1 \left(\|u_0\|_1^2 + \|u_1\|_{-1}^2 \right) \tag{11.5}$$

for all $(u_0, u_1) \in D^1 \times D^{-1}$, with a constant depending only on T.

On the other hand, the inverse inequality also holds under some necessary restrictions: we recall from [12] and [7, p. 208] the following theorem:

Theorem 11.2 *For almost all choices of $(\ell_1, \ldots, \ell_N) \in (0, \infty)^N$, the solutions of (11.1) satisfy the estimates*

$$\|u_0\|_s^2 + \|u_1\|_{s-2}^2 \le c_s \int_0^T |f(t)|^2 \, dt \tag{11.6}$$

for every $s < 1$, with a suitable constant c_s, independent of the particular choice of the initial data.

This theorem was proved by representing the solutions of (13.19) by Fourier series and by evaluating the two sides of (11.6) by applying a generalization of a classical Parseval type theorem of Ingham [6], obtained in [3,4]; see also [11]. Let us recall the form of the solutions because we will need it in the proof of Theorem 11.1 in the next section.

Setting

$$\mu_{jk} := \frac{k\pi}{\ell_j} \quad \text{and} \quad \omega_{jk} := \mu_{jk}^2$$

for brevity, the solutions of (13.19) are given by the series

$$u_j(t, x) = \sum_{k=1}^{\infty} (a_{jk} e^{i\omega_{jk}t} + b_{jk} e^{-i\omega_{jk}t}) \sin \mu_{jk} x, \quad j = 1, \ldots, N, \tag{11.7}$$

with suitable complex coefficients a_{jk} and b_{jk}.

Furthermore, using these coefficients the estimate (11.6) may be rewritten in the following equivalent form:

$$\sum_{j=1}^{N} \sum_{k=1}^{\infty} k^{2s} \left(|a_{jk}|^2 + |b_{jk}|^2 \right)$$

$$\leq c_s \int_0^T \left| \sum_{j=1}^{N} \sum_{k=1}^{\infty} \mu_{jk} a_{jk} e^{i\omega_{jk}t} + \mu_{jk} b_{jk} e^{-i\omega_{jk}t} \right|^2 dt. \tag{11.8}$$

In the rest of this section we explain the appearance of the assumption "almost all" in Theorem 11.2 (this explication will not be used in the next section). We refer to [7, pp. 207–210] for more details.

Using (11.7) and the definition of the spaces D_j^s we obtain by a straightforward computation that[2]

$$\|u_{j0}\|_s^2 + \|u_{j1}\|_{s-2}^2 \asymp \sum_{k=1}^{\infty} k^{2s} \left(|a_{jk}|^2 + |b_{jk}|^2 \right)$$

---

[2]We write $g \asymp h$ if there exist two positive constants α, β, independent of the parameters j, k, n, such that $\alpha g \leq h \leq \beta g$.

for each j, and hence

$$\|u_0\|_s^2 + \|u_1\|_{s-2}^2 \asymp \sum_{j=1}^{N}\sum_{k=1}^{\infty} k^{2s}\left(|a_{jk}|^2 + |b_{jk}|^2\right). \tag{11.9}$$

It remains to compare the right side of this relation with the integral on the right side of (11.6). For this first we observe that

$$f(t) = \sum_{j=1}^{N}\sum_{k=1}^{\infty}(\mu_{jk}a_{jk}e^{i\omega_{jk}t} + \mu_{jk}b_{jk}e^{-i\omega_{jk}t})$$

is a trigonometric type series. If the "spectrum"

$$\{\pm\omega_{jk} \ : \ j = 1,\ldots,N, \quad k = 1,2,\ldots\}$$

were uniformly discrete, then we could apply a classical generalization of Parseval's equality, due to Ingham [6], to conclude that

$$\int_0^T |f(t)|^2\, dt \asymp \sum_{j=1}^{N}\sum_{k=1}^{\infty}|\mu_{jk}|^2\left(|a_{jk}|^2 + |b_{jk}|^2\right),$$

or equivalently

$$\int_0^T |f(t)|^2\, dt \asymp \sum_{j=1}^{N}\sum_{k=1}^{\infty}k^2\left(|a_{jk}|^2 + |b_{jk}|^2\right),$$

provided T is sufficiently large. Moreover, since[3] $\omega_{j,k+1} - \omega_{j,k} \to \infty$ for each fixed j as $k \to \infty$, it can be shown by applying a method of Haraux [5] that the last estimate remains valid for *every* (even arbitrarily small) $T > 0$.

However, the spectrum is not uniformly discrete. But it has a weaker gap property, which enables us, by applying a theorem from [3] and [4], to obtain a weaker conclusion of the form

$$\int_0^T |f(t)|^2\, dt \geq c\sum_{j=1}^{N}\sum_{k=1}^{\infty}k^2\lambda_{jk}^{2N-2}\left(|a_{jk}|^2 + |b_{jk}|^2\right),$$

with a suitable positive constant c and with

$$\lambda_{jk} := \min\left\{|\omega_{jk} - \omega_{j'k'}| \ : \ (j,k) \neq (j',k')\right\}.$$

---

[3] We write more correctly j, k instead of jk for more clarity.

For each fixed j, the set $\{\pm\omega_{jk} : k = 1, 2, \ldots\}$ is uniformly discrete. On the other hand, the differences $|\omega_{jk} - \omega_{j'k'}|$ may be small when $j \neq j'$, and this makes the last estimate weaker. However, if the ratio $\ell_j/\ell_{j'}$ does not belong to some special set of zero Lebesgue measure, then a theorem on Diophantine approximation yields an estimate of the form

$$\lambda_{jk} \geq c_\beta \left|\omega_{j,k}\right|^{-\beta}$$

for each fixed $\beta > 0$, with a positive constant c_β, which is independent of j and k. Since $\omega_{j,k} \asymp k^2$, this yields the inequality

$$\int_0^T |f(t)|^2 \, dt \geq c_\beta' \sum_{j=1}^{N} \sum_{k=1}^{\infty} k^{2-(4N-4)\beta} \left(\left|a_{jk}\right|^2 + \left|b_{jk}\right|^2\right)$$

for each fixed $\beta > 0$, with another positive constant c_β'. In view of (11.9), the theorem follows by choosing $\beta > 0$ such that $2s = 2 - (4N - 4)\beta$.

11.3   Proof of Theorem 11.1

Keeping the notations

$$\mu_{jk} := \frac{k\pi}{\ell_j} \quad \text{and} \quad \omega_{jk} := \mu_{jk}^2$$

of the preceding section and setting

$$\mu_n := \frac{n\pi}{\ell} \quad \text{and} \quad \omega_{jkn} := \mu_{jk}^2 + \mu_n^2 = \omega_{jk} + \mu_n^2$$

for brevity, it is well-known that the solutions of (11.1) are given by the Fourier series

$$u_j(t, x, y) = \sum_{k=1}^{\infty} \sum_{n=1}^{\infty} (a_{jkn} e^{i\omega_{jkn}t} + b_{jkn} e^{-i\omega_{jkn}t}) \sin \mu_{jk} x \sin \mu_n y,$$

$$j = 1, \ldots, N, \qquad (11.10)$$

with suitable complex coefficients a_{jkn} and b_{jkn}.

Using these expressions we have

$$\left\| u_{j0} \right\|_s^2 = \sum_{k=1}^{\infty} \sum_{n=1}^{\infty} (k^2 + n^2)^s \left|a_{jkn} + b_{jkn}\right|^2$$

and

$$\left\| u_{j1} \right\|_{s-2}^2 = \sum_{k=1}^{\infty} \sum_{n=1}^{\infty} (k^2 + n^2)^{s-2} \omega_{jkn}^2 \left| a_{jkn} - b_{jkn} \right|^2.$$

Since $\omega_{jkn} \asymp k^2 + n^2$, we obtain the following estimate of the left side of (11.3):

$$\left\| u_{j0} \right\|_s^2 + \left\| u_{j1} \right\|_{s-2}^2 \asymp \sum_{k=1}^{\infty} \sum_{n=1}^{\infty} (k^2 + n^2)^s \left(\left| a_{jkn} \right|^2 + \left| b_{jkn} \right|^2 \right). \tag{11.11}$$

Next we evaluate the right side of (11.3). Using (13.31) we have

$$f_y(t, y) = \sum_{j=1}^{N} \sum_{k=1}^{\infty} \sum_{n=1}^{\infty} (a_{jkn} e^{i\omega_{jkn}t} + b_{jkn} e^{-i\omega_{jkn}t}) \mu_{jk} \mu_n \cos \mu_n y.$$

Using the orthogonality of the functions $\cos \mu_n y$ in $L^2(0, \ell)$, it follows that

$$\int_0^\ell |f_y(t, y)|^2 \, dy = \frac{\ell}{2} \sum_{n=1}^{\infty} \mu_n^2 \left| \sum_{j=1}^{N} \sum_{k=1}^{\infty} (a_{jkn} e^{i\omega_{jkn}t} + b_{jkn} e^{-i\omega_{jkn}t}) \mu_{jk} \right|^2.$$

Now a crucial observation is that, since $\omega_{jkn} := \omega_{jk} + \mu_n^2$, we have

$$\left| \sum_{j=1}^{N} \sum_{k=1}^{\infty} (a_{jkn} e^{i\omega_{jkn}t} + b_{jkn} e^{-i\omega_{jkn}t}) \mu_{jk} \right|^2 = \left| \sum_{j=1}^{N} \sum_{k=1}^{\infty} (a_{jkn} e^{i\omega_{jk}t} + b_{jkn} e^{-i\omega_{jk}t}) \mu_{jk} \right|^2$$

for each fixed n, and therefore

$$\int_0^\ell |f_y(t, y)|^2 \, dy = \frac{\ell}{2} \sum_{n=1}^{\infty} \mu_n^2 \left| \sum_{j=1}^{N} \sum_{k=1}^{\infty} (a_{jkn} e^{i\omega_{jk}t} + b_{jkn} e^{-i\omega_{jk}t}) \mu_{jk} \right|^2.$$

Applying the equalities (11.6) and (11.8), it follows that

$$\int_0^T \int_0^\ell |f_y(t, y)|^2 \, dy \, dt$$

$$= \frac{\ell}{2} \sum_{n=1}^{\infty} \mu_n^2 \int_0^T \left| \sum_{j=1}^{N} \sum_{k=1}^{\infty} (a_{jkn} e^{i\omega_{jk}t} + b_{jkn} e^{-i\omega_{jk}t}) \mu_{jk} \right|^2 \, dt$$

$$\geq \frac{\ell}{2c_s} \sum_{n=1}^{\infty} \sum_{j=1}^{N} \sum_{k=1}^{\infty} k^{2s} \mu_n^2 (\left| a_{jkn} \right|^2 + \left| b_{jkn} \right|^2).$$

Since $s < 1$, we have

$$k^{2s}\mu_n^2 = k^{2s}\pi^2 n^2 \ell^{-2} \geq \frac{\pi^2}{\ell^2}k^{2s}n^{2s} \geq \frac{\pi^2}{\ell^2}\left(\frac{k^2+n^2}{2}\right)^s$$

for all $k, n = 1, 2, \ldots$. We deduce therefore from the preceding estimate the following inequality:

$$\int_0^T \int_0^\ell |f_y(t, y)|^2 \, dy \, dt \geq \frac{\pi^2}{2^{s+1}\ell c_s} \sum_{n=1}^\infty \sum_{j=1}^N \sum_{k=1}^\infty (k^2 + n^2)^s \left(|a_{jkn}|^2 + |b_{jkn}|^2\right).$$

(11.12)

Combining (11.11) and (11.12), the estimate (11.3) of Theorem 11.1 follows with a suitable constant C_s.   □

References

1. Baiocchi, C., Komornik, V., Loreti, P.: Teorèmes du type Ingham et application à la théorie du contrôle. C. R. Acad. Sci. Paris Sér. I Math. **326**, 453–458 (1998)
2. Baiocchi, C., Komornik, V., Loreti, P.: Ingham type theorems and applications to control theory. Bol. Un. Mat. Ital. Serie 8 **2-B**(1), 33–63 (1999)
3. Baiocchi, C., Komornik, V., Loreti, P.: Généralisation d'un théorème de Beurling et application à la théorie du contrôle. C. R. Acad. Sci. Paris Sér. I Math. **330**(4), 281–286 (2000)
4. Baiocchi, C., Komornik, V., Loreti, P.: Ingham-Beurling type theorems with weakened gap conditions. Acta Math. Hungar. **97**(1–2), 55–95 (2002)
5. Haraux, A.: Séries lacunaires et contrôle semi-interne des vibrations d'une plaque rectangulaire. J. Math. Pures Appl. **68**, 457–465 (1989)
6. Ingham, A.E.: Some trigonometrical inequalities with applications in the theory of series. Math. Z. **41**, 367–379 (1936)
7. Komornik, V., Loreti, P.: Fourier Series in Control Theory. Springer, New York (2005)
8. Lions, J.-L.: Contrôle des systèmes distribués singuliers. Gauthier-Villars, Paris (1983)
9. Lions, J.-L.: Exact controllability, stabilizability, and perturbations for distributed systems. Siam Rev. **30**, 1–68, (1988)
10. Lions, J.-L.: Contrôlabilité exacte et stabilisation de systèmes distribués I-II. Masson, Paris (1988)
11. Loreti, P.: On some gap theorems. In: European Women in Mathematics—Marseille 2003. CWI Tract, vol. 135, pp. 39–45. Centrum Wisk. Inform., Amsterdam (2005)
12. Sikolya, E.: Simultaneous observability of networks of beams and strings. Bol. Soc. Paran. Mat. **21**(1–2), 1–11 (2003)

Chapter 12
Kernel Estimates for Nonautonomous Kolmogorov Equations with Potential Term

Markus Kunze, Luca Lorenzi, and Abdelaziz Rhandi

To the memory of Prof. Alfredo Lorenzi

Abstract Using time dependent Lyapunov functions, we prove pointwise upper bounds for the heat kernels of some nonautonomous Kolmogorov operators with possibly unbounded drift and diffusion coefficients and a possibly unbounded potential term.

12.1 Introduction

We consider nonautonomous evolution equations

$$\begin{cases} \partial_t u(t,x) = \mathscr{A}(t)u(t,x), \ (t,x) \in (s,1] \times \mathbb{R}^d, \\ u(s,x) = f(x), \qquad x \in \mathbb{R}^d, \end{cases} \tag{12.1}$$

M. Kunze
Graduiertenkolleg 1100, University of Ulm, 89069 Ulm, Germany
e-mail: markus.kunze@uni-ulm.de

L. Lorenzi (✉)
Dipartimento di Matematica e Informatica, Università degli Studi di Parma, Parco Area delle Scienze 53/A, 43124 Parma, Italy
e-mail: luca.lorenzi@unipr.it

Thank you dad, for having conveyed your love for mathematics to me!

A. Rhandi
Dipartimento di Ingegneria dell'Informazione, Ingegneria Elettrica e Matematica Applicata, Università degli Studi di Salerno, Via Ponte Don Melillo 1, 84084 Fisciano (Sa), Italy
e-mail: arhandi@unisa.it

© Springer International Publishing Switzerland 2014
A. Favini et al. (eds.), *New Prospects in Direct, Inverse and Control Problems for Evolution Equations*, Springer INdAM Series 10,
DOI 10.1007/978-3-319-11406-4_12

where the time dependent operators $\mathscr{A}(t)$ are defined on smooth functions φ by

$$\mathscr{A}(t)\varphi(x) = \sum_{ij=1}^{d} q_{ij}(t,x)D_{ij}\varphi(x) + \sum_{i=1}^{d} F_i(t,x)D_i\varphi(x) - V(t,x)\varphi(x).$$

We write $\mathscr{A}_0(t)$ for the operator $\mathscr{A}(t)+V(t)$. Throughout this article, we will always assume that the following hypothesis on the coefficients are satisfied.

Hypothesis 12.1 *The coefficients q_{ij}, F_j and V are defined on $[0,1] \times \mathbb{R}^d$ for i, $j = 1,\ldots,d$. Moreover,*

1. *there exists an $\varsigma \in (0,1)$ such that $q_{ij}, F_j, V \in C_{loc}^{\frac{\varsigma}{2},\varsigma}([0,1] \times \mathbb{R}^d)$ for all i, $j = 1,\ldots,d$. Further, $q_{ij} \in C^{0,1}((0,1) \times \mathbb{R}^d)$;*
2. *the matrix $Q = (q_{ij})$ is symmetric and uniformly elliptic in the sense that there exists a number $\eta > 0$ such that*

$$\sum_{i,j=1}^{d} q_{ij}(t,x)\xi_i\xi_j \geq \eta|\xi|^2 \quad \text{for all } \xi \in \mathbb{R}^d, \ (t,x) \in [0,1] \times \mathbb{R}^d;$$

3. *$V \geq 0$;*
4. *there exist a nonnegative function $Z \in C^2(\mathbb{R}^d)$ and a constant $M \geq 0$ such that $\lim_{|x|\to\infty} Z(x) = \infty$ and we have $\mathscr{A}(t)Z(x) \leq M$, as well as $\eta\Delta_x Z(x) + F(t,x) \cdot \nabla_x Z(x) - V(t,x)Z(x) \leq M$, for all $(t,x) \in [0,1] \times \mathbb{R}^d$;*
5. *there exists a nonnegative function $Z_0 \in C^2(\mathbb{R}^d)$ such that $\lim_{|x|\to\infty} Z_0(x) = \infty$ and we have $\mathscr{A}_0(t)Z_0(x) \leq M$, as well as $\eta\Delta_x Z_0(x) + F(t,x) \cdot \nabla_x Z_0(x) \leq M$, for all $(t,x) \in [0,1] \times \mathbb{R}^d$.*

We summarize Hypothesis 12.1(4)–(5) saying that Z (resp. Z_0) is a *Lyapunov function* for the operators $\mathscr{A}$ and $\eta\Delta + F \cdot \nabla_x - V$ (resp. for the operators $\mathscr{A}_0$ and $\eta\Delta + F \cdot \nabla_x$).

Clearly, 5 implies 4. However, for applications it will be important to differentiate between Z and Z_0.

The previous assumptions guarantee that, for any $f \in C_b(\mathbb{R}^d)$, the Cauchy problem (12.1) admits a unique solution $u \in C_b([s,1] \times \mathbb{R}^d) \cap C^{1,2}((s,1] \times \mathbb{R}^d)$. Moreover, there exists an evolution family $(G(t,s))_{(t,s)\in D} \subset \mathscr{L}(C_b(\mathbb{R}^d))$, where $D = \{(t,s) \in [0,1]^2 : t \geq s\}$, which governs Equation (12.1), i.e., $u(t,x) = (G(t,s)f)(x)$. Here and throughout the paper, the index "b" stands for boundedness.

By [2, Proposition 3.1], the operators $G(t,s)$ are given by *Green kernels* $g(t,s,\cdot,\cdot)$, i.e., we have

$$G(t,s)f(x) = \int_{\mathbb{R}^d} f(y)g(t,s,x,y)\,dy. \tag{12.2}$$

Our aim is to prove estimates for the Green kernel g. Similar results as we present here have been obtained in [10–13] for autonomous equations without potential term. The case of autonomous equations with potential term was treated in [1, 8, 9].

Recently, generalizing techniques from [4] to the parabolic situation, the authors of the present article extended these results also to nonautonomous equations and, even more importantly, allowed also unbounded diffusion coefficients, see [7]. In this article, we extend the results of [7] to also allow potential terms in the equation.

Applying our main abstract result (Theorem 12.6) in a concrete situation, we obtain the following result. In its formulation, for $s \geq 0$, we use the notation $|x|_*^s$ to denote a smooth version of the s-th power of the absolute value function, i.e., $|x|_*^s = |x|^s$ whenever $|x| \geq 1$ and the map $x \mapsto |x|_*^s$ is twice continuously differentiable in $\mathbb{R}^d$. This is done to meet the differentiability requirement in Hypothesis 12.1(1), (3) and (5) and also later differentiability requirements. If $s = 0$ or $s \geq 2$ we can choose $|x|_*^s = |x|^s$ for any $x \in \mathbb{R}^d$ as this is already twice continuously differentiable.

Theorem 12.2 *Let $k > d + 2$, $m, r \geq 0$ and $p > 1$ be given with $p > m - 1$ and $r > m - 2$. We consider the (time independent) operator $\mathscr{A}(t) \equiv \mathscr{A}$, defined on smooth functions φ by*

$$\mathscr{A}\varphi(x) = (1 + |x|_*^m)\Delta\varphi(x) - |x|^{p-1}x \cdot \nabla\varphi(x) - |x|^r\varphi(x).$$

Then we have the following estimates for the associated Green kernel g:

1. if $p \geq \frac{1}{2}(m + r)$, then for $\alpha > \frac{p+1-m}{p-1}$ and $\varepsilon < \frac{1}{p+1-m}$ we have

$$g(t, s, x, y) \leq C(t - s)^{1 - \frac{\alpha(m \vee p)k}{p+1-m}} e^{-\varepsilon(t-s)^\alpha |y|_*^{p+1-m}};$$

2. if $p < \frac{1}{2}(m + r)$, then for $\varepsilon < \frac{2}{r+2-m}$ and $\alpha > \frac{r-m+2}{r+m-2}$, if $r + m > 2$, and $\alpha > \frac{r+2-m}{2(p-1)}$, if $r + m \leq 2$, we have

$$g(t, s, x, y) \leq C(t - s)^{1 - \frac{\alpha(2m \vee 2p \vee r)k}{r+2-m}} e^{-\varepsilon(t-s)^\alpha |y|_*^{\frac{1}{2}(r+2-m)}},$$

for all $x, y \in \mathbb{R}^d$ and $s \in [0, t)$.

Here, C is a positive constant.

These bounds should be compared to the ones in [1, Example 3.3], where the case $m = 0$ was considered. We would like to note that in Theorem 12.2 we have restricted ourselves to the autonomous situation so that one can compare the results with those in [1]. Genuinely nonautonomous examples can easily be constructed along the lines of [7, Sect. 5].

12.2 Time Dependent Lyapunov Functions

In this section we introduce time dependent Lyapunov functions and prove that they are integrable with respect to the measures $g_{t,s}(x, dy) := g(t, s, x, y)dy$, where $g(t, s, \cdot, \cdot)$ is the Green kernel associated to the evolution operator $G(t, s)$, see (12.2), and $g(t, \cdot, x, \cdot) \in L^1((0, 1) \times \mathbb{R}^d)$. To do so, it is important to have information about the derivative of $G(t,s)f$ with respect to s. We have the following result, taken from [2, Lemma 3.4]. Here and in the rest of the paper, the index "c" stands for compactly supported.

Lemma 12.1 *1. For $f \in C_c^2(\mathbb{R}^d)$, $s_0 \leq s_1 \leq t$ and $x \in \mathbb{R}^d$ we have*

$$G(t,s_1)f(x) - G(t,s_0)f(x) = -\int_{s_0}^{s_1} G(t,\sigma)\mathscr{A}(\sigma)f(x)\,d\sigma. \qquad (12.3)$$

2. For $f \in C^2(\mathbb{R}^d)$, constant and positive outside a compact set and $x \in \mathbb{R}^d$, the function $G(t,\cdot)\mathscr{A}(\cdot)f(x)$ is integrable in $[0,t]$ and for $s_0 \leq s_1 \leq t$ we have

$$G(t,s_1)f(x) - G(t,s_0)f(x) \geq -\int_{s_0}^{s_1} G(t,\sigma)\mathscr{A}(\sigma)f(x)\,d\sigma.$$

We note that in the case where $V \equiv 0$ part (2) in Lemma 12.1 follows trivially from part (1), since in that situation $G(t,s)\mathbb{1} \equiv \mathbb{1}$ and $\mathscr{A}(t)\mathbb{1} = 0$ so that equation (12.3) holds for $f = \mathbb{1}$, cf. [6, Lemma 3.2].

Let us note some consequences of Lemma 12.1 for later use. First of all, part (1) of the lemma implies that $\partial_s G(t,s)f = -G(t,s)\mathscr{A}(s)f$ for $f \in C_c^2(\mathbb{R}^d)$. Arguing as in [7, Lemma 2.2], we see that for $0 \leq a \leq b \leq t$, $x \in \mathbb{R}^d$ and $\varphi \in C_c^{1,2}([a,b] \times \mathbb{R}^d)$, the function $s \mapsto G(t,s)\varphi(s)(x)$ is differentiable in $[a,b]$ and

$$\partial_s G(t,s)\varphi(s)(x) = G(t,s)\partial_s\varphi(s)(x) - G(t,s)\mathscr{A}(s)\varphi(s)(x).$$

Consequently, for such a function φ we have that

$$\int_a^b G(t,s)\big[\partial_s\varphi(s) - \mathscr{A}(s)\varphi(s)\big](x)\,ds = G(t,b)\varphi(b)(x) - G(t,a)\varphi(a)(x),$$

$$(12.4)$$

for every $x \in \mathbb{R}^d$.

As a consequence of formula (12.4) and [3, Corollary 3.11] we get the following result.

Lemma 12.2 *For any $t \in (0,1]$ and any $x \in \mathbb{R}^d$ the function $g(t,\cdot,x,\cdot)$ is continuous (actually, locally Hölder continuous) in $(0,t) \times \mathbb{R}^d$.*

We now introduce time dependent Lyapunov functions.

Definition 12.1 Let $t \in (0,1]$. A *time dependent Lyapunov function* (on $[0,t]$) is a function $0 \leq W \in C([0,t] \times \mathbb{R}^d) \cap C^{1,2}((0,t) \times \mathbb{R}^d)$ such that

1. $W(s,x) \leq Z(x)$ for all $(s,x) \in [0,t] \times \mathbb{R}^d$;
2. $\lim_{|x|\to\infty} W(s,x) = \infty$, uniformly for s in compact subsets of $[0,t)$;

3. there exists a function $0 \le h \in L^1((0, t))$ such that

$$\partial_s W(s, x) - \mathcal{A}(s)W(s) \ge -h(s)W(s) \tag{12.5}$$

and

$$\partial_s W(s) - (\eta \Delta_x W(s) + F(s) \cdot \nabla_x W(s) - V(s)W(s)) \ge -h(s)W(s), \tag{12.6}$$

on $\mathbb{R}^d$, for every $s \in (0, t)$.

Sometimes, we will say that W is a time dependent Lyapunov function *with respect to h* to emphasize the dependence on h.

Proposition 12.1 *Let W be a time dependent Lyapunov function on $[0, t]$ with respect to h. Then for $0 \le s \le t$ and $x \in \mathbb{R}^d$ the function $W(s)$ is integrable with respect to the measure $g_{t,s}(x, dy)$. Moreover, setting*

$$\zeta_W(s, x) := \int_{\mathbb{R}^d} W(s, y) g_{t,s}(x, dy)$$

we have

$$\zeta_W(s, x) \le e^{\int_s^t h(\tau) d\tau} W(t, x). \tag{12.7}$$

Proof Let us first note that by [2, Proposition 4.7] the function Z is integrable with respect to $g_{t,s}(x, dy)$. Moreover,

$$G(t, s)Z(x) := \int_{\mathbb{R}^d} Z(y) g_{t,s}(x, dy) \le Z(x) + M(t - s). \tag{12.8}$$

It thus follows immediately from domination that $W(s)$ is integrable with respect to $g_{t,s}(x, dy)$.

We now fix a sequence of functions $\psi_n \in C^\infty([0, \infty))$ such that

(i) $\psi_n(\tau) = \tau$ for $\tau \in [0, n]$;
(ii) $\psi_n(\tau) \equiv$ const. for $\tau \ge n + 1$;
(iii) $0 \le \psi_n' \le 1$ and $\psi_n'' \le 0$.

Let us also fix $0 \le s < r < t$. Note that, for any $n \in \mathbb{N}$, the function $W_n := \psi_n \circ W$ is the sum of a function in $C_c^{1,2}([0, r] \times \mathbb{R}^d)$ and a positive constant. Indeed, $W(s, \sigma) \to \infty$ as $|x| \to \infty$ uniformly on $[0, r]$. For a positive constant function, we have by Lemma 12.1(2) that

$$G(t, r)\mathbb{1} - G(t, s)\mathbb{1} \ge -\int_s^r G(t, \sigma)\mathcal{A}(\sigma)\mathbb{1} \, d\sigma = \int_s^r G(t, \sigma)\big[\partial_\sigma \mathbb{1} - \mathcal{A}(\sigma)\mathbb{1}\big] \, d\sigma.$$

Combining this with Equation (12.4), it follows that

$$G(t, r)W_n(r)(x) - G(t, s)W_n(s)(x)$$

$$\geq \int_s^r G(t, \sigma)[\partial_\sigma W_n(\sigma) - \mathscr{A}(\sigma)W_n(\sigma)](x)\, d\sigma$$

$$= \int_s^r G(t, \sigma)[\psi_n'(W(\sigma))(\partial_\sigma W(\sigma) - \mathscr{A}(\sigma)W(\sigma))](x)\, d\sigma$$

$$- \int_s^r G(t, \sigma)[V(\sigma)W(\sigma)\psi_n'(W(\sigma)) - V(\sigma)\psi_n(W(\sigma))](x)\, d\sigma$$

$$- \int_s^r G(t, \sigma)[\psi_n''(W(\sigma))(Q(\sigma)\nabla_x W(\sigma) \cdot \nabla_x W(\sigma))](x)\, d\sigma$$

$$\geq - \int_s^r G(t, \sigma)[\psi_n'(W(\sigma))h(\sigma)W(\sigma)](x)\, d\sigma, \tag{12.9}$$

for any $x \in \mathbb{R}^d$, since $G(t, s)$ preserves positivity and the condition $\psi_n'' \leq 0$ implies that $y\psi_n'(y) - \psi_n(y) \leq 0$ for any $y \geq 0$.

We next want to let $r \uparrow t$. We fix an increasing sequence $(r_k) \subset (s, t)$, converging to t as $k \to \infty$. By monotone convergence, we clearly have

$$\int_s^{r_k} G(t, \sigma)[h(\sigma)W_n(\sigma)](x)\, d\sigma \to \int_s^t G(t, \sigma)[h(\sigma)W_n(\sigma)](x)\, d\sigma$$

as $k \to \infty$. We now claim that $G(t, r_k)W_n(r_k)(x) \to G(t, t)W_n(t)(x) = W_n(t, x)$ as $k \to \infty$. To see this, we note that for $f \in C_b(\mathbb{R}^d)$, the function $(s, x) \mapsto G(t, s)f(x)$ is continuous in $[0, t] \times \mathbb{R}^d$ as a consequence of [2, Theorem 4.11]. This immediately implies that $G(t, r_k)W_n(t)(x) \to G(t, t)W_n(t)(x) = W_n(t, x)$ as $k \to \infty$. Moreover, from (12.8) it follows that

$$g_{t,s}(\mathbb{R}^d \setminus B(0, R)) \leq \frac{1}{\inf_{\mathbb{R}^d \setminus B(0,R)} Z} \int_{\mathbb{R}^d} Z(y)g_{t,s}(x, dy) \leq \frac{Z(x) + M}{\inf_{\mathbb{R}^d \setminus B(0,R)} Z}, \tag{12.10}$$

where $B(0, R) \subset \mathbb{R}^d$ denotes the open ball centered at 0 with radius R. Since the right-hand side of (12.10) converges to zero as $R \to \infty$, the set of measures $\{g_{t,s}(x, dy) : s \in [0, t]\}$ is tight.

Taking into account that $W_n(r_k)$ is uniformly bounded and converges locally uniformly to $W_n(t)$ as $k \to \infty$, it is easy to see that

$$G(t, r_k)W_n(r_k)(x) - G(t, r_k)W_n(t)(x) = \int_{\mathbb{R}^d} (W_n(r_k, y) - W_n(t, y))\, g_{t,r_k}(x, dy) \to 0$$

as $k \to \infty$. Combining these two facts, it follows that $G(t, r_k)W_n(r_k)(x) \to W_n(t, x)$ as claimed.

Thus, letting $r \uparrow t$ in (12.9), we find that

$$W_n(t, x) - G(t, s)W_n(s)(x) \geq -\int_s^t G(t, \sigma)\left[\psi_n'(W(\sigma))h(\sigma)W(\sigma)\right](x)\, d\sigma.$$

$$(12.11)$$

Note that $\psi_n'(W(\sigma))h(\sigma)W(\sigma)$ and $W_n(s)$ converge increasingly to $W(\sigma)h(\sigma)$ and $W(s)$, respectively, as $n \to \infty$, for any $\sigma \in [s, t]$. Since each operator $G(t, \sigma)$ preserves positivity, we can use monotone convergence to let $n \to \infty$ in (12.11), obtaining

$$W(t, x) - G(t, s)W(s)(x) \geq -\int_s^t h(\sigma)G(t, \sigma)W(\sigma)(x)\, d\sigma.$$

Equivalently,

$$\zeta_W(t, x) - \zeta_W(s, x) \geq -\int_s^t h(\sigma)\zeta_W(\sigma, x)\, d\sigma, \qquad x \in \mathbb{R}^d. \tag{12.12}$$

This inequality yields (12.7). Indeed, the function Φ, defined by

$$\Phi(\tau) := \left(\zeta_W(t, x) + \int_\tau^t h(\sigma)\zeta_W(\sigma, x)\, d\sigma\right)e^{\int_s^\tau h(\sigma)\, d\sigma},$$

is continuous on $[s, t]$ and increasing since its weak derivative is nonnegative by (12.12). Hence $\Phi(s) \leq \Phi(t)$, from which (12.7) follows at once if we take again (12.12) into account. □

Let us illustrate this in the situation of Theorem 12.2.

Proposition 12.2 *Consider the (time independent) operator $\mathscr{A}(t) \equiv \mathscr{A}$, defined by*

$$\mathscr{A}\varphi(x) = (1 + |x|_*^m)\Delta\varphi(x) - |x|^{p-1}x \cdot \nabla\varphi(x) - |x|^r\varphi(x),$$

where $m, r \geq 0$ and $p > 1$. Moreover, assume one of the following situations:

(i) $p > m - 1$, $\beta := p + 1 - m$ and $\delta < 1/\beta$;
(ii) $r > m - 2$, $\beta := \frac{1}{2}(r + 2 - m)$ and $\delta < 1/\beta$.

Then the following properties hold true:

1. the function $Z(x) := \exp(\delta|x|_*^\beta)$ satisfies Part (4) of Hypothesis 12.1;
2. for $0 < \varepsilon < \delta$ and $\alpha > \alpha_0$, the function $W(s, x) := \exp(\varepsilon(t - s)^\alpha|x|_*^\beta)$ is a time dependent Lyapunov function in the sense of Definition 12.1. Here, $\alpha_0 = \frac{\beta}{p-1}$ if we assume condition (ii) and additionally $m + r \leq 2$. In all other cases, $\alpha_0 = \frac{\beta}{m+\beta-2}$.

Proof In the computations below, we assume that $|x| \geq 1$ so that $|x|_*^s = |x|^s$ for $s \geq 0$. At the cost of slightly larger constants, these estimates can be extended to all of $\mathbb{R}^d$. We omit the details which can be obtained as in the proof of [7, Lemma 5.2]

1. By direct computations, we see that

$$\mathscr{A}Z(x) = \delta\beta\left[(1 + |x|^m)|x|^{\beta-2}(d + \beta - 2 + \delta\beta|x|^\beta) - |x|^{p-1+\beta} - |x|^r\right]Z(x).$$

The highest power of $|x|$ appearing in the first term is $|x|^{m+2\beta-2}$ which, in case (i) is exactly $|x|^{p-1+\beta}$, in case (ii) it is exactly $|x|^r$. In both cases, the highest power in the square brackets has a negative coefficient in front, namely $\delta\beta - 1$. Thus $\lim_{|x|\to\infty} \mathscr{A}Z(x) = -\infty$. It now follows from the continuity of $\mathscr{A}Z$ that $\mathscr{A}Z \leq M$ for a suitable constant M. Since $\eta\Delta Z + F \cdot \nabla Z - VZ \leq \mathscr{A}Z$, where $\eta = 1 + \inf_{x\in\mathbb{R}^d} |x|_*^m$. We conclude that the function $\eta\Delta Z + F \cdot \nabla Z - VZ$ is bounded from above as well.

2. We note that since $\varepsilon < \delta$, we have $W(s, x) \leq (Z(x))^{\frac{\varepsilon}{\delta}} \leq Z(x)$ for all $s \in [0, t]$ and $x \in \mathbb{R}^d$ so that (1) in Definition 12.1 is satisfied. Condition (2) is immediate from the definition of W so that it only remains to verify condition (3).

A computation shows that

$$\partial_s W(s, x) - \mathscr{A}W(s, x)$$

$$= -\varepsilon\alpha(t - s)^{\alpha-1}|x|^\beta W(s, x) - \varepsilon\beta(t - s)^\alpha W(s, x)\times \qquad (12.13)$$

$$\times \left[(1 + |x|^m)|x|^{\beta-2}(d + \beta - 2 + \varepsilon\beta(t - s)^\alpha|x|^\beta) - |x|^{p-1+\beta}\right]$$

$$+ |x|^r W(s, x)$$

$$= -\varepsilon\alpha(t - s)^{\alpha-1}|x|^\beta W(s, x) - \varepsilon\beta(t - s)^\alpha W(s, x)\times$$

$$\times \left[(1 + |x|^m)|x|^{\beta-2}(d + \beta - 2 + \delta\beta|x|^\beta) - |x|^{p-1+\beta}\right]$$

$$+ |x|^r W(s, x)$$

$$+ \varepsilon\beta^2(\delta - \varepsilon)(t - s)^\alpha(1 + |x|^m)|x|^{2\beta-2}W(s, x)$$

$$\geq \varepsilon(t - s)^{\alpha-1}|x|^\beta\left((\delta - \varepsilon)\beta^2(t - s)|x|^{m+\beta-2} - \alpha\right)W(s, x)$$

$$- \varepsilon\beta(t - s)^\alpha W(s, x)\left[(1 + |x|^m)|x|^{\beta-2}(d + \beta - 2 + \delta\beta|x|^\beta)\right.$$

$$\left. - |x|^{p-1+\beta} - |x|^r\right], \qquad (12.14)$$

where in the last inequality we took into account that $\varepsilon\beta(t - s)^\alpha < 1$.

To further estimate $\partial_s W(s) - \mathscr{A} W(s)$, we first assume that $\beta + m - 2 \geq 0$. This condition is satisfied under condition (i) and also under condition (ii) provided that $m + r > 2$. We set $C := \left[(\delta - \varepsilon)\beta^2/\alpha\right]^{-\frac{1}{\beta+m-2}}$ and distinguish two cases.

Case 1: $|x| \geq C(t - s)^{-\frac{1}{\beta+m-2}}$.

In this case $(\delta - \varepsilon)\beta^2(t - s)|x|^{\beta+m-2} \geq \alpha$ so that the first summand in (12.14) is nonnegative. Replacing C with a larger constant if necessary, we can – as in the proof of part (1) – ensure that also the second summand is positive so that overall $\partial_s W(s) - \mathscr{A} W(s) \geq 0$ in this case.

Case 2: $1 \leq |x| < C(t - s)^{-\frac{1}{\beta+m-2}}$.

In this case, we start again from estimate (12.13). We drop the terms involving $-|x|^{p-1+\beta}$ and $|x|^r$ and, using that $|x| \geq 1$, estimate further as follows:

$$W(s, x)^{-1}(\partial_s W(s, x) - \mathscr{A} W(s, x))$$
$$\geq -\varepsilon\alpha(t - s)^{\alpha-1}|x|^\beta - 2\varepsilon\beta(t - s)^\alpha |x|^{m+\beta-2}(d + \beta - 2 + \varepsilon\beta|x|^\beta)$$
$$\geq -\varepsilon\alpha(t - s)^{\alpha-1}C^\beta(t - s)^{-\frac{\beta}{m+\beta-2}} - 2\varepsilon\beta(t - s)^\alpha C^{m+\beta-2}(t - s)^{-1}\times$$
$$\times \left(d + \beta - 2 + \varepsilon\beta C^\beta(t - s)^{-\frac{\beta}{m+\beta-2}}\right)$$
$$\geq -\tilde{C}(t - s)^{\alpha-1-\frac{\beta}{m+\beta-2}} =: -h(s).$$

Note that $h \in L^1(0, t)$ since $\alpha - 1 - \frac{\beta}{m+\beta-2} > -1$ by assumption.

Suppose now that $m + \beta - 2 \leq 0$, so that $|x|^{m+\beta-2} \leq 1$ for $|x| \geq 1$. Taking again into account that $\varepsilon\beta(t - s)^\alpha < 1$ and dropping the term involving $|x|^r$, we derive from (12.13) that

$$W(s, x)^{-1}(\partial_s W(s, x) - \mathscr{A} W(s, x))$$
$$\geq -\varepsilon(t - s)^{\alpha-1}|x|^\beta \left(\alpha + 2\beta - \beta(t - s)|x|^{p-1}\right) - 2(d + \beta - 2),$$

for any $|x| \geq 1$. We can now argue as above taking $C = \left[(\alpha + 2\beta)/\beta\right]^{\frac{1}{p-1}}$ and distinguishing the cases $|x| \geq C(t - s)^{-\frac{1}{p-1}}$ and $1 \leq |x| < C(t - s)^{-\frac{1}{p-1}}$. We conclude that

$$W(s, x)^{-1}(\partial_s W(s, x) - \mathscr{A} W(s, x)) \geq -\varepsilon C^\beta(\alpha + 2\beta)(t - s)^{\alpha-1-\frac{\beta}{p-1}} =: -h(s),$$

for any $s \in (0, t)$, $|x| \geq 1$, and $h \in L^1((0, t))$ due to the condition on α.

We have thus proved (12.5) in Definition 12.1. The analogous estimate (12.6) for $\eta\Delta_x + F \cdot \nabla_x - c$ follows from observing that $\eta\Delta_x W + F \cdot \nabla_x W - cW \leq \mathscr{A} W$. □

12.3   Kernel Bounds in the Case of Bounded Diffusion Coefficients

Throughout this section, we set $R(a,b) := (a,b) \times \mathbb{R}^d$ and $\overline{R}(a,b) := [a,b] \times \mathbb{R}^d$ for any $0 \leq a < b \leq 1$. Moreover, we assume that the coefficients q_{ij} and their spatial derivatives $D_k q_{ij}$ are bounded on $R(0,b)$ for $i, j, k = 1, \ldots, d$ and every $b < 1$. We will remove this additional boundedness assumption in the next section.

Fix now $t \in [0,1]$. For $0 \leq a < b \leq t$, $x \in \mathbb{R}^d$ and $k \geq 1$, we define the quantities $\Gamma_j(k, x, a, b)$ for $j = 1, 2$ by

$$\Gamma_1(k, x, a, b) := \left(\int_{R(a,b)} |F(s, y)|^k g(t, s, x, y) \, ds \, dy \right)^{\frac{1}{k}},$$

where g is the Green kernel associated with $\mathscr{A}$, and

$$\Gamma_2(k, x, a, b) := \left(\int_{Q(a,b)} |V(s, y)|^k g(t, s, x, y) \, ds \, dy \right)^{\frac{1}{k}}.$$

We also make an additional assumption about the parabolic equation governed by the operators $\mathscr{A}_0$ without potential term. Hypothesis 12.1(5) guarantees that the Cauchy problem (12.1) with $\mathscr{A}$ being replaced by $\mathscr{A}_0$ admits a unique solution $u \in C_b(\overline{R}(s,1)) \cap C^{1,2}(R(s,1))$ for any $f \in C_b(\mathbb{R}^d)$. The associated evolution operator admits a Green kernel which we denote by g_0. In the following lemma, we will deal with the space $\mathscr{H}^{p,1}(R(a,b))$ of all functions in $W_p^{0,1}(R(a,b))$ with distributional time derivative in $(W_{p'}^{0,1}(R(a,b)))'$, where $1/p + 1/p' = 1$. We refer the reader to [5, 10] for more details on these spaces. Here, we just prove the following result which is crucial in the proof of Theorem 12.4 (cf. [10, Lemma 7.2]).

Lemma 12.3 *Let* $u \in \mathscr{H}^{p,1}(R(a,b)) \cap C_b(\overline{R}(a,b))$ *for some* $p \in (1, \infty)$. *Then, there exists a sequence* $(u_n) \subset C_c^\infty(\mathbb{R}^{d+1})$ *of smooth functions such that* u_n *tends to* u *in* $W_p^{0,1}(R(a,b))$ *and locally uniformly in* $\overline{R}(a,b)$, *and* $\partial_t u_n$ *converges to* $\partial_t u$ *weakly* in* $(W_{p'}^{0,1}(R(a,b)))'$ *as* $n \to \infty$.

Proof We split the proof in two steps: first we prove the statement with $R(a,b)$ being replaced with $\mathbb{R}^{d+1}$ and, then, using this result we complete the proof.

Step 1. Let $\vartheta \in C_c^\infty(\mathbb{R})$ be a smooth function such that $\vartheta \equiv 1$ in $(-1,1)$ and $\vartheta \equiv 0$ in $\mathbb{R} \setminus (-2,2)$. For any $\sigma > 0$, any $t \in \mathbb{R}$ and any $x \in \mathbb{R}^d$, set $\vartheta_\sigma(t, x) = \vartheta(|t|/\sigma)\vartheta(|x|/\sigma)$. Next, we define the function $u_n \in C_c^\infty(\mathbb{R}^{d+1})$ by setting

$$u_n(t, x) = n^{d+1}\vartheta_n(t, x) \int_{\mathbb{R}^{d+1}} u(s, y)\vartheta_{1/n}(t - s, x - y) \, ds \, dy$$

$$=: n^{d+1}\vartheta_n(t, x)(u \star \vartheta_{1/n})(t, x),$$

for any $(t, x) \in \mathbb{R}^{d+1}$ and any $n \in \mathbb{N}$. Clearly, u_n converges to u in $W_p^{0,1}(\mathbb{R}^{d+1})$ and locally uniformly in $\mathbb{R}^{d+1}$.

Let us fix a function $\psi \in W_{p'}^{0,1}(\mathbb{R}^{d+1})$. Applying the Fubini-Tonelli theorem and taking into account that $\vartheta_{1/n}(r,z) = \vartheta_{1/n}(-r,-z)$ for any $(r,z) \in \mathbb{R}^{d+1}$, we easily deduce that $\langle \partial_t u_n, \psi \rangle = \langle \partial_t u, \psi_n \rangle$ for any $n \in \mathbb{N}$, where $\psi_n = n^{d+1} \vartheta_{1/n} \star (\vartheta_n \psi)$ and $\langle \cdot, \cdot \rangle$ denotes the duality pairing of $W_{p'}^{0,1}(\mathbb{R}^{d+1})$ and $(W_{p'}^{0,1}(\mathbb{R}^{d+1}))'$. Since ψ_n converges to ψ in $W_{p'}^{0,1}(\mathbb{R}^{d+1})$ as $n \to \infty$, we conclude that $\langle \partial_t u, \psi_n \rangle \to \langle \partial_t u, \psi \rangle$ as $n \to \infty$. This shows that $\partial_t u_n \overset{*}{\rightharpoonup} \partial_t u$ in $(W_{p'}^{0,1}(\mathbb{R}^{d+1}))'$ as $n \to \infty$.

Step 2.   Let us now consider the general case. We extend $u \in \mathcal{H}^{p,1}(R(a,b)) \cap C_b(\overline{R}(a,b))$ to $(3a - 2b, 2b - a)$, by symmetry, first with respect to $t = b$ and then with respect to $t = a$. The so obtained function v belongs to $\mathcal{H}^{p,1}(R(3a - 2b, 2b - a)) \cap C_b(\overline{R}(3a - 2b, 2b - a))$. Proving that $v \in W_p^{0,1}(R(3a - 2b, 2b - a)) \cap C_b(\overline{R}(3a - 2b, 2b - a))$ is immediate. Hence, it remains to prove that the distributional derivative $\partial_t v$ belongs to $(W_{p'}^{0,1}(R(3a - 2b, 2b - a)))'$. To that end fix $\varphi \in C_c^\infty(R(3a - 2b, 2b - a))$ and observe that

$$\int_{R(3a-2b,2b-a)} v \partial_t \varphi \, dt \, dx = \int_{R(a,b)} u \partial_t \Phi \, dt \, dx, \tag{12.15}$$

where the function $\Phi = \varphi - \varphi(2b - \cdot, \cdot) - \varphi(2a - \cdot, \cdot) + \varphi(2a - 2b + \cdot, \cdot)$ belongs to $W_{p'}^{0,1}(R(a,b))$. It follows immediately that $\langle \partial_t v, \varphi \rangle = \langle \partial_t u, \Phi \rangle$. The density of $C_c^\infty(R(a,b))$ in $W_{p'}^{0,1}(R(a,b))$ implies that $\partial_t v \in (W_{p'}^{0,1}(R(3a - 2b, 2b - a)))'$.

We now fix a function $\zeta \in C_c^\infty((3a - 2b, 2b - a))$ such that $\zeta \equiv 1$ in $[a,b]$. Applying Step 1 to the function $(t,x) \mapsto \zeta(t)v(t,x)$, which belongs to $\mathcal{H}^{p,1}(\mathbb{R}^{d+1}) \cap C_b(\mathbb{R}^{d+1})$, we can find a sequence $(u_n) \subset C_c^\infty(\mathbb{R}^{d+1})$ converging to the function ζv locally uniformly in $\mathbb{R}^{d+1}$ and in $W_p^{0,1}(\mathbb{R}^{d+1})$, and such that $\partial_t u_n \overset{*}{\rightharpoonup} \partial_t(\zeta v)$ in $(W_{p'}^{0,1}(\mathbb{R}^{d+1}))'$. Clearly, u_n converges to u locally uniformly in $\overline{R}(a,b)$ and in $W_p^{0,1}(R(a,b))$. Moreover, fix $\varphi \in W_{p'}^{0,1}(R(a,b))$ and denote by $\overline{\varphi}$ the null extension of φ to the whole of $\mathbb{R}^{d+1}$. Clearly, $\overline{\varphi}$ belongs to $W_{p'}^{0,1}(\mathbb{R}^{d+1})$. Since

$$\int_{R(a,b)} \partial_t u_n \varphi \, dt \, dx = \int_{\mathbb{R}^{d+1}} \partial_t u_n \overline{\varphi} \, dt \, dx$$

and $\partial_t u_n \overset{*}{\rightharpoonup} \partial_t(v\zeta)$ in $(W_{p'}^{0,1}(\mathbb{R}^{d+1}))'$, from formula (12.15) and since $\zeta' \overline{\varphi} \equiv 0$ and $\zeta \overline{\varphi} \equiv \varphi$, it follows that

$$\lim_{n \to \infty} \int_{R(a,b)} \partial_t u_n \varphi \, dt \, dx = \langle \partial_t(\zeta v), \overline{\varphi} \rangle = \int_{R(a,b)} v \zeta' \overline{\varphi} \, dt \, dx + \langle \partial_t v, \zeta \overline{\varphi} \rangle$$

$$= \langle \partial_t v, \overline{\varphi} \rangle = \langle \partial_t u, \varphi \rangle.$$

This completes the proof. $\square$

Lemma 12.4 *Let $0 \le a < b < t$ and $x \in \mathbb{R}^d$. Moreover, assume that $g_0(t, \cdot, x, \cdot) \in L^\infty(R(a, b))$. Then, $g(t, \cdot, x, \cdot) \in C_b(\overline{R}(a, b))$. Moreover, if for some $q > 1$ we have $\Gamma_1(q, x, a, b) < \infty$ and $\Gamma_2(q, x, a, b) < \infty$, then $g(t, \cdot, x, \cdot) \in \mathcal{H}^{p,1}(R(\tilde{a}, \tilde{b}))$ for all $p \in (1, q)$ and any $a < \tilde{a} < \tilde{b} < b$.*

Proof By the maximum principle, $g(t, \cdot, x, \cdot) \le g_0(t, \cdot, x, \cdot)$ almost surely. Hence, $g(t, \cdot, x, \cdot) \in L^\infty(R(a, b))$. The continuity of the function $g(t, \cdot, x, \cdot)$ follows from Lemma 12.2. To infer that $g(t, \cdot, x, \cdot)$ belongs to $\mathcal{H}^{p,1}(R(\tilde{a}, \tilde{b}))$, for any $\tilde{a}$ and $\tilde{b}$ as in the statement of the lemma, we want to use [10, Lemma 3.2] (see also [7, Lemma 3.2] for the nonautonomous situation). We note that the proof of that lemma remains valid for operators with potential term, provided that both $\Gamma_1(q, x, a, b) < \infty$ and $\Gamma_2(q, x, a, b) < \infty$. Thus [7, Lemma 3.2] yields $g \in \mathcal{H}^{p,1}(R(\tilde{a}, \tilde{b}))$ for all $p \in (1, q)$. $\qquad\square$

We next establish the kernel estimates. To that end, we use time-dependent Lyapunov functions. We make the following assumptions.

Hypothesis 12.3 *Fix $0 < t \le 1$, $x \in \mathbb{R}^d$ and $0 < a_0 < a < b < b_0 < t$. Let time dependent Lyapunov functions W_1, W_2 with $W_1 \le W_2$ and a weight function $1 \le w \in C^{1,2}(R(0, t))$ be given such that*

1. *the functions $w^{-2}\partial_s w$ and $w^{-2}\nabla_y w$ are bounded on $R(a_0, b_0)$;*
2. *there exist a constant $k > d + 2$ and constants $c_1, \ldots, c_7 \ge 1$, possibly depending on the interval (a_0, b_0), such that*

$$\text{(i)} \quad w \le c_1 w^{\frac{k-2}{k}} W_1^{\frac{2}{k}}, \qquad \text{(ii)} \quad |Q\nabla_y w| \le c_2 w^{\frac{k-1}{k}} W_1^{\frac{1}{k}},$$

$$\text{(iii)} \quad |\text{Tr}(QD^2 w)| \le c_3 w^{\frac{k-2}{k}} W_1^{\frac{2}{k}}, \quad \text{(iv)} \quad |\partial_s w| \le c_4 w^{\frac{k-2}{k}} W_1^{\frac{2}{k}},$$

$$\text{(v)} \quad \Big|\sum_{i=1}^{d} D_i q_{ij}\Big| \le c_5 w^{-\frac{1}{k}} W_2^{\frac{k}{k}},$$

and

$$\text{(vi)} \quad |F| \le c_6 w^{-\frac{1}{k}} W_2^{\frac{1}{k}}, \qquad \text{(vii)} \quad V^{\frac{1}{2}} \le c_7 w^{-\frac{1}{k}} W_2^{\frac{1}{k}},$$

on $R(a_0, b_0)$;
3. *$g_0(t, \cdot, x, \cdot) \in L^\infty(R(a_0, b_0))$.*

Having fixed t and x, we write $\rho(s, y) := g(t, s, x, y)$ to simplify notation. We can now prove the main result of this section.

Theorem 12.4 *Assume Hypotheses 12.3. Then there exists a positive constant C_1, depending only on d, k and η, such that*

$$w\rho \le C_1 \Bigg[c_1^{\frac{k}{2}} \sup_{s \in (a_0, b_0)} \zeta_{W_1}(s) + \Bigg(\frac{c_1^{\frac{k}{2}}}{(b_0 - b)^{\frac{k}{2}}} + c_2^k + c_3^{\frac{k}{2}} + c_4^{\frac{k}{2}} \Bigg) \int_{a_0}^{b_0} \zeta_{W_1}(s)\, ds$$

$$+ \Bigg(c_2^{\frac{k}{2}} c_6^{\frac{k}{2}} + c_5^k + c_6^k + c_7^k \Bigg) \int_{a_0}^{b_0} \zeta_{W_2}(s)\, ds \Bigg] \tag{12.16}$$

in $R(a, b)$.

Proof We first assume that the weight function w, along with its first order partial derivatives is bounded. It follows from Hypothesis 12.3(2)(i) and (vi) that

$$\Gamma_1(k/2, x, a_0, b_0)^{\frac{k}{2}} = \int_{R(a_0,b_0)} |F(s,y)|^{\frac{k}{2}} g(t,s,x,y)\, ds\, dy$$

$$\leq \int_{R(a_0,b_0)} w(s,x)|F(s,x)|^{\frac{k}{2}} g(t,s,x,y)\, ds\, dy$$

$$\leq c_6^{\frac{2}{2}} \int_{R(a_0,b_0)} w(s,y)^{\frac{1}{2}} W_2(s,y)^{\frac{1}{2}} g(t,s,x,y)\, ds\, dy$$

$$\leq c_1^{\frac{k}{4}} c_6^{\frac{2}{2}} \int_{R(a_0,b_0)} W_2(s,y) g(t,s,x,y)\, ds\, dy < \infty,$$

as a consequence of Proposition 12.1. Moreover, using Hypothesis 12.3(2)(vii) instead, it follows that

$$\Gamma_2(k/2, x, a_0, b_0)^{\frac{k}{2}} \leq c_7^k \int_{a_0}^{b_0} \zeta_{W_2}(s,x)\, ds < \infty.$$

We thus infer from Lemma 12.4 that $g(t, \cdot, x, \cdot) \in L^\infty(R(a_0, b_0)) \cap \mathscr{H}^{p,1}(R(a_1, b_1))$ for all $p \in (1, \frac{k}{2})$, where $a_0 < a_1 < a < b < b_1 < b_0$.

Let $\vartheta : \mathbb{R} \to \mathbb{R}$ be a smooth function with $\vartheta(s) = 1$ for $s \in [a, b]$, $\vartheta(s) = 0$ for $s \geq b_1$, $0 \leq \vartheta \leq 1$ and $|\vartheta'| \leq 2(b_1 - b)^{-1}$ in $\mathbb{R}$. Given $\psi \in C_c^{1,2}(R(a_1, b_1))$, we put $\varphi(s,y) := \vartheta(s)^{\frac{k}{2}} w(s,y)\psi(s,y)$. It follows from (12.4) that

$$\int_{R(a_1,b_1)} \left[\partial_s \varphi(s,y) - \mathscr{A}(s)\varphi(s,y)\right]\rho(s,y)\, ds\, dy = 0. \tag{12.17}$$

We write $\tilde{\rho} := \vartheta^{\frac{k}{2}}\rho$ and note that $w\tilde{\rho} \in \mathscr{H}^{p,1}(R(a_1, b_1))$ for all $p \in (1, \frac{k}{2})$, since w and its derivatives are bounded. Thus with some standard computations involving integration by parts we derive from (12.17) that

$$\int_{R(a_1,b_1)} \left[\langle Q\nabla_y(w\tilde{\rho}), \nabla_y \psi\rangle - \psi\partial_s(w\tilde{\rho})\right] ds\, dy$$

$$= \int_{R(a_1,b_1)} \tilde{\rho}\left(2\sum_{i,j=1}^d q_{ij}(D_i w)(D_j\psi) - \sum_{i,j=1}^d w(D_i q_{ij})(D_j\psi) + w\langle F, \nabla_y\psi\rangle\right) ds\, dy$$

$$- \frac{k}{2}\int_{R(a_1,b_1)} \rho w\psi\, \vartheta^{\frac{k-2}{k}}\vartheta'\, ds\, dy$$

$$+ \int_{R(a_1,b_1)} \psi\left(\tilde{\rho}\mathrm{Tr}(QD^2 w) + \tilde{\rho}\langle F, \nabla_y w\rangle - \tilde{\rho}Vw - \tilde{\rho}\partial_s w\right) ds\, dy,$$

where, with a slight abuse of notation, we denote by $\int_{R(a_1,b_1)} \psi \partial_s(w\bar{\rho}) \, ds \, dy$ the pairing between $\partial_s(w\bar{\rho}) \in (W_{p'}^{0,1}(R(a_1,b_1)))'$ and $\psi \in W_{p'}^{0,1}(R(a_1,b_1))$.

We now want to apply [7, Theorem 3.7] to the function $u = w\bar{\rho}$ and infer that there exists a constant C, depending only on η, d and k (but not on $\|Q\|_\infty$), such that

$$
\|w\tilde{\rho}\|_\infty \leq C \left(\|w\tilde{\rho}\|_{\infty,2} + \|\tilde{\rho}Q\nabla_y w\|_k + \|\tilde{\rho}Fw\|_k + \sum_{j=1}^d \left\| \tilde{\rho}w \sum_{i=1}^d D_i q_{ij} \right\|_k \right.
$$

$$
+ \|\tilde{\rho}Vw\|_{\frac{k}{2}} + \frac{k}{b_1-b} \|\rho w \vartheta^{\frac{k-2}{k}}\|_{\frac{k}{2}} + \|\tilde{\rho}\mathrm{Tr}(QD^2 w)\|_{\frac{k}{2}} + \|\tilde{\rho}\partial_s w\|_{\frac{k}{2}}
$$

$$
\left. + \|\tilde{\rho}F \cdot \nabla_y w\|_{\frac{k}{2}} \right), \tag{12.18}
$$

where for $p \in [1,\infty)$ we denote by $\|f\|_p$ the usual L^p-norm of the function $f : R(a_1,b_1) \to \mathbb{R}$. Moreover, $\|f\|_{\infty,2} := \sup_{s \in (a_1,b_1)} \|f(s,\cdot)\|_{L^2(\mathbb{R}^d)}$.

Note that a major tool in the proof of that theorem is the formula

$$
\int_{R(a_1,b_1)} \vartheta (v-\ell)_+ \partial_t v \, dt \, dx = \frac{1}{2} \left[\int_{\mathbb{R}^d} \vartheta (v(b_1)-\ell)_+^2 \, dx \right.
$$

$$
\left. - \int_{\mathbb{R}^d} \vartheta (v(a_1)-\ell)_+^2 \, dx \right] \tag{12.19}
$$

satisfied by $v = w\bar{\rho} \in \mathscr{H}^{p,1}(R(a_1,b_1))$, any $\ell > 0$ and any nonnegative function $\vartheta \in C_c^\infty(\mathbb{R}^d)$, if $p > d+2$. However, formula (12.19) is satisfied also in the case $p \leq d+2$, which is our situation, if we additionally assume that $v \in C_b(\overline{R}(a_1,b_1))$ (which follows from Lemma 12.4). Its proof can be obtained arguing as in [7, Lemma 3.6] taking Lemma 12.3 into account, with slight and straightforward changes. Once formula (12.19) is established, the proof of (12.18) follows the same lines as in [7, Theorem 3.7] with no changes.

We now estimate the terms in the right-hand side of (12.18), using part (2) of Hypothesis 12.3. We have

$$
\|\tilde{\rho}Q\nabla_y w\|_k^k = \int_{R(a_1,b_1)} |\tilde{\rho}Q\nabla_y w|^k \, ds \, dy \leq c_2^k \int_{R(a_1,b_1)} \tilde{\rho}^k w^{k-1} W_1 \, ds \, dy
$$

$$
\leq c_2^k \|\tilde{\rho}w\|_\infty^{k-1} \int_{a_1}^{b_1} \zeta_{W_1}(s,x) \, ds.
$$

Let us write $M_k := \int_{a_1}^{b_1} \zeta_{W_k}(s, x)\, ds$ and $\bar{M} := \sup_{s \in (a_1, b_1)} \zeta_1(s, x)$. With similar estimates as above, we find

$$\|\tilde{\rho} F w\|_k \leq c_6 \|\tilde{\rho} w\|_\infty^{\frac{k-1}{k}} M_2^{\frac{1}{k}}, \qquad \left\| \tilde{\rho} w \sum_{i=1}^{d} D_i q_{ij} \right\|_k \leq c_5 \|\tilde{\rho} w\|_\infty^{\frac{k-1}{k}} M_2^{\frac{1}{k}},$$

$$\|\tilde{\rho} V w\|_{\frac{k}{2}} \leq c_7^2 \|\tilde{\rho} w\|_\infty^{\frac{k-2}{k}} M_2^{\frac{2}{k}}, \qquad \|\rho w \vartheta^{\frac{k-2}{2}}\|_{\frac{k}{2}} \leq c_1 \|\tilde{\rho} w\|_\infty^{\frac{k-2}{k}} M_1^{\frac{2}{k}},$$

$$\|\tilde{\rho} \mathrm{Tr}(Q D^2 w)\|_{\frac{k}{2}} \leq c_3 \|\tilde{\rho} w\|_\infty^{\frac{k-2}{k}} M_1^{\frac{2}{k}}, \qquad \|\tilde{\rho} \partial_s w\|_{\frac{k}{2}} \leq c_4 \|\tilde{\rho} w\|_\infty^{\frac{k-2}{k}} M_1^{\frac{2}{k}},$$

$$\|\tilde{\rho} F \cdot \nabla_y w\|_{\frac{k}{2}} \leq \eta^{-1} c_2 c_6 \|\tilde{\rho} w\|_\infty^{\frac{k-2}{k}} M_2^{\frac{2}{k}}, \quad \|w\tilde{\rho}\|_{\infty, 2} \leq c_1^{\frac{k}{4}} \|w\tilde{\rho}\|_\infty^{\frac{1}{2}} \bar{M}^{\frac{1}{2}}.$$

From (12.18) and the above estimates, we obtain the following inequality for $X := \|w\tilde{\rho}\|_\infty^{\frac{k}{k}}$:

$$X^k \leq \alpha X^{\frac{k}{2}} + \beta X^{k-1} + \gamma X^{k-2},$$

where $\alpha := C c_1^{\frac{k}{4}} \bar{M}^{\frac{1}{2}}$, $\beta = C \left(c_2 M_1^{\frac{1}{k}} + (c_6 + c_5 d) M_2^{\frac{1}{k}} \right)$ and

$$\gamma = C \left(\frac{c_1}{b_1 - b} + c_3 + c_4 \right) M_1^{\frac{2}{k}} + C(c_2 c_6 + c_7^2) M_2^{\frac{2}{k}}.$$

Estimating $\alpha X^{k/2} \leq \frac{1}{4} X^k + \alpha^2$, we find

$$X^k \leq \frac{4}{3} \alpha^2 + \frac{4}{3} \beta X^{k-1} + \frac{4}{3} \gamma X^{k-2}. \tag{12.20}$$

We note that the function

$$f(r) = r^k - \frac{4}{3} \beta r^{k-1} - \frac{4}{3} \gamma r^{k-2} - \frac{4}{3} \alpha^2 = r^{k-2} \left(r^2 - \frac{4}{3} \beta r - \frac{4}{3} \gamma \right) - \frac{4}{3} \alpha^2$$

$$:= r^{k-2} g(r) - \frac{4}{3} \alpha^2$$

is increasing in $\left(\frac{4}{3} \beta + \sqrt{\frac{4}{3} \gamma} + \left(\frac{4}{3} \alpha^2 \right)^{\frac{1}{k}}, \infty \right)$ since the functions $r \mapsto r^{k-2}$ and g are positive and increasing. Moreover,

$$f\left(\frac{4}{3} \beta + \sqrt{\frac{4}{3} \gamma} + \left(\frac{4}{3} \alpha^2 \right)^{\frac{1}{k}} \right) = \left(\frac{4}{3} \beta + \sqrt{\frac{4}{3} \gamma} + \left(\frac{4}{3} \alpha^2 \right)^{\frac{1}{k}} \right)^{k-2} \times$$

$$\times \left[\left(\frac{4}{3} \alpha^2 \right)^{\frac{2}{k}} + \left(\frac{4}{3} \right)^{\frac{3}{2}} \beta \gamma^{\frac{1}{2}} + 2 \left(\frac{4}{3} \right)^{\frac{k+2}{2k}} \alpha^{\frac{2}{k}} \left(\frac{\sqrt{3}}{3} \beta + \sqrt{\gamma} \right) \right] - \frac{4}{3} \alpha^2$$

$$> \left(\frac{4}{3} \alpha^2 \right)^{\frac{k-2}{k}} \left(\frac{4}{3} \alpha^2 \right)^{\frac{2}{k}} - \frac{4}{3} \alpha^2 = 0.$$

From these observations and inequality (12.20) it follows that $X \leq \frac{4}{3}\beta + \sqrt{\frac{4}{3}\gamma} + \left(\frac{4}{3}\alpha^2\right)^{\frac{1}{k}}$. Equivalently,

$$\|\tilde{\rho}w\|_\infty \leq K_1 \left(\alpha^2 + \beta^k + \gamma^{\frac{k}{2}}\right),$$

for some positive constant K_1. Taking into account that $c \geq 1$, one derives (12.16) from this by plugging in the definitions of α, β, γ and, then, letting $a_1 \downarrow a_0$ and $b_1 \uparrow b_0$.

To finish the proof of the theorem, it remains to remove the additional assumption on the weight w. To that end, we set $w_\varepsilon := \frac{w}{1+\varepsilon w}$. Using Hypothesis 12.3(1), we see that w_ε, along with its partial derivatives, is bounded. Straightforward computations show that Part (2) of Hypothesis 12.3 is satisfied with the same constants $c_1, \ldots, c_7$. Thus the first part of the proof shows that (12.18) is satisfied with w replaced with w_ε and the constants on the right-hand side do not depend on ε. Thus, upon $\varepsilon \downarrow 0$ we obtain (12.18) for the original w. $\qquad\square$

12.4   The Case of General Diffusion Coefficients

We now remove the additional boundedness assumption imposed in Sect. 12.3. We do this by approximating general diffusion coefficients with bounded ones, taking advantage of the fact that the constant C_1 obtained in Theorem 12.4 does not depend on the supremum norm of the diffusion coefficients. More precisely, we approximate the diffusion matrix Q as follows. Given a function $\varphi \in C_c^\infty(\mathbb{R})$ such that $\varphi \equiv 1$ in $(-1, 1)$, $\varphi \equiv 0$ in $\mathbb{R} \setminus (-2, 2)$ and $|t\varphi'(t)| \leq 2$ for all $t \in \mathbb{R}$, we define $\varphi_n(s, x) := \varphi(W_1(s, x)/n)$ for $s \in [0, t]$ and $x \in \mathbb{R}^d$. We put

$$q_{ij}^{(n)}(s, x) := \varphi_n(s, x)q_{ij}(s, x) + (1 - \varphi_n(s, x))\eta\delta_{ij},$$

where δ_{ij} is the Kronecker delta, and define the operators $\mathscr{A}_n(s)$ by

$$\mathscr{A}_n(s) := \sum_{i,j=1}^{d} q_{ij}^{(n)}(s)D_{ij} + \sum_{j=1}^{d} F_j(s)D_j - V(s).$$

We collect some properties of the approximating operators, omitting the easy proof.

Lemma 12.5 *Each operator $\mathscr{A}_n$ satisfies Hypothesis 12.1 in $[0, t]$, and its diffusion coefficients are bounded together with their first-order spatial derivatives. Moreover, any time dependent Lyapunov function for the operator $\partial_s - \mathscr{A}(s)$ on $[0, t]$ is a time dependent Lyapunov function for the operator $\partial_s - \mathscr{A}_n(s)$ with respect to the same h.*

It follows that the parabolic equation (12.1) with $\mathscr{A}$ replaced with $\mathscr{A}_n$ is wellposed and the solution is given through an evolution family $(G_n(r,s))_{0 \leq s \leq r \leq t}$. Moreover, for $s < r$ the operator $G_n(r,s)$ is given by a Green kernel $g_n(r,s,\cdot,\cdot)$. We write $\mathscr{A}_n^0 := \mathscr{A}_n + V$ and denote the Green kernel associated to the operators $\mathscr{A}_n^0$ by g_n^0.

We make the following assumptions.

Hypothesis 12.5 *Fix* $0 < t \leq 1$, $x \in \mathbb{R}^d$ *and* $0 < a_0 < a < b < b_0 < t$ *and assume we are given time dependent Lyapunov functions* W_1, W_2 *with* $W_1 \leq W_2 \leq c_0 Z^{1-\sigma}$ *for some constants* $c_0 > 0$ *and* $\sigma \in (0,1)$ *and a weight function* $1 \leq w \in C^2(\mathbb{R}^d)$ *such that*

1. *Hypotheses 12.3(1)–(2) are satisfied;*
2. $|\Delta_y w| \leq c_8 w^{\frac{k-2}{k}} W_1^{\frac{2}{k}}$ *and* $|Q \nabla_y W_1| \leq c_9 w^{-\frac{1}{k}} W_1 W_2^{\frac{1}{k}}$ *on* $[a_0, b_0] \times \mathbb{R}^d$, *for certain constants* $c_8, c_9 \geq 1$;
3. *for* $n \in \mathbb{N}$ *we have* $g_n^0(t, \cdot, x, \cdot) \in L^\infty((a,b) \times \mathbb{R}^d)$.

In order to prove kernel estimates for the Green kernel g, we apply Theorem 12.4 to the operators $\mathscr{A}_n$ and then let $n \to \infty$. To do so, we have to show that the operators $\mathscr{A}_n$ satisfy Hypothesis 12.3.

Lemma 12.6 *The operator* $\mathscr{A}_n$ *satisfies Hypothesis 12.3 with the same constants* c_1, c_4, c_6, c_7 *and with* c_2, c_3 *and* c_5 *being replaced, respectively, by* $2c_2$, $c_3 + \eta c_8$ *and* $c_5 + 4c_9$.

Proof Since part (1) is obvious and part (3) follows directly from part (3) in Hypothesis 12.5, we only need to check part (2) of Hypothesis 12.3. Here, the estimates (i), (iv), (vi) and (vii) are obvious, as they do not depend on the diffusion coefficients. Let us next note that

$$|\nabla_y w| = |Q^{-1} Q \nabla_y w| \leq \eta^{-1} c_2 w^{\frac{k-1}{k}} W_1^{\frac{1}{k}},$$

so that

$$|Q_n \nabla_y w| = |\varphi_n Q \nabla_y w + (1-\varphi_n)\eta \nabla_y w| \leq |Q \nabla_y w| + \eta |\nabla_y w| \leq 2c_2 w^{\frac{k-1}{k}} W_1^{\frac{1}{k}}.$$

This gives (ii) for Q_n. As for (iii), we have

$$|\text{Tr}(Q_n D^2 w)| \leq |\text{Tr}(Q D^2 w)| + \eta |\Delta w| \leq (c_3 + \eta c_8) w^{\frac{k-2}{k}} W_1^{\frac{2}{k}}.$$

It remains to check (v). We note that

$$\sum_{i=1}^d D_i q_{ij}^{(n)} = \varphi_n \sum_{i=1}^d D_i q_{ij} + \frac{\varphi'(W_1/n)}{n} \left[(Q \nabla_y W_1)_j - \eta D_j W_1 \right].$$

As $|t\varphi'(t)| \leq 2$, it follows that

$$\left|\frac{\varphi'(W_1/n)}{n}[(Q\nabla_y W_1)_j - \eta D_j W_1]\right| \leq \frac{2}{W_1}(|Q\nabla_y W_1| + \eta|\nabla_y W_1|).$$

Consequently,

$$\left|\sum_{i=1}^{d} D_i q_{ij}^{(n)}\right| \leq \left|\sum_{i=1}^{d} D_i q_{ij}\right| + \frac{2}{W_1}(|Q\nabla_y W_1| + \eta|\nabla_y W_1|) \leq (c_5 + 4c_9)w^{-\frac{1}{k}}W_2^{\frac{1}{k}}.$$

This finishes the proof. $\qquad\square$

We shall need the following convergence result for the Green kernels.

Lemma 12.7 *Fix* $r \leq t$ *and* $x \in \mathbb{R}^d$ *and define* $\rho_n(s,y) := g_n(r,s,x,y)$ *and* $\rho(s,y) := g(r,s,x,y)$ *for* $s \in [0,r]$ *and* $y \in \mathbb{R}^d$. *Then* $\rho_n \to \rho$, *locally uniformly in* $(0,r) \times \mathbb{R}^d$.

Proof The proof is obtained as that of [7, Proposition 2.9]. We give a sketch. Using Schauder interior estimates and a diagonal argument, one shows that for any $f \in C_c^{2+\varsigma}(\mathbb{R}^d)$ $G_n(\cdot,s)f$ converges to $G(\cdot,s)f$ locally uniformly. This implies that the measure $\rho_n(s,y)\,dsdy$ converges weakly to the measure $\rho(s,y)\,dsdy$.

On the other hand, [3, Corollary 3.11] implies that for a compact set $K \subset \mathbb{R}^d$ and a compact interval $J \subset (0,r)$ we have $\|\rho_n\|_{C^\gamma(J \times K)} \leq C$ for certain constants $C > 0$ and $\gamma \in (0,1)$ independent of n. Thus, by compactness, a subsequence converges locally uniformly to some continuous function ψ which, by the above, has to be ρ. $\qquad\square$

We can now state and prove our main result.

Theorem 12.6 *Assume Hypothesis 12.5. Then there exists a positive constant* C_1, *depending only on* d,k *and* η, *such that*

$$w\rho \leq C_1\left[c_1^{\frac{k}{2}}\sup_{s\in(a_0,b_0)}\zeta_{W_1}(s) + \left(\frac{c_1^{\frac{k}{2}}}{(b_0-b)^{\frac{k}{2}}} + c_2^k + c_3^{\frac{k}{2}} + c_4^{\frac{k}{2}} + c_8^{\frac{k}{2}}\right)\int_{a_0}^{b_0}\zeta_{W_1}(s)\,ds\right.$$

$$\left. + \left(c_2^{\frac{k}{2}}c_6^{\frac{k}{2}} + c_5^k + c_6^k + c_7^k + c_9^k\right)\int_{a_0}^{b_0}\zeta_{W_2}(s)\,ds\right] \qquad (12.21)$$

in $(a,b) \times \mathbb{R}^d$.

Proof We apply Theorem 12.4 to the operators $\mathscr{A}_n$. Taking Lemma 12.6 into account, we obtain

$$w\rho_n \leq C_1 \left[c_1^{\frac{k}{2}} \sup_{s \in (a_0, b_0)} \zeta_{1,n}(s) + \left(c_2^{\frac{k}{2}} c_6^{\frac{k}{2}} + (c_5 + 4c_9)^k + c_6^k + c_7^k \right) \int_{a_0}^{b_0} \zeta_{2,n}(s)\, ds \right.$$

$$\left. + \left(\frac{c_1^{\frac{k}{2}}}{(b_0 - b)^{\frac{k}{2}}} + (2c_2)^k + (c_3 + \eta c_8)^{\frac{k}{2}} + c_4^{\frac{k}{2}} \right) \int_{a_0}^{b_0} \zeta_{1,n}(s)\, ds \right],$$

$$(12.22)$$

in (a, b), where $\zeta_{j,n}(s) := \int_{\mathbb{R}^d} W_j(x, y) g_n(t, s, x, y)\,dy$. Note that $\zeta_{j,n}$ is well defined by Proposition 12.1, since W_j is also a time dependent Lyapunov function for $\mathscr{A}_n$ by Lemma 12.5. Since $\rho_n \to \rho$ locally uniformly by Lemma 12.7, Estimate (12.21) follows from (12.22) upon $n \to \infty$ once we prove that the right-hand sides also converge.

To that end, it suffices to prove that $\zeta_{j,n}$ converges to ζ_{W_j} uniformly on (a_0, b_0). Using the estimate $W_j \leq c_0 Z^{1-\sigma}$ and Hölder's inequality, we find

$$|\zeta_{j,n}(s) - \zeta_{W_j}(s)| \leq \int_{\mathbb{R}^d} W_j(s)|\rho_n(s) - \rho(s)|\,dy$$

$$\leq \int_{B(0,R)} W_j(s)|\rho_n(s) - \rho(s)|\,dy$$

$$+ \int_{\mathbb{R}^d \setminus B(0,R)} W_j(s)\rho_n(s)\,dy + \int_{\mathbb{R}^d \setminus B(0,R)} W_j(s)\rho(s)\,dy$$

$$\leq \|W_j\|_{L^\infty((a_0,b_0) \times B(0,R))} \|\rho_n - \rho\|_{L^\infty((a_0,b_0) \times B(0,R))} |B(0,R)|$$

$$+ c_0 \left(\int_{\mathbb{R}^d \setminus B(0,R)} Z(y) g_n(t, s, x, y)\,dy \right)^{1-\sigma} (g_n(t, s, x, \mathbb{R}^d \setminus B(0,R)))^\sigma$$

$$+ c_0 \left(\int_{\mathbb{R}^d \setminus B(0,R)} Z(y) g(t, s, x, y)\,dy \right)^{1-\sigma} (g(t, s, x, \mathbb{R}^d \setminus B(0,R)))^\sigma,$$

$$(12.23)$$

where $|B(0, R)|$ denotes the Lebesgue measure of the ball $B(0, R)$. We first note that, as a consequence of Equation (12.8) (which is also valid if G is replaced with G_n since Z is also a Lyapunov function for $\mathscr{A}_n$), the integrals $\int_{\mathbb{R}^d} Z(y) g_n(t, s, x, y)\,dy$ are uniformly bounded. Arguing as in the proof of (12.10), it is easy to check that the measures $\{g_n(t, s, x, y)\,dy : s \in [0, t]\}$ are tight. Therefore, the last two terms in (12.23) can be bounded by any given $\varepsilon > 0$ if R is chosen large enough. Since $\rho_n \to \rho$ locally uniformly, given R, also the first term in (12.23) can be bounded by ε if n is large enough. Thus, altogether $\zeta_{j,n} \to \zeta_{W_j}$ uniformly on $[a_0, b_0]$. This finishes the proof. □

12.5 Proof of Theorem 12.2

Let us come back to the example from Theorem 12.2. We start by observing that the same computations as in the proof of Proposition 12.2 show that the function $Z_0(x) = \exp(\delta|x|_*^{p+1-m})$ is a Lyapunov function for both the operators $\mathscr{A}_0$ and $\eta\Delta_x - F \cdot \nabla_x$.

To obtain estimates for the Green kernel associated with the operator $\mathscr{A}$, we want to apply Theorem 12.6. We assume that we are in the situation of Proposition 12.2 and pick $0 < \varepsilon_0 < \varepsilon_1 < \varepsilon_2 < \delta$, where $\delta < 1/\beta$, and $\alpha > \frac{\beta}{m+\beta-2}$. For $\beta \geq 2$, we define the functions $w, W_1, W_2 : [0, t] \times \mathbb{R}^d$ by

$$w(s, y) := e^{\varepsilon_0(t-s)^\alpha |y|_*^\beta} \quad \text{and} \quad W_j(s, y) := e^{\varepsilon_j(t-s)^\alpha |y|_*^\beta}.$$

Let us check the conditions of Theorem 12.6. As a consequence of Proposition 12.2, W_1 and W_2 are time dependent Lyapunov functions which obviously satisfy $W_1 \leq W_2 \leq Z^{1-\sigma}$ for suitable σ, where $Z(y) := \exp(\delta|y|_*^\beta)$. We have to verify that with this choice of w, W_1 and W_2 Hypothesis 12.5 is satisfied. As before, we make only computations assuming that $|x| \geq 1$, omitting the details concerning the neighborhood of the origin.

We now fix arbitrary $a_0, b_0 \in (0, t)$ with $a_0 < b_0$. Note that $w(s, y)^{-2}\partial_s w(s, y) = -\varepsilon_0\alpha(t - s)^{\alpha-1}|y|^\beta e^{-\varepsilon_0(t-s)^\alpha|y|^\beta}$. This is clearly bounded. Similarly, one sees that $w^{-2}\nabla_y w$ is bounded.

Let us now turn to part (2) of Hypotheses 12.3 and 12.5. Since $w \leq W_1$, clearly (2)(i) is satisfied with $c_1 = 1$. As for (2)(ii), we have

$$\frac{|Q(s, y)\nabla_y w(s, y)|}{w(s, y)^{1-1/k} W_1(s, y)^{1/k}} = \varepsilon_0\beta(t - s)^\alpha |y|^{\beta-1}(1 + |y|^m)e^{-\frac{1}{k}(\varepsilon_1-\varepsilon_0)(t-s)^\alpha|y|^\beta}.$$

To bound this expression, we note that for $\tau, \gamma, z > 0$, we have

$$z^\gamma e^{-\tau z^\beta} = \tau^{-\frac{\gamma}{\beta}}(\tau z^\beta)^{\frac{\gamma}{\beta}}e^{-\tau z^\beta} \leq \tau^{-\frac{\gamma}{\beta}}\left(\frac{\gamma}{\beta}\right)^{\frac{\gamma}{\beta}}e^{-\frac{\gamma}{\beta}} =: \tau^{-\frac{\gamma}{\beta}}C(\gamma, \beta),$$

which follows from the fact that the maximum of the function $t \mapsto t^p e^{-t}$ on $(0, \infty)$ is attained at the point $t = p$. Applying this estimate in the case where $z = |y|$, $\tau = k^{-1}(\varepsilon_1 - \varepsilon_0)(t - s)^\alpha$, $\beta = \beta$ and $\gamma = \beta - 1 + m$, we get

$$\frac{|Q(s, y)\nabla_y w(s, y)|}{w(s, y)^{1-1/k} W_1(s, y)^{1/k}}$$

$$\leq 2\varepsilon_0\beta(t - s)^\alpha \left(\frac{\varepsilon_1 - \varepsilon_0}{k}\right)^{-\frac{\beta-1+m}{\beta}} (t - s)^{-\alpha\frac{\beta-1+m}{\beta}} C(\beta - 1 + m, \beta)$$

$$=: \bar{c}(t - s)^{-\frac{\alpha(m-1)}{\beta}} \leq \bar{c}(t - b_0)^{\frac{-\alpha(m-1)_+}{\beta}},$$

for a certain constant $\bar{c}$.

Thus we can choose the constant c_2 as $\bar{c}(t - b_0)^{-\frac{\alpha(m-1)_+}{\beta}}$, where $\bar{c}$ is a universal constant. Note that c_2 depends on the interval (a_0, b_0) only through the factor $(t - b_0)^{-\gamma_2}$. As it turns out, similar estimates show that also for (2)(iii)–(vii) in Hypothesis 12.3 and in Part (2) of Hypothesis 12.5 we can choose constants $c_3, \ldots, c_9$ of this form, however with different exponents $\gamma_3, \ldots, \gamma_9$. We now determine the exponents we can choose. To simplify the presentation, we drop constants from our notation and write $\lesssim$ to indicate an estimate involving a constant which merely depends on $d, m, p, r, k, \varepsilon_0, \varepsilon_1, \varepsilon_2$.

As for (iii) we find

$$\frac{|\mathrm{Tr}(QD^2w(s, y))|}{w(s, y)^{1-2/k} W_1(s, y)^{2/k}}$$

$$\lesssim \left[(t - s)^\alpha |y|^{\beta-2+m} + (t - s)^{2\alpha} |y|^{2\beta-2+m}\right] e^{-\frac{2}{k}(\varepsilon_1-\varepsilon_0)(t-s)^\alpha |y|^\beta}$$

$$\lesssim (t - s)^{2\alpha} (t - s)^{-\alpha \frac{2\beta-2+m}{\beta}} \leq (t - b_0)^{-\frac{\alpha(m-2)_+}{\beta}},$$

so that here $\gamma_3 = \frac{(m-2)_+}{\beta}$. The estimates

$$\frac{|\partial_s w(s, y)|}{w(s, y)^{1-2/k} W_1(s, y)^{2/k}} \lesssim (t - s)^{\alpha-1} |y|^\beta e^{-\frac{2}{k}(\varepsilon_1-\varepsilon_0)(t-s)^\alpha |y|^\beta}$$

$$\lesssim (t - s)^{\alpha-1}(t - s)^{-\alpha} \leq (t - b_0)^{-1},$$

$$\frac{w(s, y)^{1/k} |\sum_{i=1}^d D_i q_{ij}(s, y)|}{W_2(s, y)^{1/k}} \lesssim |y|^m e^{-\frac{1}{k}(\varepsilon_2-\varepsilon_0)(t-s)^\alpha |y|^\beta} \lesssim (t - s)^{-\frac{\alpha m}{\beta}}$$

$$\leq (t - b_0)^{-\frac{\alpha m}{\beta}}$$

and

$$\frac{w(s, y)^{1/k} |F(s, y)|}{W_2(s, y)^{1/k}} = |y|^p e^{-\frac{1}{k}(\varepsilon_2-\varepsilon_0)(t-s)^\alpha |y|^\beta} \lesssim (t - s)^{-\frac{\alpha p}{\beta}} \leq (t - b_0)^{-\frac{\alpha p}{\beta}},$$

show that in (iv), resp. (v), resp. (vi) we can choose $\gamma_4 = 1$, resp. $\gamma_5 = \frac{\alpha m}{\beta}$ resp. $\gamma_6 = \frac{\alpha p}{\beta}$.

A similar estimate as for (vi) shows that in (vii) we can choose $\gamma_7 = \frac{\alpha r}{2\beta}$.

Concerning part (2) of Hypothesis 12.5, we note that repeating the computations for Hypothesis 12.3(2)(ii)–(iii) with $m = 0$, we see that in the estimate for $|\Delta_y w|$ and $|Q \nabla_y W_1|$ we can pick $c_8 = c_9 = \bar{c}$.

Finally for part (3) of Hypothesis 12.3, we note that in this special situation the boundedness of the Green kernel for the associated operators without potential term can also be established using time dependent Lyapunov functions. This has been done in [7].

We may thus invoke Theorem 12.6. To that end, given $s \in (0, t)$, we choose $a_0 := \max\{s - (t-s)/2, s/2\}$ and $b_0 := s + (t-s)/2$ so that $t - b_0 = (t-s)/2$ and $b_0 - a_0 \le t - s$. Let us note that, as a consequence of Proposition 12.1,

$$\zeta_{W_j}(s, x) \le \exp\left(\int_s^t h(\tau) \, d\tau\right) W_j(t, x) = \exp\left(\int_s^t h(\tau) \, d\tau\right).$$

Thus, recalling the form of h from the proof of Proposition 12.2, we see that there exists a constant H, depending only on α, β and m, hence independent of (a_0, b_0), such that

$$\int_{a_0}^{b_0} \zeta_{W_j}(s) \, ds \le H(b_0 - a_0) \le H(t-s).$$

Thus, by Theorem 12.6, we find that, for a certain constant C, we have

$$w\rho \le C\left((t-s)^{1-\frac{k}{2}} + (t-s)^{1-\frac{\alpha}{2\beta}((m-1)_+ + p)k} + (t-s)^{1-\frac{\alpha}{\beta}\Lambda k}\right), \tag{12.24}$$

where $\Lambda = m \vee p \vee \frac{r}{2}$. To simplify this further, we note first that

$$\Lambda \ge \frac{1}{2}((m-1)_+ + p).$$

Now, let us assume that both $p > m - 1$ and $r > m - 2$ so that we can either assume (i) or (ii) in Proposition 12.2. Note that in case (i), we have, by the choice of α, that

$$\frac{\alpha\Lambda}{\beta} \ge \frac{\alpha p}{\beta} > \frac{p}{m + \beta - 2} = \frac{p}{p-1} > \frac{1}{2}.$$

In case (ii), we distinguish the cases $r + m > 2$ and $r + m \le 2$. If $r + m > 2$ we have

$$\frac{\alpha\Lambda}{\beta} \ge \frac{\alpha r}{2\beta} > \frac{r}{2(m + \beta - 2)} = \frac{r}{r + m - 2} > \frac{1}{2},$$

since $r > m - 2$. On the other hand, if $r + m \le 2$, then

$$\frac{\alpha\Lambda}{\beta} \ge \frac{\alpha p}{\beta} > \frac{p}{p-1} > \frac{1}{2},$$

Thus, the right-hand side of (12.24) can be estimated by a constant times $(t - s)^{1-\frac{\alpha}{\beta}\Lambda k}$.

Therefore, if $p \geq \frac{1}{2}(m + r)$, we pick $\beta = p + 1 - m$. We have, for $\alpha > \frac{p+1-m}{p-1}$, $\varepsilon < \frac{1}{p+1-m}$,

$$g(t, s, x, y) \leq C(t - s)^{1 - \frac{\alpha(m \vee p)k}{p+1-m}} e^{-\varepsilon(t-s)^{\alpha}|y|_*^{p+1-m}},$$

for a certain constant C. On the other hand, for $p < \frac{1}{2}(m + r)$, we pick $\beta = \frac{1}{2}(r + 2 - m)$. So, we obtain

$$g(t, s, x, y) \leq C(t - s)^{1 - \frac{\alpha(2m \vee 2p \vee r)}{2(r+2-m)}k} e^{-\varepsilon(t-s)^{\alpha}|y|_*^{\frac{1}{2}(r+2-m)}},$$

for $\varepsilon < \frac{2}{r+2-m}$ and $\alpha > \frac{r-m+2}{r+m-2}$ if $r + m > 2$, and $\alpha > \frac{r+2-m}{2(p-1)}$ if $r + m \leq 2$, where, again, C is a positive constant independent of t and s. This finishes the proof of Theorem 12.2.

References

1. Aibeche, A., Laidoune, K., Rhandi, A.: Time dependent Lyapunov functions for some Kolmogorov semigroups perturbed by unbounded potentials. Arch. Math. (Basel) **94**, 565–577 (2010)
2. Angiuli, L., Lorenzi, L.: Compactness and invariance properties of evolution operators associated with Kolmogorov operators with unbounded coefficients. J. Math. Anal. Appl. **379**, 125–149 (2011)
3. Bogachev, V.I., Krylov, N.V., Röckner, M.: On regularity of transition probabilities and invariant measures of singular diffusions under minimal conditions. Commun. Partial Differ. Equ. **26**, 2037–2080 (2001)
4. Fornaro, S., Fusco, N., Metafune, G., Pallara, D.: Sharp upper bounds for the density of some invariant measures. Proc. R. Soc. Edinburgh Sect. A **139**, 1145–1161 (2009)
5. Krylov, N.V.: Some properties of traces for stochastic and deterministic parabolic weighted Sobolev spaces. J. Funct. Anal. **183**, 1–41 (2001)
6. Kunze, M., Lorenzi, L., A. Lunardi, A.: Nonautonomous Kolmogorov parabolic equations with unbounded coefficients. Trans. Am. Math. Soc. **362**, 169–198 (2010)
7. Kunze, M., Lorenzi, L., Rhandi, A.: Kernel estimates for nonautonomous Kolmogorov equations (submitted). ArXiv:1308.1926 (2013)
8. Laidoune, K., Metafune, G., Pallara, D., Rhandi, A.: Global properties of transition kernels associated to second order elliptic operators. In: Escher, J. et al. (eds.) Progress in Nonlinear Differential Equations and Their Applications, vol. 60. Springer, Basel (2011)
9. Lorenzi, L., Rhandi, A.: On Schrödinger type operators with unbounded coefficients: generation and heat kernel estimates (submitted). ArXiv:1203.0734 (2012)
10. Metafune, G., Pallara, D., Rhandi, A.: Global properties of transition probabilities of singular diffusions. Theory Probab. Appl. **54**, 68–96 (2010)
11. Metafune, G., Spina, C.: Kernel estimates for a class of Schrödinger semigroups. J. Evol. Equ. **7**, 719–742 (2007)
12. Metafune, G., Spina, C., Tacelli, C.: Elliptic operators with unbounded diffusions and drift coefficients in L^p spaces. Adv. Differ. Equ. **19**, 473–526 (2014)
13. Spina, C.: Kernel estimates for a class of Kolmogorov semigroups. Arch. Math. **91**, 265–279 (2008)

Chapter 13
On Some Inverse Problems of the Calculus of Variations for Second Order Differential Equations with Deviating Arguments and Partial Derivatives

Galina Kurina

This paper is dedicated to the memory of Professor
A. Lorenzi
I met Alfredo Lorenzi in Italy at the international meetings devoted to differential equations, inverse problems and control theory which were organized by him and Professor A. Favini. Alfredo spoke in Russian very well. We had numerous discussions on different problems which were extremely interesting and useful. I will always remember the friendliness of Alfredo, his exceptional professionalism in mathematics. The death of A. Lorenzi is irreparable loss for the mathematics and mathematicians.

Abstract The paper is devoted to the study of solvability conditions for inverse problems of the calculus of variations for two types of one-dimensional second order differential equations with respect to functions of two variables with deviating arguments. We obtain explicit formulae for the integrand of the inverse problem solution depending on deviating arguments and partial derivatives of the first order. For this purpose we integrate functions standing before the second order derivatives in the given equation with respect to some of their arguments. This method is compared with the standard approach for recovering a functional from a given equation by using the integration of a bilinear form with respect to an auxiliary parameter. We also demonstrate by presenting explicit examples that sometimes the proposed approach is more preferable than the traditional one.

G. Kurina (✉)
Voronezh State University, 1 Universitetskaya pl., 394006 Voronezh, Russia

Voronezh Institute of Law and Economics, 119-A Leninskii pr., 394042 Voronezh, Russia
e-mail: kurina@math.vsu.ru

© Springer International Publishing Switzerland 2014
A. Favini et al. (eds.), *New Prospects in Direct, Inverse and Control Problems for Evolution Equations*, Springer INdAM Series 10,
DOI 10.1007/978-3-319-11406-4_13

13.1 Introduction

The inverse problem of the calculus of variations consists in the following. Let a second order differential equation with deviating arguments and partial derivatives be given. It is required to know whether there exists a functional defined by an integral for which this equation is a necessary extremum condition for the functional. If such a functional exists, then we need to find it. The inverse problem of the calculus of variations is closely connected with the variational method of the differential equations research. It is very important for this method to obtain a solution of the inverse problem in the form of an integral with the integrand containing derivatives of smaller order than in the given equation (see, for instance [2]). If an one-parameter group of variational symmetries is known, then we can find a solution of Euler's equation of the second order in quadratures (see, e.g., [9]). The survey [3] is devoted to various approaches and results concerning inverse problems of the calculus of variations.

Differential equations with deviating arguments have numerous applications in automatic control theory, in the theory of self-oscillating systems, in the study of duct-burning problems in rocketry. They occur in problems of long-term forecasting in economics, in various biophysical problems, etc. The reason for the occurrence of delays in variational problems in control theory is sometimes related to time delays incurred in signal transmission. However, usually it is due to simplifying assumptions that reduce the action of intermediate transmitting and amplifying devices in the system to delays in the transmission of signals [1].

To the best of my knowledge, if inverse problems of the calculus of variations are solvable then two approaches to solve these problems can be applied. The most known method uses the integration with respect to an auxiliary parameter of a bilinear form depending on a given equation (see, for instance [2]). The second approach allows us to obtain the explicit expression for the integrand of a functional depending on the first order derivatives by integrating functions standing before the second order derivatives in the given equation with respect to some of their arguments [11]. The solution of the inverse problem of the calculus of variations in the explicit form for the second order ordinary differential equation with deviating arguments has been found by the first approach in [10] for linear systems with symmetrical conditions and in [12] by the second approach for non-linear equations with asymmetrical conditions for solutions. Note that the first approach has been also used for partial differential difference equations (see, for instance [8]).

The present paper deals with the inverse problem of the calculus of variations for two types of second order differential equations with respect to functions of two variables with deviating arguments. Using the second approach the solvability conditions and the explicit formulae for the integrand of the inverse problem solution, depending on functions with deviating arguments and partial derivatives of the first order, are obtained. We compare two approaches to inverse problems of the calculus of variations by considering illustrative examples.

The paper is organized as follows. In Sect. 13.2 we present the results devoted to inverse problems in the symmetrical case where a solution of the given equation satisfies a prescribed condition on the boundary of the domain of function definition. Section 13.3 deals with an asymmetrical case in which a solution of a given equation satisfies conditions of various forms on parts of the rectangle being a domain of function definition. Examples illustrating the solving of considered problems are given. In Sect. 13.4 we compare two approaches to inverse problems of the calculus of variations by considering two examples.

13.2 Symmetrical Problems

13.2.1 Euler's Equation

Let Q be an open bounded connected subset in $I\!R^n$ with a sufficiently smooth boundary ∂Q. We begin with one result from [7] concerning the necessary extremum condition for the functional

$$J(z) = \int_Q F(t, z(t), z_{t_1}(t), \ldots, z_{t_n}(t), z(\omega_1(t)), z_{t_1}(\omega_1(t)), \ldots, z_{t_n}(\omega_1(t)), \ldots,$$

$$z(\omega_m(t)), z_{t_1}(\omega_m(t)), \ldots, z_{t_n}(\omega_m(t))) \, dt$$

(13.1)

under the given boundary condition

$$z(t) = \varphi(t), \ t \in \partial Q.$$

(13.2)

Here $m \geq 0$ is fixed, the function F is assumed to be twice continuously differentiable and the mapping $\omega_k : Q \to Q$ is surjective, $\omega_k \in C^2(\bar{Q})$, and its inverse $\gamma_k = \omega_k^{-1} \in C^2(\bar{Q})$ for all $k = 0, \ldots, m$. For notational convenience we set $\omega_0(t) = t$. The notation $z_{t_k}(\omega)$ stands for the partial derivative of z with respect to t_k at the point ω. A function z providing an extremum for functional (13.1) is sought among functions from $C^2(Q) \cap C^1(\bar{Q})$ satisfying the boundary condition (13.2).

Set

$$\Phi := \sum_{k=0}^{m} F|_{t=\gamma_k(t)} I_k,$$

(13.3)

where $I_k = |\gamma_k'(t)|$ and $\gamma_k'(t)$ is the Jacobian $\frac{D(\gamma_{k_1}, \ldots, \gamma_{k_n})}{D(t_1, \ldots, t_n)}$. Note that $I_0 = 1$.

If the functional (13.1) under the boundary conditions (13.2) attains an extremum on a function $z(t)$, then this function satisfies the following equation

$$\Phi_{z(t)} - \sum_{i=1}^{n} \frac{\partial}{\partial t_i} \Phi_{z_{t_i}(t)} = 0.$$

(13.4)

The proof of this fact can be found in [7]. Note that relation (13.4) is a generalization of Euler's equation to the case of the functional (13.1).

A similar result for the case $n = 2$ and $m = 1$ is presented in [4]. We will consider this particular situation. Let $t = (x, y)$. Assume also that F does not depend on partial derivatives at the point $w_1(t) = w(t)$, that is, relation (13.1) has the form

$$J(z) = \int_Q F(x, y, z(x, y), z_x(x, y), z_y(x, y), z(\omega(x, y)))dxdy, \qquad (13.5)$$

where $\omega(x, y) = (\xi(x, y), \eta(x, y))$. We assume the same conditions on Q and F as before. For instance, let $Q = \{(x, y) \in I\!R^2 : x^2 + y^2 < r^2\}$ and ω be a rotation. In this case, (13.3) becomes

$$\Phi := F + \tilde{F},$$

where

$$\tilde{F} := F(\alpha(t), \beta(t), z(\alpha(t), \beta(t)), z_x(\alpha(t), \beta(t)), z_y(\alpha(t), \beta(t)), z(t))I,$$

$$t = (x, y) \in Q, \qquad (\alpha(x, y), \beta(x, y)) = \omega^{-1}(x, y), \qquad I = \left| \frac{D(\alpha, \beta)}{D(x, y)} \right|.$$

Using (13.4), we obtain Euler's equation for functional (13.5)

$$F_z - \frac{\partial}{\partial x} F_{z_x} - \frac{\partial}{\partial y} F_{z_y} + \tilde{F}_z = 0.$$

Let us write the last equality in the expanded form

$$- F_{z_x z_x} z_{xx} - 2 F_{z_x z_y} z_{xy} - F_{z_y z_y} z_{yy} + F_z - F_{z_x x} - F_{z_x z} z_x - F_{z_x z(\omega)} z_\omega \omega_x -$$

$$- F_{z_y y} - F_{z_y z} z_y - F_{z_y z(\omega)} z_\omega \omega_y + \tilde{F}_z = 0. \qquad (13.6)$$

13.2.2   The Inverse Problem of the Calculus of Variations

Taking into account the form of Euler's equation (13.6), let us consider the following equation with deviating arguments and partial derivatives

$$A z_{xx}(x, y) + B z_{xy}(x, y) + C z_{yy}(x, y) + D = 0. \qquad (13.7)$$

Here $D := D_1 + D_2$, the functions A, B, C, and D_1 depend on six variables

$$x, y, z(x, y), z_x(x, y), z_y(x, y), z(\xi(x, y), \eta(x, y)), \qquad (13.8)$$

and the function D_2 depends on variables (13.8) and

$$z_x(\xi(x, y), \eta(x, y)), z_y(\xi(x, y), \eta(x, y)), \alpha(x, y), \beta(x, y), z(\alpha(x, y), \beta(x, y)),$$
$$z_x(\alpha(x, y), \beta(x, y)), z_y(\alpha(x, y), \beta(x, y)).$$

All functions in (13.7) are continuously differentiable with respect to the arguments. It is also assumed that admissible functions z satisfy (13.2).

Consider the inverse problem of the calculus of variations for equation (13.7). Namely, we seek for a functional of the form (13.5) such that the corresponding Euler's equation coincides with equation (13.7).

The next result provides necessary solvability conditions for the considered inverse problem.

Theorem 13.1 *If the inverse problem of the calculus of variations for equation (13.7) has a solution in the form of the functional (13.5) with the function F, which is three times continuously differentiable, then the functions A, B, C, and D satisfy the following conditions:*

$$A_{z_y} = \frac{1}{2} B_{z_x}, \tag{13.9}$$

$$C_{z_x} = \frac{1}{2} B_{z_y}, \tag{13.10}$$

$$D_{z_x} = A_x + A_z z_x + A_{z(\omega)} z_\omega \omega_x + \frac{1}{2} B_y + \frac{1}{2} B_z z_y + \frac{1}{2} B_{z(\omega)} z_\omega \omega_y, \tag{13.11}$$

$$D_{z_y} = \frac{1}{2} B_x + \frac{1}{2} B_z z_x + \frac{1}{2} B_{z(\omega)} z_\omega \omega_x + C_y + C_z z_y. \tag{13.12}$$

Proof If the inverse problem of the calculus of variations has a solution, then there is a function F such that equation (13.7) is Euler's equation for the functional (13.5) with the integrand F. Hence, the left-hand side of equality (13.7) coincides with the left-hand side of (13.6) for the function F. Differentiating successively with respect to z_{xx}, z_{xy}, and z_{yy} we have

$$A = -F_{z_x z_x}, \tag{13.13}$$

$$B = -2F_{z_x z_y}, \tag{13.14}$$

$$C = -F_{z_y z_y}, \tag{13.15}$$

$$D = D_1 + D_2 = F_z - F_{z_x x} - F_{z_x z} z_x - F_{z_x z(\omega)} z_\omega \omega_x - F_{z_y y}$$
$$- F_{z_y z} z_y - F_{z_y z(\omega)} z_\omega \omega_y + \tilde{F}_z. \tag{13.16}$$

Using the differentiation, we arrive at conditions (13.9)–(13.12).   □

Remark 13.1 If (13.7) does not contain deviating arguments, then the solvability conditions (13.9)–(13.12) correspond to conditions (6.6)–(6.9) from [11].

Analogously [11, p. 104], where an equation without deviating arguments has been considered, we obtain from (13.13)

$$F = -\frac{1}{2} \int \int B \, dz_y \, dz_x + \int U(x, y, z, z_x, z(\omega)) dz_x + V(x, y, z, z_y, z(\omega)),$$

(13.17)

where the functions U and V are unknown. In the last formula and further, the notation of the form $\int \int f(x, y) dx dy$ means an iterated integral with respect to x and y. Taking into account (13.13) and (13.15), we get

$$A = \frac{1}{2} \int B_{z_x} dz_y - U_{z_x}, \quad C = \frac{1}{2} \int B_{z_y} dz_x - V_{z_y z_y}.$$

Since we have no aim to find the general form of F, in view of the last two relations we can take

$$U = \frac{1}{2} \int \int B_{z_x} \, dz_y \, dz_x - \int A \, dz_x,$$

(13.18)

$$V_{z_y} = \frac{1}{2} \int \int B_{z_y} \, dz_x \, dz_y - \int C \, dz_y,$$

(13.19)

and

$$V = \frac{1}{2} \int \int \int B_{z_y} \, dz_x \, dz_y \, dz_y - \int \int C \, dz_y \, dz_y + W,$$

(13.20)

where the function $W = W(x, y, z, z(\omega))$ is unknown. Note that W is a three times continuously differentiable function of four variables x, y, $z(x, y)$, and $z(\omega(x, y))$. We seek for a non-general form of F. Therefore, we can find W from the equality $F_z = D_1$ and relations (13.17), (13.18), and (13.20)

$$W = \int \left(D_1 + \frac{1}{2} \int \left(\int B_z dz_x - \int \int B_{z_x z} \, dz_x \, dz_x - \int \int B_{zz_y} \, dz_x \, dz_y \right) dz_y \right.$$

$$\left. + \int \int A_z \, dz_x \, dz_x + \int \int C_z \, dz_y \, dz_y \right) dz.$$

(13.21)

In order to ensure (13.16) we must require the validity of the equality

$$D_2 = -F_{z_x x} - F_{z_x z} z_x - F_{z_x z(\omega)} z_\omega \omega_x - F_{z_y y} - F_{z_y z} z_y - F_{z_y z(\omega)} z_\omega \omega_y + \tilde{F}_z,$$

(13.22)

where F is defined by (13.17), (13.18), (13.20), (13.21).

We will also assume that

$$(D_1)_{z_x} = -\frac{1}{2} \int \left(B_z - \int B_{z_xz} dz_x - \int B_{zz_y} dz_y \right) dz_y - \int A_z dz_x - \int \int C_{zz_x} dz_y \, dz_y,$$
(13.23)

$$(D_1)_{z_y} = -\frac{1}{2} \left(\int B_z dz_x - \int \int B_{z_xz} \, dz_x \, dz_x - \int \int B_{zz_y} \, dz_x \, dz_y \right)$$

$$- \int \int A_{zz_y} \, dz_x \, dz_x - \int C_z dz_y. \quad (13.24)$$

Theorem 13.2 *If the twice continuously differentiable functions A, B, C, and D in (13.7) satisfy (13.9), (13.10), and (13.22)–(13.24), then the function F, defined by (13.17), (13.18), (13.20), and (13.21), determines the solution (13.5) of the inverse problem of the calculus of variations for equation (13.7).*

Proof In view of (13.9), (13.10), (13.23), and (13.24) it follows from (13.18), (13.19), (13.21), respectively, that $U_{z_y} = 0$, $V_{z_yz_x} = 0$, $W_{z_x} = 0$, $W_{z_y} = 0$. Hence, $V_{z_x} = 0$. Substitute the expression for the function F, defined by (13.17), (13.18), (13.20), and (13.21), into the left-hand side in (13.6). Using the given conditions, we obtain the left-hand side in (13.7). □

Remark 13.2 The obtained formulae show that the differentiability conditions for the functions in (13.7) can be relaxed for some functions.

13.2.3 Example 1

Let $Q = \{(x, y) \in I\!R^2 : x^2 + y^2 < r^2\}$ and

$$\xi(x, y) = \frac{x - y}{\sqrt{2}}, \quad \eta(x, y) = \frac{x + y}{\sqrt{2}}, \quad \alpha(x, y) = \frac{x + y}{\sqrt{2}}, \quad \beta(x, y) = \frac{y - x}{\sqrt{2}}.$$

Consider the equation of form (13.7) with

$$A = \frac{1}{2}B = C = -D_1 = -\exp\left\{z(x, y) + z_x(x, y) + z_y(x, y) + z\left(\frac{x - y}{\sqrt{2}}, \frac{x + y}{\sqrt{2}}\right)\right\}$$

$$D_2 = -D_1 \times \left(z_x(x, y) + z_y(x, y) + \sqrt{2}z_y\left(\frac{x - y}{\sqrt{2}}, \frac{x + y}{\sqrt{2}}\right)\right)$$
(13.25)

$$+ \exp\left\{z\left(\frac{x + y}{\sqrt{2}}, \frac{y - x}{\sqrt{2}}\right) + z_x\left(\frac{x + y}{\sqrt{2}}, \frac{y - x}{\sqrt{2}}\right) + z_y\left(\frac{x + y}{\sqrt{2}}, \frac{y - x}{\sqrt{2}}\right) + z(x, y)\right\}.$$

Let us solve the inverse problem of the calculus of variations. By (13.17), (13.18), (13.20), and (13.21) we obtain

$$U = W = V = 0,$$

$$F = \exp\left\{z(x, y) + z_x(x, y) + z_y(x, y) + z\left(\frac{x - y}{\sqrt{2}}, \frac{x + y}{\sqrt{2}}\right)\right\}. \quad (13.26)$$

Conditions (13.9), (13.10), and (13.22)–(13.24) from Theorem 13.2 are valid in this case. Therefore, F in (13.26) defines the required functional (13.5). It is not difficult to verify directly that Eq. (13.7) with functions (13.25) is really Euler's equation for the functional (13.5) with the obtained function (13.26).

13.3 Asymmetrical Problems

Let $Q = (x_0, x_1) \times (y_0, y_1)$. Denote by $L_2(Q)$ the space of square integrable functions and by $H(Q)$ the space of functions $z(x, y)$ which are differentiable with respect to x almost everywhere on Q and such that $z \in L_2(Q)$ and $z_x \in L_2(Q)$. The norm in $H(Q)$ is defined by $\|z\|_H = (\|z\|_{L_2(Q)}^2 + \|z_x\|_{L_2(Q)}^2)^{1/2}$.

Consider the problem of an extremum of the functional

$$J(z) = \int_{x_0}^{x_1} dx \int_{y_0}^{y_1} F\left(x, y, z(x, y - h), z(x, y), z_x(x, y - h), z_x(x, y)\right) dy \quad (13.27)$$

with respect to functions of two variables (see [5, 6]). Here $0 < h < y_1 - y_0$ and the function F is assumed to be twice continuously differentiable on Q. Admissible functions belong to the space $H(Q)$ and, moreover, satisfy the following boundary conditions

$$z(x, y) = \varphi(x, y), \ (x, y) \in E_0; \quad z(x, y) = \chi(x), \ (x, y) \in E_1; \quad (13.28)$$

$$z(x, y) = \mu(y), \ (x, y) \in G_0; \quad z(x, y) = \nu(y), \ (x, y) \in G_1. \quad (13.29)$$

Here $E_0 = \{(x, y) : x \in [x_0, x_1], y \in [y_0 - h, y_1)\}$, $E_1 = \{(x, y) : x \in [x_0, x_1], y = y_1\}$, $G_0 = \{(x, y) : x = x_0, y \in (y_0, y_1)\}$, and $G_1 = \{(x, y) : x = x_1, y \in (y_0, y_1)\}$. The functions φ, χ, μ, and ν defined on E_0, E_1, G_0, and G_1 respectively are given (see Fig. 13.1). It is also assumed that $\varphi \in H(E_0)$ and the functions μ and ν are integrable and essentially bounded in G_0 and G_1 respectively.

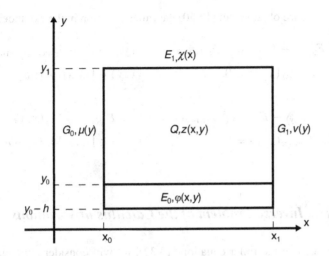

Fig. 13.1 The domains of definition of functions φ, χ, μ, and ν

13.3.1 Euler's Equation

Define

$$\tilde{F} := F(x, y + h, z(x, y), z(x, y + h), z_x(x, y), z_x(x, y + h)),$$

$\tilde{F}_z$ and $\tilde{F}_{z_x}$ denote the partial derivatives of $\tilde{F}$ with respect to the third and to the fifth arguments respectively. Introduce the following notation

$$\Psi = F + \tilde{F}, \quad y \in [y_0, y_1 - h]; \qquad \Psi = F, \quad y \in (y_1 - h, y_1].$$

Now we are formulating the necessary condition for an extremum of functional (13.27).

Theorem 13.3 ([5, 6]) *Let an extremum of functional (13.27) be attained at z under boundary conditions (13.28) and (13.29) in the space $H(Q)$. Then almost everywhere on Q the function z satisfies the equation*

$$\Psi_{z(x,y)} - \frac{d}{dx}\Psi_{z_x(x,y)} = 0, \tag{13.30}$$

which is an analog of Euler's equation for the considered problem.

Further, we will consider the function F in (13.27) which does not depend on $z_x(x, y - h)$, i.e. we will deal with the functional of the form

$$J(z) = \int_{x_0}^{x_1} dx \int_{y_0}^{y_1} F(x, y, z(x, y - h), z(x, y), z_x(x, y))\, dy. \tag{13.31}$$

For this case, we obtain from (13.30) the Euler equation in the expanded form

$$F_{z(x,y)} + \tilde{F}_{z(x,y)} - F_{z_x(x,y)x} - F_{z_x(x,y)z(x,y-h)}z_x(x,y-h) - F_{z_x(x,y)z(x,y)}z_x(x,y)$$
$$- F_{z_x(x,y)z_x(x,y)}z_{xx}(x,y) = 0, \qquad (x,y) \in [x_0,x_1] \times [y_0,y_1-h];$$

$$F_{z(x,y)} - F_{xz_x(x,y)} - F_{z_x(x,y)z(x,y-h)}z_x(x,y-h) - F_{z_x(x,y)z(x,y)}z_x(x,y)$$
$$- F_{z_x(x,y)z_x(x,y)}z_{xx}(x,y) = 0, \qquad (x,y) \in [x_0,x_1] \times (y_1-h, y_1].$$
$$(13.32)$$

13.3.2   The Inverse Problem of the Calculus of Variations

Taking into account the Euler equation (13.32), we will consider the equations of the form

$$A z_{xx}(x,y) + B + C z_x(x,y-h) + \tilde{D} = 0, \quad (x,y) \in [x_0,x_1] \times [y_0, y_1 - h],$$
$$(13.33)$$

$$A z_{xx}(x,y) + B + C z_x(x,y-h) = 0, \quad (x,y) \in [x_0,x_1] \times (y_1-h, y_1], \quad (13.34)$$

where $\tilde{D} = D(x, y+h, z(x,y), z(x,y+h), z_x(x,y+h))$ and the functions A, B, C, and D depend on x, y, $z(x,y-h)$, $z(x,y)$, and $z_x(x,y)$. We assume that they are twice continuously differentiable. In fact, the smoothness assumptions can be weakened (see formulae below). It follows from (13.33) that it is sufficiently to define the function D on $[x_0,x_1] \times [y_0+h, y_1]$.

Let us refine the settings of the inverse problem. For given equations (13.33) and (13.34) it is required to find a functional of the form (13.31) for which the left-hand side of Euler's equation coincides with the left-hand side of equations (13.33) and (13.34) for all twice continuously differentiable functions $z(\cdot, \cdot)$ satisfying conditions (13.28) and (13.29).

Theorem 13.4 *The inverse problem of the calculus of variations for equations (13.33) and (13.34) has a solution if and only if the functions in (13.33) and (13.34) satisfy the following conditions*

$$C_{z_x(x,y)} - A_{z(x,y-h)} = 0, \quad (x,y) \in Q, \qquad (13.35)$$

$$B_{z_x(x,y)} - A_x - z_x(x,y) A_{z(x,y)} = 0, \quad (x,y) \in Q, \qquad (13.36)$$

$$D_{z_x(x,y)} + C = 0, \quad (x, y) \in [x_0, x_1] \times (y_0 + h, y_1], \tag{13.37}$$

$$D_{z(x,y)} + \int_0^{z_x(x,y)} C_{z(x,y)} dz_x(x, y) - G_{z(x,y-h)z(x,y)} = 0, \quad (x, y) \in [x_0, x_1] \times (y_0 + h, y_1], \tag{13.38}$$

where

$$G = G(x, y, z(x, y-h), z(x, y)) = \int_0^{z(x,y)} \left(B - \int_0^{z_x(x,y)} B_{z_x(x,y)} dz_x(x, y) + E_x \right) dz(x, y) + K, \tag{13.39}$$

$$K = \begin{cases} 0, & (x, y) \in [x_0, x_1] \times (y_1 - h, y_1], \\ \int_0^{z(x,y)} \left(\tilde{D} + \int_0^{z_x(x,y+h)} \tilde{C} dz_x(x, y + h) - \tilde{G}_{z(x,y)} \right) dz(x, y), & (13.40) \\ (x, y) \in [x_0, x_1] \times [y_0, y_1 - h], \end{cases}$$

$$E = E(x, y, z(x, y-h), z(x, y)) = \int_0^{z(x,y-h)} \left(\int_0^{z_x(x,y)} A_{z(x,y-h)} dz_x(x, y) - C \right) dz(x, y-h),$$

$$(x, y) \in Q, \tag{13.41}$$

the function K for $x \in [x_0, x_1]$, $y \in [y_0, y_1 - h]$ is determined successively on the domains

$$[x_0, x_1] \times (y_1 - kh, y_1 - (k-1)h], \quad k = 2, \ldots : y_1 - kh \geq y_0.$$

The function F, which determines the solution of the form (13.31) of the inverse problem of the calculus of variations, can be expressed as follows:

$$F = - \int_0^{z_x(x,y)} \left(\int_0^{z_x(x,y)} A \, dz_x(x, y) \right) dz_x(x, y) + Ez_x(x, y) + G, \quad (x, y) \in Q. \tag{13.42}$$

Remark 13.3 If it turns out that one of the integrals in the statement of Theorem 13.4 does not exist, then we can replace the lower limit of integration by any number such that this integral does exist.

Proof Necessity. Suppose that the inverse problem has a solution. Then the left-hand sides of (13.32) coincide with the left-hand sides of (13.33) and (13.34) respectively.

Therefore, the following equalities (13.43)–(13.46) must be valid

$$- F_{z_x(x,y)z_x(x,y)} = A, \quad (x, y) \in [x_0, x_1] \times [y_0, y_1], \tag{13.43}$$

$$- F_{z_x(x,y)z(x,y-h)} = C, \quad (x, y) \in [x_0, x_1] \times [y_0, y_1], \tag{13.44}$$

$$F_{z(x,y)} + \tilde{F}_{z(x,y)} - F_{z_x(x,y)x} - F_{z_x(x,y)z(x,y)}z_x(x, y) = B + \tilde{D}, \ (x, y) \in [x_0, x_1]$$
$$\times [y_0, y_1 - h], \tag{13.45}$$

$$F_{z(x,y)} - F_{z_x(x,y)x} - F_{z_x(x,y)z(x,y)}z_x(x, y) = B, \ (x, y) \in [x_0, x_1] \times (y_1 - h, y_1], \tag{13.46}$$

From (13.43), we obtain

$$F_{z_x(x,y)} = - \int_0^{z_x(x,y)} A \, dz_x(x, y) + E(x, y, z(x, y - h), z(x, y)), \tag{13.47}$$

and

$$F = - \int_0^{z_x(x,y)} (\int_0^{z_x(x,y)} A \, dz_x(x, y)) \, dz_x(x, y) + E(x, y, z(x, y - h), z(x, y))z_x(x, y)$$
$$+ G(x, y, z(x, y - h), z(x, y)), \tag{13.48}$$

where $(x, y) \in [x_0, x_1] \times [y_0, y_1]$. Here E and G are sufficiently smooth functions. Substituting (13.47) for $F_{z_x(x,y)}$ into (13.44), we have

$$\int_0^{z_x(x,y)} A_{z(x,y-h)} \, dz_x(x, y) - C = E_{z(x,y-h)}, \quad (x, y) \in [x_0, x_1] \times [y_0, y_1]. \tag{13.49}$$

The function E must not explicitly depend on $z_x(x, y)$, hence, after differentiating the last equality with respect to $z_x(x, y)$, we obtain (13.35). If relation (13.35) holds, then the left-hand side of (13.49) does not explicitly depend on $z_x(x, y)$.

Substituting (13.48) into (13.46), we obtain

$$
- \int_0^{z_x(x,y)} \left(\int_0^{z_x(x,y)} A_{z(x,y)} \, dz_x(x,y) \right) dz_x(x,y) + \int_0^{z_x(x,y)} A_x \, dz_x(x,y)
$$

$$
+ z_x(x,y) \int_0^{z_x(x,y)} A_{z(x,y)} \, dz_x(x,y) - B = E_x - G_{z(x,y)} \qquad (13.50)
$$

for all $(x, y) \in [x_0, x_1] \times (y_1 - h, y_1]$. Differentiating the last equality with respect to $z_x(x, y)$ and noting that the functions E and G are explicitly independent of $z_x(x, y)$, we have (13.36) for $(x, y) \in [x_0, x_1] \times (y_1 - h, y_1]$. If (13.36) holds, then the left hand side of (13.50) does not explicitly depend on $z_x(x, y)$.

Substituting (13.48) into (13.45), we obtain

$$
- \int_0^{z_x(x,y)} \left(\int_0^{z_x(x,y)} A_{z(x,y)} \, dz_x(x,y) \right) dz_x(x,y) + \int_0^{z_x(x,y)} A_x \, dz_x(x,y) + z_x(x,y) \cdot
$$

$$
\int_0^{z_x(x,y)} A_{z(x,y)} \, dz_x(x,y) - \int_0^{z_x(x,y+h)} \left(\int_0^{z_x(x,y+h)} \tilde{A}_{z(x,y)} \, dz_x(x,y+h) \right) dz_x(x,y+h) - B - \tilde{D}
$$

$$
= E_x - \tilde{E}_{z(x,y)} z_x(x, y + h) - G_{z(x,y)} - \tilde{G}_{z(x,y)}, \quad (x, y) \in [x_0, x_1] \times [y_0, y_1 - h].
$$

It follows from (13.35) that $\tilde{A}_{z(x,y)} = \tilde{C}_{z_x(x,y+h)}$. Therefore, using the last equality and (13.49), we have

$$
- \int_0^{z_x(x,y)} \left(\int_0^{z_x(x,y)} A_{z(x,y)} \, dz_x(x,y) \right) dz_x(x,y) + \int_0^{z_x(x,y)} A_x \, dz_x(x,y)
$$

$$
+ z_x(x,y) \int_0^{z_x(x,y)} A_{z(x,y)} \, dz_x(x,y) - B - E_x - \tilde{D} - \int_0^{z_x(x,y+h)} \tilde{C} \, dz_x(x, y + h)
$$

$$
= -G_{z(x,y)} - \tilde{G}_{z(x,y)}, \quad (x, y) \in [x_0, x_1] \times [y_0, y_1 - h].
$$

$$(13.51)$$

Differentiating this equality with respect to $z_x(x, y)$, we obtain (13.36) for $(x, y) \in [x_0, x_1] \times [y_0, y_1 - h]$. If (13.36) holds, then the left-hand side of (13.51) does not explicitly depend on $z_x(x, y)$. Using the differentiation of (13.51) with respect to $z_x(x, y + h)$, we obtain (13.37). If (13.37) holds, then the left-hand side of (13.51) does not explicitly depend on $z_x(x, y + h)$.

Taking into account (13.36), from (13.50) and (13.51) we obtain the equalities

$$
B - \int_0^{z_x(x,y)} B_{z_x(x,y)} \, dz_x(x, y) + E_x = G_{z(x,y)}, \qquad (x, y) \in [x_0, x_1] \times (y_1 - h, y_1];
$$

$$
B - \int_0^{z_x(x,y)} B_{z_x(x,y)} \, dz_x(x, y) + E_x + \tilde{D} + \int_0^{z_x(x,y+h)} \tilde{C} \, dz_x(x, y + h)
$$

$$
= G_{z(x,y)} + \tilde{G}_{z(x,y)}, \qquad (x, y) \in [x_0, x_1] \times [y_0, y_1 - h].
$$

$$(13.52)$$

Noting that the functions B, E, and G are independent of $z(x, y + h)$ and differentiating (13.52) with respect to $z(x, y + h)$, we obtain condition (13.38). If (13.38) holds, then the expression for $G_{z(x,y)}$ obtained from (13.52) does not explicitly depend on $z(x, y + h)$.

Sufficiency. We will show that under conditions (13.35)–(13.38) functional (13.31) with the function F from (13.42) is a solution of the inverse problem of the calculus of variations for equations (13.33) and (13.34).

Recall that the problem of determining the general form of the function F is not posed here. In view of condition (13.35), the expressions for E and G do not explicitly depend on $z_x(x, y)$ and in view of conditions (13.37) and (13.38), the expression for the function G is independent of $z_x(x, y + h)$ and $z(x, y + h)$ respectively.

We substitute into the second equation in (13.32) the expressions for partial derivatives found from (13.42) regarding the explicit independence of E and G of $z_x(x, y)$.

$$
- \int_0^{z_x(x,y)} \left(\int_0^{z_x(x,y)} A_{z(x,y)} \, dz_x(x, y) \right) dz_x(x, y) + E_{z(x,y)} z_x(x, y) + G_{z(x,y)}
$$

$$
+ \int_0^{z_x(x,y)} A_x \, dz_x(x, y) - E_x + \left(\int_0^{z_x(x,y)} A_{z(x,y)} \, dz_x(x, y) - E_{z(x,y)} \right) z_x(x, y)
$$

$$
+ \left(\int_0^{z_x(x,y)} A_{z(x,y-h)} \, dz_x(x, y) - E_{z(x,y-h)} \right) z_x(x, y - h) + A z_{xx}(x, y) = 0,
$$

$$(x, y) \in [x_0, x_1] \times (y_1 - h, y_1].$$

From here, in view of (13.39)–(13.41), and (13.36), we obtain (13.34).

Further, we substitute the found from (13.42) expressions for the partial derivatives of the function F and $\tilde{F}$ into the first equation in (13.32).

$$
- \int_0^{z_x(x,y)} (\int_0^{z_x(x,y)} A_{z(x,y)}\, dz_x(x,y))\, dz_x(x,y) + E_{z(x,y)}z_x(x,y) + G_{z(x,y)} -
$$

$$
- \int_0^{z_x(x,y+h)} (\int_0^{z_x(x,y+h)} \tilde{A}_{z(x,y)}dz_x(x,y+h))dz_x(x,y+h) + \tilde{E}_{z(x,y)}z_x(x,y+h) + \tilde{G}_{z(x,y)} +
$$

$$
+ \int_0^{z_x(x,y)} A_x\, dz_x(x,y) - E_x + (\int_0^{z_x(x,y)} A_{z(x,y)}\, dz_x(x,y) - E_{z(x,y)})z_x(x,y) +
$$

$$
+(\int_0^{z_x(x,y)} A_{z(x,y-h)}\, dz_x(x,y) - E_{z(x,y-h)})z_x(x,y-h) + Az_{xx}(x,y) = 0,
$$

$$(x,y) \in [x_0, x_1] \times [y_0, y_1 - h].$$

Taking into account (13.35), (13.36), and (13.39)–(13.41), we obtain (13.33) from
the last equality. Note that in transforming the two last expressions we have used the
easily verified identity $\int_0^x (\int_0^x f(x)dx)\, dx + \int_0^x f(x)\, dx = x \int_0^x f(x)dx.$ $\square$

13.3.3   Example 2

Let us solve the inverse problem of the calculus of variations for equations of the
form

$$M(x, y, z(x, y-h), z(x, y), z_x(x, y-h), z_x(x, y), z_{xx}(x, y))$$

$$= -(2x^2 - z(x, y-h))z_{xx}(x, y) + 24z(x, y) - 4xz_x(x, y) + z_x(x, y)z_x(x, y-h) = 0,$$

$$(x, y) \in [x_0, x_1] \times (y_1 - h, y_1], \qquad (13.53)$$

$$M(x, y, z(x, y-h), z(x, y), z_x(x, y-h), z_x(x, y), z_{xx}(x, y)) - \frac{1}{2}(z_x(x, y+h))^2 = 0,$$

$$(x, y) \in [x_0, x_1] \times [y_0, y_1 - h]. \qquad (13.54)$$

Here $h \in (0, y_1 - y_0)$ is a parameter.

Comparing (13.53) and (13.54) with (13.34) and (13.33) respectively, we get

$$A = -2x^2 + z(x, y - h), B = 24z(x, y) - 4xz_x(x, y), C = z_x(x, y),$$

$$D = -\frac{1}{2}(z_x(x, y))^2.$$

It is not difficult to check that the solvability conditions (13.35)–(13.38) obtained in Theorem 13.4 are valid in this case.

Using (13.39)–(13.42), we obtain

$$E = K = 0, \ G = 12(z(x, y))^2,$$

$$F = 12(z(x, y))^2 + x^2(z_x(x, y))^2 - \frac{1}{2}z(x, y - h)(z_x(x, y))^2.$$

The last function F determines functional (13.31). It is easily to verify immediately that (13.53), (13.54) is really Euler's equation for the functional of form (13.31) with the obtained function F.

13.4 The Comparison of Two Approaches

In this section, we will compare two methods for finding solutions of the inverse problems of the calculus of variations using some examples. We consider equations of form (13.7). It is assumed that $(x, y) \in Q$, Q is an open circle with the center in the origin of coordinates, ω is a rotation, $\gamma(x, y) = \omega^{-1}(x, y) = (\alpha(x, y), \beta(x, y))$, the boundary function $\varphi = 0$.

Example 3 Let us consider the equation

$$z_{xx}(x, y) + z_{xy}(x, y) + z_{yy}(x, y) + z^\mu(x, y)z^\nu(\omega(x, y))$$

$$+ \frac{\nu}{\mu + 1}z^{\mu+1}(\gamma(x, y))z^{\nu-1}(x, y) = 0, \ \mu \neq -1. \tag{13.55}$$

Here

$$A = B = C = 1, \quad D_1 = z^\mu(x, y)z^\nu(\omega(x, y)), \quad D_2 = \frac{\nu}{\mu + 1}z^{\mu+1}(\gamma(x, y))z^{\nu-1}(x, y).$$

Conditions (13.9), (13.10), and (13.22)–(13.24) are satisfied. Therefore, we obtain from Theorem 13.2 an integrand of a solution of the inverse problem

$$F = -\frac{1}{2}(z_x^2(x, y) + z_x(x, y)z_y(x, y) + z_y^2(x, y)) + \frac{z^{\mu+1}(x, y)}{\mu + 1}z^\nu(\omega(x, y)).$$

$$(13.56)$$

Applying the traditional approach for finding a solution of the inverse problem for equation (13.55), we have the following expression

$$\int_{Q} \int_{0}^{1} (s(z_{xx}(x, y) + z_{xy}(x, y) + z_{yy}(x, y)) + s^{\mu+\nu}(z^\mu(x, y)z^\nu(\omega(x, y))$$

$$+ \frac{\nu}{\mu + 1}z^{\mu+1}(\gamma(x, y))z^{\nu-1}(x, y)))z(x, y) \, ds \, dx \, dy.$$

$$(13.57)$$

If $\mu + \nu > -1$, then we integrate with respect to s. Further, in the integrals of three first addends we pass to iterated integrals and use the formula of integration by parts. In the integral of the fifth addend we make the change of the variables $u = \alpha(x, y), v = \beta(x, y)$. By taking into account the zero boundary condition for z and the form of ω, we obtain a solution of the inverse problem with the integrand of the form (13.56). Note that using the formula for the solution of the inverse problem from Theorem 13.2 is simpler then (13.57).

For $\mu + \nu \leq -1$ we have in (13.57) the divergent integral with respect to s, therefore in this case we cannot find the solution of the inverse problem for equation (13.55) using the traditional approach.

Example 4 Solve the inverse problem of the calculus of variations for the equation

$$z_{xx}(x, y) + z_{xy}(x, y) + z_{yy}(x, y) + \exp(z(x, y)z(\omega(x, y)))$$

$$+ \exp(z(x, y)z(\gamma(x, y)))(z(x, y)z(\gamma(x, y)) - 1)/z^2(x, y) = 0. \quad (13.58)$$

Here

$$A = B = C = 1, \quad D_1 = \exp(z(x, y)z(\omega(x, y))),$$

$$D_2 = \exp(z(x, y)z(\gamma(x, y)))(z(x, y)z(\gamma(x, y)) - 1)/z^2(x, y).$$

Conditions (13.9), (13.10), and (13.22)–(13.24) from Theorem 13.2 are satisfied here. Hence, we obtain from (13.17), (13.18), (13.20), and (13.21) the integrand of a solution of the inverse problem

$$F = -\frac{1}{2}(z_x^2(x, y) + z_x(x, y)z_y(x, y) + z_y^2(x, y)) + \frac{\exp(z(x, y)z(\omega(x, y)))}{z(\omega(x, y))}.$$

Applying the traditional approach for finding a solution of the inverse problem for equation (13.58), we have the following expression

$$\int_Q \int_0^1 (s(z_{xx}(x, y) + z_{xy}(x, y) + z_{yy}(x, y)) + \exp(s^2 z(x, y)z(\omega(x, y)))$$

$$+ \exp(s^2 z(x, y)z(\gamma(x, y)))(s^2 z(x, y)z(\gamma(x, y)) - 1)/(s^2 z^2(x, y)))z(x, y) \, ds \, dx \, dy.$$

We cannot obtain from here the explicit presentation for a solution of the inverse problem in the form (13.5), since the last relation contains the integrals with respect to s which cannot be found in elementary functions.

Acknowledgements The author is grateful to Professors V. G. Zadorozhnii and A. L. Skubachevskii for useful discussions.

References

1. El'sgol'ts, L.E.: Qualitative Methods in Mathematical Analysis. Tranlations of Mathematical Monographs, vol. 12. American Mathematical Society (AMS), Providence (1964)
2. Filippov, V.M.: Variational Principles for Nonpotential Operators, Translations of Mathematical Monographs, vol. 77. American Mathematical Society (AMS), Providence (1989)
3. Fillipov, V.M., Savchin, V.M., Shorohov, S.G.: Variational principles for nonpotential operators. J. Math. Sci. **68**, 275–398 (1994)
4. Kamenskii, G.A.: Variational and boundary-value problems with deviating argument. Differ. Equ. **6**, 1349–1358 (1970) [in Russian]
5. Kamenskii, G.A.: Asymmetrical variational problem for nonlocal functionals. Funct. Differ. Equ. **10**, 485–492 (2003)
6. Kamenskii, G.: On boundary value problems connected with variational problems for nonlocal functionals. Funct. Differ. Equ. **12**, 245–270 (2005)
7. Kamenskii, G.A, Skubachevskii, A.L.: Extrema of Functionals with Deviating Arguments. MAI, Moscow (1979) [in Russian]
8. Kolesnikova I.A., Popov A.M., Savchin V.M.: On variational formulations for functional differential equations. Funct. Spaces Appl. **5**, 89–101 (2007)
9. Olver, P.J.: Applications of Lie Groups to Differential Equations. Springer-Verlag, Berlin/Heidelberg/New York/Tokio (1986)
10. Popov, A.M.: Inverse problem of the calculus of variations for systems of differential-difference equations of second order. Math. Notes **72**, 687–691 (2002)
11. Zadorozhnii, V.G.: Methods of Variational Analysis. Institut kompyuternyh issledovanii, Moscow/Izhevsk (2006) [in Russian]
12. Zadorozhnii, V.G., Kurina, G. A.: Inverse problem of the variational calculus for differential equations of second order with deviating argument. Math. Notes **90**, 218–226 (2011)

Chapter 14
Intrinsic Decay Rate Estimates for Semilinear Abstract Second Order Equations with Memory

Irena Lasiecka and Xiaojun Wang

Abstract Semilinear abstract second order equation with a memory is considered. The memory kernel $g(t)$ is subject to a general assumption, introduced for the first time in Alabau-Boussouira and Cannarsa (C. R. Acad. Sci. Paris Ser. I **347**, 867–872, 2009), $g' \leq -H(g)$, where the function $H(\cdot) \in C^1(R^+)$ is positive, increasing and convex with $H(0) = 0$. The corresponding result announced in Alabau-Boussouira and Cannarsa (C. R. Acad. Sci. Paris Ser. I **347**, 867–872, 2009) (with a brief idea about the proof) provides the decay rates expressed in terms of the relaxation kernel in the case relaxation kernel satisfies *the equality $g' = -H(g)$* (Theorem 2.2 in Alabau-Boussouira and Cannarsa, C. R. Acad. Sci. Paris Ser. I **347**, 867–872, 2009). In the case of *inequality $g' \leq -H(g)$*, Alabau-Boussouira and Cannarsa (C. R. Acad. Sci. Paris Ser. I **347**, 867–872, 2009) claims uniform decay of the energy without specifying the rate (Theorem 2.1 in Alabau-Boussouira and Cannarsa, C. R. Acad. Sci. Paris Ser. I **347**, 867–872, 2009). The result presented in this paper establishes the decay rate estimates for the general case of *inequality $g' \leq -H(g)$*. The decay rates are expressed (Theorem 2) in terms of the solution to a given nonlinear dissipative ODE governed by $H(s)$. Applications to semilinear elasto-dynamic systems with memory are also provided.

I. Lasiecka (✉)
Department of Mathematical Sciences, University of Memphis, Memphis, TN 38152, USA

IBS, Polish Academy of Sciences, Warsaw
e-mail: lasiecka@memphis.edu

X. Wang
Department of Mathematical Sciences, University of Memphis, Memphis, TN 38152, USA
e-mail: xwang13@memphis.edu

© Springer International Publishing Switzerland 2014
A. Favini et al. (eds.), *New Prospects in Direct, Inverse and Control Problems for Evolution Equations*, Springer INdAM Series 10,
DOI 10.1007/978-3-319-11406-4_14

14.1 Introduction

14.1.1 The Model

This paper is concerned with the existence and uniform decay rates of solutions to the following semilinear abstract second order in time equation:

$$\begin{cases} u_{tt} + Au - g * Au = f_s(u), & \text{for } t > 0, \\ u(0) = u_0 \in D(A^{\frac{1}{2}}), \ u_t(0) = u_1 \in \mathcal{H}. \end{cases} \tag{14.1}$$

Here $A : \mathcal{D}(A) \to \mathcal{H}$ is a self-adjoint positive operator defined on a real Hilbert space $\mathcal{H}$. The scalar function $g(t)$ is the relaxation kernel of convolution operator $\mathcal{A} : \mathcal{H} \to \mathcal{H}$:

$$\mathcal{A}(z) \triangleq (g * z)(t) = \int_0^t g(t - s)z(s)\, ds,$$

and $f_s : \mathcal{D}(A^{1/2}) \to \mathcal{H}$ is a nonlinear operator representing the source terms. It is assumed that f_s is locally Lipschitz with the property that its potential function $F : \mathcal{D}(A^{1/2}) \to R$ (i.e. $\nabla F(\cdot) = f_s(\cdot)$) satisfies the dissipativity condition $F(s) \leq 0$.

We assume here $g : \mathbb{R}_+ \to \mathbb{R}_+$ is a $C^1(\mathbb{R}_+)$ function,[1] monotone decreasing and such that

$$g(0) > 0 \text{ and } 1 - \int_0^\infty g(s)\, ds = l > 0, . \tag{14.2}$$

System of type (14.1) has been used to model the viscoelastic materials with memory effects and extensively investigated in the literature, see [28, 30, 31]. Under the assumptions imposed later on one shows without difficulties that the system (14.1) has a unique "finite energy" solution $(u, u_t) \in C(0, \infty, \mathcal{D}(A^{1/2} \times \mathcal{H})$, which is "regular" for regular initial data. Problems related to long time behavior, including asymptotic behavior have occupied central role in the literature on viscoelastic models [16, 30, 31] including references therein and more recently [1, 2, 5, 17, 25, 26, 28, 32] among others Our goal is to study the decay rates of the energy associated with the system (14.1). Precise definition of the energy is given below after introducing notation used throughout the manuscript.

Notations: We work on a Hilbert space $\mathcal{H}$, on which we denote the inner product and norm by $(\cdot, \cdot)$ and $\| \cdot \|$ respectively. Throughout this work, $C, C_1, C_2, c_0, c_1, c_2,$ etc., are some generic constants possibly different at different occurrences. For a

---

[1]In this work, we use $\mathbb{R}_+$ for $[0, \infty)$, which is different from $(0, \infty)$. Hence $C^1(\mathbb{R}_+)$ is for $C^1([0, \infty))$, meaning that it contains functions whose first derivatives can be continuously extended to the boundary. Similar to $C^2(\mathbb{R}_+)$ and so on.

given function $H(s)$ we shall also use the notation $\hat{H}(s) \triangleq c_1 H(c_2 s)$ for some positive constants c_1, c_2. In other words, $\hat{H}$ will denote the rescale of function H.

We introduce a quantity

$$(g \circ v)(t) \triangleq \int_0^t g(t-s)||v(t)-v(s)||^2 ds, \forall v \in \mathscr{H}$$

which plays a critical role in the study of memory problems and has been used in the literature.

The decay of a solution to (14.1) is described by the decay of the energy function given by

$$E(t) \triangleq \frac{1}{2}||u_t||^2 + \frac{1}{2}(1 - \int_0^t g(s)ds)||A^{\frac{1}{2}}u||^2 + \frac{1}{2}(g \circ A^{\frac{1}{2}}u)(t) - F(u(t)). \quad (14.3)$$

With the above setup one has that $E(t) = E_k(t) + E_p(t)$, where the potential energy $E_p(t)$ is topologically equivalent to $||A^{1/2}u||$.

14.1.2 Canonical Examples

We shall present several canonical examples illustrating the abstract model introduced above. Let $\Omega \subset R^n$ be a bounded domain with a smooth boundary Γ.

14.1.2.1 Semilinear Wave Equation

$$u_{tt} - \Delta u - g * \Delta u + |u|^{p-1}u = 0, in \ \Omega \times (0, \infty)$$

$$u = 0 \ on \ \Gamma \times (0, \infty). \quad (14.4)$$

Let $\mathscr{H} = L_2(\Omega)$. Then $\mathscr{D}(A^{1/2}) = H_0^1(\Omega)$. We identify $f_s(u) = -|u|^{p-1}u$ which is locally Lipschitz from $H^1(\Omega)$ to $L_2(\Omega)$, for all $2p \leq p^*$, $p^* = \frac{2n}{n-2}$. The latter follows from Sobolev's embeddings $H^1(\Omega) \subset L_{p*}(\Omega)$. Thus the source function complies with the hypotheses. In this case we have $F(u) = -(p+1)^{-1} \int_\Omega |u|^{p+1} dx$. F is locally Lipschitz $H^1(\Omega) \rightarrow \mathbb{R}_-$.

14.1.2.2 Berger's Plate with Clamped Boundary Conditions

$$u_{tt} + \Delta^2 u - g * \Delta^2 u - \Delta u \int_\Omega |\nabla u|^2 dx = 0, \ in \ \Omega \times (0, \infty)$$

$$u = 0, \partial_\nu u = 0 \ on \ \Gamma \times (0, \infty). $$

$$(14.5)$$

Let $\mathscr{H} = L_2(\Omega)$. Then $f_s(u) = \Delta u \|\nabla u\|^2$ and $F(u) = -\frac{1}{4}\|\nabla u\|^4$. The operator A is given by $Au = \Delta^2 u$ with

$$\mathscr{D}(A) = H_0^2(\Omega) \cap H^4(\Omega).$$

We have

$$\mathscr{D}(A^{1/2}) = H_0^2(\Omega)$$

thus f_s is locally Lipschitz from $\mathscr{D}(A^{1/2}) \to \mathscr{H}$.

Next example exploits plate equation with free boundary conditions represented by shear forces, torques and the moments.

14.1.2.3   Semilinear Plate with Free Boundary Conditions

$$u_{tt} + \Delta^2 u - g * \Delta^2 u + f(u) = 0, in \ \Omega \times (0, \infty)$$

$$\partial_\nu \Delta u + (1 - \mu)\partial_\tau \partial_\nu \partial_\tau u = 0, on \ \Gamma$$

$$\partial_\nu^2 u + \nu \partial_\tau{}^2 u + \mu \ \mathrm{div} \ \nu \partial_\nu u = 0, \ on \ \Gamma. \tag{14.6}$$

Here $\partial_\nu, \partial_\tau$ denote normal and tangential derivatives. $\mu \in (0, 1/2)$ denotes Poisson's modulus. Let $\mathscr{H} = L_2(\Omega)$ with $\Omega \subset R^2$. Then assuming that $f(s)s \geq 0$ we have $F(u) = -(\int_0^1 f(su)ds, u)_\Omega$, so that $F(u) \leq 0$. The operator A is defined as $Au = \Delta^2 u$ with

$$\mathscr{D}(A) = \{u \in H^4(\Omega), \partial_\nu \Delta u + (1-\nu)\partial_\tau \partial_\nu \partial_\tau u = 0, \partial_\nu^2 u + \nu \partial_\tau{}^2 + \mu \ \mathrm{div} \ \nu \partial_\nu u = 0, \ on \ \Gamma\}.$$

It is well known by now that A is selfadjoint, nonnegative and $\mathscr{D}(A^{1/2}) = H^2(\Omega)$. Since $H^2(\Omega) \subset C(\Omega)$, the nonlinear source $f(u)$ and $F(u)$ comply with all the requirements.

14.1.3   About the Problem Studied

Decay rates for the energy function associated with dissipative dynamics have occupied considerable attention in the literature. This quantitative piece of information allows for a better understanding of the properties of damping mechanism, hence it is useful in designing suitable materials and devices in order to meet preassigned targets. Both frictional and viscoelastic damping mechanism draw prime attention in this area. In the case of frictional damping the situation is well understood and the results available in the literature are fairly complete and general. Less is known in the area of viscoelasticity where the damping is caused by the past history of the dynamics. In fact, majority of the results pertain to rather specific structures of relaxation kernels, most notably the exponential ones that lead to exponential decay rates or for very special configurations of decaying kernels, see references in [7–14, 18, 19, 29]. Not surprisingly, more recent efforts are directed toward ability to consider more *general* relaxation functions. In order to understand and fully appreciate the meaning of "more general", it is instructive to step back and consider the same issue within the context of *frictional damping*. The first result in this direction for a *frictional damping* was given in [22] see also later developments in [3, 24]. We shall explain briefly the main idea.

For the abstract wave equation with frictional damping of the form

$$u_{tt} + Au + g(u_t) = f_s(u), \qquad \text{for } t > 0, \tag{14.7}$$

$$u(0) = u_0 \in D(A^{\frac{1}{2}}), \ u_t(0) = u_1 \in \mathcal{H}, \tag{14.8}$$

where $g(s)$ is a monotone, increasing function, zero at the origin it was shown that the decay rates of the energy are driven by an ODE, i.e., $E(t) \leq s(t), t > T_0$, where $s(t)$ satisfies the following differential equation.

$$s_t + \hat{H}(s) = 0, \tag{14.9}$$

where $\hat{H}$ is a rescaled variant of $H(s)$ i.e., $\hat{H}(s) = c_1 H(c_2 s)$ with $H(s)$ determined from

$$s^2 + g^2(s) \leq H^{-1}(sg(s)), |s| \leq 1. \tag{14.10}$$

In view of monotonicity of $g(s)$ such a function $H(s)$, which is continuous, increasing, convex, zero at the origin can always be constructed. Thus, (14.10) is not an assumption but the property, see [22].

The goal of the present work is to construct a similar theory for an abstract second order equation with a *viscoelastic* damping. In fact, following [2] we shall formulate a hypothesis which is meant to play the role of condition (14.10) in the context of *viscoelastic* damping. The relaxation kernel is assumed (see [2]) to obey the following differential inequality:

$$g'(t) + H(g(t)) \leq 0 \ \text{for } t > 0, \tag{14.11}$$

and the controlling function $H(s)$ is assumed convex, continuous and strictly increasing with $H(0) = 0$. The assumption (14.11) was introduced for the first time in this form in [2] for the treatment of viscoelastic damping. While conditions (14.10) and (14.11) share some similarities in the sense that damping mechanisms [frictional in (14.10) and viscoelastic in (14.11)] are characterized by convex, increasing functions $H(s)$, the real nature of both conditions is different. Inequality in (14.10) is not an additional condition imposed on $g(s)$. It is rather a property of a continuous, increasing function $g(s)$. As shown in [22], such a function $H(s)$ can be always constructed for a given $g(s)$ which is continuous, increasing and zero at the origin. Instead, condition (14.11) is a real restriction imposed on a monotone decreasing function $g(s)$ (for instance, it requires convexity of $g(s)$). However, condition (14.11) introduced in [2] is still very general and allows for many parallels and borrows from the analysis of the frictional damping satisfying (14.10). In particular, the role of convexity of $H(s)$ exploited via appropriate use of Jensen's inequality and a connection to dissipative ODE solutions are paramountly common in both cases.

Thus, the goal of this paper is to relate the asymptotic decay rates for the energy of PDE weak solutions to the decay rates driven by a dissipative ODE governed by a rescaled variant of function $H(s)$ in (14.11). While there are many papers in the literature which provide decay rates for the energy with a "specified" (linear, polynomial) behavior of $H(s)$, the problem of finding general argument which gives the correct answer for an *un-quantitized* behavior of $H(s)$ at the origin (critical region) is relatively novel, particularly in the case of memory damping. The first result in this direction for viscoelastic damping has been announced in [2]. Alabau-Boussouira and Cannarsa [2] asserts the decay rates expressed in terms of the relaxation kernel (Theorem 2.2 in [2]) for the case of relaxation kernel satisfying *the equality* $g' = -H(g)$. In the case of *inequality* $g' \leq -H(g)$, [2] claims an uniform decay of the energy without specifying the rate (Theorem 2.1). The announcement of the result in [2] attributes its proof to convexity arguments along with suitable weighted inequalities developed in [3, 4].

The result presented in this paper provides decay rate estimates for the general case of *inequality* $g' \leq -H(g)$. The decay rates are expressed in terms of a function bounded by a solution to a given nonlinear dissipative ODE governed by $H(s)$. Our aim is to develop an elementary method, based on ODE methodology introduced in [22], which is capable to provide specified decay estimates for solutions corresponding to general structures of relaxation kernels defined by the inequality (14.11). The nonlocal character of the damping provides a new set of mathematical difficulties to be dealt with. Starting with general assumption (14.11) (introduced in [2]) we adjust convexity arguments of [22]) in order to incorporate the nonlocal terms into the multipliers estimates via multiple applications of Jensen's inequality (used heavily in [22]). The method is intrinsic and is based on an adaptation of the main idea introduced in [22] for the treatment of frictional damping. It relies on an exploitation of convexity of function $H(s)$, via Jensen's inequality, along with a suitable comparison theory for ODE's. The final result is that the decay rates for PDE solutions are given by ODE solutions governed by rescales of function $H(s)$.

Our paper is organized as follows: In Sect. 14.2, we introduce the notations, state the assumptions and the main results. Section 14.3 is devoted to the proofs. Section 14.3.1 provides some preliminary material and estimates which are critical for the development. These include generalized Jensen's inequality, derivation of the energy balance, and the iterative estimates involving convolutions (so called α-Sequences) which are used critically later on. Sections 14.3.2 and 14.3.3 deal with the main core of the proofs.

14.2 Main Results

14.2.1 Preliminaries

Our main task is to study optimal decay rates, after the global existence of weak (mild or even strong actually) solution is asserted. The former is accomplished by showing local existence plus the non-blowup property. The latter requires the investigation of the energy function $E(t)$. To this end, we introduce the damping term $D(t)$ defined by

$$D(t) \triangleq -\frac{1}{2}(g' \circ A^{\frac{1}{2}}u)(t) + \frac{1}{2}g(t)||A^{\frac{1}{2}}u(t)||^2 dx \geq 0. \tag{14.12}$$

It is standard by now ([1] and references therein) to derive from (14.1) the energy balance equality

$$E(t) + \int_s^t D(\tau)d\tau = E(s), \forall s \leq t, \tag{14.13}$$

where the energy function $E(t)$ is given by (14.3). Thus, the energy is decreasing and we always have $E(t) \leq E(s), s < t$. This answers global existence part provided, however, one has local solutions along with a priori bounds. Precise formulation of this statement is given below.

14.2.2 Wellposedness of Finite Energy Solutions

Assumptions (A):

1. there exists a constant $c_0 > 0$ such that the operator A satisfies the following generalized Poincare inequality:

$$||u|| \leq c_0||A^{\frac{1}{2}}u||, \forall u \in D(A^{\frac{1}{2}});$$

2. $f_s : D(A^{\frac{1}{2}}) \to \mathcal{H}$ is locally Lipschitz with $f_s(0) = 0$ and $(f_s(u), u) \leq 0$.

As a consequence of Assumption (A) with $F(u) = \int_0^1 (f_s(tu), u)dt, u \in D(A^{\frac{1}{2}})$ we have

- $F(0) = 0, F(u) \leq 0$ and $f_s(\cdot) = \nabla F(\cdot)$ on $D(A^{\frac{1}{2}})$;
- $\frac{d}{dt} F(u(t)) = (f_s(u), u_t)$ and $|F(u)| \leq \phi(\|A^{\frac{1}{2}}u\|)\|\|A^{\frac{1}{2}}u\|^2$, where $\phi(u)$ is a bounded function for bounded arguments.

Assumption (A) constitutes a set of standard hypotheses imposed on semi linear potential function F in the context of wellposedness of local and global solutions corresponding to abstract second order semi linear evolutions.. Part 2 in Assumption (A) reflects local Lipschitz continuity and it corresponds to Hypothesis (H1) in [1]. The last bullet in Assumption (A) states that nonlinear part of potential energy is locally bounded see Assumption 3.1 p. 486 in [15]. This is stated in the same form as (14.25) in Assumption (H2) in [1]. The following wellposedness result is known and standard by now [1] and references therein.

Theorem 14.1 *Let* $(u_0, u_1) \in D(A^{\frac{1}{2}}) \times \mathcal{H}$ *be given. Assume that the Assumption (A) is satisfied, then the problem (1.1) has a unique global (weak) solution*

$$u \in C(\mathbb{R}_+; D(A^{\frac{1}{2}})) \cap C^1(\mathbb{R}_+; \mathcal{H}).$$

Moreover, if $(u_0, u_1) \in D(A) \times D(A^{\frac{1}{2}})$, *then the solution satisfies*

$$u \in C(\mathbb{R}_+; D(A)) \cap C^1(\mathbb{R}_+; D(A^{\frac{1}{2}})) \cap C^2(\mathbb{R}_+; \mathcal{H}).$$

14.2.3 Uniform Decay Rates for the Energy Function

By (14.13), we know that the solution decreases in time. We want to find out if *it decreases to zero and how fast it decreases*. In order to provide a quantitative answer, the following assumption is needed.

Assumptions (B):

1. assume that there exists a convex function $H \in C^1(\mathbb{R}_+)$, which is strictly increasing with $H(0) = 0$ such that

$$g'(t) + H(g(t)) \leq 0, \forall t > 0; \tag{14.14}$$

2. let $y(t)$ be a solution of the following ODE:

$$y'(t) + H(y(t)) = 0, y(0) = g(0) > 0$$

and assume that there exists $\alpha_0 \in (0, 1)$ such that $y^{1-\alpha_0} \in L_1(\mathbb{R}_+)$;
3. in addition to $H \in C^1(\mathbb{R}_+)$, assume there exists an interval $[0, \bar{\delta}], \bar{\delta} > 0$ such that $H(\cdot) \in C^2(0, \infty)$ and $\liminf_{x \to 0^+} \{x^2 H''(x) - x H'(x) + H(x)\} \geq 0$,

Remark 14.1 • The inequality in Assumption (B)-1 along with convexity requirement is the same as inequality (7) introduced in [2].

• Note that $H(\cdot) \in C^2(\mathbb{R}_+)$ means $H''(x)$ can be continuously extended to
0, for instance $H(x) = x(x + 1)^{p-1}, x \geq 0, 1 < p < 2$. However, a
$C^2(0, \infty)$ function may have second derivative unbounded at zero, for instance,
$H(x) = |x|^{p-1}x, x \in \mathbb{R}^n_+$. Thus, this case requires a different treatment. Instead
of assuming $H(\cdot) \in C^2(\mathbb{R}_+)$ or $H(x) \in C^2(0, \infty)$, we shall consider a limited
regularity $H(\cdot) \in C^1(\mathbb{R}_+)$ only, supplemented with an additional assumption
that all the functions defined by $H^{(k)}(x) \triangleq x^{1-\frac{1}{\alpha_k}} H(x^{\frac{1}{\alpha_k}}), k = 0, 1, \ldots m$, are
convex on $[0, \bar{\delta}]$, where $\alpha_k = (k + 1)\alpha_0$ and $m + 1$ is the first integer to make
$k\alpha_0 \geq 1$. The assumption stated in (B3) guarantees this to happen.

We begin with the following preliminary uniform decay result.

Proposition 14.1 *Under the Assumptions (A), (B1), (B2), the energy $E(t)$ of the
solutions associated to weak solutions to problem (14.1) decay to zero uniformly
with respect to the topology of finite energy space. More specifically, $E(t) \leq
S(t), t > T_0$, with some $T_0 > 0$ where $S(t)$ satisfies the nonlinear ODE*

$$S_t + \hat{H}_{\alpha_0}(S) = 0, S(0) = E(0)$$

where $\hat{H}_\alpha = c_1 H(c_2 s^{\frac{1}{\alpha}})$ for some constants $c_1, c_2 > 0$.

The result stated above is suboptimal since $\alpha_0 < 1$. However it constitutes a
necessary first step for the analysis. The second and main result provides sharp
decay rates where $\hat{H}_\alpha$ is used with $\alpha = 1$.

Theorem 14.2 *Let Assumption (A)–(B) be in force. Then, the decay rates obtained
for the energy of weak solutions obey the estimate for some $T_0 > 0$*

$$E(t) \leq S(t) \text{ for } t > T_0$$

*where $S(t)$ satisfies $S_t + \hat{H}(S) = 0, S(0) = E(0)$, with $\hat{H}$ given by $\hat{H}(S) =
c_1 H(c_2 S)$ with some suitable constants $c_1, c_2 > 0$. The constants c_1, c_2 and T_0 do
not depend on a particular solution but they may depend on $E(0)$*

Remark 14.2 • Theorem 14.2 is a viscoelastic analogue of a related result obtained
for a frictional damping in [22].

• A related result has been announced in [2]. More specifically, under the hypothesis that $g(t)$ satisfies the differential *equality* $g'(t) + H(g(t)) = 0, \forall t > 0, [2]$
announces that the decay rates for the linear system (14.1) with $f_s = 0$ obey the
inequality $E(t) \leq C(E_u(0))g(t)$. In the case of *inequality* $g'(t) + H(g(t)) \leq
0, \forall t > 0$, Theorem 2.1 in [2] announces uniformity of the decay but without
specifying the rates. It should be noted that techniques used in [2] and the present
manuscript are very different. While [2] refers to weighted energy inequalities
developed in [3–5], the present work is based on the ODE method developed
in [22]. We believe that this latter method provides a more effective framework

when dealing with the case of *inequality* characterization of the relaxation kernel. Such characterization provides incomplete information on the kernel itself, thus any claim comparing the decay rates of the PDE with the actual decay rates of a given kernel is not reasonable to expect. The only ground for comparison is the ODE which simultaneously characterizes both quantities involved.

Remark 14.3 A good testing ground asserting significance of Theorem 14.2 is the polynomial case when $H(s) = |s|^{p-1}s$ with $p \in (1,2)$. This corresponds to relaxation kernel $g(t) \leq \frac{1}{t^{1/(p-1)}}$ four t large. It is easy to verify that all the conditions in Assumption B are satisfied. Indeed, $\alpha_0 < \frac{p-2}{p-1}$ satisfies the requirement in Assumption B(2). As for B(3), the variant (b) applies. For $x \to 0+$ we have

$$x^2 H'' - xH'(x) + H(x) = (p-1)^2 x^p$$

which is nonnegative. Thus, all the hypotheses apply for the maximal range of the parameter $p \in [1,2)$ and we obtain sharp decay rates for the energy $E(t) \sim t^{-\frac{1}{p-1}}$ for $p > 1$. The exponential decay rates for $p = 1$ correspond to a linear function $H(s)$—being well understood in the past.

In order to point out the novelty and significance of the result in Theorem 14.2, it suffices to mention that almost all papers in the field when considering this model impose the following restriction on the range of parameters $p \in [1, 3/2)$. This also includes [1] where sharp polynomial decay rates are proved for $p \in [1, 3/2)$ see (24), Assumption (H2) in [1]. This unnatural restriction is removed by using the methodology presented in this paper. In fact, sub-optimality of the results in the polynomial case was a motivation for introducing new method in [20], which not only generalizes previous theories, but it also provides results in the cases which were explicitly ruled out in previous treatments. However, the proof in [20] is inductive and it requires verification of a "dynamic" hypothesis (we note that the inductive hypothesis introduced in [20] is very different from "successive energy decay estimates or "boot strap" arguments used in [1, 13]. Instead, in the present framework, the optimal result is obtained in one shot. One should also mention that iterative method already used in [13] (and later in [1], when applied to the polynomial case, still required restrictions on the parameters $p \in [1, 3/2)$. Instead [2] states the result as valid for the range $p \in (1, 2)$, however since there is no proof given in [2] it is difficult to discern the technicalities responsible for this particular point (iterative method used before is simply not enough).

Remark 14.4 The result stated in Theorem 14.2 applies directly to canonical examples of nonlinear waves and plates introduced in Sect. 14.2. In particular, one shows that the decays obtained for the energy function are sharp with respect to reconstructing properties of relaxation function $g(s)$. In the polynomial case, with semi linear terms in the viscoelastic equation, [1] proves the decay rates which are sharp but applicable only to restricted range of $p \in [1, 3/2)$.

14.2.4 Discussion of the Results and of the Method Used in the Context of the Past Literature

As already mentioned, decay rates for the energy function of viscoelastic abstract models have been considered in past literature rather frequently. While vast majority of the work deals with well quantized relaxation kernels or with kernels specifying "linear type" of differential inequalities characterized by $g'(t) + \xi(t)g(t) \leq 0$ with suitable assumptions imposed on time dependent function $\xi(t)$ [7–9, 12, 18, 19, 25, 26, 32], there have been few attempts to reach the level of generality which does not require such specifications. In this context we should mention [2, 20, 32]. All these works follow a general idea of employing convex analysis (as in [22]) in order to obtain estimates independent on specific characterization of the kernel. Paper [2] announces (without a proof—with only brief sketch of the proof) sharp decay result for *linear* problem under the assumption that the relaxation kernel satisfies differential *equation* rather than *inequality* Theorem 2.2. In the case of *inequality*, [2] provides only (Theorem 2.1) the statement on uniformity of decay rates without specifying the rates. Paper [32] studies coupled second order evolution equations where one equation is subject to memory effects with relaxation kernel satisfying *equality* differential relation, in addition to other technical assumptions. In particular Assumption (2.3) in [32] requires total subordination of the operator B_i (coupling term) with respect to the square roots of the generators—which is a typical assumption in coupled dynamics but not natural for a single dynamics. Manuscript [27] is the first one which derives the decay rates for the case of *differential inequality*. However, these rates are *non-optimal*. Non-optimality is already manifested in the case of algebraically decaying kernels where the method used forces the restricted range of parameters, as explained in Remark 14.3 above. Lasiecka et al. [20] achieves sharp decay rates for the case of inequality, however the most general result requires an inductive-dynamic hypothesis to be satisfied. The contribution of the present work is that it removes the limitation of [20] and there is no need for the dynamic hypothesis.

It should also be mentioned that from the point of view of methodology, the methods used in [2, 27, 32] rely on weighted energy inequalities, while the method used now and in [20] pursues the approach of [22] which is ODE based. It appears that ODE based method is more effective when formulating the results for relaxation kernels satisfying *differential inequalities*-rather than equalities. This is due to the fact that relaxation kernels quantitated by inequalities may be poorly characterized—thus decay rates given in terms of the relaxation kernel itself may not be possible at all. Instead, ODE approach exploits the maximum information available in this case—with ODE being an accurate description. In addition, there are several generalizations developed in the present work where the methodology employed allows to eliminate numerous technical restrictions. Thus, the novel contribution of this work is from the point of view of both the *results* obtained and the *methodology* developed.

We shall briefly explain the approach used in this paper. Let $u(t)$ be a solution to (14.1). We rewrite $(g \circ A^{\frac{1}{2}}u)(t) = \int_0^t g(t - s)f^2(t,s)ds$ for notational convenience, where f is defined by

$$f(t,s) = ||A^{\frac{1}{2}}(u(t) - u(s))||. \tag{14.15}$$

For $0 < \alpha \leq 1$ we pay attention to the important parameter

$$c(t,\alpha) \triangleq \int_0^t g(t-s)^{1-\alpha} f^2(t,s)ds, \tag{14.16}$$

which is related the α-Sequence in Proposition (14.2). This parameter was already used in [20, 27]. Given the Assumption (B)(2) along with the bound $f^2(t,s) \leq 2E(0)$, we know that for some $\alpha_0 \in (0,1)$,

$$0 < c_{\alpha_0} \triangleq \sup_{t>0} c(t,\alpha_0) \leq C \int_0^\infty y^{1-\alpha_0} dt < \infty. \tag{14.17}$$

The above allows [20] to construct function $\hat{H}_{T,\alpha_0} \triangleq H(c_T s^{\alpha_0^{-1}})$, where T is a sufficiently large -but finite and c_T is a suitable constant. The crux of the matter is showing that the validity of (14.17) implies the estimate

$$E((n+1)T) + \hat{H}_{T,\alpha_0}\{E((n+1)T)\} \leq E(nT) \tag{14.18}$$

for all $T > T_0$ and uniformly in $n = 1,2,\ldots$.

Note that the new function $\hat{H}_{T,\alpha_0}$ is more "convex" than the original one $H(s)$ - thus it will produce "slower" decays. By exploiting boundedness in (14.17), Lemma 3.3 in [22] the method in [20] shows that the decay of the energy are driven by ODE

$$S_t + \hat{H}_{T,\alpha_0}(S) = 0,$$

which gives uniform decays (non-optimal though, since $\alpha_0 < 1$) to zero asserted in Proposition 14.1. In order to improve the decay rates we use a bootstrap argument by applying new piece of information regarding the decay of function $f(t,s)$ and thus improving index $\alpha_0 \to \alpha_1$ where we have strict inequality $\alpha_0 < \alpha_1$. Continuing the process iteratively, for optimality one needs to show the limit of α_n is 1, since $\alpha = 1, H_\alpha = H$ corresponds to the optimal value. Typically an infinite sequence of α_n is generated and the condition (14.16) is tested dynamically.

In contrast with the above procedure developed in [20] we will be constructing iterations with a uniform step [exploiting the Assumption (B)(3)], thus the limiting argument is no longer necessary. The final result is quantitatively the same, namely that the decay rates are driven by a rescale of function H.

14.3 Proofs of the Results

14.3.1 Preliminary Inequalities

In this section we shall enlist several estimates used through the course of the proofs. We will begin with Jensen's inequality which was important devise used since [22] where for the first time unquantified damping in the convex framework was considered. For readers' convenience we recall this inequality stated in the exact form to be used in the arguments to follow.

Let F be a convex increasing function on $[a, b]$, $f : \Omega \to [a, b]$ and h are integrable functions on Ω such that $h(x) \geq 0$, and $\int_\Omega h(x)dx = k > 0$, then generalized Jensen's inequality states that

$$F[\frac{1}{k} \int_\Omega f(x)h(x)dx] \leq \frac{1}{k}[\int_\Omega F[f(x)]h(x)dx]. \tag{14.19}$$

The above version of Jensen's inequality is well known in the literature, see [6, 23] and has been also critically used in [20, 22, 27].

For readers convenience we shall also recall derivation of the energy balance in (14.13). In fact, this derivation is helpful in motivating particular structure of the energy function.

Lemma 14.1 *The strong form of energy equality takes the form:*

$$\frac{d}{dt}\{\frac{1}{2}||u_t||^2 + \frac{1}{2}(1 - \int_0^t g(s)ds)||A^{\frac{1}{2}}u||^2 + \frac{1}{2}(g \circ A^{\frac{1}{2}}u)(t) - F(u(t))\}$$

$$= \frac{1}{2}\int_0^t g'(t - s)||A^{\frac{1}{2}}(u(t) - u(s))||^2 ds - \frac{1}{2}g(t)||A^{\frac{1}{2}}u(t)||^2.$$

The derivation of this equality is standard-once we have regular local in time solutions (guaranteed in our case Theorem 14.1). To recall the calculations we proceed as follows.

Taking the inner product of (14.1) and u_t on $\mathscr{H}$, we have

$$(u_{tt} + Au - g * Au, u_t) = (f_s(u), u_t)$$

$$\Rightarrow \frac{1}{2}\frac{d}{dt}||u_t||^2 + \frac{1}{2}\frac{d}{dt}||A^{\frac{1}{2}}u||^2 - \int_0^t g(t - s)(Au(s), u_t(t))ds = \frac{1}{2}\frac{d}{dt}F(u).$$

Now the third term on the left

$$-\int_0^t g(t-s)(Au(s), u_t(t))ds$$

$$= \int_0^t g(t-s)(Au(t) - Au(s), u_t(t))ds - \int_0^t g(t-s)(Au(t), u_t(t))ds$$

$$= \frac{1}{2}\int_0^t g(t-s)\frac{d}{dt}||A^{\frac{1}{2}}(u(t) - u(s))||^2 ds - \frac{1}{2}\int_0^t g(s)ds\frac{d}{dt}||A^{\frac{1}{2}}u(t)||^2$$

$$= \frac{1}{2}\frac{d}{dt}\int_0^t g(t-s)||A^{\frac{1}{2}}(u(t) - u(s))||^2 ds - \frac{1}{2}\int_0^t g'(t-s)||A^{\frac{1}{2}}(u(t) - u(s))||^2 ds$$

$$-\frac{1}{2}\frac{d}{dt}[\int_0^t g(s)ds||A^{\frac{1}{2}}u(t)||^2] + \frac{1}{2}g(t)||A^{\frac{1}{2}}u(t)||^2.$$

We end up with energy equality

$$\frac{d}{dt}\left\{\frac{1}{2}||u_t||^2 + \frac{1}{2}(1 - \int_0^t g(s)ds)||A^{\frac{1}{2}}u||^2 + \frac{1}{2}(g \circ A^{\frac{1}{2}}u)(t) - F(u(t))\right\}$$

$$= \frac{1}{2}\int_0^t g'(t-s)||A^{\frac{1}{2}}(u(t) - u(s))||^2 ds - \frac{1}{2}g(t)||A^{\frac{1}{2}}u(t)||^2.$$

Thus it is natural to define the energy

$$E(t) \triangleq \frac{1}{2}||u_t||^2 + \frac{1}{2}(1 - \int_0^t g(s)ds)||A^{\frac{1}{2}}u||^2 + \frac{1}{2}(g \circ A^{\frac{1}{2}}u)(t) - F(u(t)).$$

The following integral inequality will be used frequently in the derivation of energy estimate in Sect. 14.3.

Lemma 14.2 *Let u be a solution of (14.1) and $\psi \in L^1(0, \infty)$, $\psi \geq 0$, a.e.. Then,*

$$\left|\left|\int_0^t \psi(t-s)(u(t) - u(s))\,ds\right|\right|^2 \leq ||\psi||_{L^1(0,\infty)}(\psi \circ u)(t). \qquad (14.20)$$

Proof The proof is given in [20, 27]. For readers' convenience we repeat the arguments. We have, making using of Cauchy-Schwarz inequality and Fubini's theorem,

$$\left|\left|\int_0^t \psi(t-s)(u(t) - u(s))\,ds\right|\right|^2$$

$$= \int_\Omega \left(\int_0^t \sqrt{\psi(t-s)}\sqrt{\psi(t-s)}(u(t) - u(s))\,ds\right)^2 dx$$

$$\leq \int_{\Omega} \left(\int_0^t \psi(\xi)\, d\xi \right) \int_0^t \psi(t-s)(u(t)-u(s))^2\, ds dx$$

$$\leq \|\psi\|_{L^1(0,\infty)} \int_0^t \psi(t-s)\|(u(t)-u(s))\|^2\, dx ds,$$

which proves the lemma. $\qquad\qquad\qquad\qquad\qquad\qquad\qquad\qquad\qquad\qquad$ $\square$

Another critical estimate is used to "boot strap" the decay rate of the energy. To achieve the optimal decay rate, we reiterate this estimate and generate a finite sequence of parameters, which we call the α-*Sequence*, whose boundedness is in turn required to carry out the iteration.

Proposition 14.2 (α-Sequence) *Let $\alpha_0 \in (0,1)$ and $y(t) \in C[0,\infty)$ be a positive function decreasing to zero. Moreover*

$$\int_0^{\infty} y^{1-\alpha_0}(t)\, dt = L < \infty.$$

Let m be a positive integer such that $m\alpha_0 < 1$ and $(m+1)\alpha_0 \geq 1$. A finite sequence of parameters, in form of definite integrals on $[0,t]$, are generated in the following way:

$$I_k(t) = \int_0^t y^{1-k\alpha_0}(t-s)y^{(k-1)\alpha_0}(s)\, ds, \quad k = 1, 2, \dots, m, \qquad (14.21)$$

$$I_{m+1}(t) = \int_0^t y^{m\alpha_0}(s)\, ds. \qquad\qquad\qquad\qquad\qquad (14.22)$$

Then each $I_k(t), k = 1, \dots, m+1$, is bounded, uniformly in t, namely,

$$\sup_{t>0} I_k(t) < \infty.$$

Proof We first assume $y(0) \leq 1$.

For $k = 1, \dots, m$, we have

$$I_k = \int_0^t y^{1-k\alpha_0}(t-s)y^{(k-1)\alpha_0}(s)\, ds$$

$$= \int_0^{t/2} y^{1-k\alpha_0}(t-s)y^{(k-1)\alpha_0}(s)\, ds + \int_{t/2}^t y^{1-k\alpha_0}(t-s)y^{(k-1)\alpha_0}(s)\, ds$$

$$\leq \int_0^{t/2} y^{1-k\alpha_0}(s)y^{(k-1)\alpha_0}(s)\, ds + \int_{t/2}^t y^{1-k\alpha_0}(t-s)y^{(k-1)\alpha_0}(t-s)\, ds$$

$$= \int_0^{t/2} y^{1-\alpha_0}(s)ds + \int_{t/2}^t y^{1-\alpha_0}(t-s)ds$$

$$= 2 \int_0^{t/2} y^{1-\alpha_0}(s)ds \leq 2L < \infty.$$

Above we use the fact that $y(t)$, hence $y^{1-k\alpha_0}(t)$, is a decreasing function and on each subinterval we have either $s < t - s$ or $s > t - s$.

For I_{m+1}, we have

$$I_{m+1} = \int_0^t y^{m\alpha_0}(s)ds \leq \int_0^t y^{1-\alpha_0}(s)ds \leq L < \infty,$$

since $(m+1)\alpha_0 > 1 \Rightarrow m\alpha_0 > 1 - \alpha_0$ and $y(t) \leq y(0) \leq 1$.

For y with $y(0) > 1$, we can always find a time t_0 so that $y(t) \leq 1, t > t_0$ since y decreases to zero. The fact $y \in C[0, \infty)$ implies the only possibility to stop the integrals from being finite is the asymptotic behavior of y at infinity. So y being larger than 1 on a finite interval $[0, t_0]$ does not bear influence on our result.   □

14.3.2   Proof of Proposition 14.1

This follows directly from [20]. Indeed, since $c(t, \alpha_0) < \infty$ we construct a suboptimal function $H_{\alpha_0}(s) = H(s^{\frac{1}{\alpha_0}})$ where its rescale $\hat{H}_\alpha$ is also continuous, convex on $[0, \infty)$, strictly increasing and zero at the origin. It is shown in [20] that the inequality in (14.18) holds. Thus solution driven by the ODE $s_t(t) + c_1 H((c_2 s)^{\frac{1}{\alpha_0}}) = 0$, provides the decay for the energy $E(t)$—see [20]. Since H is increasing, $s' < -c < 0 \Rightarrow s(t) \to 0 \Rightarrow E(t) \leq s(t) \to 0$, where the constants c_1 and c_2 depend on $||A^{1/2}u(0)|||$ but not on a specific solution. This proves the uniform convergence to zero of the energy function which are quantified by α_0.

Our next step is to improve the decay rates by making these eventually independent on α_0.

14.3.3   Proof of Theorem 14.2

Given the result in Proposition 14.1, Theorem 14.2 is a consequence of the following sequence of lemmas. The proofs of these results are given later.

Lemma 14.3 *There exists $T > 0$ and the positive constants C_{1T}, C_{2T} such that for $n = 1, 2 \ldots$*

$$E((n+1)T) \leq C_{1T} \int_{nT}^{(n+1)T} (g \circ A^{\frac{1}{2}}u)(t)dt + C_{2T} \int_{nT}^{(n+1)T} D(t)dt. \qquad (14.23)$$

In what follows we shall use another function -H related -defined as

$$H_{1,\alpha}(s) \triangleq \alpha s^{1-\frac{1}{\alpha}} H_{\alpha}(s) = \alpha s^{1-\frac{1}{\alpha}} H(s^{\frac{1}{\alpha}}). \qquad (14.24)$$

As before we shall denote the rescale of these functions by $\hat{H}_{1,\alpha}$. Also note that all three functions $H, H_{\alpha}, H_{1,\alpha}$ coincide for the optimal value of the parameter $\alpha = 1$. For $\alpha < 1$, H_{α} and $H_{1,\alpha}$ represent additional convexification which then leads to compromised decay rates for the energy function.

Lemma 14.4 *Under the Assumptions (B)(3), there exists an interval* $(0, \delta), 0 < \delta < \bar{\delta}$ *on which we have*

1. $H_{1,\alpha}(0) = 0$ *and* $H_{1,\alpha}(s)$ *is increasing and convex;*
2. *moreover, if* $c_{\alpha} \triangleq \sup_{t>0} c(\alpha, t) = \sup_{t>0} \int_0^t g^{1-\alpha}(t-s)f^2(t,s)ds < \infty$, *then there exist constants* ϑ *so that*

$$H_{1,\alpha}[\vartheta(g \circ A^{\frac{1}{2}}u)](t) \leq \alpha \vartheta D(t), \ for \ t \in [nT, (n+1)T], \ n = 1, 2 \ldots, \qquad (14.25)$$

with $\vartheta \in (0, 1)$ *independent on* n.

Lemma 14.5 *Given the results of lemmas above, we have*

$$E((n+1)T) + \hat{H}_{1,\alpha}\{E((n+1)T)\} \leq E(nT) \qquad (14.26)$$

holds for all $T > T_0$, *where* $\hat{H}_{1,\alpha}$ *is defined by*

$$\hat{H}_{1,\alpha}^{-1}(x) = \frac{C_{1T}}{\vartheta} TH_{1,\alpha}^{-1}\left[\frac{\alpha \vartheta}{T}x\right] + C_{2T}x, \forall x \in \mathbf{R}.$$

Moreover, we can show that $\hat{H}_{1,\alpha}(s)$ *is a convex, continuous increasing and zero at the origin function. Here* $C_{1T}, C_{2T}, \vartheta$ *depends only on* α, T, *but not on* n;

Lemma 14.5 leads to the following decay result proved originally in [22] with more explicit estimates in [21] (see also [3] for results on decay rates of the energy expressed in terms of integral inequalities):

Lemma 14.6 *Given the Lemma 14.5 and condition (14.25) we have the following decay rates for the energy function*

$$E(t) \leq s(t), \forall t > T$$

with

$$S_t + \hat{H}_{1,\alpha}(s) = 0, s(0) = E(T),^2 \qquad (14.27)$$

Lemma 14.7 *Comparison principle.*
  Given $y(t)$ satisfying

$$y_t + H(y) = 0, y(0) = y_0 > 0, \qquad (14.28)$$

and $s(t)$ satisfying

$$s_t + \alpha s^{1-\frac{1}{\alpha}} H(s^{\frac{1}{\alpha}}) = 0, s(0) = s_0 > 0. \qquad (14.29)$$

Both function y and s are positive, decreasing, Then we have

$$s(t) = y^{\alpha}(t + c), \qquad (14.30)$$

for some constant c, meaning the decay rate of $s(t)$ is identical to $y^{\alpha}(t)$ up to a finite time delay.

Lemma 14.8 *Iteration for optimality.*
  We show the following decay rates for the energy function

$$E(t) \le S(t), \text{for } t > T$$

with

$$S_t + \hat{H}(S) = 0, S(0) = E(T) \qquad (14.31)$$

where $\hat{H}(x) = c_1 H(c_2 x)$ is a convex, continuous increasing and zero at the origin function.

Remark 14.5 Heuristic for the proof of the main result. Lemma 14.3 gives the energy inequality, while Lemmas 14.4–14.6 show the energy is driven by damping function $H_{1,\alpha}$ which has "weaker" decay rate than H. Lemma 14.7 discovers the explicit relation between the energy decay rate caused by $H_{1,\alpha}$ and that by H, namely between $E(t)$ and $y(t)$. In view of the critical quantity $\int_0^t g^{1-\alpha}(t - s) f^2(t,s) ds \le 2 \int_0^t y^{1-\alpha}(t-s) E(s) ds$, this explicit relation gives an extra bit room for $y^{1-\alpha}(t-s)$, hence $g^{1-\alpha}(t-s)$, to improve. More precisely, a larger α is reached while $\int_0^t g^{1-\alpha}(t - s) f^2(t,s) ds$ still being bounded uniformly in t. Lemma 14.8 adopts the α-sequence property and completes the proof of Theorem 14.2.

---

[2] Here we can take $s(0) = E(0)$ for notational convenience, since $E(T) \le E(0)$.

Proof (of Lemma 14.3)

We seek global integral energy estimates for (i) potential energy $||A^{\frac{1}{2}}u||^2$, (ii) the kinetic energy $||u_t||^2$ and (iii) viscoelastic energy $g \circ A^{1/2}u$; This is accomplished by using standard multipliers used in viscoelasticity [1, 10–14] with the difference that our estimates are carried out over the discrete time segments $[nT, (n + 1)T]$ without any weights, rather than over the full time intervals $(0, \infty)$ with the assigned weights. This difference is related to the fact that our method (based on [22]) seeks a comparison with a discrete version of ODE.

1. Estimate on the potential energy $\int_{nT}^{(n+1)T} ||A^{\frac{1}{2}}u(t)||^2 dt$.
Taking the inner product of (14.1) and u, we have

$$(u_{tt} + Au - g * Au, u) = (f_s(u), u).$$

Since $(u_{tt}, u) = \frac{d}{dt}(u_t, u) - ||u_t||^2$ and $(Au, u) = ||A^{\frac{1}{2}}u||^2$, we get

$$\frac{d}{dt}(u_t, u) - ||u_t||^2 + ||A^{\frac{1}{2}}u||^2 - (f_s(u), u) = (g * Au, u).$$

The following are easy to check

$$|(g * Au, u)| = |\int_0^t g(t - s)(A^{\frac{1}{2}}u(s), A^{\frac{1}{2}}u(t))ds|$$

$$\leq |\int_0^t g(t-s)(A^{\frac{1}{2}}[u(s) - u(t)], A^{\frac{1}{2}}u(t))ds| + |\int_0^t g(t-s)(A^{\frac{1}{2}}u(t), A^{\frac{1}{2}}u(t))ds|$$

$$\leq (1 - l)(1 + \epsilon)||A^{\frac{1}{2}}u(t)||^2 + C_\epsilon(g \circ A^{\frac{1}{2}}u)(t),$$

and

$$|(u_t, u)| \leq \frac{||u||^2}{2} + \frac{||u_t||^2}{2} \leq C(||A^{\frac{1}{2}}u||^2 + ||u_t||^2) \leq CE(t)$$

which implies $-(u_t, u)|_{nT}^{(n+1)T} \leq C[E(nT) + E((n + 1)T)]$.
After Integrating on time interval $[nT, (n + 1)T]$, we arrive at

$$\int_{nT}^{(n+1)T} ||A^{\frac{1}{2}}u||^2 dt - \int_{nT}^{(n+1)T} (f_s(u), u)dt \tag{14.32}$$

$$\leq C \int_{nT}^{(n+1)T} ||u_t||^2 dt + C \int_{nT}^{(n+1)T} (g \circ A^{\frac{1}{2}}u)(t)dt - (u_t, u)|_{nT}^{(n+1)T}$$

$$\leq C \int_{nT}^{(n+1)T} ||u_t||^2 dt + C \int_{nT}^{(n+1)T} (g \circ A^{\frac{1}{2}}u)(t)dt + C[E(nT) + E((n + 1)T)].$$

Here the generic constant C depends on ϵ, l, but not on n;

2. *Estimate on the kinetic energy.* Now we want to bound term $\int_{nT}^{(n+1)T} ||u_t||^2 dt$. We take the inner product of each term in (14.1) with term $\int_0^t g(t-s)(u(t) - u(s))ds$ and have

$$\int_{nT}^{(n+1)T} (u_{tt}(t), \int_0^t g(t-s)(u(t)-u(s))ds)dt \qquad (14.33)$$

$$= (u_t, \int_0^t g(t-s)(u(t)-u(s)))|_{nT}^{(n+1)T}$$

$$- \int_{nT}^{(n+1)T} (u_t(t), \int_0^t g'(t-s)(u(t)-u(s))dsdt$$

$$- \int_{nT}^{(n+1)T} \int_0^t g(s)ds||u_t(t)||^2 dt$$

$$\int_{nT}^{(n+1)T} (Au(t), \int_0^t g(t-s))(u(t)-u(s))ds)dt \qquad (14.34)$$

$$= \int_{nT}^{(n+1)T} (A^{\frac{1}{2}}u(t), \int_0^t g(t-s)A^{\frac{1}{2}}(u(t)-u(s))ds)dt$$

$$\int_{nT}^{(n+1)T} (\int_0^t g(t-s)Au(s)ds, \int_0^t g(t-s)(u(t)-u(s))ds)dt$$

$$= \int_{nT}^{(n+1)T} (\int_0^t g(t-s)A^{\frac{1}{2}}u(s)ds, \int_0^t g(t-s)A^{\frac{1}{2}}(u(t)-u(s))dsdt$$

$$= - \int_{nT}^{(n+1)T} ||\int_0^t g(t-s)A^{\frac{1}{2}}(u(s)-u(t))ds||^2 dt \quad (14.35)$$

$$+ \int_{nT}^{(n+1)T} (\int_0^t g(s)ds)A^{\frac{1}{2}}u(t), \int_0^t g(t-s)A^{\frac{1}{2}}(u(t)-u(s))dsdt.$$

Combining the equalities in (14.33)–(14.35) leads to

$$(u_t, \int_0^t g(t-s)(u(t)-u(s))ds)|_{nT}^{(n+1)T} - \int_{nT}^{(n+1)T} (u_t(t), \int_0^t g'(t-s)(u(t)-u(s))dsdt$$

$$- \int_{nT}^{(n+1)T} \int_0^t g(s)ds||u_t(t)||^2 dt + \int_{nT}^{(n+1)T} (A^{\frac{1}{2}}u(t), \int_0^t g(t-s)A^{\frac{1}{2}}(u(t)-u(s))dsdt$$

$$= - \int_{nT}^{(n+1)T} ||\int_0^t g(t-s)A^{\frac{1}{2}}(u(s)-u(t))ds||^2 dt$$

$$+ \int_{nT}^{(n+1)T} \int_0^t g(s)ds(A^{\frac{1}{2}}u(t), \int_0^t g(t-s)A^{\frac{1}{2}}(u(t)-u(s))ds)dt$$

$$+ \int_{nT}^{(n+1)T} (f_s(u), \int_0^t g(t-s)(u(t)-u(s))ds)dt.$$

We rewrite it into

$$\int_{nT}^{(n+1)T} \int_0^t g(s)ds||u_t(t)||^2 dt$$

$$= (u_t, \int_0^t g(t-s)(u(t)-u(s))ds)|_{nT}^{(n+1)T}$$

$$- \int_{nT}^{(n+1)T} (u_t(t), \int_0^t g'(t-s)(u(t)-u(s))ds)dt$$

$$+ \int_{nT}^{(n+1)T} (A^{\frac{1}{2}}u(t), \int_0^t g(t-s)A^{\frac{1}{2}}(u(t)-u(s))ds)dt$$

$$+ \int_{nT}^{(n+1)T} ||\int_0^t g(t-s)A^{\frac{1}{2}}(u(s)-u(t))ds||^2 dt$$

$$- \int_{nT}^{(n+1)T} \int_0^t g(s)ds(A^{\frac{1}{2}}u(t), \int_0^t g(t-s)A^{\frac{1}{2}}(u(t)-u(s))ds)dt$$

$$- \int_{nT}^{(n+1)T} (f_s(u), \int_0^t g(t-s)(u(t)-u(s))ds)dt$$

$$= I_1 + I_2 + I_3 + I_4 + I_5 + I_6.$$

Now we need the estimates on the right hand side
For the *first term*, we show that for the

$$I_1 = (u_t, \int_0^t g(t-s)(u(t)-u(s))ds)|_{nT}^{(n+1)T} \le C[E(nT) + E((n+1)T)],$$

$$(14.36)$$

where the constant C does not depend on T.
Indeed since

$$||\int_0^t g(t-s)(u(t)-u(s))ds||^2$$

$$\le ||g||_{L^1(\mathbb{R}_+)} \int_0^t g(t-s)||u(t)-u(s)||^2 ds$$

$$\le c_0||g||_{L^1(\mathbb{R}_+)} \int_0^t g(t-s)||A^{\frac{1}{2}}(u(t)-u(s))||^2 ds$$

$$\le C(g \circ A^{\frac{1}{2}}u)(t),$$

we have

$$\left(u_t, \int_0^t g(t-s)(u(t)-u(s))ds\right)$$

$$\leq \frac{1}{2}||u_t(t)||^2 + \frac{1}{2}||\int_0^t g(t-s)(u(t)-u(s))ds||^2$$

$$\leq CE(t)$$

which implies (14.36).
Now for the rest of the terms appearing in the expression
The second term:

$$I_2 = \int_{nT}^{(n+1)T} \left(u_t(t), \int_0^t g'(t-s)(u(t)-u(s))ds\right)dt$$

$$\leq \epsilon \int_{nT}^{(n+1)T} ||u_t(t)||^2 dt + C_\epsilon ||g'||_{L^1} \int_{nT}^{(n+1)T} (|g'| \circ A^{\frac{1}{2}}u)(t)dt$$

$$\leq \epsilon \int_{nT}^{(n+1)T} ||u_t(t)||^2 dt - C \int_{nT}^{(n+1)T} (g' \circ A^{\frac{1}{2}}u)(t)dt,$$

since $g' < 0$ and $||g'||_{L^1} = g(0) < \infty$.
The third term:

$$I_3 = \int_{nT}^{(n+1)T} \left(A^{\frac{1}{2}}u(t), \int_0^t g(t-s)A^{\frac{1}{2}}(u(t)-u(s))ds\right)dt$$

$$\leq \int_{nT}^{(n+1)T} ||A^{\frac{1}{2}}u(t)|| ||\int_0^t g(t-s)A^{\frac{1}{2}}(u(t)-u(s))ds||dt$$

$$\leq \epsilon \int_{nT}^{(n+1)T} ||A^{\frac{1}{2}}u(t)||^2 dt + C_\epsilon \int_{nT}^{(n+1)T} (g \circ A^{\frac{1}{2}}u)(t)dt.$$

The fourth term:

$$I_4 = \int_{nT}^{(n+1)T} ||\int_0^t g(t-s)A^{\frac{1}{2}}(u(s)-u(t))ds||^2 dt$$

$$\leq C \int_{nT}^{(n+1)T} (g \circ A^{\frac{1}{2}}u)(t)dt.$$

The fifth term:

$$I_5 = -\int_{nT}^{(n+1)T} \int_0^t g(s)ds(A^{\frac{1}{2}}u(t), \int_0^t g(t-s)A^{\frac{1}{2}}(u(t)-u(s))ds)dt$$

$$\leq \int_{nT}^{(n+1)T} |\int_0^t g(s)ds|||A^{\frac{1}{2}}u(t)|||| \int_0^t g(t-s)A^{\frac{1}{2}}(u(t)-u(s))ds||dt$$

$$\leq \epsilon \int_{nT}^{(n+1)T} ||A^{\frac{1}{2}}u(t)||^2 dt + C_\epsilon \int_{nT}^{(n+1)T} (g \circ A^{\frac{1}{2}}u)(t)dt.$$

The final term:

$$I_6 = -\int_{nT}^{(n+1)T} (f_s(u), \int_0^t g(t-s)(u(t)-u(s))ds)dt$$

$$\leq C_\epsilon \int_{nT}^{(n+1)T} ||\int_0^t g(t-s)|(u(t)-u(s))|ds||^2 dt + \epsilon \int_{nT}^{(n+1)T} ||A^{1/2}u(t)||^2$$

$$\leq C \int_{nT}^{(n+1)T} (g \circ A^{\frac{1}{2}}u)(t)dt + \epsilon \int_{nT}^{(n+1)T} ||A^{1/2}u(t)||^2$$

where the last inequality results from the fact that $f_s(u)$ is locally Lipschitz.
Indeed, at 0 we have $||f_s(u) - 0|| \leq C_{||A^{1/2}u||}||A^{1/2}u - 0||$, while we know
$||A^{1/2}u(t)||^2 \leq E(t) \to 0$.
Combining the above estimates of $I1 - I5$, we arrive at

$$\int_{nT}^{(n+1)T} [\int_0^t g(s)ds]||u_t(t)||^2 dt \leq \epsilon \int_{nT}^{(n+1)T} ||u_t(t)||^2 dt + \epsilon \int_{nT}^{(n+1)T} ||A^{\frac{1}{2}}u(t)||^2$$

$$+ C_\epsilon \int_{nT}^{(n+1)T} (g \circ A^{\frac{1}{2}}u)(t)dt$$

$$- C_\epsilon \int_{nT}^{(n+1)T} (g' \circ A^{\frac{1}{2}}u)(t)dt$$

$$+ C[E(nT) + E((n+1)T)],$$

that is

$$\int_{nT}^{(n+1)T} \{\int_0^t g(s)ds - \epsilon\}||u_t(t)||^2 dt \leq \epsilon \int_{nT}^{(n+1)T} ||A^{\frac{1}{2}}u(t)||^2$$

$$+C_\epsilon \int_{nT}^{(n+1)T} (g \circ A^{\frac{1}{2}}u)(t)dt - C_\epsilon \int_{nT}^{(n+1)T} (g' \circ A^{\frac{1}{2}}u)(t)dt$$

$$+C[E(nT) + E((n+1)T)],$$

We can pick $T > 0$ so that $\{\int_0^t g(s)ds - \epsilon\} > \delta_0 > 0$, so that we have

$$\int_{nT}^{(n+1)T} ||u_t(t)||^2 dt \le \epsilon \int_{nT}^{(n+1)T} ||A^{\frac{1}{2}}u(t)||^2 \tag{14.37}$$

$$+ C_\epsilon \int_{nT}^{(n+1)T} (g \circ A^{\frac{1}{2}}u)(t)dt - C_\epsilon \int_{nT}^{(n+1)T} (g' \circ A^{\frac{1}{2}}u)(t)dt$$

$$+ C[E(nT) + E((n+1)T)]$$

Now combining with (14.32), selecting ϵ small enough so that the term of $||A^{\frac{1}{2}}u||^2$ is absorbed by the term on the left, we get

$$\int_{nT}^{(n+1)T} ||A^{\frac{1}{2}}u||^2 dt \le C \int_{nT}^{(n+1)T} (g \circ A^{\frac{1}{2}}u)(t)dt - C \int_{nT}^{(n+1)T} (g' \circ A^{\frac{1}{2}}u)(t)dt$$

$$+ C[E(nT) + E((n+1)T)],$$

which in turn by (14.32) gives

$$\int_{nT}^{(n+1)T} ||u_t||^2 dt \le C \int_{nT}^{(n+1)T} (g \circ A^{\frac{1}{2}}u)(t)dt - C \int_{nT}^{(n+1)T} (g' \circ A^{\frac{1}{2}}u)(t)dt$$

$$+ C[E(nT) + E((n+1)T)];$$

3. *Estimate on the total energy $E(t)$.*
The above estimates give

$$\int_{nT}^{(n+1)T} ||A^{\frac{1}{2}}u||^2 dt + \int_{nT}^{(n+1)T} ||u_t||^2 dt \le C \int_{nT}^{(n+1)T} (g \circ A^{\frac{1}{2}}u)(t)dt$$

$$-C \int_{nT}^{(n+1)T} (g' \circ A^{\frac{1}{2}}u)(t)dt + C[E(nT) + E((n+1)T)]. \tag{14.38}$$

Noting that by Assumption A.2 we have

$$\int_{nT}^{(n+1)T} F(u)dt \le \int_{nT}^{(n+1)T} \phi(||A^{\frac{1}{2}}u||)||A^{\frac{1}{2}}u||^2 dt$$

$$\le \phi(\sqrt{E_0}) \int_{nT}^{(n+1)T} ||A^{\frac{1}{2}}u||^2 dt$$

$$\le C \int_{nT}^{(n+1)T} ||A^{\frac{1}{2}}u||^2 dt$$

which leads to the following observability inequality:

$$\int_{nT}^{(n+1)T} E(t)dt \leq C \int_{nT}^{(n+1)T} (g \circ A^{\frac{1}{2}}u)(t)dt$$

$$-C \int_{nT}^{(n+1)T} (g' \circ A^{\frac{1}{2}}u)(t)dt + C[E(nT) + E((n+1)T)]. \quad (14.39)$$

Energy identity gives

$$E((n+1)T) + \frac{1}{2} \int_{nT}^{(n+1)T} \left(g(t)\|A^{\frac{1}{2}}u(t)\|^2 - (g' \circ A^{\frac{1}{2}}u)(t) \right)dt = E(nT). \tag{14.40}$$

Recalling $D(t) = \frac{1}{2}g(t)\|A^{\frac{1}{2}}u(t)\|^2 - \frac{1}{2}(g' \circ A^{\frac{1}{2}}u)(t)$, we write

$$E((n+1)T) + \int_{nT}^{(n+1)T} D(t)dt = E(nT).$$

From (14.39)

$$\int_{nT}^{(n+1)T} E(t)dt \leq C \int_{nT}^{(n+1)T} (g \circ A^{\frac{1}{2}}u)(t)dt$$

$$+ C \int_{nT}^{(n+1)T} D(t)dt + CE((n+1)T).$$

Since

$$\int_{nT}^{(n+1)T} E(t)dt \geq TE((n+1)T),$$

we have

$$TE((n+1)T) \leq C \int_{nT}^{(n+1)T} (g \circ A^{\frac{1}{2}}u)(t)dt + C \int_{nT}^{(n+1)T} D(t)dt + CE((n+1)T).$$

Selecting $T > C$ gives

$$E((n+1)T) \leq \frac{C_1 T}{\vartheta} \int_{nT}^{(n+1)T} (g \circ A^{\frac{1}{2}}u)(t)dt + C_{2T} \int_{nT}^{(n+1)T} D(t)dt.$$

The constants $C_i, i = 1, 2$ depends only on T and not on n. The same argument applies to all intervals $[nT, (n+1)T], n = 1, 2 \ldots$.

$\square$

Proof (of Lemma 14.4)

Since $H \in C^1[0, \infty)$, by a simple variable substitution and L'Hospital rule we can see that function $H_{1,\alpha}(s) = \alpha s^{1-\frac{1}{\alpha}} H(s^{\frac{1}{\alpha}}), 0 < \alpha < 1$ is well-defined with $H_{1,\alpha}(0) = 0$.

1. To show that there exist a $\delta > 0$ such that function $H_{1,\alpha}(s)$ is increasing and convex on $(0, \delta)$, we perform some simple calculations. Let $k = \frac{1}{\alpha} \geq 1$. We have $H_{1,\alpha}(s) = \frac{1}{k} s^{1-k} H(s^k)$ and

$$H'_{1,\alpha}(s) = \frac{(1-k)}{k} s^{-k} H(s^k) + s^{1-k} H'(s^k) s^{k-1}$$

$$= \frac{1}{k} s^{-k} H(s^k) + [H'(s^k) - s^{-k} H(s^k)] > 0, \forall s > 0.$$

The last inequality results from the properties of $H(s)$:

$$H(x) \leq x H'(x), x > 0. \tag{14.41}$$

Indeed, this follows from geometric interpretation of convexity of $H(x)$ and the fact that $H(0) = 0$ which then gives $H'(x)x - H(x) > 0, \forall x > 0$. Thus we show that $H_{1,\alpha}(x)$ is increasing on the positive half line.

For the second derivative on $(0, \infty)$, we have

$$H''_{1,\alpha}(s) = (1-k)(-1)s^{-k-1} H(s^k) + (1-k)s^{-k} H'(s^k)s^{k-1} + ks^{k-1} H''(s^k)$$

$$= \frac{1}{s^{k+1}} [kx^2 H''(x) - (k-1)xH'(x) + (k-1)H(x)] \tag{14.42}$$

with $x = s^k$.

We first assume Assumption (B)(3)(a) holds, namely $H(x) \in C^2(\mathbb{R}_+)$. Noting $H(0) = 0$, simple Taylor expansion gives

$$H(x) = H'(0)x + H''(0)\frac{x^2}{2} + o(x^2)$$

$$H'(x) = H'(0) + H''(0)x + o(x)$$

$$H''(x) = H''(0) + o(1), x \to 0.$$

Then (14.42) implies

$$H''_{1,\alpha}(s) = \frac{1}{s^{k+1}} [kx^2 H''(x) - (k-1)xH'(x) + (k-1)H(x)]$$

$$= \frac{1}{s^{k+1}} [\frac{(k+1)}{2} x^2 H''(0) + o(x^2)].$$

So long as $H''(0) > 0$, which is the case, there exist an interval $[0, \delta], \delta > 0$ on which $H_{1,\alpha}(s) = \alpha s^{1-\frac{1}{\alpha}} H(s^{\frac{1}{\alpha}})$ is increasing and convex.

For the case $H(x) \in C^2(0, \infty)$, but $H''(0)$ might be undefined, we rewrite

$$H''_{1,\alpha}(s) = \frac{1}{s^{k+1}}[kx^2 H''(x) - (k-1)xH'(x) + (k-1)H(x)]$$

$$= \frac{1}{s^{k+1}}\left(x^2 H''(x) + (k-1)[x^2 H''(x) - xH'(x) + H(x)]\right).$$

By Assumption (B)(3)(b) we have $H''_{1,\alpha}(s) \geq 0, \forall s, 0 < s < \delta$. Hence we conclude the proof for part 1, namely, there exist an interval $(0, \bar\delta), 0 < \delta < \bar\delta$ on which $H_{1,\alpha}(s) = \alpha s^{1-\frac{1}{\alpha}} H(s^{\frac{1}{\alpha}})$ is increasing and convex.

2. Now we prove that $H_{1,\alpha}(s) = \alpha s^{1-\frac{1}{\alpha}} H(s^{\frac{1}{\alpha}})$ satisfies

$$H_{1,\alpha}[\vartheta(g \circ A^{\frac{1}{2}})(t)] \leq \alpha\vartheta D(t) \tag{14.43}$$

(or equivalently $(g \circ A^{\frac{1}{2}})(t) \leq \frac{1}{\vartheta} H_{1,\alpha}^{-1}(\alpha\vartheta D(t)))$, under the assumption

$$0 < c_\alpha = \sup_{t>0} c(\alpha, t) = \sup_{t>0} \int_0^t g^{1-\alpha}(t-s)f^2(t,s)ds < \infty.$$

Noting that $(g \circ A^{\frac{1}{2}})(t) = \int_0^t g(t-s)f^2(t,s)ds$ and $c(\alpha, t) = \int_0^t g^{1-\alpha}(s)f^2(t,s)ds < \infty$, also noting $H_{1,\alpha}(s) = \alpha s^{1-\frac{1}{\alpha}} H(s^{\frac{1}{\alpha}})$ is convex, by Jensen's Inequality we have

$$H_{1,\alpha}\left[\vartheta \int_0^t g(t-s)f^2(t,s)ds\right] = H_{1,\alpha}\left[\int_0^t \vartheta g^\alpha(t-s)g^{1-\alpha}(s)f^2(t,s)ds\right]$$

$$= H_{1,\alpha}\left[\frac{1}{c(t,\alpha)} \int_0^t \vartheta c(t,\alpha)g^\alpha(t-s)g^{1-\alpha}(t-s)f^2(t,s)ds\right]$$

$$\leq \frac{1}{c(t,\alpha)} \int_0^t H_{1,\alpha}[\vartheta c(t,\alpha)g^\alpha(t-s)]g^{1-\alpha}(t-s)f^2(t,s)ds$$

$$= \frac{1}{c(t,\alpha)} \int_0^t \alpha[\vartheta c(t,\alpha)g^\alpha(t-s)]^{1-\frac{1}{\alpha}} H([\vartheta c(t,\alpha)g^\alpha(t-s)]^{\frac{1}{\alpha}})g^{1-\alpha}(t-s)f^2(t,s)ds$$

$$= \frac{\alpha\vartheta^{1-\frac{1}{\alpha}}}{c(t,\alpha)^{\frac{1}{\alpha}}} \int_0^t H([\vartheta^{\frac{1}{\alpha}}c(t,\alpha)^{\frac{1}{\alpha}}g(t-s)])f^2(t,s)ds.$$

Now we can take ϑ such that $\vartheta^{\frac{1}{\alpha}}c(t,\alpha)^{\frac{1}{\alpha}} \leq 1$, e.g. $\vartheta = \frac{1}{c_\alpha}$, thus we have

$$H\left(\left[\vartheta^{\frac{1}{\alpha}}c(t,\alpha)^{\frac{1}{\alpha}}g(t-s)\right]\right) \leq \vartheta^{\frac{1}{\alpha}}c(t,\alpha)^{\frac{1}{\alpha}}H(g(t-s)),$$

because $H(0) = 0$ and $H(x)$ is convex.

So we have

$$H_{1,\alpha}\left[\vartheta \int_0^t g(t-s)f^2(t,s)ds\right] \leq \alpha\vartheta \int_0^t H\left[g(t-s)\right]f^2(t,s)ds$$

$$\leq \alpha\vartheta \int_0^t \left[-g'(t-s)\right]f^2(t,s)ds \leq \alpha\vartheta D(t).$$

This completes the proof of (14.43). $\square$

Proof (of Lemma 14.5) From (14.23)–(14.25), we have

$$E((n+1)T) \leq C_{1T} \int_{nT}^{(n+1)T} (g \circ A^{\frac{1}{2}}u)(t)dt + C_{2T} \int_{nT}^{(n+1)T} D(t)dt$$

$$\leq \frac{C_{1T}}{\vartheta} \int_{nT}^{(n+1)T} H_{1,\alpha}^{-1}(\alpha\vartheta D(t))dt + C_{2T} \int_{nT}^{(n+1)T} D(t)dt$$

$$\leq \frac{C_{1T}}{\vartheta} TH_{1,\alpha}^{-1}\left[\frac{\alpha\vartheta}{T} \int_{nT}^{(n+1)T} D(t)dt\right] + C_{2T} \int_{nT}^{(n+1)T} D(t)dt$$

$$\leq \hat{H}_{1,\alpha}^{-1}\left[\int_{nT}^{(n+1)T} D(t)dt\right].$$

Here we define $\hat{H}_{1,\alpha}^{-1}$, as the inverse of a function $\hat{H}_{1,\alpha}$, by

$$\hat{H}_{1,\alpha}^{-1}(x) = \frac{C_{1T}}{\vartheta} TH_{1,\alpha}^{-1}\left[\frac{\alpha\vartheta}{T}x\right] + C_{2T}x.$$

It is easy to see that $\hat{H}_{1,\alpha}^{-1}$ is increasing and concave with $\hat{H}_{1,\alpha}^{-1}(0) = 0$. Hence $\hat{H}_{1,\alpha}$ is increasing, convex and through origin. Furthermore,

$$E((n+1)T) \leq \hat{H}_{1,\alpha}^{-1}\left[\int_{nT}^{(n+1)T} D(t)dt\right].$$

$$\Rightarrow \hat{H}_{1,\alpha}[E((n+1)T)] \leq \left[\int_{nT}^{(n+1)T} D(t)dt\right].$$

$$\Rightarrow \hat{H}_{1,\alpha}[E((n+1)T)] \leq E(nT) - E((n+1)T).$$

$$\Rightarrow E((n+1)T) + \hat{H}_{1,\alpha}[E((n+1)T)] \leq E(nT). \qquad \square$$

Proof (of Lemma 14.6)

We define $\underline{H}_\alpha = I - (I + \hat{H}_{1,\alpha})^{-1}$. By Lemma 3.3 in [22] and Lemma 14.5 $E(t) \leq s(t)$ where $s(t)$ satisfies the ODE.

$$s_t + \underline{H}_\alpha(s) = 0, s(0) = E(T).$$

See Lemma 3.3 in [22] for the detailed proof. On the other hand, $\underline{H}_\alpha(s)$ has the same asymptotic behavior at the origin as $\hat{H}_{1,\alpha}$, note the following

$$\underline{H}_\alpha = I - (I + \hat{H}_{1,\alpha})^{-1} = (I + \hat{H}_{1,\alpha}) \circ (I + \hat{H}_{1,\alpha})^{-1} - (I + \hat{H}_{1,\alpha})^{-1}$$

$$= \hat{H}_{1,\alpha} \circ (I + \hat{H}_{1,\alpha})^{-1}.$$

Since $\hat{H}_{1,\alpha} \in C^1[0, \infty)$, we have $\hat{H}_{1,\alpha}(x) = O(x)$ at the origin. We use $A \approx B$ to represent that A and B have the same end behaviors. Then

$$(I + \hat{H}_{1,\alpha}) \approx I \Rightarrow \hat{H}_\alpha \circ (I + \hat{H}_\alpha)^{-1} \approx \hat{H}_{1,\alpha} \circ I \Rightarrow \underline{H}_\alpha \approx \hat{H}_{1,\alpha}, x \to 0.$$

For a detailed discussion, see [21, Corollary 1. p. 1770]. By similar arguments, from the definition

$$\hat{H}_{1,\alpha}^{-1}(x) = \frac{C_{1T}}{\vartheta} TH_{1,\alpha}^{-1}\left[\frac{\alpha\vartheta}{T}x\right] + C_{2T}x,$$

and the fact that they are both convex at zero, we can see $\underline{H}_\alpha \approx \hat{H}_{1,\alpha}(x)$ for $x > 0$ in the neighborhood of 0. $\qquad \square$

Proof (of Lemma 14.7)

Let $\mathcal{H}(y) = \int_y^\infty \frac{dx}{H(x)}, y > 0$. It is easy to see $\mathcal{H}$ is positive and decreasing function on $[0, \infty)$ with $\frac{d}{dy}\mathcal{H}(y) = -\frac{dy}{H(y)}$. Thus

$$y_t + H(y) = 0 \Rightarrow \frac{dy}{\mathcal{H}(y)} = -dt$$

$$\Rightarrow d\mathcal{H}(y) = dt$$

$$\Rightarrow \mathcal{H}(y) - \mathcal{H}(y_0) = t,$$

$$\Rightarrow y(t) = \mathcal{H}^{-1}(\mathcal{H}(y_0) + t).$$

Similarly,

$$s_t + \alpha s^{1-\frac{1}{\alpha}} H(s^{\frac{1}{\alpha}}) = 0 \Rightarrow (s^{\frac{1}{\alpha}})_t + H(s^{\frac{1}{\alpha}}) = 0,$$

$$\Rightarrow s^{\frac{1}{\alpha}}(t) = \mathcal{H}^{-1}(\mathcal{H}(s_0^{\frac{1}{\alpha}}) + t) = y(t + c),$$

$$\Rightarrow s(t) = y^{\alpha}(t + c),$$

with $c = \mathcal{H}(s_0^{\frac{1}{\alpha}}) - \mathcal{H}(y_0)$.   □

Proof (of Lemma 14.8) The above lemmas show that the energy is bounded by a function $s_0(t) \approx y^{\alpha_0}(t)$. It might contain a delay, but it makes no difference since we are considering the asymptotic behavior.

To conclude the proof of Theorem 14.2, we adopt an iteration process to find a sequence $\alpha_k, k = 1, \ldots, m + 1$ such that $\alpha_0 < \alpha_1 < \ldots < \alpha_m < 1 = \alpha_{m+1}$ and the corresponding controlling function, $s_k(t) \approx y^{\alpha_k}(t)$.

Under Assumption (B)(1)–(2), we have

$$c(\alpha_0, t) = \int_0^t g^{1-\alpha_0}(t - s) f^2(t, s) ds$$

$$\leq 2E(0) \int_0^t g^{1-\alpha_0}(t - s) ds = 2E(0) \int_0^t g^{1-\alpha_0}(s) ds$$

$$\leq 2E(0) \int_0^\infty y^{1-\alpha_0}(t) dt < \infty.$$

This is the critical estimate to initialize the second part of Lemma 14.4. Then by Lemma 14.6, $E(t)$ is driven by a function $s(t)$ which satisfies the differential equation

$$s_t + \alpha_0 s^{1-\frac{1}{\alpha_0}} H(s^{\frac{1}{\alpha_0}}) = 0.$$

Thus Lemma 14.7 implies

$$E(t) \leq s(t) \leq C y^{\alpha_0}(t).$$

So we have the first improvement on our energy estimate. Getting the optimal decay $E(t) \leq Cy(t)$ is equivalent to say

$$s_t + \alpha_0 s^{1-\frac{1}{\alpha_l}} H(s^{\frac{1}{\alpha_l}}) = 0 \qquad (14.44)$$

with

$$\alpha_l = 1,$$

or to have

$$\int_0^t g^{1-\alpha_l}(t-s)f^2(t,s)ds < \infty, \tag{14.45}$$

with $\alpha_l = 1$, since (14.45) implies (14.44).

This can be done in finitely many steps. First we expect to pick α_1 so that $\alpha_0 < \alpha_1 \leq 1$ and

$$c(\alpha_1, t) = \int_0^t g^{1-\alpha_1}(t-s)f^2(t,s)dt < \infty.$$

Since $f^2(t,s) \leq E(t) + E(s) \leq 2E(s)$, it suffices to have $\int_0^t y^{1-\alpha_1}(t)E(s)dt < \infty$, or to have $\int_0^t y^{1-\alpha_1}(t)y^{\alpha_0}(s)dt < \infty$. Now Proposition 14.2 takes over and the generated finite sequence $I_k(t)$ which bound $c(\alpha_k, t)$ all the way until $\alpha_m = 1$. That completes the proof.

The iteration will generate sequence $\alpha_1 = 2\alpha_0, \alpha_2 = 3\alpha_0, \alpha_m = (m+1)\alpha_0$ until we have $\alpha_m = (m+1)\alpha_0 > 1$ for some m. Since $\alpha_0 > 0$, such a finite m exists, thus we can reach the optimality by letting $\alpha_m = 1$ in m steps. □

Corollary 14.1 Assume that $H(s) \sim |s|^{p-1}s, p \in [1, 2)$ and consider Examples given in Sect. 14.2. Then there exist positive constant $c = c(E(T)), c_1 = c_1(E(T))$ such that

$$E(t) \sim s(t)$$

where $s(t)$ satisfies the following nonlinear ODE with some $T_0 > 0$

$$s_t(t) + cH(c_1 s(t)) = 0, t > T_0, s(0) = E(0). \tag{14.46}$$

The above result recovers optimal range of the parameter $p \in [1, 2)$ for the decay rates corresponding to the algebraic decay rates of relaxation kernels.

Acknowledgements Research of I. Lasiecka and X. Wang partially supported by DMS Grant 0104305 and AFOSR Grant FA9550-09-1-0459.

References

1. Alabau-Boussouira, F., Cannarsa, P., Sforza, D.: Decay estimates for second order evolution equations with memory. J. Funct. Anal. **254**, 1342–1372 (2008)
2. Alabau-Boussouira, F., Cannarsa, P.: A general method for proving sharp energy decay rates for memory-dissipative evolution equations. C. R. Acad. Sci. Paris Ser. I **347**, 867–872 (2009)
3. Alabau-Boussouira, F.: Convexity and weighted integral inequalities for energy decay rates of nonlinear dissipative hyperbolic systems. Appl. Math. Optim. **51**, 61–105 (2005)

4. Alabau-Boussouira, F.: A unified approach via convexity for optimal energy decay rates of finite and infinite dimensional vibrating damped systems with applications to semi-discretized vibrating damped systems. J. Differ. Equ. **248**, 1473–1517 (2010)
5. Alabau-Boussouira, F.: On some recent advances on stabilization for hyperbolic equations. Lecture Note in Mathematics, CIME Foundation Subseries, Control of Partial Differential Equations, vol. 2048, pp.1–100. Springer, New York (2012)
6. Arnold, V.I.: Mathematical Methods of Classical Mechanics. Springer, New York (1989)
7. Barreto, R., Lapa, E.C., Munoz Rivera, J.E.: Decay rates for viscoelastic plates with memory. J. Elast. **44**(1), 61–87 (1996)
8. Berrimi, S., Messaoudi, S.A.: Exponential decay of solutions to a viscoelastic equation with nonlinear localized damping. Electronic J. Differ. Equ. **88**, 1–10 (2004)
9. Berrimi S., Messaoudi, S. A.: Existence and decay of solutions of a viscoelastic equation with a nonlinear source. Nonlinear Anal. **64**, 2314–2331 (2006)
10. Barreto, R., Munoz Rivera, J.E.: Uniform rates of decay in nonlinear viscoelasticity for polynomial decaying kernels. Appl. Anal. **60**, 263–283 (1996)
11. Cavalcanti, M.M., Domingos Cavalcanti, V.N., Martinez, P.: General decay rate estimates for viscoelastic dissipative systems. Nonlinear Anal. **68**(1), 177–193 (2008)
12. Cavalcanti, M.M., Domingos Cavalcanti, V.N., Soriano, J.A.: Exponential decay for the solution of semilinear viscoelastic wave equations with localized damping. Electronic J. Differ. Equ. **44**, 1–14 (2002)
13. Cavalcanti, M.M., Oquendo, H.P.: Frictional versus viscoelastic damping in a semilinear wave equation. SIAM J. Control Optim. **42**(4), 1310–1324 (2003)
14. Cabanillas, E.L., Munoz Rivera, J.E.: Decay rates of solutions of an anisotropic inhomogeneous n-dimensional viscoelastic equation with polynomial decaying kernels. Commun. Math. Phys. **177**, 583–602 (1996)
15. Chueshov, I., Lasiecka, I.: Attracors for second order evolution equations with nonlinear damping. J. Dyn. Differ. Equ. **16**(2), 469–512 (2004)
16. Dafermos, C.M.: Asymptotic stability in viscoelasticity. Arch. Rational Mech. Anal. **37**, 297–308 (1970)
17. Fabrizio, M., Polidoro, S.: Asymptotic decay for some differential systems with fading memory. Appl. Anal. **81**(6), 1245–1264 (2002)
18. Han, X., Wang, M.: General decay of energy for a viscoelastic equation with nonlinear damping. Math. Methods Appl. Sci. **32**(3), 346–358 (2009)
19. Han, X., Wang, M.: General decay rates of energy for the second order evolutions equations with memory. Acta Appl. Math. **110**, 195–207 (2010)
20. Lasiecka, I., Messaoudi, S. A., Mustafa, M.I.: Note on intrinsic decay rates for abstract wave equations with memory. J. Math. Phys. **54**, 031504 (2013). doi:10.1063/1.4793988
21. Lasiecka, I., Toundykov, D.: Energy decay rates for the semilinear wave equation with nonlinear localized damping and source terms. Nonlinear Anal. **64**(8), 1757–1797 (2006)
22. Lasiecka, I., Tataru, D.: Uniform boundary stabilization of semilinear wave equation with nonlinear boundary dissipation. Differ. Integral Equ. **6**, 507–533 (1993)
23. Lieb, E.H., Loss, M.: Analysis, 2nd edn. Graduate Studies in Mathematics, vol. 14. American Mathematical Society, Providence (2001)
24. Martinez, P.: A new method to obtain decay rate estimates for dissipatve systems. ESAIM Control Optim. Calc. **4**, 419–444 (1999)
25. Messaoudi, S.A.: General decay of solutions of a viscoelastic equation. J. Math. Anal. Appl. **341**, 1457–1467 (2008)
26. Messaoudi, S.A.: General decay of the solution energy in a viscoelastic equation with a nonlinear source. Nonlinear Anal. **69**, 2589–2598 (2008)
27. Messaoudi, S., Mustafa, S.: General stability result for viscoelastic wave equations. J. Math. Phys. **53**, (2012)
28. Muñoz Rivera, J., Peres Salvatierra A.: Asymptotic behaviour of the energy in partially viscoelastic materials. Q. Appl. Math. **59**, 557–578 (2001)

29. Muñoz Rivera, J., Lapa E.C., Barreto, R.: Decay rates for viscoelastic plates with memory. J. Elast. **44**(1), 61–87 (1996)
30. Prüss, J.: Evolutionary Integral Equations and Applications. Monography Math. vol 87. Birkhäuser (1993)
31. Renardy, M., Hrusa, W., Nohel, J.: Mathematical problems in viscoelasticity. Pitman Monographs and Surveys in Pure and Applied Mathematics, vol. 35. Longman Scientific & Technical, Harlow. Wiley, New York (1987)
32. Xiao, T., Liang, J.: Coupled second order semilinear evolution equations indirectly damped via memory effects. J. Differ. Equ. **254**(5), 2128–2157 (2013)

Chapter 15
Inverse Problem for a Linearized Jordan–Moore–Gibson–Thompson Equation

Shitao Liu and Roberto Triggiani

In memory of Alfredo Lorenzi: scholar, collaborator, friend

Abstract We consider an inverse problem for the linearized Jordan–Moore–Gibson–Thompson equation, which is a third-order (in time) PDE in the original unknown u that arises in nonlinear acoustic waves modeling high-intensity ultrasound. Both canonical recovery problems are investigated: (i) uniqueness and (ii) stability, by use of just one boundary measurement. Our approach relies on the dynamical decomposition of the Jordan–Moore–Gibson–Thompson equation given in Marchand et al. (Math. Methods Appl. Sci. **35**, 1896–1929, 2012), which identified 3 distinct models in the new variable z. By using now z-model ♯3, we weaken by two units the regularity requirements on the data of the original u-dynamics over our prior effort Liu and Triggiani (J. Inverse Ill-Posed Probl. **21**, 825–869, 2013), which instead employed z-model ♯1.

15.1 Physical Motivation of the Model

Classical models of nonlinear acoustics are the Kuznetsov's equation [15, 21], the Westervelt equation [14, 15, 17], and the Kokhlov–Zabolotskaya–Kuznetsov (KZK) equation. Nonlinear acoustic wave propagation encompasses a wide range of applications for medical and industrial use, such as high-intensity focused ultrasound (HIFU) in lithotripsy, thermotherepy, ultrasound cleaning, sonochemistry, etc. The aforementioned mathematical models are second order in time

S. Liu
Department of Mathematical Sciences, O-110 Martin Hall, Clemson University, Clemson, SC 29634, USA
e-mail: liul@clemson.edu

R. Triggiani (✉)
Department of Mathematical Sciences, 373 Dunn Hall, University of Memphis, Memphis, TN 38152, USA
e-mail: rtrggani@memphis.edu

© Springer International Publishing Switzerland 2014
A. Favini et al. (eds.), *New Prospects in Direct, Inverse and Control Problems for Evolution Equations*, Springer INdAM Series 10,
DOI 10.1007/978-3-319-11406-4_15

and characterized by the presence of a viscoelastic damping term. These models incorporate the classical Fourier Law for the heat flux. As is well known, this law leads to the paradox of infinite speed of propagation of the solutions, which may not be acceptable in several applications. Accordingly, to avoid this pathology, several other constitutive relations for the heat flux have been proposed within the derivation of the corresponding nonlinear acoustic model [12, 13]. Among these is the Maxwell–Cattaneo Law, which has recently resurfaced in the modeling of several physical phenomena. In the present context of nonlinear acoustic waves modeling high-intensity ultrasound, replacement of the Fourier law by the Maxwell–Cattaneo's law yields a third order (in time) nonlinear equation, which is referred to as the Jordan–Moore–Gibson–Thompson equation [12, 13]. Its linearized version is reported below.

Mathematical Model of the Linearized J–M–G–T Equation Let Ω be an open bounded domain in $\mathbb{R}^n$, $n \geq 2$, with smooth boundary Γ (subject to further assumptions to be specified below). The *linearized* third-order Jordan–Moore–Gibson–Thompson PDE-equation—which arises in high-intensity ultrasound—is as follows:

$$\begin{cases} \tau u_{ttt} + \alpha(x)u_{tt} - c^2 \Delta u - b \Delta u_t = 0 \text{ in } Q = \Omega \times [0, T] & (15.1a) \\[2mm] u(\cdot, \frac{T}{2}) = u_0, \ u_t(\cdot, \frac{T}{2}) = u_1, \ u_{tt}(\cdot, \frac{T}{2}) = u_2 \text{ in } \Omega & (15.1b) \\[2mm] \text{either } \frac{\partial u}{\partial \nu}|_\Sigma = 0 \quad \text{in } \Sigma = \Gamma \times [0, T] & (15.1c) \\[2mm] \text{or else } u|_\Sigma = 0 \quad \text{in } \Sigma = \Gamma \times [0, T]. & (15.1d) \end{cases}$$

Here $u(x, t)$ is the velocity potential of the acoustic phenomenon; c (positive constant) is the speed of sound; τ is a positive constant accounting for relaxation; $b = \delta + \tau c^2$, where δ is the diffusivity of sound, is a positive constant. Henceforth, without further mention, we *normalize* τ and set $\tau = 1$. Instead, the positive coefficient $\alpha(x)$ is allowed to depend on the space variable $x \in \Omega$.

Mathematical Review A mathematical analysis of the u-problem (15.1a-b-c) or (15.1a-b-d) was recently given in [16], in fact, for the fully nonlinear model, by use of energy methods; and in [48] by use of a semi-group approach, where all coefficients are *constant*. These references investigate the following preliminary issues: well-posedness in natural function spaces, in particular on the space of finite energy; asymptotic behavior; and, in the case of constant coefficients, a structural decomposition as well as spectral analysis of the basic semi-group generator, with sharp decay rate in terms of the data.

Goal of the Present Paper With the foundational issues being settled, in the present paper we seek to investigate an inverse problem. This will amount to the recovery (existence of recovery and stability of recovery) of the coefficient $\gamma(x) = \alpha(x) - \frac{c^2}{b}$ of both physical and mathematical relevance. To this end, we shall employ only one suitable boundary measurement, following our first effort [47]. In addition, in the present paper, we shall seek optimal (minimal) assumptions on the initial data. Indeed, our present requirements on the data weaken by two units those imposed in [47]. This will be achieved by employing z-model ♯3 in (15.28), (15.29), rather than z-model ♯1 in (15.25a-d), as anticipated in [47, Remark 5.1, p. 850]. See however Theorem 15.5 below.

Remark 15.1 *In model (15.1a-d) we regard* $t = \frac{T}{2}$ *as initial time. This is not essential, because the change of independent variable* $t \to t - \frac{T}{2}$ *transforms* $t = \frac{T}{2}$ *to* $t = 0$. *However, this present choice is convenient in order to apply the Carleman estimates established in [35].*

Basic Assumption on the Coefficient $\alpha(x)$ **to be Identified** In reference [48] the coefficient $\alpha(x)$ was assumed to be a constant $\alpha(x) \equiv const > \frac{c^2}{b}$. This assumption, or even just $\alpha(x) \equiv const$, is however *not* needed in order to claim semigroup-wellposedness and related regularity results as stated in [48]. The full assumption $\alpha(x) \equiv const > \frac{c^2}{b}$ was only needed in carrying out the analysis of (sharp) exponential decay of said semigroup, explicitly in terms of the coefficients $\{\alpha, c, b\}$. In this paper the coefficient $\alpha(x)$ is unknown and is the object of our proposed inverse problem. Accordingly, we shall need to make throughout on $\alpha(x)$ the following assumption (to be explained in Remark 15.2 below):

$$
\begin{cases}
\alpha(\cdot) \text{ is a multiplier } \mathcal{D}(\mathscr{A}^{\frac{m}{2}}) \to \mathcal{D}(\mathscr{A}^{\frac{m}{2}}), \text{ or} \\
\alpha \in M(\mathcal{D}(\mathscr{A}^{\frac{m}{2}}) \to \mathcal{D}(\mathscr{A}^{\frac{m}{2}})), \ [49], \ m > \dfrac{\dim\Omega}{2}.
\end{cases}
\tag{15.2}
$$

Here, the operator $\mathscr{A}$ is $(-\Delta)$ with either Neumann boundary condition (B.C.) or else with Dirichlet B.C. in (15.1a-d). Accordingly, condition (15.2) can be reformulated as follows:

$$
\begin{cases}
\alpha \in M(H^m(\Omega) \to H^m(\Omega)), \ m > \dfrac{\dim\Omega}{2}, \\
\text{plus appropriate boundary compatibility conditions.}
\end{cases}
\tag{15.3}
$$

Explicit conditions of this type were already encountered and exploited in [46] and the authors' paper on which they are based such as [44, 45]. Regarding [46] we may quote the following spots: Remark 5.5, Eq. (5.69b); and above all Eq. (5.136) which is precisely condition (15.2) in the case of Dirichlet B.C. for the operator $\mathscr{A}$. More specifically, in the aforementioned case where $\mathscr{A}$ includes Dirichlet B.C., then boundary compatibility conditions to associate to (15.3) are the following:

$$
\left. \frac{\partial \alpha}{\partial \nu} \right|_\Gamma = 0 \text{ for } dim\Omega = 2; \text{ and in addition } \left. \frac{\partial(\Delta \alpha)}{\partial \nu} \right|_\Gamma = 0 \text{ for } dim\Omega = 3.
\tag{15.4}
$$

Remark 15.2 *For bounded domain* Ω *of class* $C^{0,1}$, *a most useful and enlightening description of the spaces of multipliers* $M(H^m(\Omega) \to H^m(\Omega))$ *is available [49, Sect. 6.3.3, p. 251]. In particular, we invoke [49, Theorem 3, p. 252] in our* *present context of Eq. (15.3), with the following parameters/notation:* $p = 2$, $pm = 2m > dim\Omega$ *(a condition a-fortiori satisfied by assumption (15.2)),* $l = m$, $W_2^m(\Omega)$ *(defined in [49, p. 8]), so that* $W_2^m(\Omega) = H^m(\Omega)$ *in our notation. We thus obtain that the space of multipliers* $M(H^m(\Omega) \to H^m(\Omega))$ *in (15.3) coincides with the space* $W_2^m(\Omega) = H^m(\Omega)$:

$$M(H^m(\Omega) \to H^m(\Omega)) \equiv H^m(\Omega), \quad 2m > dim\Omega. \tag{15.5a}$$

Thus, assumption $\alpha \in M(H^m(\Omega) \to H^m(\Omega))$ *in (15.3) becomes now checkable in our present setting and then implies by (15.5a) that*

$$\alpha \in H^m(\Omega), \text{ moreover that } \alpha \in C(\Omega) \tag{15.5b}$$

by the usual embedding. Furthermore, a sufficient checkable condition for (15.3) to hold true is

$$\alpha \in W^{m,\infty}(\Omega), \quad m > \frac{dim\Omega}{2}. \tag{15.5c}$$

Here below we shall focus mostly on the case of Neumann B.C. in (15.1a-c), though relevant technicalities of the case of Dirichlet B.C. will also be noted explicitly. Accordingly, in our treatment we write simply $\mathscr{A}$ to indicate the Neumann case as in (15.17) below.

Main Results: Neumann B.C. (15.1a-b-c)

Theorem 15.1 (Uniqueness of Inverse Problem for u-system (15.1a-b-c) with Neumann B.C.) *Assume the preliminary geometric and analytic assumptions (A.1) and (A.2) stated in Sect. 15.3 below and given in terms of a scalar function* $d(x)$. *Let*

$$T > T_0 \equiv 2\sqrt{\max_{x\in\overline{\Omega}} d(x)}. \tag{15.6}$$

With reference to the u-problem (15.1a-c) with Neumann B.C., assume that the unknown term $\alpha(\cdot)$ *satisfies assumption (15.2), equivalently assumption (15.3), which is made checkable and explicit in Remark 15.2. Furthermore, let the initial data possess the following regularity properties*

$$u_0, u_1, u_2 \in \mathscr{D}(\mathscr{A}^{\frac{m+2}{2}}) \times \mathscr{D}(\mathscr{A}^{\frac{m+1}{2}}) \times \mathscr{D}(\mathscr{A}^{\frac{m}{2}}) \subset H^{m+2}(\Omega) \times H^{m+1}(\Omega) \times H^m(\Omega),$$

$$m > \frac{dim\Omega}{2}, \tag{15.7}$$

with m non-necessarily integer, and in addition satisfy the following condition

$$\frac{c^2}{b} u_0 + u_1 = 0, \quad |u_2(x)| = \left| u_{tt}\left(x, \frac{T}{2}\right) \right| \geq r_0 > 0, \ x \in \Omega \tag{15.8}$$

for some $r_0 > 0$. Finally, if the solution to problem (15.1a-c) satisfies the additional homogeneous boundary trace condition

$$\frac{c^2}{b}u(x,t) + u_t(x,t) = 0, \ x \in \Gamma_1, \ t \in [0,T], \ \Gamma_1 = \Gamma \setminus \Gamma_0 \quad (15.9)$$

over the observed part Γ_1 (specified in assumption (A.1) of Sect. 15.3) of the boundary Γ and over the time interval T as in (15.6), then in fact

$$\gamma(x) = \alpha(x) - \frac{c^2}{b} \equiv 0, \ i.e. \ \alpha(x) \equiv \frac{c^2}{b}, \ in \ x \in \Omega. \quad (15.10)$$

Remark 15.3 *The regularity assumptions (15.2) and (15.7) imply that the solution of the u-problem (15.1a-c) (with $\tau \equiv 1$) has the regularity properties $u, u_t, u_{tt} \in L^\infty(Q)$ which is needed in the proof of Theorem 15.1. More precisely, (15.2) and (15.7) imply (see Sect. 15.2)*

$$\{u, u_t, u_{tt}\} \in C([0,T]; H^{m+2}(\Omega) \times H^{m+1}(\Omega) \times H^m(\Omega)), \quad (15.11)$$

continuously, where then the following embedding holds: $H^m(\Omega) \hookrightarrow C(\Omega) \subset L^\infty(\Omega)$ since $m > \frac{dim\Omega}{2}$.

Next, we provide the stability result of recovering $\gamma(\cdot)$. This appears to require an additional (though, in the context, reasonable) assumption: the coefficient $\alpha(\cdot)$, equivalently $\gamma(\cdot)$, is a-priori in a ball of the space $H^m(\Omega)$ of fixed but arbitrary radius R:

$$\|\gamma(\cdot)\|_{H^m(\Omega)} \leq R, \quad m > \frac{dim\Omega}{2}, \text{ hence } \gamma \text{ (and } \alpha \text{) in } C(\Omega). \quad (15.12)$$

The reason will be noted below in Remark 15.4.

Theorem 15.2 (Stability of Inverse Problem for u-system (15.1a-c) with Neumann B.C.) *With reference to the u-problem (15.1a-c), assume properties (15.2), (15.7), (15.8), (15.12) and let $T > T_0$. Then, there exists a constant C depending on the problem data and on the constant R in (15.12) but not on the unknown coefficient $\alpha(x)$ (or $\gamma(x) = \alpha(x) - \frac{c^2}{b}$), such that with $\Gamma_1 = \Gamma \setminus \Gamma_0$, Γ_0 as in (A.1) of Sect. 15.3, we have*

$$\|\gamma\|_{L^2(\Omega)} \leq C\left(\left\|\left(\frac{c^2}{b}u + u_t\right)_t\right\|_{L^2(\Gamma_1 \times [0,T])} + \left\|\left(\frac{c^2}{b}u + u_t\right)_{tt}\right\|_{L^2(\Gamma_1 \times [0,T])}\right). \quad (15.13)$$

Main Results: Dirichlet B.C. (15.1a-b-d)

Theorem 15.3 (Uniqueness of Inverse Problem for u-system (15.1a-b-d) with Dirichlet B.C.) *Assume the same hypotheses of Theorem 15.1, except that:*

(a) (15.46b) replaces now (15.46a) in assumption (A.1); and (b) the additional
Dirichlet homogeneous boundary trace condition (15.9) is replaced now by its
Neumann counterpart

$$\frac{c^2}{b}\frac{\partial u}{\partial \nu}(x,t) + \frac{\partial u_t}{\partial \nu}(x,t) = 0,\ x \in \Gamma_1,\ t \in [0,T],\ \Gamma_1 = \Gamma \setminus \Gamma_0. \quad (15.14)$$

Then, the same uniqueness conclusion (15.10) holds true.

Variations of the proof of Theorem 15.3 over that of Theorem 15.1 are noted in
Remark 15.8 at the end of Sect. 15.5.

**Theorem 15.4 (Stability of Inverse Problem for u-system (15.1a-b-d) with
Dirichlet B.C.)** *Assume the same hypotheses of Theorem 15.2 except that (15.46b)
replaces now (15.46a) in assumption (A.1). Then the following double estimate
holds true with $\Gamma_1 = \Gamma \setminus \Gamma_0$:*

$$\|\gamma\|_{H_0^\theta(\Omega)} \le C \left\|\frac{\partial}{\partial \nu}\left(\frac{c^2}{b}u + u_t\right)_t\right\|_{H^\theta(0,T;L^2(\Gamma_1))} \quad (15.15a)$$

*for all $\gamma \in H_0^\theta(\Omega)$, $0 < \theta \le 1$, $\theta \ne \frac{1}{2}$, where the constant C depends on the
problem data and on the constant R in (15.12) but not on the unknown coefficient
$\alpha(x)$ (or $\gamma(x) = \alpha(x) - \frac{c^2}{b}$);*

$$c\left\|\frac{\partial}{\partial \nu}\left(\frac{c^2}{b}u + u_t\right)_t\right\|_{H^\theta(0,T;L^2(\Gamma_1))} \le \|\gamma\|_{H^\theta(\Omega)} \quad (15.15b)$$

for all $\gamma \in H^\theta(\Omega)$, $0 \le \theta \le 1$, and all $T > 0$, now with $m \ge 2$.

Variations of the proof of Theorem 15.4 over that of Theorem 15.2 are noted in
Remark 15.9 at the end of Sect. 15.6.

Remark 15.4 (Orientation on the need of assumption (15.12)) *Following the
change of variable and decomposition of the original Jordan-Moore-Gibson-
Thompson u-equation (15.1a), introduced in a critical way in [48], we shall reduce
the original inverse problem for such u-problem (15.1a) to an inverse problem for a
second-order hyperbolic equation in the new z-variable, $z = \frac{c^2}{b}u + u_t$, see (15.23).
In the latter z-equation, however—unlike a 'typical' inverse problem—the RHS of
the equation (whether in Model ♯1, Eq. (15.24); or Model ♯2, Eq. (15.26); or Model
♯3, Eq. (15.28)) depends on $u_{tt}(\gamma)$, $u_t(\gamma)$, $u(\gamma)$, respectively, rather than being
independent terms with respect to z. Thus, in particular, these RHS terms in the z-
equation and the initial conditions (I.C.) are linked (rather than being independent
as in a 'typical' inverse problem). This then complicates the inverse problem as
the map from the coefficient γ to the solution of the z-problem is no longer linear
(see Sects. 15.5 and 15.6). This creates additional difficulties, particularly in the
compactness-uniqueness argument of Proposition 15.1, steps (iv) and (v), needed*

for the stability Theorem 15.2. It is to obtain (15.129), (15.130) on $z(\gamma_n) - z(\gamma_0)$
and (15.134), (15.135) on $z_t(\gamma_n) - z_t(\gamma_0)$ that assumption (15.12) is invoked in step
(iv), Eq. (15.124), to obtain $\gamma_0 \in C(\Omega)$, hence (15.127).

Literature   To handle such z-problem, with either Neumann B.C. or Dirichlet B.C., we shall adapt and modify the techniques for both uniqueness and stability, just under one suitable boundary measurement, that were employed in the book chapter [46], while overcoming additional challenges for the stability result by means of the additional assumption (15.12), as noted in Remark 15.4, by means also multiplier theory [49]. These, in turn, were an improvement of [40] (Neumann) and [44] (Dirichlet), where the object was the recovery (both uniqueness and stability) of the damping coefficient of the equation. Instead, in references [39,45], the aim was to recover, in one shot, both damping and potential coefficients, again by use of just one boundary measurement. For recovery of the coefficients from strongly coupled hyperbolic and Schrödinger systems, we refer to Liu and Triggiani [38,41–43].

All of the aforementioned references are critically based on a few basic technical points: (1) use of the sharp Carleman estimates as in [35]; (2) optimal/sharp interior and boundary regularity theory of second-order hyperbolic equations of Dirichlet or Neumann type [22–24, 26–28, 54]; (3) use of multiplier theory [49]; (4) a suitable post-Carleman estimate device introduced in [9, Theorem 8.2.2, p. 231] (not available in the first edition [8]); (5) sharp controllability inequalities for either the Neumann or the Dirichlet problem from [35], using also critically the boundary trace result of [32] (generalizing [5,25,56,57], in the canonical case).

There are various forms of 'Carleman estimates'. They were introduced in [4] for a 2-variable problem. The monographs [6, 7, 55] consider Carleman estimates for compactly supported solutions. These are inadequate for control or inverse problems, where a key role is played precisely by the *traces* of the solutions on the boundary. Here we use the ones in [35] which, unlike prior literature, include *explicit* boundary terms and apply to $H^{1,1}$-solutions with suitable L^2-boundary traces, rather than $H^{2,2}$-solutions as in past literature. The use of Carleman estimates for inverse problems was introduced in the pioneering work [3], to be followed by [18], [20]. Pointwise Carleman estimates for unique continuation over-determined problems have been studied intensely by the Novosibirski school, see [36]. The pointwise Carleman estimates in [35] were inspired by Lavrentev et al. [36] in the study primarily of the Neumann-control problems. With motivation coming from control theory (continuous observability inequalities, stabilization inequalities, global unique continuation for over-determined problems), Carleman estimates for non-compactly supported solutions were introduced in [51–53] (with lower order terms) in an abstract evolution setting, which stimulated the ad-hoc studies in [30] (still with lower order terms) for second-order hyperbolic equations; [31] for first-order hyperbolic equations, [29] for shells, etc.

Additional references for inverse problems based on Carleman estimates include [9, 10] (Neumann), [2] (Dirichlet), [11] (Dirichlet), [59] (Dirichlet), the recent review paper [19] and the references therein. Carleman estimates generalizing [35] to Riemannian wave equations are given in [58], extending also the treatment of

[33, 34] for control theory problems. For numerical treatments of inverse problems we refer to the established text [20] as well as the review paper [19] and the recent book [1].

A Generalization of the Global Uniqueness Result The following more general global uniqueness result of any damping coefficient ($\alpha(\cdot)$, or $\gamma(\cdot)$; not only zero as in Theorems 15.1 and 15.3) can be given. It starts with solutions $u(\gamma_1)$ and $u(\gamma_2)$, corresponding to two damping coefficients $\gamma_1(\cdot)$ and $\gamma_2(\cdot)$, of problem (15.1a-d), possessing the same I.C. $\{u_0, u_1, u_2\}$, the same B.C. (either Dirichlet or Neumann, respectively) and enjoying the same inverse problems data (either Neumann or Dirichlet boundary traces on $\Gamma_1 \times [0, T]$, respectively). It then concludes that, in fact, $\gamma_1(x) \equiv \gamma_2(x)$ in Ω. The proof is a minor modification of the proof of Theorem 15.1 or Theorem 15.3 (more precisely, of Theorem 15.3 or Theorem 1.6 in [47]). This is so since it uses the counterpart of model ♯1. Accordingly, it requires initial data $\{u_0, u_1, u_2\}$ more regular by two units over the regularity (15.7) (that is, it requires the same regularity as in [47, Eq. (1.7)]) and also the coefficients $\gamma_1(\cdot)$ and $\gamma_2(\cdot)$ will be sought in a class smoother by two units (that is, in the same class of [47, Eq. (1.9)]). The authors wish to thank a referee for suggesting extending the investigation to include the problem of Theorem 15.5 below.

Theorem 15.5 (Uniqueness of inverse problem for the u-system (15.1a-d) with $\tau \equiv 1$) *Assume the preliminary geometric and analytic assumptions (A.1) in the form (15.46a) or (15.46b) respectively, and (A.2) stated in Sect. 15.3. Let T be as in (15.6). Assume that the damping coefficients $\alpha_1(\cdot)$ and $\alpha_2(\cdot)$ (equivalently $\gamma_i = \alpha_i - \frac{c^2}{b}$, $i = 1, 2$) be in the class $H^m(\Omega)$ however now for $m > \frac{\dim\Omega}{2} + 2$. Let $u(\gamma_1)$, $u(\gamma_2)$ be the corresponding solutions with same I.C.*

$$u(\gamma_i)|_{\frac{T}{2}} = u_0, \; u_t(\gamma_i)|_{\frac{T}{2}} = u_1, \; u_{tt}(\gamma_i)|_{\frac{T}{2}} = u_2, \; i = 1, 2,$$

and satisfying (15.8) and the regularity in (15.7) however now with $m > \frac{\dim\Omega}{2} + 2$. Let $u(\gamma_1)$, $u(\gamma_2)$ satisfy the same B.C.

$$\frac{\partial u(\gamma_1)}{\partial \nu}\Big|_\Sigma = \frac{\partial u(\gamma_2)}{\partial \nu}\Big|_\Sigma \; (resp. \; u(\gamma_1)|_\Sigma = u(\gamma_2)|_\Sigma)$$

as well as the same corresponding additional inverse theory boundary conditions

$$\frac{c^2}{b} u(\gamma_1) + u_t(\gamma_1) = \frac{c^2}{b} u(\gamma_2) + u_t(\gamma_2) \text{ on } \Gamma_1 \times [0, T];$$

$$(resp. \; \frac{c^2}{b} \frac{\partial u(\gamma_1)}{\partial \nu} + \frac{\partial u_t(\gamma_1)}{\partial \nu} = \frac{c^2}{b} \frac{\partial u(\gamma_2)}{\partial \nu} + \frac{\partial u_t(\gamma_2)}{\partial \nu} \text{ on } \Gamma_1 \times [0, T]).$$

Then in fact

$$\gamma_1(x) \equiv \gamma_2(x) \ in \ \Omega.$$

The proof is given in Sect. 15.7.

15.2 First Step in the Proof: Reduction to a More Convenient z-problem by a Change of Variable: Well-Posedness

Problem (15.1a-c) (with $\tau = 1$) can be embedded into the following abstract system on the Hilbert space $\mathscr{H}$:

$$u_{ttt} + \alpha(x)u_{tt} + c^2 \mathscr{A} u + b \mathscr{A} u_t = 0, \quad \text{on } \mathscr{H} \tag{15.16}$$

along with the initial condition (I.C.) (15.1a-b), where $\mathscr{A}$ is a non-negative (unbounded) self-adjoint operator on $\mathscr{H}$. Namely,

$$\mathscr{H} = L^2(\Omega), \ \mathscr{A} = -\Delta, \ \mathscr{D}(\mathscr{A}) = \{f \in H^2(\Omega) : \frac{\partial f}{\partial \nu}|_\Gamma = 0\}. \tag{15.17}$$

Then the u-problem (15.16) can be rewritten as a first-order problem in the variables $\{u, u_t, u_{tt}\}$ as

$$\frac{d}{dt} \begin{bmatrix} u \\ u_t \\ u_{tt} \end{bmatrix} = \begin{bmatrix} 0 & I & 0 \\ 0 & 0 & I \\ -c^2 \mathscr{A} & -b \mathscr{A} & -\alpha(\cdot) \end{bmatrix} \begin{bmatrix} u \\ u_t \\ u_{tt} \end{bmatrix} = G_\alpha \begin{bmatrix} u \\ u_t \\ u_{tt} \end{bmatrix}. \tag{15.18}$$

Semigroup Well-Posedness and Regularity of Problem (15.18)

Theorem 15.6 *Assume (15.2) on $\alpha(\cdot)$. Then the operator G_α in (15.18) generates an s.c. group $e^{G_\alpha t}$ on the space $S_m \equiv \mathscr{D}(\mathscr{A}^{\frac{m+2}{2}}) \times \mathscr{D}(\mathscr{A}^{\frac{m+1}{2}}) \times \mathscr{D}(\mathscr{A}^{\frac{m}{2}}), \ m = 0, 1, 2 \ldots$. In particular, we have*

$$\{u_0, u_1, u_2\} \in \mathscr{D}(\mathscr{A}^{\frac{m+2}{2}}) \times \mathscr{D}(\mathscr{A}^{\frac{m+1}{2}}) \times \mathscr{D}(\mathscr{A}^{\frac{m}{2}})$$

$$\Rightarrow \{u, u_t, u_{tt}\} = e^{G_\alpha t}\{u_0, u_1, u_2\} \in C([0, T]; \mathscr{D}(\mathscr{A}^{\frac{m+2}{2}}) \times \mathscr{D}(\mathscr{A}^{\frac{m+1}{2}}) \times \mathscr{D}(\mathscr{A}^{\frac{m}{2}}))$$

$$\Rightarrow \{u_t, u_{tt}, u_{ttt}\} = G_\alpha e^{G_\alpha t}\{u_0, u_1, u_2\} \in$$

$$C([0, T]; \mathscr{D}(\mathscr{A}^{\frac{m+1}{2}}) \times \mathscr{D}(\mathscr{A}^{\frac{m}{2}}) \times \mathscr{D}(\mathscr{A}^{\frac{m-1}{2}}))$$

$$\Rightarrow \{u_{tt}, u_{ttt}, u_{tttt}\} = G_\alpha^2 e^{G_\alpha t}\{u_0, u_1, u_2\} \in$$

$$C([0, T]; \mathscr{D}(\mathscr{A}^{\frac{m}{2}}) \times \mathscr{D}(\mathscr{A}^{\frac{m-1}{2}}) \times \mathscr{D}(\mathscr{A}^{\frac{m-2}{2}}))$$

$$\tag{15.19}$$

continuously. As a matter of fact, the above relationships (15.19) hold true also for m real positive. Henceforth, accordingly, m may be taken to be a real positive number, in order to get sharp/optimal results.

Proof **Step 1.**   Assume first that $\alpha(x) \equiv const.$ Then, Theorem 15.6 (for $m = 0, 1$) is given in [48, Theorems 2.1 and 2.2]: the operator G_α in (15.18) generates a s.c. group in either of the spaces (recalling also the notation of [48])

$$S_0 \equiv U_3 = \mathscr{D}(\mathscr{A}) \times \mathscr{D}(\mathscr{A}^{\frac{1}{2}}) \times \mathscr{H}; \ S_1 \equiv U_4 = \mathscr{D}(\mathscr{A}^{\frac{3}{2}}) \times \mathscr{D}(\mathscr{A}) \times \mathscr{D}(\mathscr{A}^{\frac{1}{2}})$$
(15.20)

(and, in fact, also on the spaces $U_1 \equiv \mathscr{D}(\mathscr{A}^{\frac{1}{2}}) \times \mathscr{D}(\mathscr{A}^{\frac{1}{2}}) \times \mathscr{H}$ and $U_2 \equiv \mathscr{D}(\mathscr{A}) \times \mathscr{D}(\mathscr{A}) \times \mathscr{D}(\mathscr{A}^{\frac{1}{2}})$, see [48, Eqs. (2.30a) and (2.33a)]).

Step 2.   The operator G_0 obtained from G_α by setting $\alpha(x) \equiv 0$ satisfies Step 1. Next, the operator

$$P_\alpha = \begin{bmatrix} 0 & 0 & 0 \\ 0 & 0 & 0 \\ 0 & 0 & \alpha(\cdot) \end{bmatrix} \text{ is a bounded perturbation } S_m \to S_m$$
(15.21)

under the multiplier assumption (15.2) on $\alpha(\cdot)$, so that the first statement in (15.19) holds true, from which the other two follow readily.   □

It is also convenient to rewrite (15.16) as follows

$$(u_t + \alpha(\cdot)u)_{tt} + b\mathscr{A}\left(\frac{c^2}{b}u + u_t\right) = 0$$
(15.22)

which then suggests–as in [48]–to introduce a new variable z:

$$z = \frac{c^2}{b}u + u_t = (\alpha u + u_t) - \gamma(\cdot)u; \ z_t = \frac{c^2}{b}u_t + u_{tt}; \ \gamma(\cdot) = \alpha(\cdot) - \frac{c^2}{b}.$$
(15.23)

By means of (15.23)—to be used in (15.22)—the original u-problem (15.16) is transformed into the following z-problems. Three models can be extracted.

Model #1:

$$z_{tt} = -b\mathscr{A}z - \gamma(\cdot)u_{tt}(\gamma)$$
(15.24)

where $u(\gamma)$ intends to emphasize that the solution u depends on the coefficient γ. The PDE-version is then, recalling (15.1a-c):

$$
\begin{cases}
z_{tt} = b \Delta z - \gamma(x) u_{tt}(\gamma) \text{ in } \Omega \times [0, T] & (15.25a) \\[2mm]
z(\cdot, \dfrac{T}{2}) = \dfrac{c^2}{b} u_0 + u_1 \text{ in } \Omega & (15.25b) \\[2mm]
z_t(\cdot, \dfrac{T}{2}) = \dfrac{c^2}{b} u_1 + u_2 \text{ in } \Omega & (15.25c) \\[2mm]
\dfrac{\partial z}{\partial \nu}\Big|_\Sigma = 0 \quad \text{in } \Sigma = \Gamma \times [0, T]. & (15.25d)
\end{cases}
$$

The homogeneous Neumann boundary condition (B.C.) in (15.25a-d) follows from the original B.C. for u in (15.1a-d), via (15.23) on z.

Model ♯2 (Use $u_{tt} = z_t - \frac{c^2}{b} u_t$ from (15.23) in (15.24))

$$
z_{tt} = -b \mathscr{A} z - \gamma(\cdot) z_t + \gamma(\cdot) \frac{c^2}{b} u_t(\gamma) \tag{15.26}
$$

whose PDE-version is then

$$
\begin{cases}
z_{tt} = b \Delta z - \gamma(\cdot) z_t + \gamma(\cdot) \frac{c^2}{b} u_t(\gamma) \text{ in } \Omega \times [0, T] \\
\text{along with Eqs. (15.25b-c-d).}
\end{cases} \tag{15.27}
$$

Model ♯3 (Use $u_t = z - \frac{c^2}{b} u$ from (15.23) in (15.26))

$$
z_{tt} = -b \mathscr{A} z - \gamma(\cdot) z_t + \gamma(\cdot) \frac{c^2}{b} z - \gamma(\cdot) \left(\frac{c^2}{b} \right)^2 u(\gamma) \tag{15.28}
$$

whose PDE-version is then

$$
\begin{cases}
z_{tt} = b \Delta z - \gamma(x) z_t + \gamma(x) \frac{c^2}{b} z - \gamma(x) (\frac{c^2}{b})^2 u(\gamma) \text{ in } \Omega \times [0, T] \\
\text{along with Eq. (15.25b-c-d).}
\end{cases} \tag{15.29}
$$

Our analysis and corresponding results will depend on the model ♯1, ♯2, ♯3 used. In [47], we have considered the inverse problem corresponding to model ♯1. In this paper, however, following [47, Remark 5.1, p. 851], we shall concentrate on model ♯3. Model ♯3, Eq. (15.29) has the advantage over Model ♯1, Eq. (15.25a), that the 'forcing term' involves $u(\gamma)$ rather than $u_{tt}(\gamma)$. Thus, the corresponding inverse theory results of the present paper lowers the regularity requirements on the data of the original u-dynamics by two space units: $m > \frac{\dim \Omega}{2}$ now as in (15.7), rather than $m > \frac{\dim \Omega}{2} + 2$ as in [47, (1.7)]. The presence of the terms z and z_t

in (15.29) are 'benign' with respect to the Carleman estimate of Sect. 15.4. as noted in Remark 15.5 below.

Remark 15.5 *We note that in passing from (15.25a) to (15.27) to (15.29), new 2-terms such as z_t and z are introduced, while the RHS term $\gamma(\cdot)u_{tt}(\gamma)$ reduces its time derivatives to $\gamma(\cdot)u_t(\gamma)$ and $\gamma(\cdot)u(\gamma)$. From the point of view of the Carleman estimates to be employed below, terms such as z, z_t are accommodated and handled at no extra effort. Then, model $\sharp 3$ invoking only $u(\gamma)$ would appear at least at the outset, to be the one imposing minimal requirements of regularity of the original u-dynamics (15.1a-d) [47, Remark 5.1, p. 851].*

At any rate the above change of variable $u \to z$ has led to a second-order (in time) equation, indeed the wave equation problem (15.29). An important feature thereof is that: the 'forcing term'– say $-\gamma(x)(\frac{c^2}{b})^2 u(\gamma)$ in (15.29); or $\gamma(x)\frac{c^2}{b}u_t(\gamma)$ in (15.27); or $-\gamma(x)u_{tt}(\gamma)$ in (2.10a) is *not* independent on the z-problem, in particular on its I.C., but actually forcing terms and z, hence I.C. are linked. This then implies that the solution of the z-problems depends *nonlinearly* on γ, see Eq. (15.76e). This fact is a complication in the present inverse problem study, with respect to the usual case of 'linear inverse problems', where forcing terms are chosen freely and I.C. are independent, in fact the latter are homogeneous (zero).

In Sect. 15.6, we shall invoke (part of) the following result.

Theorem 15.7 *Let $\alpha_n(\cdot)$, $n = 1, 2, \ldots$ be a sequence of coefficients for problem (15.16) or (15.18) satisfying assumption (15.2) as well as (15.12). Furthermore, assume that*

$$\alpha_n \to \alpha_0 \quad \text{in } L^2(\Omega). \tag{15.30}$$

Let the I.C. satisfy (recall (15.20))

$$[u_0, u_1, u_2] \in U_3 \equiv S_0 = \mathscr{D}(\mathscr{A}) \times \mathscr{D}(\mathscr{A}^{\frac{1}{2}}) \times \mathscr{H} \tag{15.31}$$

(case $m = 0$ implied by (15.7)). Then, with reference to (15.18), the s.c. groups $e^{G_{\alpha_n}t}$ and $e^{G_{\alpha_0}t}$, guaranteed on $S_0 \equiv U_3$ by Theorem 15.6, satisfy:

$$e^{G_{\alpha_n}t} \begin{bmatrix} u_0 \\ u_1 \\ u_2 \end{bmatrix} \to e^{G_{\alpha_0}t} \begin{bmatrix} u_0 \\ u_1 \\ u_2 \end{bmatrix}, \quad \begin{bmatrix} u_0 \\ u_1 \\ u_2 \end{bmatrix} \in U_3 \equiv S_0 \tag{15.32}$$

uniformly in t in bounded intervals; explicitly

$$\|\{u(t;\alpha_n), u_t(t;\alpha_n), u_{tt}(t;\alpha_n)\} - \{u(t;\alpha_0), u_t(t;\alpha_0), u_{tt}(t;\alpha_0)\}\|_{U_3} \to 0, \tag{15.33a}$$

in particular

$$\|u(t;\alpha_n) - u(t;\alpha_0)\|_{\mathscr{D}(\mathscr{A})} \to 0, \quad \|u_t(t;\alpha_n) - u_t(t;\alpha_0)\|_{\mathscr{D}(\mathscr{A}^{\frac{1}{2}})} \to 0 \qquad (15.33b)$$

uniformly in t in bounded intervals.

Proof We shall invoke the approximation Theorem as e.g. in [50, Theorem 4.5, p. 88]. To this end, recalling (15.18) and (15.21) we write by (15.30)

$$G_{\alpha_n} = G_0 + P_{\alpha_n}; \quad G_{\alpha_0} = G_0 + P_0; \text{ on state space } U_3 \equiv S_0 = \mathscr{D}(\mathscr{A}) \times \mathscr{D}(\mathscr{A}^{\frac{1}{2}}) \times \mathscr{H} :$$

$$\|P_{\alpha_n} x\|_{S_0} = \|\alpha_n x_3\|_{\mathscr{H} = L^2(\Omega)} \leq \|\alpha_n\|_{L^\infty(\Omega)} \|x_3\|_{\mathscr{H}} \leq p \|x\|_{S_0} \qquad (15.34)$$

recalling (15.21) and assumption (15.12): $\|\alpha_n\|_{H^m(\Omega)} \leq R$, $2m > dim\Omega$, so that $H^m(\Omega) \hookrightarrow C(\Omega)$ and $\|\alpha_n\|_{C(\Omega)} \leq C\|\alpha_n\|_{H^m(\Omega)} \leq CR$. If $\|e^{G_0 t}\|_{\mathscr{L}(U_3)} \leq M_0 e^{w_0 t}$, then

$$\|e^{G_{\alpha_n} t}\|_{\mathscr{L}(U_3)} \leq M_0 e^{(w_0 + M_0 p)t}, \quad \|P_{\alpha_n}\|_{\mathscr{L}(U_3)} \leq p, \qquad (15.35)$$

(uniformly in n) by [50, Theorem 1.1, p. 76]. Finally we apply [50, Theorem 4.5, p. 88] with D there as $D = S_m$, so that D is dense, then by the definition (15.21) in the state space $U_3 \equiv S_0 = \mathscr{D}(\mathscr{A}) \times \mathscr{D}(\mathscr{A}^{\frac{1}{2}}) \times \mathscr{H}$ as required. Thus, if $x = [x_1, x_2, x_3] \in D$, then we estimate and obtain by (15.30)

$$\|G_{\alpha_n} x - G_{\alpha_0} x\|_{U_3} = \|P_{\alpha_n} x - P_{\alpha_0} x\|_{U_3} = \|(\alpha_n - \alpha_0)x_3\|_{\mathscr{H} = L^2(\Omega)}$$
$$\leq C \|\alpha_n - \alpha_0\|_{L^2(\Omega)} \|x_3\|_{L^\infty(\Omega)} \to 0 \qquad (15.36)$$

since $x_3 \in \mathscr{D}(\mathscr{A}^{\frac{m}{2}}) \hookrightarrow C(\Omega)$ for m as in (15.2). The other assumptions in [50, Theorem 4.5, p. 88] are satisfied and then conclusion (15.32) follows, of which (15.33b) is a restatement via (15.18). □

Theorem 15.8 *Assume hypotheses (15.2) and (15.12). Then on the space $S_m = \mathscr{D}(\mathscr{A}^{\frac{m+2}{2}}) \times \mathscr{D}(\mathscr{A}^{\frac{m+1}{2}}) \times \mathscr{D}(\mathscr{A}^{\frac{m}{2}})$, $2m > dim\Omega$, we have:*

(a) the operator P_{α_n} in (15.21) satisfies

$$\|P_{\alpha_n}\|_{\mathscr{L}(S_m)} \leq R, \quad \text{uniformly in } n; \qquad (15.37)$$

(b) consequently, the s.c. group on S_m generated by $G_{\alpha_n} = G_0 + P_{\alpha_n}$ satisfies for some constants $M_1, \tilde{w}_1 > 0$:

$$\|e^{G_{\alpha_n} t}\|_{\mathscr{L}(S_m)} \leq M_1 e^{\tilde{w}_1 t}, \quad t \geq 0, \quad \text{uniformly in } n. \qquad (15.38)$$

Equivalently, with reference to problem (15.18), $\gamma_n = \alpha_n - \frac{c^2}{b}$,

$$\|\{u(t, \gamma_n), u_t(t, \gamma_n), u_{tt}(t, \gamma_n)\}\|_{S_m} \le C_T \|\{u_0, u_1, u_2\}\|_{S_m}, \quad t \in [0, T];$$
(15.39)

in particular, since $\mathscr{D}(\mathscr{A}^{\frac{m}{2}}) \subset H^m(\Omega) \hookrightarrow C(\Omega)$ under present assumption $2m > \dim\Omega$,

$$\|u(\cdot, \gamma_n)\|_{L^\infty(Q)}, \|u_t(\cdot, \gamma_n)\|_{L^\infty(Q)}, \|u_{tt}(\cdot, \gamma_n)\|_{L^\infty(Q)} \le C_{T, u_0, u_1, u_2, R}.$$
(15.40)

Proof (a) For $x = \{x_1, x_2, x_3\} \in S_m$, we return to (15.21) and estimate

$$\|P_{\alpha_n} x\|_{S_m} = \|\alpha_n x_3\|_{\mathscr{D}(\mathscr{A}^{\frac{m}{2}})} \le \|\alpha_n\|_{M(\mathscr{D}(\mathscr{A}^{\frac{m}{2}}) \to \mathscr{D}(\mathscr{A}^{\frac{m}{2}}))} \|x_3\|_{\mathscr{D}(\mathscr{A}^{\frac{m}{2}})} \quad (15.41)$$

$$= \|\alpha_n\|_{H^m(\Omega)} \|x_3\|_{\mathscr{D}(\mathscr{A}^{\frac{m}{2}})} \le R \|x_3\|_{\mathscr{D}(\mathscr{A}^{\frac{m}{2}})} \quad (15.42)$$

$$\le R \|x\|_{S_m}, \quad \text{uniformly in } n. \quad (15.43)$$

In (15.41), we have invoked assumption (15.2) on α_n, and in going to (15.42) we have invoked consequence (15.5a), from which assumption (15.12) yields (15.43). Thus (15.43) is established.

(b) As in the proof of Theorem 15.7, however now on the space S_m rather than the space S_0, if $\|e^{G_0 t}\|_{\mathscr{L}(S_m)} \le M_1 e^{w_1 t}$, then [50, Theorem 1.1, p. 76] yields via (15.37)

$$\|e^{G_{\alpha_n} t}\|_{\mathscr{L}(S_m)} \le M_1 e^{(w_1 + M_1 R)t}, \quad t \ge 0 \quad (15.44)$$

and (15.38) is established. Returning to problem (15.18), we obtain

$$\left\| \begin{bmatrix} u(t, \gamma_n) \\ u_t(t, \gamma_n) \\ u_{tt}(t, \gamma_n) \end{bmatrix} \right\|_{S_m} = \left\| G_{\alpha_n t} \begin{bmatrix} u_0 \\ u_1 \\ u_2 \end{bmatrix} \right\|_{S_m} \le M_1 e^{\tilde{w}_1 t} \left\| \begin{bmatrix} u_0 \\ u_1 \\ u_2 \end{bmatrix} \right\|_{S_m} \quad (15.45)$$

which yields then the desired uniform bound (15.39).   □

15.3   General Setting: Main Geometrical Assumptions

Following [56, Sect. 5], [25, 35], throughout this paper, we make the following assumptions:

(A.1) Given the triple $\{\Omega, \Gamma_0, \Gamma_1\}$, $\partial\Omega = \overline{\Gamma_0 \cup \Gamma_1}$, there exists a strictly convex (real-valued) non-negative function $d : \overline{\Omega} \to \mathbb{R}^+$, of class $C^3(\overline{\Omega})$, such that, if we introduce the (conservative) vector field $h(x) = [h_1(x), \ldots, h_n(x)] \equiv \nabla d(x)$, $x \in \Omega$, then the following two properties hold true:

$$
\begin{cases}
\dfrac{\partial d}{\partial v}\Big|_{\Gamma_0} = \nabla d \cdot v = h \cdot v = 0 \text{ on } \Gamma_0; \text{ in the Neumann case} & (15.46a) \\[2mm]
\dfrac{\partial d}{\partial v}\Big|_{\Gamma_0} = \nabla d \cdot v = h \cdot v \leq 0 \text{ on } \Gamma_0; \text{ in the Dirichlet case}; & (15.46b)
\end{cases}
$$

(ii) the (symmetric) Hessian matrix $\mathcal{H}_d$ of $d(x)$ [i.e., the Jacobian matrix J_h of $h(x)$] is strictly positive definite on $\overline{\Omega}$: there exists a constant $\rho > 0$ such that for all $x \in \overline{\Omega}$:

$$
\mathcal{H}_d(x) = J_h(x) =
\begin{bmatrix}
d_{x_1 x_1} & \cdots & d_{x_1 x_n} \\
\vdots & & \vdots \\
d_{x_n x_1} & \cdots & d_{x_n x_n}
\end{bmatrix}
=
\begin{bmatrix}
\dfrac{\partial h_1}{\partial x_1} & \cdots & \dfrac{\partial h_1}{\partial x_n} \\
\vdots & & \vdots \\
\dfrac{\partial h_n}{\partial x_1} & \cdots & \dfrac{\partial h_n}{\partial x_n}
\end{bmatrix}
\geq \rho I;
$$

(15.47)

(A.2) $d(x)$ has no critical point on $\overline{\Omega}$:

$$
\inf_{x \in \Omega} |h(x)| = \inf_{x \in \Omega} |\nabla d(x)| = s > 0. \tag{15.48}
$$

Remark 15.6 *Assumption (A.1) =(15.46a) is due to the Neumann B.C. of the hyperbolic problem to follow. It was introduced in [Tr.1, Sect. 5]. For the corresponding Dirichlet problem the condition $h \cdot v = 0$ in Γ_0 in (15.46a) can be relaxed to $h \cdot v \leq 0$ in Γ_0 as in (15.46b). Assumption (A.2) is needed for the validity of the pointwise Carleman estimate in Sect. 15.4 below (it will imply that the constant β be positive, $\beta > 0$, in estimate (15.56)–(15.57) below, [35, (1.1.15b), (4.19)]). Assumption (A.2) can, in effect, be entirely dispensed with [35, Sect. 10] by use of two vector fields. For sake of keeping the exposition simpler, we shall not exploit this substantial generalization. Assumptions (A.1) and (A.2) hold true for large classes of triples $\{\Omega, \Gamma_0, \Gamma_1\}$. One canonical case is that Γ_0 be flat: here then we can take $d(x) = |x - x_0|^2$, with x_0 collocated on the hyperplane containing Γ_0 and outside Ω. Then $h(x) = \nabla d(x) = 2(x - x_0)$ is radial. Another case is where Γ_0 is either convex or concave and subtended by a common point; more precisely see [35, Theorem A.4.1]; in which case, the corresponding required $d(\cdot)$ can also be explicitly constructed. Other classes are given in [35]. See illustrative configurations in the appendix of the present paper.*

15.4 First Basic Step of Proofs: A Carleman Estimate and Continuous Observability/Regularities Inequalities at the $H^1 \times L^2$-level [22, 35]

We recall from [35] a Carleman estimate and a continuous observability inequality (COI), both at the $H^1 \times L^2$-level, that play key roles in the proofs.

Pseudo-Convex Function. [35, p. 230] Having chosen, on the strength of assumption (A.1), a strictly convex potential function $d(x)$ satisfying the preliminary scaling condition $\min_{x \in \overline{\Omega}} d(x) = m > 0$, we next introduce the pseudo-convex function $\varphi(x, t) : \overline{\Omega} \times \mathbb{R} \to \mathbb{R}^+$ of class C^3 by setting for $T > T_0$, as in (15.6), and $0 < a < b$,

$$\varphi(x, t) = d(x) - \frac{a}{b}\left(t - \frac{T}{2}\right)^2; \quad x \in \overline{\Omega}, \ t \in [0, T]; \quad T_0^2 \equiv 4 \max_{x \in \overline{\Omega}} d(x). \tag{15.49a}$$

By (15.49a), with $T > T_0$, there exists $\delta > 0$ such that for a suitable constant $a < b$

$$T^2 > 4 \max_{x \in \overline{\Omega}} d(x) + 4\delta, \quad \text{and thus} \quad \frac{a}{b} T^2 > 4 \max_{x \in \overline{\Omega}} d(x) + 4\delta. \tag{15.49b}$$

Henceforth, let $\varphi(x, t)$ be defined by (15.49a) with T and a chosen as above, unless otherwise explicitly noted. Such function $\varphi(x, t)$ has the following properties [35, p. 230]:

(a) For the constant $\delta > 0$ fixed in (15.49b), we have

$$\begin{cases} \varphi(x, 0) \equiv \varphi(x, T) \leq \max_{x \in \overline{\Omega}} d(x) - \frac{a}{b}\frac{T^2}{4} \leq -\delta \\ \qquad\qquad \text{uniformly in} x \in \overline{\Omega}; \tag{15.50a} \\ \varphi(x, t) \leq \varphi\left(x, \frac{T}{2}\right), \quad \text{for any } t > 0 \text{ and any } x \in \overline{\Omega}. \tag{15.50b} \end{cases}$$

(b) There are t_0 and t_1, with $0 < t_0 < \frac{T}{2} < t_1 < T$, say, chosen symmetrically around $\frac{T}{2}$, such that

$$\min_{x \in \overline{\Omega}, t \in [t_0, t_1]} \varphi(x, t) \geq \sigma, \quad \text{where} \quad 0 < \sigma < m = \min_{x \in \overline{\Omega}} d(x), \tag{15.51}$$

since $\varphi\left(x, \frac{T}{2}\right) = d(x) \geq m > 0$, under present choice. Moreover, let $Q(\sigma)$ be the subset of $\Omega \times [0, T] \equiv Q$ defined by [35, Eq. (1.1.19), p. 232]

$$Q(\sigma) = \{(x, t) : \varphi(x, t) \geq \sigma > 0, x \in \Omega, 0 \leq t \leq T\}. \tag{15.52}$$

The following important property of $Q(\sigma)$ [35, Eq. (1.1.20), p. 232] will be needed later in Sect. 15.5:

$$\Omega \times [t_0, t_1] \subset Q(\sigma) \subset \Omega \times [0, T]. \tag{15.53}$$

Carleman Estimate at the $H^1 \times L^2$-level. We next consider the following general second-order hyperbolic equation

$$y_{tt} - \Delta y = F(y) + f(x,t) \quad \text{in } Q = \Omega \times [0, T] \tag{15.54a}$$

$$F(y) = q_1(x,t)y + q_2(x,t)y_t + q_3(x,t) \cdot \nabla y, \quad f \in L^2(Q) \tag{15.54b}$$

$q_1, q_2, |q_3| \in L^\infty(Q)$, so that

$$|F(y)|^2 \le C_T[y^2 + y_t^2 + |\nabla y|^2], \quad (x,t) \in Q. \tag{15.54c}$$

The y-equation (15.54a) is at first considered without the imposition of boundary conditions. We shall consider initially solutions $y(x,t)$ of (15.54a) in the class

$$y \in H^{2,2}(Q) \equiv L^2(0, T; H^2(\Omega)) \cap H^2(0, T; L^2(\Omega)). \tag{15.55}$$

For these solutions the following Carleman estimate was established in [35, Theorem 5.1, p. 255].

Theorem 15.9 *Assume hypotheses (A.1) and (A.2) in Sect. 15.3. Let $\varphi(x,t)$ be defined by (15.49a) and $f \in L^2(Q)$ as in (15.54b).*

(a) *([35, p. 255]) Let y be a solution of Eq. (15.54a) in the class (15.55). Then, the following one-parameter family of estimates hold true, with $\rho > 0$ as in (15.47), $\beta > 0$ a suitable constant (β is positive by virtue of (A.2), see Remark 15.6), for all $\tau > 0$ sufficiently large, $\epsilon > 0$ small, and $E_y(\cdot)$ defined in (15.59) below:*

$$BT_y|_\Sigma + 2 \int_0^T \int_\Omega e^{2\tau\varphi} |f|^2 dQ + C_{1,T} e^{2\tau\sigma} \int_0^T \int_\Omega y^2 dQ$$

$$\ge C_{1,\tau} \int_Q e^{2\tau\varphi} [y_t^2 + |\nabla y|^2] dQ$$

$$+ C_{2,\tau} \int_{Q(\sigma)} e^{2\tau\varphi} y^2 dxdt - c_T \tau^3 e^{-2\tau\delta} [E_y(0) + E_y(T)] \tag{15.56}$$

$$C_{1,\tau} = \tau\epsilon\rho - 2C_T, \quad C_{2,\tau} = 2\tau^3\beta + \mathcal{O}(\tau^2) - 2C_T, \quad \beta > 0. \tag{15.57}$$

Here $\delta > 0$, $\sigma > 0$ and $\sigma > -\delta$ are the constants in (15.50a), (15.51), C_T, c_T and $C_{1,T}$ are positive constants depending on T as well as d (but not on τ). In addition, the boundary terms $BT_y|_\Sigma$, $\Sigma = \Gamma \times [0, T]$, are given explicitly by

$$BT_y|_\Sigma = 2\tau \int_0^T \int_\Gamma e^{2\tau\varphi} \left(y_t^2 - |\nabla y|^2 \right) h \cdot v \, d\Gamma \, dt$$

$$+ 2\tau \int_0^T \int_\Gamma e^{2\tau\varphi} \left[2\tau^2 \left(|h|^2 - 4c^2 \left(t - \frac{T}{2} \right)^2 \right) \right.$$

$$\left. + \tau(\eta - \Delta d - 2c) \right] y^2 h \cdot v \, d\Gamma \, dt$$

$$+ 8c\tau \int_0^T \int_\Gamma e^{2\tau\varphi} \left(t - \frac{T}{2} \right) y_t \frac{\partial y}{\partial v} \, d\Gamma \, dt$$

$$+ 4\tau \int_0^T \int_\Gamma e^{2\tau\varphi} (h \cdot \nabla y) \frac{\partial y}{\partial v} \, d\Gamma \, dt$$

$$+ 4\tau^2 \int_0^T \int_\Gamma e^{2\tau\varphi} \left[|h|^2 - 4c^2 \left(t - \frac{T}{2} \right)^2 + \frac{\eta}{2\tau} \right] y \frac{\partial y}{\partial v} \, d\Gamma \, dt$$

$$(15.58)$$

where $h(x) = \nabla d(x)$, $\eta(x) = \Delta d(x) - 2c - 1 + k$ for $0 < k < 1$ a constant. Moreover, $Q(\sigma)$ is the set defined in (15.52). The energy function $E_y(t)$ is defined as

$$E_y(t) = \int_\Omega [y^2(x,t) + y_t^2(x,t) + |\nabla y(x,t)|^2] d\Omega. \qquad (15.59)$$

(b) ([35, Theorem 8.2, p. 266]) The validity of the Carleman estimate (15.56)–(15.58) in (a) can be extended to the following class of finite energy solutions of (15.54b)

$$\begin{cases} y \in H^{1,1}(Q) = L^2(0, T; H^1(\Omega)) \cap H^1(0, T; L^2(\Omega)); & (15.60a) \\ y_t \in L^2(0, T; L^2(\Gamma)), \dfrac{\partial y}{\partial v} \in L^2(\Sigma) \equiv L^2(0, T; L^2(\Gamma)). & (15.60b) \end{cases}$$

Neumann B.C.: Continuous Observability Inequality at the $H^1 \times L^2$-level. Consider the following Neumann problem, with F as in (15.54b):

$$\begin{cases} y_{tt}(x,t) - \Delta y(x,t) = F(y) + f(x,t) \text{ in } Q = \Omega \times [0, T]; & (15.61a) \\[2mm] y\left(\cdot, \dfrac{T}{2}\right) = y_0(x); \quad y_t\left(\cdot, \dfrac{T}{2}\right) = y_1(x) \text{ in } \Omega; & (15.61b) \\[2mm] \dfrac{\partial y}{\partial \nu}(x,t)|_\Sigma = 0 \text{ in } \Sigma = \Gamma \times [0, T]. & (15.61c) \end{cases}$$

Theorem 15.10 *[35, Theorem 9.2, p. 269] Assume (A.1), (A.2) of Sect. 15.3. Let f, F satisfy (15.54b). Then, the following continuous observability inequality holds true:*

$$C_T\left(\|y_0\|^2_{H^1(\Omega)} + \|y_1\|^2_{L^2(\Omega)}\right) \le \int_0^T \int_{\Gamma_1} (y^2 + y_t^2) d\Gamma_1 dt + \|f\|^2_{L^2(Q)}, \quad (15.62)$$

wherever the right-hand side is finite. Here $T > T_0$, with T_0 defined by (15.6), and $C_T > 0$ is a positive constant depending on T and Γ_1 is defined by $\Gamma_1 = \Gamma \setminus \Gamma_0$ with Γ_0 in (15.46a).

For the COI with $F \equiv 0$, we also refer to Lasiecka and Triggiani [25].

Dirichlet B.C.: Continuous Observability and Regularity Estimates at the $H^1 \times L^2$-level. Proceeding analogously, we obtain Theorem 15.11 below. Consider the following problem

$$\begin{cases} y_{tt}(x,t) = \Delta y(x,t) + F(y) + f(x,t) \text{ in } Q = \Omega \times [0, T]; & (15.63a) \\[2mm] y\left(\cdot, \dfrac{T}{2}\right) = y_0(x); \quad y_t\left(\cdot, \dfrac{T}{2}\right) = y_1(x) \text{ in } \Omega; & (15.63b) \\[2mm] y(x,t)|_\Sigma = 0 \text{ in } \Sigma = \Gamma \times [0, T], & (15.63c) \end{cases}$$

with initial conditions $y_0 \in H_0^1(\Omega)$, $y_1 \in L^2(\Omega)$. F and f as in (15.54b). Then, its solution satisfies

$$y \in C([0, T]; H_0^1(\Omega)), \quad y_t \in C([0, T]; L^2(\Omega)), \quad \text{a-fortiori} \quad y \in H^{1,1}(Q). \quad (15.64)$$

Theorem 15.11 (Counterpart of [35, Theorem. 9.2, p. 269], [22–24]) *Assume hypothesis (A.1), (A.2) of Sect. 15.3. For problem (15.63a-c), the following continuous observability/regularity inequalities hold true:*

$$c_T\left(\|y_0\|^2_{H_0^1(\Omega)} + \|y_1\|^2_{L^2(\Omega)}\right) \le \int_0^T \int_{\Gamma_1} \left(\frac{\partial y}{\partial \nu}\right)^2 d\Gamma_1\, dt + \|f\|^2_{L^2(Q)}, \quad T > T_0 \quad (15.65)$$

$$\int_0^T \int_{\Gamma_1} \left(\frac{\partial y}{\partial \nu}\right)^2 d\Gamma_1 dt \leq C_T \left(\|y_0\|_{H_0^1(\Omega)}^2 + \|y_1\|_{L^2(\Omega)}^2\right) + \|f\|_{L^1(0,T;L^2(\Omega))}^2, \forall T > 0.$$
(15.66)

Here, T_0 is defined by (15.6) for the first inequality (15.65) (the second inequality (15.66) holds for all $T > 0$); Γ_1, is the controlled or observed portion of the boundary, with $\Gamma_0 = \Gamma \backslash \Gamma_1$ satisfying (3.1_D), and c_T, C_T are positive constants depending on T.

For the regularity inequality (15.66) we refer to Lasiecka and Triggiani [23,24] and [22, Theorem 2.1, p. 151]. For the COI (15.65) we refer, in the present context, also to [5,37,56] with $F \equiv 0$.

Remark 15.7 *The COI (15.62) in the Neumann case and (15.65) in the Dirichlet case, respectively may be interpreted also as follows: if problem (15.61a-c), respectively (15.63a-c) has non-homogeneous forcing term $f \in L^2(Q)$ and Dirichlet boundary traces $y, y_t \in L^2(\Sigma_1)$, respectively Neumann boundary trace $\frac{\partial y}{\partial \nu}|_{\Sigma_1} \in L^2(\Sigma_1)$, then necessarily the I.C. $\{y_0, y_1\}$ must lie in $H^1(\Omega) \times L^2(\Omega)$, respectively $H_0^1(\Omega) \times L^2(\Omega)$. This will be used in Step 3 in Sect. 15.5 below, in connection with the u_{tt}-overdetermined problems.*

15.5 Uniqueness of Linear Inverse Problem for the *u*-problem (15.1a-c): Proof of Theorem 15.1

Step 1. We return to the *z*-mixed problem (15.29) supplemented by the additional assumed B.C. (15.9), that is,

$$\begin{cases} z_{tt} = b\Delta z - \gamma(x)z_t + \frac{c^2}{b}\gamma(x)z - \left(\frac{c^2}{b}\right)^2\gamma(x)u(\gamma) \text{ in } \Omega \times [0, T]; & (15.67a) \\ z\left(\cdot, \frac{T}{2}\right) = \frac{c^2}{b}u_0 + u_1 = 0; \quad z_t\left(\cdot, \frac{T}{2}\right) = \frac{c^2}{b}u_1 + u_2 \text{ in } \Omega; & (15.67b) \\ \frac{\partial z}{\partial \nu}|_{\Sigma} = 0, \quad z|_{\Sigma_1} = 0 \text{ in } \Sigma, \Sigma_1 & (15.67c) \end{cases}$$

under assumptions (15.7), (15.8) and Remark 15.2, which we rewrite for convenience:

$$\begin{cases} \alpha(x), \text{ hence } \gamma(x) \in M(\mathscr{D}(\mathscr{A}^{\frac{m}{2}}) \to \mathscr{D}(\mathscr{A}^{\frac{m}{2}})); \ u, u_t, u_{tt} \in L^\infty(Q) \\ \frac{c^2}{b}u_0 + u_1 = 0; \text{ in particular } \gamma \in L^\infty(\Omega) \text{ (Remark 15.2)} \end{cases}$$
(15.68)

(We recall that a sufficient checkable condition for (15.68) to hold is: $\alpha \in W^{m,\infty}(\Omega)$, $m > \frac{\dim\Omega}{2}$, along with appropriate boundary compatibility conditions). As a consequence, the COI (15.62) applied to the over-determined z-problem (15.67a-c) with $RHS = (\frac{c^2}{b})^2\gamma(x)u(\gamma) \in L^2(Q)$ by (15.68) implies that

$$z_t\left(\cdot, \frac{T}{2}\right) = \frac{c^2}{b}u_1 + u_2 \in L^2(\Omega); \tag{15.69}$$

and then the over-determined problem (15.67a-c) satisfies a-fortiori the properties

$$z \in H^{1,1}(Q) = L^2(0, T; H^1(\Omega)) \cap H^1(0, T; L^2(\Omega)); \quad BT_z|_\Sigma = 0 \tag{15.70}$$

by recalling (15.67a-c) and assumption $h \cdot v = 0$ on Γ_0 by (15.46a) in (15.58) to obtain $BT_z|_\Sigma = 0$. Thus, the Carleman estimates (15.56) are applicable to the z-problem (15.67a-c) with $F(z) = -\gamma(x)z_t + \frac{c^2}{b}\gamma(x)z$ and $f(x,t) = -(\frac{c^2}{b})^2\gamma(x)u(\gamma)$ and yield via (15.70) on $BT_z|_\Sigma = 0$:

$$C_{1,\tau}\int_0^T\int_\Omega e^{2\tau\varphi}[z_t^2 + |\nabla z|^2]dQ + C_{2,\tau}\int_{Q(\sigma)} e^{2\tau\varphi}z^2 dx\,dt$$

$$\leq C_{b,c}\int_0^T\int_\Omega e^{2\tau\varphi}|\gamma u|^2 dQ + C_{T,z}\,e^{2\tau\sigma} + c_T\tau^3 e^{-2\tau\delta}[E_z(0) + E_z(T)], \tag{15.71}$$

where we have set via (15.56) for fixed $z = z(\gamma)$: $C_{T,z} = C_{1,T}\int_0^T\int_\Omega z^2 dQ$.

Step 2. In this step, we differentiate in time the z-mixed problem (15.67a-c). Thus obtaining by use of the IC, hence $\Delta z(\cdot, \frac{T}{2}) = 0$, $u(\cdot, \frac{T}{2}) = u_0$:

$$\begin{cases} (z_t)_{tt} = b\Delta(z_t) - \gamma(x)(z_t)_t + \dfrac{c^2}{b}\gamma(x)z_t \\[2mm] \qquad -(\dfrac{c^2}{b})^2\gamma(x)u_t(\gamma) \text{ in } \Omega \times [0, T]; & (15.72a) \\[3mm] z_t(\cdot, \dfrac{T}{2}) = \dfrac{c^2}{b}u_1 + u_2 \text{ in } \Omega; & (15.72b) \\[3mm] (z_t)_t\left(\cdot, \dfrac{T}{2}\right) = -\gamma(x)\left[\dfrac{c^2}{b}(\dfrac{c^2}{b}u_0 + u_1) + u_2\right] \\[2mm] \qquad = -\gamma(x)u_2 \text{ in } \Omega; & (15.72c) \\[3mm] \dfrac{\partial(z_t)}{\partial v}\bigg|_\Sigma = 0, \quad z_t|_{\Sigma_1} = 0 \text{ in } \Sigma, \ \Sigma_1. & (15.72d) \end{cases}$$

This was noted in [47, Eq. (5.24), p. 850]. Eq. (15.72c) uses (15.8). Again the COI (15.62) applied to the over-determined problem (15.72a-d) with the *RHS* = $-(\frac{c^2}{b})^2\gamma(x)u_t(\gamma) \in L^2(Q)$ by (15.68) implies now for the I.C. that

$$\frac{c^2}{b}u_1 + u_2 \in H^1(\Omega); \quad \gamma(x)u_2 \in L^2(\Omega) \tag{15.73}$$

so that the over-determined problem (15.72a-d) satisfies the properties, counterpart of (15.70):

$$z_t \in H^{1,1}(Q); \quad BT_{z_t}|_\Sigma = 0. \tag{15.74}$$

Thus, the Carleman estimates (15.56) are applicable to the (z_t)-problem (15.72a-d) with $F(z_t) = -\gamma(x)(z_t)_t + \frac{c^2}{b}\gamma(x)z_t$ and $f(x,t) = -(\frac{c^2}{b})^2\gamma(x)u_t(\gamma)$ this time and yield the counterpart of (15.71):

$$C_{1,\tau}\int_0^T\int_\Omega e^{2\tau\varphi}[z_{tt}^2 + |\nabla z_t|^2]\,dQ + C_{2,\tau}\int_{Q(\sigma)}e^{2\tau\varphi}z_t^2\,dx\,dt$$

$$\leq C_{b,c}\int_0^T\int_\Omega e^{2\tau\varphi}|\gamma u_t|^2 dQ + C_{T,z_t}\,e^{2\tau\sigma} + c_T\tau^3 e^{-2\tau\delta}[E_{z_t}(0) + E_{z_t}(T)]. \tag{15.75}$$

Step 3. In this step, we differentiate in time, the z_t-mixed problem (15.72a-d) one more time, thus obtaining after two cancellations to obtain:

$$\begin{cases}
(z_{tt})_{tt} = b\Delta(z_{tt}) - \gamma(x)(z_{tt})_t + \dfrac{c^2}{b}\gamma(x)z_{tt} \\[4pt]
\qquad\qquad -(\dfrac{c^2}{b})^2\gamma(x)u_{tt}(\gamma) \text{ in } Q; & (15.76a) \\[10pt]
(z_{tt})(\cdot, \dfrac{T}{2}) = -\gamma(x)u_2 \text{ in } \Omega; & (15.76b) \\[10pt]
(z_{tt})_t\left(\cdot, \dfrac{T}{2}\right) = b\Delta(\dfrac{c^2}{b}u_1 + u_2) - \gamma(x)u_3(\gamma) \text{ in } \Omega; & (15.76c) \\[10pt]
\dfrac{\partial(z_{tt})}{\partial\nu}|_\Sigma = 0, \quad z_{tt}|_{\Sigma_1} = 0 \text{ in } \Sigma, \ \Sigma_1; & (15.76d) \\[10pt]
z_{ttt}(\cdot, \dfrac{T}{2}) = b\Delta(\dfrac{c^2}{b}u_1 + u_2) + \gamma(\gamma u_2 + \dfrac{c^2}{b}u_2). & (15.76e)
\end{cases}$$

To obtain (15.76e) we have used $u_{ttt}(\cdot, \frac{T}{2}) = u_3(\gamma) = -(\gamma + \frac{c^2}{b})u_2 + b\Delta(\frac{c^2}{b}u_0 + u_1)$ from (1.1) with $\tau = 1$, evaluated at $t = \frac{T}{2}$, and (15.8). Again the COI (15.62) applied to the over-determined problem (15.76a-d), with

$RHS = -(\frac{c^2}{b})^2 \gamma(x) u_{tt}(\gamma) \in L^2(Q)$ by (15.68) implies for the I.C., hence for the solution that

$$\left\{ (z_{tt}\left(\cdot, \frac{T}{2}\right), (z_{tt})_t \left(\cdot, \frac{T}{2}\right) \right\} \in H^1(\Omega) \times L^2(\Omega); \ z_{tt} \in H^{1,1}(Q), \quad BT_{z_{tt}}|_\Sigma = 0,$$
(15.77)

counterpart of (15.74). Thus, the Carleman estimates (15.56) are applicable to the (z_{tt})-problem (15.76a-d) with $F(z_{tt}) = -\gamma(x)(z_{tt})_t + \frac{c^2}{b}\gamma(x)z_{tt}$ and $f(x,t) = -(\frac{c^2}{b})^2 \gamma(x) u_{tt}(\gamma)$ this time and yield the counterpart of (15.71), (15.75):

$$C_{1,\tau} \int_0^T \int_\Omega e^{2\tau\varphi}[z_{ttt}^2 + |\nabla z_{tt}|^2]dQ + C_{2,\tau} \int_{Q(\sigma)} e^{2\tau\varphi} z_{tt}^2 \, dx \, dt$$

$$\leq C_{b,c} \int_0^T \int_\Omega e^{2\tau\varphi}|\gamma u_{tt}|^2 dQ + C_{T,z_{tt}} e^{2\tau\sigma} + c_T \tau^3 e^{-2\tau\delta}[E_{z_{tt}}(0) + E_{z_{tt}}(T)].$$
(15.78)

Step 4.   Adding up (15.71), (15.75), (15.78) together yields the combined inequality

$$C_{1,\tau} \int_Q e^{2\tau\varphi} \left[z_t^2 + z_{tt}^2 + z_{ttt}^2 + |\nabla z|^2 + |\nabla z_t|^2 + |\nabla z_{tt}|^2 \right] dQ$$

$$+ C_{2,\tau} \int_{Q(\sigma)} e^{2\tau\varphi}[z^2 + z_t^2 + z_{tt}^2]dx\,dt$$

$$\leq C_{b,c} \int_0^T \int_\Omega e^{2\tau\varphi} \left[|\gamma u|^2 + |\gamma u_t|^2 + |\gamma u_{tt}|^2 \right] dQ + [C_{T,z} + C_{T,z_t} + C_{T,z_{tt}}] e^{2\tau\sigma}$$

$$+ c_T \tau^3 e^{-2\tau\delta} [E_z(0) + E_{z_t}(0) + E_{z_{tt}}(0) + E_z(T) + E_{z_t}(T) + E_{z_{tt}}(T)].$$
(15.79)

We note that the energy terms $E_{z_t}(0)$, $E_{z_{tt}}(0)$, $E_{z_t}(T)$ and $E_{z_{tt}}(T)$ above in (15.79) actually also depend on γ. Next, we invoke one more time properties $u, u_t, u_{tt} \in L^\infty(Q)$ from (15.68) (already critically used to arrive at (15.79)):

$$|\gamma(x)u| \leq \|u(\gamma)\|_{L^\infty(Q)}|\gamma(x)|; \ |\gamma(x)u_t| \leq \|u_t(\gamma)\|_{L^\infty(Q)}|\gamma(x)|;$$

$$|\gamma(x)u_{tt}| \leq \|u_{tt}(\gamma)\|_{L^\infty(Q)}|\gamma(x)| \qquad (15.80)$$

in Q. Using (15.80) in (15.79) we arrive at

$$C_{1,\tau} \int_Q e^{2\tau\varphi} \left[z_t^2 + z_{tt}^2 + z_{ttt}^2 + |\nabla z|^2 + |\nabla z_t|^2 + |\nabla z_{tt}|^2 \right] dQ$$

$$+ C_{2,\tau} \int_{Q(\sigma)} e^{2\tau\varphi} \left[z^2 + z_t^2 + z_{tt}^2 \right] dx\, dt \leq \tilde{C}_{b,c,u,T,z} \left\{ \int_Q e^{2\tau\varphi} |\gamma|^2 dQ + e^{2\tau\sigma} \right.$$

$$\left. + \tau^3 e^{-2\tau\delta} [E_z(0) + E_{z_t}(0) + E_{z_{tt}}(0) + E_z(T) + E_{z_t}(T) + E_{z_{tt}}(T)] \right\}$$
(15.81)

where $\tilde{C}_{b,c,u,T,z}$ is a positive constant depending on b, c, T, z, u, hence on γ.

Step 5. In this step, we follow the idea of a strategy proposed in [9, Theorem 8.2.2, p. 231]. We evaluate (15.67a) at the initial time $t = \frac{T}{2}$, and (for the first time) use the vanishing $z(\cdot, \frac{T}{2}) = 0$, to obtain via the positivity hypothesis in (15.8) on $u_{tt}(\cdot, \frac{T}{2})$:

$$\left| z_{tt} \left(\cdot, \frac{T}{2} \right) \right| = |\gamma(x) u_2| \geq r_0 |\gamma(x)|,$$
(15.82)

$$|\gamma(x)| \leq \frac{1}{r_0} \left| z_{tt} \left(x, \frac{T}{2} \right) \right| \quad x \in \overline{\Omega}.$$
(15.83)

Claim: Using (15.83) in the first integral term on the RHS of (15.81) yields,

$$\int_Q e^{2\tau\varphi} |\gamma|^2 dQ = \int_0^T \int_\Omega e^{2\tau\varphi} |\gamma|^2 d\Omega\, dt \leq \frac{T}{r_0^2} \int_\Omega |z_{tt}(x,0)|^2 d\Omega$$

$$+ \frac{T}{r_0^2} (2c\tau T + 1) \int_\Omega \int_0^{T/2} e^{2\tau\varphi(x,s)} |z_{tt}(x,s)|^2 ds\, d\Omega$$

$$+ \frac{T}{r_0^2} \int_\Omega \int_0^{T/2} e^{2\tau\varphi(x,s)} |z_{ttt}(x,s)|^2 ds\, d\Omega.$$
(15.84)

A proof of (15.84) can be found, for example, in [40, (4.22), p. 1648] or in [46, Eq. (4.41)] and we omit the details here.

Step 6. We substitute (15.84) for the first integral term on the RHS of (15.81) and obtain, after obvious majorizations,

$$C_{1,\tau} \int_Q e^{2\tau\varphi} \left[z_t^2 + z_{tt}^2 + z_{ttt}^2 + |\nabla z|^2 + |\nabla z_t|^2 + |\nabla z_{tt}|^2 \right] dQ$$

$$+ C_{2,\tau} \int_{Q(\sigma)} e^{2\tau\varphi} \left[z^2 + z_t^2 + z_{tt}^2 \right] dx\, dt$$

$$\leq \tilde{C}_{b,c,u,T,z}\left\{\left[\left(\frac{T}{r_0^2}\right)(2Tc\tau+1)\int_Q e^{2\tau\varphi}|z_{tt}|^2 dQ\right.\right.$$

$$\left.+\int_Q e^{2\tau\varphi}|z_{ttt}|^2 dQ + \int_\Omega |z_{tt}(x,0)|^2\right]$$

$$\left.+ e^{2\tau\sigma} + \tau^3 e^{-2\tau\delta}[E_z(0) + E_{z_t}(0) + E_{z_{tt}}(0) + E_z(T) + E_{z_t}(T) + E_{z_{tt}}(T)]\right\}.$$

$$(15.85)$$

We note again that the energy terms $E_{z_t}(0)$, $E_{z_{tt}}(0)$, $E_{z_t}(T)$ and $E_{z_{tt}}(T)$ above in (15.85) actually also depend on γ.

Step 7.   From here, the proof proceeds as in past cases (see e.g. [40, Step 7, 8, p. 1650] or [46, Eq. (4.46) through (4.53)]) using $e^{2\tau\varphi} < e^{2\tau\sigma}$ on $Q \setminus Q(\sigma)$ by (15.52), the property $C_{1,\tau} \sim \tau$, $C_{2,\tau} \sim \tau^3$ in (15.57) for large enough τ to have

$$\tau^2 \int_{Q(\sigma)} [z^2 + z_t^2 + z_{tt}^2]dxdt \leq C_{data},$$

where C_{data} is a constant only depend on the problem data but not on τ. Therefore by taking $\tau \to \infty$ we finally obtain first

$$z = z_t = z_{tt} \equiv 0 \text{ in } Q(\sigma). \tag{15.86}$$

Finally, invoking the property (15.53) that $\Omega \times [t_0, t_1] \subset Q(\sigma) \subset Q$, and that $\frac{T}{2} \in [t_0, t_1]$, we obtain in particular from (15.72c) or (15.76a-b) and (15.86) that

$$z_{tt}\left(x, \frac{T}{2}\right) = -\gamma(x)u_{tt}\left(x, \frac{T}{2}\right) \equiv 0, \quad \text{for all } x \in \Omega. \tag{15.87}$$

Recalling for the second time that by (15.8) $|u_{tt}(x, \frac{T}{2})| \geq r_0, x \in \Omega$, we conclude that

$$\gamma(x) \equiv 0 \quad \text{i.e.} \quad \alpha(x) \equiv \frac{c^2}{b} \text{ in } \Omega \tag{15.88}$$

as desired. Thus, with this step the proof of Theorem 15.1 is complete.   □

Remark 15.8 *(Variations of the above proof of Theorem 15.1 to obtain the proof of Theorem 15.3). In Step 1, the counterpart of problem (15.67a-c) is now:*

$$
\begin{cases}
z_{tt} = b\Delta z - \gamma(x)z_t + \dfrac{c^2}{b}\gamma(x)z - (\dfrac{c^2}{b})^2\gamma(x)u(\gamma) \text{ in } \Omega \times [0, T]; & (15.89a) \\[2mm]
z(\cdot, \dfrac{T}{2}) = \dfrac{c^2}{b}u_0 + u_1 = 0; \quad z_t\left(\cdot, \dfrac{T}{2}\right) = \dfrac{c^2}{b}u_1 + u_2 \text{ in } \Omega; & (15.89b) \\[2mm]
z|_\Sigma = 0, \quad \dfrac{\partial z}{\partial \nu}|_{\Sigma_1} = 0 \text{ in } \Sigma, \ \Sigma_1. & (15.89c)
\end{cases}
$$

Next, instead of the COI (15.62) for the Neumann case, one invokes the COI (15.65) of the Dirichlet case, as applied to problem (15.89a-c) and thus obtains the same conclusion (15.70). Consequently, the same regularity $z \in H^{1,1}(Q)$ as in (15.70) holds true now, while the boundary terms now satisfies $BT_z|_\Sigma \leq 0$, which is the inequality in the right direction while invoking the Carleman estimate. This then yields the same inequality (15.71). Similarly, in subsequent steps, one obtains: $BT_{z_t}|_\Sigma \leq 0$, $BT_{z_{tt}}|_\Sigma \leq 0$, rather than the conditions $BT_{z_t}|_\Sigma = 0$, $BT_{z_{tt}}|_\Sigma = 0$, as in (15.74) and (15.75). No other changes are needed.

15.6 Stability of Inverse Problem for the u-problem (15.1a-d). Proof of Theorem 15.2

Step 1. Let $u = u(\gamma)$ be the solution of problem in (15.1a-d) and let $z = z(\gamma) = \frac{c^2}{b}u + u_t$ be the corresponding solution of problem (15.29), subject to assumption (15.8) on $z(\cdot, \frac{T}{2}) = \frac{c^2}{b}u_0 + u_1 = 0$, that is,

$$
\begin{cases}
z_{tt} = b\Delta z - \gamma(x)z_t + \dfrac{c^2}{b}\gamma(x)z - (\dfrac{c^2}{b})^2\gamma(x)u(\gamma) \text{ in } Q; & (15.90a) \\[2mm]
z\left(\cdot, \dfrac{T}{2}\right) = 0; \quad z_t\left(\cdot, \dfrac{T}{2}\right) = \dfrac{c^2}{b}u_1 + u_0 \text{ in } \Omega; & (15.90b) \\[2mm]
\dfrac{\partial z}{\partial \nu}|_\Sigma = 0 \text{ in } \Sigma & (15.90c)
\end{cases}
$$

to be repeatedly invoked below. Here $\gamma \in L^\infty(\Omega)$, a-fortiori, see Remark 15.2. We set

$$
v = v(\gamma) = z_t(\gamma)
$$

so that v satisfies the z_t-problem obtained from differentiating problem (15.90a-c), that is (this is (15.72a-d) without the over-determined B.C.)

$$
\begin{cases}
v_{tt} = b\Delta v - \gamma(x)v_t + \dfrac{c^2}{b}\gamma(x)v - (\dfrac{c^2}{b})^2\gamma(x)u_t(\gamma) \text{ in } Q; & (15.91a) \\[2ex]
v\left(\cdot, \dfrac{T}{2}\right) = \dfrac{c^2}{b}u_1 + u_2; \quad v_t\left(\cdot, \dfrac{T}{2}\right) = -\gamma(x)u_2(x) \text{ in } \Omega; & (15.91b) \\[2ex]
\dfrac{\partial v}{\partial \nu}\Big|_\Sigma = 0 \text{ in } \Sigma. & (15.91c)
\end{cases}
$$

By linearity we split v into two components

$$v = \psi + \zeta \tag{15.92}$$

where ψ satisfies the same problem as v, however with homogeneous 'forcing term'

$$
\begin{cases}
psi_{tt} = b\Delta\psi - \gamma(x)\psi_t + \dfrac{c^2}{b}\gamma(x)\psi \text{ in } Q & (15.93a) \\[2ex]
\psi\left(\cdot, \dfrac{T}{2}\right) = v\left(\cdot, \dfrac{T}{2}\right) = \dfrac{c^2}{b}u_1 + u_2 \text{ in } \Omega & (15.93b) \\[2ex]
\psi_t\left(\cdot, \dfrac{T}{2}\right) = v_t\left(\cdot, \dfrac{T}{2}\right) = -\gamma(x)u_2(x) \text{ in } \Omega & (15.93c) \\[2ex]
\dfrac{\partial \psi}{\partial \nu}\Big|_\Sigma = 0 \text{ in } \Sigma & (15.93d)
\end{cases}
$$

where ζ satisfies the same problem as v, with homogeneous initial conditions

$$
\begin{cases}
\zeta_{tt} = b\Delta\zeta - \gamma(x)\zeta_t + \dfrac{c^2}{b}\gamma(x)\zeta - (\dfrac{c^2}{b})^2\gamma(x)u_t(\gamma) \text{ in } Q & (15.94a) \\[2ex]
\zeta\left(\cdot, \dfrac{T}{2}\right) = 0; \quad \zeta_t\left(\cdot, \dfrac{T}{2}\right) = 0 \text{ in } \Omega & (15.94b) \\[2ex]
\dfrac{\partial \zeta}{\partial \nu}\Big|_\Sigma = 0 \text{ in } \Sigma. & (15.94c)
\end{cases}
$$

Remark 15.9 *With I.C. for the ψ-problem as assumed in (15.7), we can apply the COI (15.62) and obtain a-fortiori: there exists a constant $C_T > 0$ (independent of ψ) such that*

$$
r_0^2\|\gamma\|_{L^2(\Omega)}^2 \leq \|\gamma(\cdot)u_2(\cdot)\|_{L^2(\Omega)}^2 \leq C_T^2 \int_0^T \int_{\Gamma_1} [\psi^2 + \psi_t^2]\, d\Sigma_1 \tag{15.95}
$$

(wherever the RHS is finite), where in the LHS of (15.95) we have recalled the positivity assumption (15.8). Recalling (15.92) in (15.95), we then obtain by the triangle inequality, with $C_{T,r_0} = \frac{C_T}{r_0}$, via also (15.92):

$$\|\gamma\|_{L^2(\Omega)}$$

$$\leq C_{T,r_0}\left(\|\psi\|_{L^2(\Gamma_1\times[0,T])} + \|\psi_t\|_{L^2(\Gamma_1\times[0,T])}\right)$$

$$\leq C_{T,r_0}\left(\|v-\zeta\|_{L^2(\Gamma_1\times[0,T])} + \|v_t-\zeta_t\|_{L^2(\Gamma_1\times[0,T])}\right)$$

$$\leq C_{T,r_0}\left(\|v\|_{L^2(\Gamma_1\times[0,T])} + \|v_t\|_{L^2(\Gamma_1\times[0,T])} + \|\zeta\|_{L^2(\Gamma_1\times[0,T])} + \|\zeta_t\|_{L^2(\Gamma_1\times[0,T])}\right)$$

$$\leq C_{T,r_0}\left(\|z_t\|_{L^2(\Gamma_1\times[0,T])} + \|z_{tt}\|_{L^2(\Gamma_1\times[0,T])} + \|\zeta\|_{L^2(\Gamma_1\times[0,T])} + \|\zeta_t\|_{L^2(\Gamma_1\times[0,T])}\right).$$

$$(15.96)$$

Via (15.23), inequality (15.96) is the desired, sought-after stability esti- mate (15.13) of Theorem 15.2, modulo (polluted by) the ζ- and ζ_t-terms. Such terms, which are nonlinear in $\gamma(\cdot)$, will be next omitted by a compactness- uniqueness argument, via a direct proof, in Proposition 15.1 below, of the corresponding estimate (15.96) without such terms ζ and ζ_t. To carry this through we need the following lemma.

Step 2. Lemma 15.12 *Consider the ζ-system (15.94a-c) with data $\gamma(x) \in L^\infty(\Omega)$, see Remark 15.1, $u_t, u_{tt} \in L^\infty(Q)$ as showed in Remark 15.3 as a consequence of assumptions (15.2), (15.7). Define the following nonlinear operators K and K_1*

$$(K\gamma)(x,t) = \zeta(x,t)|_{\Sigma_1} : L^2(\Omega) \to L^2(\Gamma_1 \times [0,T]) \qquad (15.97)$$

$$(K_1\gamma)(x,t) = \zeta_t(x,t)|_{\Sigma_1} : L^2(\Omega) \to L^2(\Gamma_1 \times [0,T]) \qquad (15.98)$$

where ζ is the unique solution of problem (15.94a-c) depending nonlinearly in $\gamma(\cdot)$. Then

$$\text{both } K \text{ and } K_1 \text{ are compact operators.} \qquad (15.99)$$

Proof Preliminaries.

(a) We shall invoke sharp (Dirichlet) trace theory results [26, 27], reproduced in Appendix 2, for the Neumann hyperbolic problem (15.94a-c). More precisely, regarding the ζ-problem (15.94a-c) with zero I.C. and forcing term $(\frac{c^2}{b})^2\gamma(x)u_t(\gamma)$, the following Dirichlet trace result hold true:

From assumptions $\gamma(x) \in L^2(\Omega)$, $u_t \in L^\infty(Q)$ as implied from the assumption (15.7) = (15.68), we get via part II of Theorem, Appendix 2:

$$\left(\frac{c^2}{b}\right)^2 \gamma(x)u_t(x,t) \in L^2(Q) \Rightarrow \zeta|_\Sigma \in H^\beta(\Sigma) \text{ continuously;}$$

(15.100)

(b) similarly, we consider now the ζ_t-problem, obtained from differentiating problem (15.94a-c) in time, thus with RHS $= -(\frac{c^2}{b})^2 \gamma(x)u_{tt}(\gamma)$, $\zeta_t(\cdot, \frac{T}{2}) = 0$, and $(\zeta_t)_t(\cdot, \frac{T}{2}) = -(\frac{c^2}{b})^2 \gamma(x)u_2 \in L^2(\Omega)$. With $\gamma(x) \in L^2(\Omega)$ and $u_t, u_{tt} \in L^\infty(Q)$ as implied from (15.7) = (15.68), and

$$(\frac{c^2}{b})^2\gamma(x)u_{tt}(x,t) \in L^2(Q), \ (\frac{c^2}{b})^2\gamma(x)u_t(x,t) \in H^1(0,T;L^2(\Omega)),$$

we get

$$\gamma u_2 \in L^2(\Omega), \ \left(\frac{c^2}{b}\right)^2 \gamma(x)u_t(x,t) \in H^1(0,T;L^2(\Omega))$$

$$\Rightarrow D_t^1\zeta|_\Sigma = \zeta_t|_\Sigma \in H^\beta(\Sigma),$$

(15.101)

continuously with β the following constant [26, 27]:

$$\beta = \frac{3}{5}, \text{ for a general } \Omega; \ \beta = \frac{2}{3}, \text{ if } \Omega \text{ is a sphere;}$$

$$\beta = \frac{3}{4} - \epsilon, \text{ if } \Omega \text{ is a parallelepiped.}$$

(15.102)

After these preliminaries, we can now draw the desired conclusions on the compactness of the operators K and K_1 defined in (15.97) and (15.98);

(c) *Compactness of K.* According to (15.100), it suffices to have $u_t \in L^\infty(Q)$ in order to have that the nonlinear map

$$\gamma \in L^2(\Omega) \to K\gamma|_\Sigma = \zeta|_\Sigma \in H^{\beta-\epsilon}(\Sigma) \text{ is compact,}$$

(15.103)

$\forall \ \epsilon > 0$ sufficiently small, for then $(\frac{c^2}{b})^2\gamma(x)u_t(x,t) \in L^2(Q)$ as required, by (15.100).

Compactness of K_1. According to (15.101), it suffices to have $u_{tt} \in L^\infty(Q)$ in order to have that the nonlinear map

$$\gamma \in L^2(\Omega) \to K_1\gamma|_\Sigma = \zeta_t|_\Sigma \in H^{\beta-\epsilon}(\Sigma) \text{ is compact,}$$

(15.104)

$\forall \epsilon > 0$, sufficiently small, for then $(\frac{c^2}{b})^2 \gamma(x) u_{tt}(x,t) \in L^2(Q)$ as required by (15.101). Hence Lemma 15.12 is proved. $\square$

Step 3.   Lemma 15.12 will allow us to absorb (omit) the terms (nonlinear in $\gamma(\cdot)$)

$$\|K\gamma = \zeta\|_{L^2(\Gamma_1 \times [0,T])} \quad \text{and} \quad \|K_1\gamma = \zeta_t\|_{L^2(\Gamma_1 \times [0,T])} \tag{15.105}$$

in the RHS of estimate (15.96) by a compactness-uniqueness argument, as noted below (15.96).

Proposition 15.1 *Consider the z-problem (15.29) with $T > T_0$ as in (15.6) under assumptions (15.7), (15.8), so that estimate (15.96) as well as Lemma 15.12 hold true. Then, the terms $K\gamma = \zeta|_{\Sigma_1}$ and $K_1\gamma = \zeta_t|_{\Sigma_1}$ measured in the $L^2(\Gamma_1 \times [0,T])$-norm can be omitted from the RHS of inequality (15.96) (for a suitable constant $C_{T,data}$), so that the desired conclusion, equation (15.13) of Theorem 15.2, holds true*

$$\|\gamma\|^2_{L^2(\Omega)} \leq C_{T,data} \left\{ \int_0^T \int_{\Gamma_1} [z_t^2 + z_{tt}^2] d\Gamma_1 dt \right\}$$

$$\leq C_{T,data} \left\{ \int_0^T \int_{\Gamma_1} \left[\left(\frac{c^2}{b} u_t + u_{tt} \right)^2 + \left(\frac{c^2}{b} u_{tt} + u_{ttt} \right)^2 \right] d\Gamma_1 dt \right\}$$

$$\tag{15.106}$$

for all $\gamma \in L^2(\Omega)$, with $C_{T,data}$ independent of γ.

Proof Step (i).   Suppose, by contradiction, that inequality (15.106) is false. Then, there exists a sequence $\{\gamma_n\}_{n=1}^\infty$, $\gamma_n \in L^2(\Omega)$, such that

$$
\begin{cases}
\text{(i)} \quad \|\gamma_n\|_{L^2(\Omega)} \equiv 1, \quad n = 1, 2, \ldots; & \tag{15.107a} \\
\text{(ii)} \quad \lim_{n\to\infty} \left(\|z_t(\gamma_n)\|_{L^2(\Gamma_1 \times [0,T])} + \|z_{tt}(\gamma_n)\|_{L^2(\Gamma_1 \times [0,T])} \right) = 0, & \tag{15.107b}
\end{cases}
$$

where $z(\gamma_n)$ solves problem (15.90a-c) with $\gamma = \gamma_n$:

$$
\begin{cases}
(z(\gamma_n))_{tt} = b\Delta z(\gamma_n) - \gamma_n(x) z_t(\gamma_n) & \tag{15.108a} \\
\qquad + \frac{c^2}{b} \gamma_n(x) z(\gamma_n) - (\frac{c^2}{b})^2 \gamma_n(x) u(\gamma_n) \text{ in } Q; & \tag{15.108b} \\
z(\gamma_n)(\cdot, \frac{T}{2}) = 0; \quad (z(\gamma_n))_t(\cdot, \frac{T}{2}) = \frac{c^2}{b} u_1 + u_2 \text{ in } \Omega; & \tag{15.108c} \\
\dfrac{\partial z(\gamma_n)(x,t)}{\partial \nu}\Big|_{\Sigma} = 0 \text{ in } \Sigma. & \tag{15.108d}
\end{cases}
$$

In view of (15.107a), there exists a subsequence, still denoted by γ_n, such that:

$$\gamma_n \text{ converges weakly in } L^2(\Omega) \text{ to some } \gamma_0 \in L^2(\Omega). \tag{15.109}$$

Moreover, since the (nonlinear) operators K and K_1 are both compact by the Lemma 15.12, it then follows by (15.109) that we have strong convergence

$$\lim_{m,n\to+\infty} \|K\gamma_n - K\gamma_m\|_{L^2(\Gamma_1\times[0,T])} = 0; \tag{15.110a}$$

$$\lim_{m,n\to+\infty} \|K_1\gamma_n - K_1\gamma_m\|_{L^2(\Gamma_1\times[0,T])} = 0. \tag{15.110b}$$

Step (ii).   On the other hand, consider the $z(\gamma_n)$ and $z(\gamma_m)$ problems corresponding to (15.108a-d) and define the difference of their corresponding time derivatives

$$w = z_t(\gamma_n) - z_t(\gamma_m) = v(\gamma_n) - v(\gamma_m). \tag{15.111}$$

Then w in (15.111) satisfies the following problem via (15.91a-c) on $v = z_t$:

$$
\begin{cases}
w_{tt} = b\Delta w - \gamma_n w_t + \dfrac{c^2}{b}\gamma_n w - (\gamma_n - \gamma_m)(z_t)_t(\gamma_m) & (15.112a) \\[2mm]
\quad + \dfrac{c^2}{b}(\gamma_n - \gamma_m)z_t(\gamma_m) \\[2mm]
\quad -(\dfrac{c^2}{b})^2(\gamma_n u_t(\gamma_n) - \gamma_m u_t(\gamma_m)) \text{ in } Q; & (15.112b) \\[2mm]
w(\cdot, \dfrac{T}{2}) = 0; \quad w_t(\cdot, \dfrac{T}{2}) = -(\gamma_n(x) - \gamma_m(x))u_2(x) \text{ in } \Omega; & (15.112c) \\[2mm]
\dfrac{\partial w(x,t)}{\partial \nu}\Big|_\Sigma = 0 \text{ in } \Sigma. & (15.112d)
\end{cases}
$$

As in Step 1 we then decompose w into components $w = w_1 + w_2$ where w_1 satisfies the same problem as w with homogeneous forcing term

$$
\begin{cases}
w_{1tt} = b\Delta w_1 - \gamma_n(w_1)_t + \dfrac{c^2}{b}\gamma_n w_1 \text{ in } Q; & (15.113a) \\[2mm]
w_1(\cdot, \dfrac{T}{2}) = 0; \quad w_{1t}(\cdot, \dfrac{T}{2}) = -(\gamma_n(x) - \gamma_m(x))u_2(x) \text{ in } \Omega; & (15.113b) \\[2mm]
\dfrac{\partial w_1(x,t)}{\partial \nu}\Big|_\Sigma = 0 \text{ in } \Sigma & (15.113c)
\end{cases}
$$

while w_2 satisfies the same problem as w, with homogeneous initial conditions

$$
\begin{cases}
w_{2tt} = b \Delta w_2 - \gamma_n (w_2)_t + \dfrac{c^2}{b} \gamma_n w_2 - (\gamma_n - \gamma_m)(z_t)_t (\gamma_m) & \text{(15.114a)} \\[2ex]
\quad + \dfrac{c^2}{b} (\gamma_n - \gamma_m) z_t (\gamma_m) & \\[2ex]
\quad - (\dfrac{c^2}{b})^2 (\gamma_n u_t (\gamma_n) - \gamma_m u_t (\gamma_m)) \text{ in } Q; & \text{(15.114b)} \\[2ex]
w_2 (\cdot, \dfrac{T}{2}) = w_{2t} (\cdot, \dfrac{T}{2}) = 0 \text{ in } \Omega; & \text{(15.114c)} \\[2ex]
\dfrac{\partial w_2 (x, t)}{\partial \nu} \big|_\Sigma = 0 \text{ in } \Sigma. & \text{(15.114d)}
\end{cases}
$$

Again we apply the COI (15.62) for the w_1-problem and get by the positivity assumption (15.8)

$$
\| \gamma_n - \gamma_m \|^2_{L^2(\Omega)} \leq C^2_{T, r_0} \int_0^T \int_{\Gamma_1} [w_1^2 + w_{1t}^2] d\Sigma_1. \tag{15.115}
$$

On the other hand, we readily see by comparing the ζ-problem (15.94a-c) with the w_2-problem (15.114a-d)

$$
\begin{cases}
w_2 = \zeta(\gamma_n) - \zeta(\gamma_m); \; w_2|_{\Sigma_1} = \zeta(\gamma_n)|_{\Sigma_1} - \zeta(\gamma_m)|_{\Sigma_1} & \text{(15.116a)} \\[1ex]
w_2|_{\Sigma_1} = K \gamma_n - K \gamma_m; \; w_{2t}|_{\Sigma_1} = K_1 \gamma_n - K_1 \gamma_m & \text{(15.116b)}
\end{cases}
$$

from the definition of K and K_1 in (15.97), (15.98). Therefore by the triangle inequality we have starting from (15.115), and invoking $w = w_1 + w_2$, (15.111), (15.116a-b):

$$
\| \gamma_n - \gamma_m \|_{L^2(\Omega)} \leq C_{T, r_0} (\| w_1 \|_{L^2(\Sigma_1)} + \| w_{1t} \|_{L^2(\Sigma_1)})
$$

$$
\leq C_{T, r_0} (\| w \|_{L^2(\Sigma_1)} + \| w_t \|_{L^2(\Sigma_1)} + \| w_2 \|_{L^2(\Sigma_1)} + \| w_{2t} \|_{L^2(\Sigma_1)})
$$

$$
\leq C_{T, r_0} (\| z_t (\gamma_n) - z_t (\gamma_m) \|_{L^2(\Sigma_1)} + \| z_{tt} (\gamma_n) - z_{tt} (\gamma_m) \|_{L^2(\Sigma_1)})
$$

$$
+ C_{T, r_0} \| K \gamma_n - K \gamma_m \|_{L^2(\Sigma_1)} + C_{T, r_0} \| K_1 \gamma_n - K_1 \gamma_m \|_{L^2(\Sigma_1)}
$$

$$
\leq C_{T, r_0} \| z_t (\gamma_n) \|_{L^2(\Sigma_1)} + C_{T, r_0} \| z_t (\gamma_m) \|_{L^2(\Sigma_1)}
$$

$$
+ C_{T, r_0} \| z_{tt} (\gamma_n) \|_{L^2(\Sigma_1)} + C_{T, r_0} \| z_{tt} (\gamma_m) \|_{L^2(\Sigma_1)}
$$

$$
+ C_{T, r_0} \| K \gamma_n - K \gamma_m \|_{L^2(\Sigma_1)} + C_{T, r_0} \| K_1 \gamma_n - K_1 \gamma_m \|_{L^2(\Sigma_1)}.
$$

$$
\tag{15.117}
$$

It then follows from (15.107b) and (15.110) as applied to the RHS of (15.117) that

$$\lim_{m,n \to +\infty} \|\gamma_n - \gamma_m\|_{L^2(\Omega)} = 0. \tag{15.118}$$

Therefore, $\{\gamma_n\}$ is a Cauchy sequence in $L^2(\Omega)$. By uniqueness of the limit, recall (15.109), it then follows that

$$\lim_{n \to \infty} \|\gamma_n - \gamma_0\|_{L^2(\Omega)} = 0. \tag{15.119}$$

Thus, in view of (15.107a), then (15.119) implies

$$\|\gamma_0\|_{L^2(\Omega)} = 1. \tag{15.120}$$

Step (iii).   Now consider the $z(\gamma_n)$-problem (15.108a-d) and its corresponding $z(\gamma_0)$-problem, with γ_0 as in (15.119), (15.120). Let

$$y(\gamma_n) = z(\gamma_n) - z(\gamma_0). \tag{15.121}$$

Then, via (15.108a-d), y satisfies the following problem

$$
\begin{cases}
y_{tt} = b\Delta y - \gamma_n y_t + \dfrac{c^2}{b}\gamma_n y - (\gamma_n - \gamma_0)z_t(\gamma_0) & \text{(15.122a)} \\[2mm]
\quad + \dfrac{c^2}{b}(\gamma_n - \gamma_0)z(\gamma_0) - \left(\dfrac{c^2}{b}\right)^2 (\gamma_n u(\gamma_n) - \gamma_0 u(\gamma_0)) \text{ in } Q; & \text{(15.122b)} \\[2mm]
y(\cdot, \dfrac{T}{2}) = y_t(\cdot, \dfrac{T}{2}) = 0 \text{ in } \Omega; & \text{(15.122c)} \\[2mm]
\dfrac{\partial y(x,t)}{\partial \nu}\Big|_{\Sigma} = 0 \text{ in } \Sigma. & \text{(15.122d)}
\end{cases}
$$

As $\gamma_n \in H^m(\Omega)$ and $2m > dim\Omega$, by the usual embedding $\gamma_n \in L^\infty(\Omega)$, as noted in Remark 15.2. Moreover, as $u, u_t, u_{tt} \in L^\infty(Q)$ as noted in Remark 15.3 and Sect. 15.2, we have $z = \frac{c^2}{b}u + u_t \in L^\infty(Q)$ and $z_t = \frac{c^2}{b}u_t + u_{tt} \in L^\infty(Q)$. Thus we have since also $\gamma_0 \in L^2(\Omega)$

$$(\gamma_n - \gamma_0)z_t(\gamma_0) + \frac{c^2}{b}(\gamma_n - \gamma_0)z(\gamma_0) - \left(\frac{c^2}{b}\right)^2 (\gamma_n u(\gamma_n) - \gamma_0 u(\gamma_0)) \in L^2(Q)$$

and hence we can again apply the sharp trace theory results [26, 27] (recalled in Appendix 2) for the y-problem (15.122a-d) as done in (15.100) for the

ζ-problem (15.94a-c) (both with zero I.C.) and obtain

$$\|y(\gamma_n)\|_{H^\beta(\Sigma)} \leq C_{T,data} \left(\|(\gamma_n - \gamma_0)z_t(\gamma_0)\|_{L^2(Q)} + \|(\gamma_n - \gamma_0)z(\gamma_0)\|_{L^2(Q)} \right.$$

$$\left. + \|\gamma_n(\cdot)u(\gamma_n) - \gamma_0(\cdot)u(\gamma_0)\|_{L^2(Q)} \right). \quad (15.123)$$

Step (iv). By assumption (15.12) we have that the elements γ_n are in a fixed ball of (arbitrary) radius R, in the space $H^m(\Omega)$:

$$\|\gamma_n\|_{H^m(\Omega)} \leq R, \quad n = 1, 2, \cdots \quad (15.124)$$

Hence, there exists a subsequence γ_{n_j} which is weakly convergent to some $\gamma_0^* \in H^m(\Omega)$:

$$\gamma_{n_j} \to \gamma_0^* \text{ weakly in } H^m(\Omega); \ \gamma_0^* \in H^m(\Omega). \quad (15.125)$$

By uniqueness of the limit, we then have

$$\gamma_0^* = \gamma_0 \in H^m(\Omega), \text{ hence } \gamma_0 \in C(\Omega) \quad (15.126)$$

as $2m > dim\Omega$ by assumption (15.12). A-fortiori, γ_0 is a multiplier $L^2(\Omega) \to L^2(\Omega)$: $\gamma_0 \in M(L^2(\Omega) \to L^2(\Omega))$.

Step (v). We return to inequality (15.123). After adding and subtracting, and recalling (15.126), we estimate

$$\|y(\gamma_n)\|_{H^\beta(\Sigma)} \leq C_{T,data} \left(\|(\gamma_n - \gamma_0)\|_{L^2(\Omega)} \|z_t(\gamma_0)\|_{L^\infty(Q)} \right.$$

$$+ \|(\gamma_n - \gamma_0)\|_{L^2(\Omega)} \|z(\gamma_0)\|_{L^\infty(Q)} + \|\gamma_n - \gamma_0\|_{L^2(\Omega)} \|u(\gamma_n)\|_{L^\infty(Q)}$$

$$\left. + \|\gamma_0(\cdot)\|_{L^\infty(\Omega)} \|u(\gamma_n) - u(\gamma_0)\|_{L^2(Q)} \right). \quad (15.127)$$

(We could have used the sharper norm $\|\gamma_0(\cdot)\|_{M(L^2(\Omega) \to L^2(\Omega))}$ instead of $\|\gamma_0(\cdot)\|_{L^\infty(\Omega)}$ recalling the line below (15.126)).

We now recall the results of Sect. 15.2: the uniform $L^\infty(Q)$-bound on $u(\cdot, \gamma_n)$ from (15.40) of Theorem 15.8, and the $L^2(Q)$-convergence result of Theorem 15.7, via (15.119). For the third and fourth term on the RHS of (15.127), they yield, respectively

$$\|u(\gamma_n)\|_{L^\infty(Q)} \leq C_{T,u_0,u_1,u_2}; \quad \lim_{n \to \infty} \|u(\gamma_n) - u(\gamma_0)\|_{L^2(Q)} = 0. \quad (15.128)$$

Using (15.119) and (15.128) on the RHS of (15.127), and recalling (15.121), we conclude first that

$$\lim_{n\to\infty} \|y(\gamma_n)\|_{H^\beta(\Sigma_1)} = \lim_{n\to\infty} \|z(\gamma_n)|_{\Sigma_1} - z(\gamma_0)|_{\Sigma_1}\|_{H^\beta(\Sigma_1)} = 0, \qquad (15.129)$$

next that

$$\lim_{n\to\infty} \|z(\gamma_n)|_{\Sigma_1} - z(\gamma_0)|_{\Sigma_1}\|_{C([0,T];L^2(\Gamma_1))} = 0, \qquad (15.130)$$

since $\beta > \frac{1}{2}$, so that $H^\beta(0,T)$ embeds in $C[0,T]$.

Step (vi). We now consider the y_t-problem, obtained from differentiating problem (6.33) in time, thus with RHS $= -(\gamma_n - \gamma_0)z_{tt}(\gamma_0) + \frac{c^2}{b}(\gamma_n - \gamma_0)z_t(\gamma_0) - (\frac{c^2}{b})^2(\gamma_n u_t(\gamma_n) - \gamma_0 u_t(\gamma_0))$ and I.C. $y_t(\cdot,\frac{T}{2}) = 0$, $(y_t)_t(\cdot,\frac{T}{2}) = (\gamma_0(x) - \gamma_n(x))u_2 \in L^2(\Omega)$ by recalling (15.121), (15.76a-b), (15.107a), (15.120). Applying to such y_t-problem Part II of Theorem, Appendix 2, we obtain

$$\|y_t(\gamma_n)\|_{H^\beta(\Sigma)} \le C_{T,data}\left(\|(\gamma_n - \gamma_0)z_{tt}(\gamma_0)\|_{L^2(Q)} + \|(\gamma_n - \gamma_0)z_t(\gamma_0)\|_{L^2(Q)}\right.$$
$$\left. +\|\gamma_n(\cdot)u_t(\gamma_n) - \gamma_0(\cdot)u_t(\gamma_0)\|_{L^2(Q)} + \|(\gamma_n - \gamma_0)u_2\|_{L^2(\Omega)}\right) \qquad (15.131)$$

counterpart of (15.101) for ζ_t. Proceeding now as in going from (15.123) to (15.127), that is, adding and subtracting, we obtain (using again (15.126)):

$$\|y_t(\gamma_n)\|_{H^\beta(\Sigma)} \le C_{T,data}\left(\|(\gamma_n - \gamma_0)\|_{L^2(\Omega)}\|z_{tt}(\gamma_0)\|_{L^\infty(Q)}\right.$$

$$+\|(\gamma_n - \gamma_0)\|_{L^2(\Omega)}\|z_t(\gamma_0)\|_{L^\infty(Q)} + \|\gamma_n - \gamma_0\|_{L^2(\Omega)}\|u_t(\gamma_n)\|_{L^\infty(Q)}$$

$$\left. +\|\gamma_0(\cdot)\|_{L^\infty(\Omega)}\|u_t(\gamma_n) - u_t(\gamma_0)\|_{L^2(Q)} + \|\gamma_n - \gamma_0\|_{L^2(\Omega)}\|u_2\|_{L^\infty(\Omega)}\right). \qquad (15.132)$$

We again recall the results of Theorem 15.7 and 15.8 for the third and fourth term on the RHS of (15.132), they yield, respectively

$$\|u_t(\gamma_n)\|_{L^\infty(Q)} \le C_{T,u_1,u_2,u_3}; \quad \lim_{n\to\infty} \|u_t(\gamma_n) - u_t(\gamma_0)\|_{L^2(Q)} = 0. \qquad (15.133)$$

From (15.119), (15.133) used on the RHS of (15.132), and recalling (15.121), we first conclude that

$$\lim_{n\to\infty} \|y_t(\gamma_n)\|_{H^\beta(\Sigma_1)} = \lim_{n\to\infty} \|z_t(\gamma_n)|_{\Sigma_1} - z_t(\gamma_0)|_{\Sigma_1}\|_{H^\beta(\Sigma_1)} = 0, \qquad (15.134)$$

and next that

$$\lim_{n\to\infty} \|z_t(\gamma_n)|_{\Sigma_1} - z_t(\gamma_0)|_{\Sigma_1}\|_{C([0,T];L^2(\Gamma_1))} = 0, \tag{15.135}$$

since $\beta > \frac{1}{2}$ as in (15.130), as $H^\beta(0,T)$ embeds in $C[0,T]$. Then, by virtue of (15.107b), combined with (15.135), we obtain that

$$z_t(\gamma_0)|_{\Sigma_1} \equiv 0; \text{ or } z(\gamma_0)|_{\Sigma_1} \equiv \text{function of } x \in \Gamma_1, \text{constant in } t \in [0,T]. \tag{15.136}$$

Step (vii).   We return to problem (15.108a-d): with $\gamma_n \in L^2(\Omega)$, $z, z_t \in L^\infty(Q)$ as noted below (15.122a) and $u \in L^\infty(Q)$, Remark 15.3, thus forcing term in $L^2(Q)$ and initial 'velocity' $\frac{c^2}{b}u_1 + u_2 \in L^2(\Omega)$. We then obtain the following regularity results from [26, 27], recalled in Appendix 2, continuously:

$$\{z(\gamma_n), (z(\gamma_n))_t\} \in C([0,T]; H^1(\Omega) \times L^2(\Omega)); \tag{15.137}$$

$$z(\gamma_n)|_\Sigma \in H^\beta(\Sigma), \tag{15.138}$$

where β is the constant in (15.102).

As a consequence of (15.119), we also have via (15.137), (15.138) and (15.130), (15.134):

$$\{z(\gamma_n), (z(\gamma_n))_t\} \to \{z(\gamma_0), (z(\gamma_0))_t\} \text{ in } C([0,T]; H^1(\Omega)\times L^2(\Omega)); \tag{15.139}$$

$$z(\gamma_n)|_\Sigma \to z(\gamma_0)|_\Sigma \text{ in } H^\beta(\Sigma). \tag{15.140}$$

On the other hand, recalling the initial condition (15.108c), we have

$$z(\gamma_n)\left(x, \frac{T}{2}\right) \equiv 0, \ x \in \overline{\Omega}, \text{ hence } z(\gamma_n)\left(x, \frac{T}{2}\right) \equiv 0, \quad x \in \Gamma_1, \tag{15.141}$$

in the sense of trace in $H^{\frac{1}{2}}(\Gamma_1)$. Then (15.141) combined with (15.129), (15.130) yields *a-fortiori*

$$z(\gamma_0)\left(x, \frac{T}{2}\right) \equiv 0, \quad x \in \Gamma_1, \tag{15.142}$$

and next, by virtue of (15.136), the desired conclusion,

$$z(\gamma_0)|_{\Sigma_1} \equiv 0. \tag{15.143}$$

By (15.119), (15.139), (15.140) applied to (15.108a-d), we have that $z(\gamma_0)$ satisfies weakly the limit problem:

$$\begin{cases} z_{tt}(\gamma_0) = b\Delta z(\gamma_0) - \gamma_0(x)z_t(\gamma_0) + \dfrac{c^2}{b}\gamma_0(x)z(\gamma_0) & \text{(15.144a)} \\[2mm] \quad -(\dfrac{c^2}{b})^2\gamma_0(x)u(\gamma_0) \text{ in } Q; & \text{(15.144b)} \\[2mm] z(\gamma_0)\left(\cdot,\dfrac{T}{2}\right) = 0; \quad z_t(\gamma_0)\left(\cdot,\dfrac{T}{2}\right) = \dfrac{c^2}{b}u_1 + u_2 \text{ in } \Omega; & \text{(15.144c)} \\[2mm] \dfrac{\partial z(\gamma_0)}{\partial\nu}\Big|_\Sigma = 0 \text{ and } z(\gamma_0)|_{\Sigma_1} = 0 \text{ in } \Sigma,\ \Sigma_1, & \text{(15.144d)} \end{cases}$$

via also (15.143), where $u_{tt}, u_{ttt} \in L^\infty(Q)$, by assumption (15.7) = (15.68), as noted in Remark 15.3 and Sect. 15.2. Moreover, the positivity assumption (15.8) holds. Thus, the uniqueness Theorem 15.1 applies and yields

$$\gamma_0(x) \equiv 0, \quad x \in \Omega. \tag{15.145}$$

Then (15.145) contradicts (15.120). Thus, assumption (15.107a-b) is false and inequality (15.115) holds true. Proposition 15.1, as well as Theorem 15.2 are then established. □

Remark 15.10 (*Variations of above proof of Theorem 15.2 to obtain the proof of Theorem 15.4*). *In Step 1, the counterpart of problem (15.91a-c) is now*

$$\begin{cases} v_{tt} = b\Delta v - \gamma(x)v_t + \dfrac{c^2}{b}\gamma(x)v - (\dfrac{c^2}{b})^2\gamma(x)u_t(\gamma) \text{ in } Q; & \text{(15.146a)} \\[2mm] v\left(\cdot,\dfrac{T}{2}\right) = \dfrac{c^2}{b}u_1 + u_2; \quad v_t\left(\cdot,\dfrac{T}{2}\right) = -\gamma(x)u_2(x) \text{ in } \Omega; & \text{(15.146b)} \\[2mm] v|_\Sigma = 0 \text{ in } \Sigma & \text{(15.146c)} \end{cases}$$

while, still with $z_t = v = \psi + \zeta$ as in (15.92), the counterpart of problems (15.93a-d) and (15.94a-c) are

$$\begin{cases} \psi_{tt} = b\Delta\psi - \gamma(x)\psi_t + \dfrac{c^2}{b}\gamma(x)\psi \text{ in } Q & \text{(15.147a)} \\[2mm] \psi\left(\cdot,\dfrac{T}{2}\right) = v\left(\cdot,\dfrac{T}{2}\right) = \dfrac{c^2}{b}u_1 + u_2 \text{ in } \Omega & \text{(15.147b)} \\[2mm] \psi_t\left(\cdot,\dfrac{T}{2}\right) = v_t\left(\cdot,\dfrac{T}{2}\right) = -\gamma(x)u_2(x) \text{ in } \Omega & \text{(15.147c)} \\[2mm] \psi|_\Sigma = 0 \text{ in } \Sigma & \text{(15.147d)} \end{cases}$$

and

$$
\begin{cases}
\zeta_{tt} = b\Delta\zeta - \gamma(x)\zeta_t + \dfrac{c^2}{b}\gamma(x)\zeta - \left(\dfrac{c^2}{b}\right)^2\gamma(x)u_t(\gamma) \text{ in } Q & (15.148a) \\[2mm]
\zeta\left(\cdot,\dfrac{T}{2}\right) = 0; \quad \zeta_t\left(\cdot,\dfrac{T}{2}\right) = 0 \text{ in } \Omega & (15.148b) \\[2mm]
\zeta|_\Sigma = 0 \text{ in } \Sigma. & (15.148c)
\end{cases}
$$

In Step 2, we now invoke the COI (15.65) for the ψ-problem (15.147a-d), with I.C. as in (15.7) and $\gamma \in L^\infty(\Omega)$ a fortiori from assumption (15.2). We obtain the counterpart of (15.95):

$$
r_0^2\|\gamma\|_{L^2(\Omega)}^2 \le \|\gamma(\cdot)u_2(\cdot)\|_{L^2(\Omega)}^2 \le C_T^2 \int_0^T \int_{\Gamma_1} \left(\frac{\partial\psi}{\partial\nu}\right)^2 d\Sigma_1, \tag{15.149}
$$

recalling the positivity assumption (15.8) on the LHS. Then (15.149) implies the counterpart of (15.96)

$$
\|\gamma\|_{L^2(\Omega)} \le C_{T,r_0}\left(\left\|\frac{\partial z_t}{\partial\nu}\right\|_{L^2(\Gamma_1\times[0,T])} + \left\|\frac{\partial\zeta}{\partial\nu}\right\|_{L^2(\Gamma_1\times[0,T])}\right)
$$

$$
= C_{T,r_0}\left(\left\|\frac{\partial}{\partial\nu}\left(\frac{c^2}{b}u + u_t\right)_t\right\|_{L^2(\Gamma_1\times[0,T])} + \left\|\frac{\partial\zeta}{\partial\nu}\right\|_{L^2(\Gamma_1\times[0,T])}\right). \tag{15.150}
$$

Inequality (15.150) is the sought-after right-hand side inequality of Theorem 15.4 for $\theta = 0$, modulo (polluted by) the $\frac{\partial\zeta}{\partial\nu}|_{\Sigma_1}$-term. It is at the level of absorbing this term by a compactness-uniqueness argument that one needs to restrict the final result to $\gamma \in H_0^\theta(\Omega)$, $0 < \theta \le 1$, $\theta \ne \frac{1}{2}$ in order to use compactness as the map $\gamma \in H^\theta(\Omega) \to \frac{\partial\zeta}{\partial\nu} \in H^\theta(\Sigma_1)$, $0 \le \theta \le 1$, $\theta \ne \frac{1}{2}$ is only continuous, but not compact for the ζ-problem (15.148a-c) with the Dirichlet B.C.. This fact is a distinguished obstacle of the stability analysis for the Dirichlet B.C. case as opposed to the Neumann B.C. case. For details we refer to [46, Sect. 5.3].

Sketch of proof of inequality (15.15b).

(a) Case $\theta = 0$. With reference to the variable $z = \frac{c^2}{b}u + u_t$ in (15.43), we shall show that

$$
\left\|\frac{\partial z_t}{\partial\nu}\right\|_{L^2(0,T;L^2(\Gamma_1))} \le C_{T,R_1}\|\{u_0,u_1,u_2\}\|_m \|\gamma\|_{L^2(\Omega)} \tag{15.151}
$$

where $\{u_0,u_1,u_2\}$ is measured in the $H^{m+2}(\Omega) \times H^{m+1}(\Omega) \times H^m(\Omega)$-norm as in (15.7), denoted here by $\|\{u_0,u_1,u_2\}\|_m$.

Proof of (15.151). Consider the z_t-problem in (15.72a-c) in the present Dirichlet case, i.e. with B.C. $z_t|_\Sigma = 0$ only, where by assumption

$$\|\gamma\|_{C(\Omega)} \le c_1\|\gamma\|_{H^m(\Omega)} \le c_1 R = R_1, \ m > \frac{\dim \Omega}{2} \ge 1. \tag{15.152}$$

(i) We estimate the forcing term in (15.72a-d)

$$\left\| \left(\frac{c^2}{b}\right)^2 \gamma(\cdot) u_t(\gamma) \right\|_{L^2(0,T;L^2(\Omega))} \le \left(\frac{c^2}{b}\right)^2 \|\gamma\|_{L^2(\Omega)} \|u_t(\gamma)\|_{L^\infty(Q)} \tag{15.153}$$

$$\le C_{b,c,R_1} \|\gamma\|_{L^2(\Omega)} \|\{u_0, u_1, u_2\}\|_m \tag{15.154}$$

where $\{u_0, u_1, u_2\}$ is measured in the $H^{m+2}(\Omega) \times H^{m+1}(\Omega) \times H^m(\Omega)$-norm, by appealing to Remark 15.3;

(ii) we next estimate the initial position in (15.72a-d), with m as in (15.152)

$$\left\| z_t\left(\cdot, \frac{T}{2}\right) \right\|_{H^1(\Omega)} = \left\| \frac{c^2}{b} u_1 + u_2 \right\|_{H^1(\Omega)} \le \frac{c^2}{b} \|\{u_1, u_2\}\|_{H^{m+1}(\Omega) \times H^m(\Omega)};$$
$$\tag{15.155}$$

(iii) finally we estimate the initial velocity in (15.72c)

$$\left\| (z_t)_t\left(\cdot, \frac{T}{2}\right) \right\|_{L^2(\Omega)} = \|\gamma(\cdot)u_2\|_{L^2(\Omega)} \le \|\gamma\|_{L^2(\Omega)} \|u_2\|_{L^\infty(\Omega)}$$

$$\le c\|\gamma\|_{L^2(\Omega)} \|u_2\|_{H^m(\Omega)} \tag{15.156}$$

m as in (15.152). Finally, armed with estimates (15.154), (15.155), (15.156) on the data, we invoke the regularity inequality in [22], [27, Theorem 10.5.3.1, p. 946] to the aforementioned z_t problem with γ uniformly bounded in $L^\infty(\Omega)$ as in (15.152) and obtain inequality (15.151).

(b) Case $\theta = 1$. We now show that with $m \ge 2$

$$\left\| \frac{\partial z_{tt}}{\partial \nu} \right\|_{L^2(0,T;L^2(\Gamma_1))} \le C_{T,R_1} \|\{u_0, u_1, u_2\}\|_m \|\gamma\|_{H^1(\Omega)} \tag{15.157}$$

in the same $H^{m+2}(\Omega) \times H^{m+1}(\Omega) \times H^m(\Omega)$-norm for $\{u_0, u_1, u_2\}$.

Proof of (15.157). We respect the same argument this time as applied to the z_{tt}-problem (15.76a-c) with B.C. only $z_{tt}|_\Sigma = 0$.

(i) as to the forcing term, we estimate

$$\left\| \left(\frac{c^2}{b} \right)^2 \gamma(\cdot) u_{tt}(\gamma) \right\|_{L^2(0,T;L^2(\Omega))} \leq \left(\frac{c^2}{b} \right)^2 \|\gamma\|_{L^2(\Omega)} \|u_{tt}(\gamma)\|_{L^\infty(Q)}$$

$$\leq \left(\frac{c^2}{b} \right)^2 C_{R_1} \|\gamma\|_{L^2(\Omega)} \|\{u_0, u_1, u_2\}\|_m$$

$$(15.158)$$

in the $H^{m+2}(\Omega) \times H^{m+1}(\Omega) \times H^m(\Omega)$-norm for $\{u_0, u_1, u_2\}$, see Remark 15.3;

(ii) as to the initial position in (15.76a-b) we estimate via (15.5a)

$$\left\| z_{tt}\left(\cdot, \frac{T}{2} \right) \right\|_{H^1(\Omega)} = \|\gamma(\cdot) u_2\|_{H^1(\Omega)}$$

$$\leq \|\gamma\|_{H^1(\Omega)} \|u_2\|_{\mathcal{M}(H^1 \to H^1)}$$

$$\leq c \|\gamma\|_{H^1(\Omega)} \|u_2\|_{H^m(\Omega)}; \qquad (15.159)$$

(iii) for the initial velocity in (15.76e)

$$(z_{tt})_t \left(\cdot, \frac{T}{2} \right) = b \Delta z_t \left(\cdot, \frac{T}{2} \right) + \gamma^2 u_2 + \frac{c^2}{b} \gamma u_2. \qquad (15.160)$$

We estimate first

$$\left\| \Delta z_t \left(\cdot, \frac{T}{2} \right) \right\|_{L^2(\Omega)} \leq \tilde{c} \left\| z_t \left(\cdot, \frac{T}{2} \right) \right\|_{H^2(\Omega)} = \tilde{c} \left\| \frac{c^2}{b} u_1 + u_2 \right\|_{H^2(\Omega)}$$

$$\leq \tilde{c} \frac{c^2}{b} \|\{u_1, u_2\}\|_{H^{m+1}(\Omega) \times H^m(\Omega)} \qquad (15.161)$$

$m \geq 2$; next in (15.160)

$$\left\| \gamma^2 u_2 + \frac{c^2}{b} \gamma u_2 \right\|_{L^2(\Omega)} \leq C_{b,c} \|\gamma\|_{L^2(\Omega)} \|\gamma\|_{L^\infty(\Omega)} \|u_2\|_{L^\infty(\Omega)}$$

$$\leq C_{b,c} \|\gamma\|_{L^2(\Omega)} R_1 \|u_2\|_{H^m(\Omega)}. \qquad (15.162)$$

Finally, estimates (15.158), (15.159), (15.162) in the data yield the final estimate (15.157) by invoking again [22], [27, Theorem 10.5.3.1, p. 946] with γ uniformly bounded in $L^\infty(\Omega)$ by (15.152). We finally interpolate between the estimate (15.151) for $\theta = 0$ and the estimate (15.157) for $\theta = 1$, the latter with $m \geq 2$.

15.7 Proof of Uniqueness Theorem 15.5

Step 1. We start from the two problems

$$u_{ttt}(\gamma_i) + \alpha(\cdot)u_{tt}(\gamma_i) + c^2\mathscr{A}u(\gamma_i) + b\mathscr{A}u_t(\gamma_i) = 0, \ i = 1, 2; \qquad (15.163a)$$

$$u(\gamma_i)|_{\frac{T}{2}} = u_0, \ u_t(\gamma_i)|_{\frac{T}{2}} = u_1, \ u_{tt}(\gamma_i)|_{\frac{T}{2}} = u_2, \ i = 1, 2 \qquad (15.163b)$$

where $\mathscr{A}$ is $(-\Delta)$ with either Neumann or Dirichlet B.C.. Set

$$w = u(\gamma_1) - u(\gamma_2). \qquad (15.164)$$

Then, by adding and subtracting $\alpha_2 u_{tt}(\gamma_1)$, one obtains that w satisfies

$$w_{ttt} + \alpha_2(\cdot)w_{tt} + c^2\mathscr{A}w + b\mathscr{A}w_t = (\alpha_2(\cdot) - \alpha_1(\cdot))u_{tt}(\gamma_1) \qquad (15.165)$$

rewritten as

$$(w_t + \alpha_2 w)_{tt} + b\mathscr{A}\left(\frac{c^2}{b}w + w_t\right) = (\alpha_2 - \alpha_1)u_{tt}(\gamma_1). \qquad (15.166)$$

Setting

$$z = \frac{c^2}{b}w + w_t = (\alpha_2 w + w_t) - \gamma_2 w, \quad \gamma_2 = \alpha_2 - \frac{c^2}{b} \qquad (15.167)$$

we obtain after a cancellation

$$z_{tt} + c\mathscr{A}z = \gamma_2(\cdot)u_{tt}(\gamma_2) - \gamma_1(\cdot)u_{tt}(\gamma_1). \qquad (15.168)$$

Equation (15.167) is the counterpart of model $\sharp 1$ in the present case of non-zero damping γ_2.

Step 2. Under the assumed regularity properties, the argument of Sect. 15.5 can be carried out (This was done in [47, Sect. 5] for model $\sharp 1$). It leads to the corresponding result

$$z = z_t = z_{tt} \equiv 0 \text{ in } Q(\sigma) \qquad (15.169)$$

in particular at $\frac{T}{2}$, recalling (15.163b):

$$0 = z_{tt}|_{\frac{T}{2}} + \mathscr{A}(z|_{\frac{T}{2}}) = \gamma_2 u_{tt}(\gamma_2)|_{\frac{T}{2}} - \gamma_1 u_{tt}(\gamma_1)|_{\frac{T}{2}} = (\gamma_2 - \gamma_1)u_2. \qquad (15.170)$$

By (15.8) as assumed, $|u_2(x)| \geq r_0 > 0$ in Ω, hence (15.170) yields the desired conclusion

$$\gamma_1(x) = \gamma_2(x) \text{ in } \Omega.$$

Theorem 15.5 is proved. □

Acknowledgements The authors wish to thank a referee for suggesting to extend their original version to include the global uniqueness of Theorem 15.5 of any damping coefficient. S.L. was supported by a startup grant from Clemson University and R.T. was supported by the National Science Foundation under Grant DMS-0104305 and by the Air Force Office of Scientific Research under Grant FA9550-09-1-0459.

Appendix 1: Admissible Geometrical Configurations in the Neumann B.C. Case

Here we present some examples in connection to the main geometrical assumptions (A.1), (A.2). We refer to Lasiecka et al. [35] for more details.

Example #1 (Any dimension ≥ 2): Γ_0 is flat.

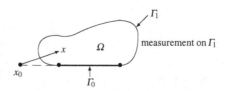

Let $x_0 \in$ hyperplane containing Γ_0, then.

$$d(x) = \|x - x_0\|^2; \quad h(x) = \nabla d(x) = 2(x - x_0).$$

Example #2 (A ball of any dimension ≥ 2): $d(x)$ in [35, Theorem. A.4.1, p. 301].

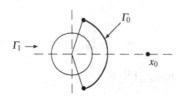

Measurement on $\Gamma_1 > \frac{1}{2}$ circumference (as in the Dirichlet case), same as for controllability.

Example #3 (Generalizing Ex #2: a domain Ω of any dimension ≥ 2 with unobserved portion Γ_0 convex, subtended by a common point x_0): $d(x)$ in [35, Theorem. A.4.1, p. 301].

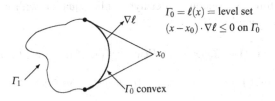

$$\Gamma_0 = \ell(x) = \text{level set}$$
$$(x - x_0) \cdot \nabla \ell \leq 0 \text{ on } \Gamma_0$$

Example #4 (A domain Ω of any dimension ≥ 2 with unobserved portion Γ_0 concave, subtended by a common point x_0): $d(x)$ in [35, Theorem. A.4.1, p. 301].

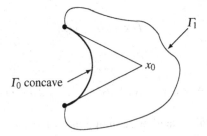

Example #5 (dim $= 2$): Γ_0 neither convex nor concave. Γ_0 is described by graph

$$y = \begin{cases} f_1(x), x_0 \leq x \leq x_1, y \geq 0; \\ f_2(x), x_2 \leq x \leq x_1, y < 0, \end{cases}$$

f_1, f_2 *logarithmic concave* on $x_0 < x < x_1$, e.g., $\sin x, -\frac{\pi}{2} < x < \frac{\pi}{2}$; $\cos x, 0 < x < \pi$

as the function $\sin x + 1$ and $\cos x + 1$ are *lobarithmic concave* on the corresponding x-intervals. See [35, Fig. A.3, p. 290]. The corresponding function $d(x)$ is given in [35, (A.2.7), p. 289].

Appendix 2: Sharp Regularity Theory for Second-Order Hyperbolic Equations of Neumann Type

Consider the following second-order hyperbolic equation with non-homogeneous Neumann B.C g and I.C. $\{w_0, w_1\}$:

$$\begin{cases} w_{tt}(x,t) - \Delta w(x,t) = F(w) + f(x,t) \text{ in } Q = \Omega \times [0,T]; \\[2mm] w(\cdot, 0) = w_0(x); \quad w_t(\cdot, 0) = w_1(x) \text{ in } \Omega; \\[2mm] \dfrac{\partial w}{\partial \nu}(x,t)|_\Sigma = g(x,t) \qquad\qquad \text{ in } \Sigma = \Gamma \times [0,T] \end{cases}$$

where the forcing term

$$f(x,t) \in L^2(Q),$$

and $F(w)$ is given by

$$F(w) = q_1(x,t)w + q_2(x,t)w_t + q_3(x,t) \cdot \nabla w,$$

subject to the following standing assumption on the coefficients: q_1, q_2, $|q_3| \in L^\infty(Q)$, so that the following pointwise estimate holds true:

$$|F(w)| \le C_T[w^2 + w_t^2 + |\nabla w|^2], \quad (x,t) \in Q.$$

We first define the parameters α and β to be the following values:

$$\begin{cases} \alpha = \frac{3}{5} - \epsilon, \ \beta = \frac{3}{5} : \text{ for a general smooth, bounded domain } \Omega; \\[2mm] \alpha = \beta = \frac{2}{3} : \text{ for a sphere domain } \Omega; \\[2mm] \alpha = \beta = \frac{3}{4} - \epsilon : \text{ for a parallelepiped domain } \Omega \end{cases}$$

where $\epsilon > 0$ is arbitrary. Then we have the following sharp regularity results:

Theorem ([27, Theorem 1.2 (ii), (iii), 1.3, p. 290]) *With reference to the above w-mixed problem, the following regularity results hold true, with α and β defined above:*

(i) [27, Theorem 2.0, (15.21), (2.9), p. 123; Theorem A, p. 117; Theorem 2.1, p. 124]. Suppose we have $f = 0$, $\{w_0, w_1\} \in H^1(\Omega) \times L^2(\Omega)$ and $g \in L^2(\Sigma)$. Then we have the unique solution w satisfies

$$w \in H^\alpha(Q) = C([0,T]; H^\alpha(\Omega)) \cap H^\alpha(0,T; L^2(\Omega)); \quad w|_\Sigma \in H^{2\alpha-1}(\Sigma).$$

(ii) [27, Theorem 5.1, (5.4), (5.5), p. 149; Theorem C, p. 118; Theorem 7.1, p. 158]. Suppose now $f \in L^2(Q)$, $\{w_0, w_1\} \in H^1(\Omega) \times L^2(\Omega)$ and $g = 0$. Then we have

$$w \in C([0, T]; H^1(\Omega)), \ w_t \in C([0, T]; L^2(\Omega)); \quad w|_\Sigma \in H^\beta(\Sigma).$$

References

1. Beilina, L., Klibanov, M.: Approximate Global Convergence and Adaptivity for Coefficient Inverse Problems. Springer, New York (2012)
2. Bukhgeim, A., Cheng, J., Isakov. V., Yamamoto, M.: Uniqueness in determining damping coefficients in hyperbolic equations. In: Saburou Saitoh, Nakao Hayashi, Masahiro Yamamoto (eds.), Analytic Extension Formulas and Their Applications, pp. 27–46. Kluwer, Dordrecht (2001)
3. Bukhgeim, A., Klibanov, M.: Global uniqueness of a class of multidimensional inverse problem. Sov. Math. Dokl. **24**, 244–257 (1981)
4. Carleman, T.: Sur un problème d'unicité pour les systèmes d'équations aux derivées partielles à deux variables independantes. Ark. Mat. Astr. Fys.**2B**, 1–9 (1939)
5. Ho, L. F.: Observabilite frontiere de l'equation des ondes. Comptes Rendus de l'Academie des Sciences de Paris **302**, 443–446 (1986)
6. Hörmander, L.: The Analysis of Linear Partial Differential Operators I. Springer, Berlin/ New York (1985)
7. Hörmander, L.: The Analysis of Linear Partial Differential Operators II. Springer, Berlin/ New York (1985)
8. Isakov, V.: Inverse Problems for Partial Differential Equations, 1st edn. Springer, New York (1998)
9. Isakov, V.: Inverse Problems for Partial Differential Equations, 2nd edn. Springer, New York (2006)
10. Isakov, V., Yamamoto, M.: Carleman estimate with the Neumann boundary condition and its application to the observability inequality and inverse hyperbolic problems. Contemp. Math. **268**, 191–225 (2000)
11. Isakov, V., Yamamoto, M.: Stability in a wave source problem by Dirichlet data on subboundary. J. Inverse Ill-Posed Probl. **11**, 399–409 (2003)
12. Jordan, P.M.: An analytic study of the Kuznetsov's equation: diffusive solitons, shock formation, and solution bifurcation. Phys. Lett. A **326**, 77–84 (2004)
13. Jordan, P.M.: Nonlinear acoustic phenomena in viscous thermally relaxing fluids: shock bifurcation and the emergence of diffusive solitions (A) (Lecture). The 9th International Conference on Theoretical and Computational Acoustics (ICTCA 2009), Dresden, Germany. J. Acoust. Soc. Am. **124**, 2491–2491 (2008)
14. Kaltenbacher, B., Lasiecka, I.: Global existence and exponential decay rates for the Westervelt equation. DCDS Ser. S **2**, 503–525 (2009)
15. Kaltenbacher, B., Lasiecka, I.: Well-posedness of the Westervelt and the Kuznetsov equations with non homogeneous Neumann boundary conditions. DCDS Suppl., 763–773 (2011)
16. Kaltenbacher, B., Lasiecka, I., Marchand, R.: Wellposedness and exponential decay rates for the Moore–Gibson–Thompson equation arising in high intensity ultrasound. Control Cybern. (2011)
17. Kaltenbacher, B., Lasiecka, I., Veljovic, S.: Well-posedness and exponential decay of the Westervelt equation with inhomogeneous Dirichlet boundary data. Progress in Nonlinear Differential Equations and Their Applications, vol. 60. Springer, Basel (2011)
18. Klibanov, M.: Inverse problems and Carleman estimates. Inverse Probl. **8**, 575–596 (1992)

19. Klibanov, M.: Carleman estimates for global uniqueness, stability and numerical methods for coefficient inverse problems. J. Inverse Ill-Posed Probl. **21**(2), (2013)

20. Klibanov, M., Timonov, A.: Carleman Estimates for Coefficient Inverse Problems and Numerical Applications. VSP, Utrecht (2004)

21. Kuznetsov, V.P.: Equations of nonlinear acoustics. Sov. Phys. **16**, 467–470 (1971)

22. Lasiecka, I., Lions, J.L., Triggiani, R.: Non-homogeneous boundary value problems for second-order hyperbolic operators. J. Math. Pures Appl. **65**, 149–192 (1986)

23. Lasiecka, I., Triggiani, R.: A cosine operator approach to modeling $L_2(0, T; L_2(\Omega))$ boundary input hyperbolic equations. Appl. Math. Optim. **7**, 35–83 (1981)

24. Lasiecka, I., Triggiani, R.: Regularity of hyperbolic equations under $L_2(0, T; L_2(\Gamma))$-Dirichlet boundary terms. Appl. Math. Optim. **10**, 275–286 (1983)

25. Lasiecka, I., Triggiani, R.: Exact controllability of the wave equation with Neumann boundary control. Appl. Math. Optim. **19**, 243–290 (1989)

26. Lasiecka, I., Triggiani, R.: Sharp regularity theory for second-order hyperbolic equations of Neumann type Part I: L_2 non-homogeneous data. Ann. Mat. Pura Appl. (IV) **CLVII**, 285–367 (1990)

27. Lasiecka, I., Triggiani, R.: Regularity theory of hyperbolic equations with non-homogeneous Neumann boundary conditions II: General boundary data. J. Differ. Equ. **94**, 112–164 (1991)

28. Lasiecka, I., Triggiani, R.: Recent advances in regularity of second-order hyperbolic mixed problems and applications. Dynamics Reported, vol. 3, pp. 104–158. Springer, New York (1994)

29. Lasiecka, I., Triggiani, R.: Carleman estimates and uniqueness for the system of strong coupled PDE's of spherical shells. Special volume of Zeits. Angerwandte Math. Mech. vol. 76, pp.277–280. Akademie, Berlin (1996)

30. Lasiecka, I., Triggiani, R.: Carleman estimates and exact controllability for a system of coupled, nonconservative second-order hyperbolic equations. Lect. Notes Pure Appl. Math. **188**, 215–245 (1997)

31. Lasiecka, I., Triggiani, R.: Exact boundary controllability of a first-order nonlinear hyperbolic equation with non-local in the integral term arising in epidemic modeling. In: Gilbert, R.P., Kajiwara, J., Xu, Y. (eds.) Direct and Inverse Problems of Mathematical Physics, pp. 363–398. ISAAC'97, The First International Congress of the International Society for Analysis, Its Applications and Computations. Kluwer (2000)

32. Lasiecka, I., Triggiani, R.: Uniform stabilization of the wave equation with Dirichlet or Neumann-feedback control without geometrical conditions. Appl. Math. Optim. **25**, 189–224 (1992)

33. Lasiecka, I., Triggiani, R., Yao, P.F.: Exact controllability for second-order hyperbolic equations with variable coefficient-principal part and first-order terms. Nonlinear Anal. **30**(1), 111–222 (1997)

34. Lasiecka, I., Triggiani, R., Yao, P.F.: Inverse/observability estimates for second-order hyperbolic equations with variable coefficients. J. Math. Anal. Appl. **235**(1), 13–57 (1999)

35. Lasiecka, I., Triggiani, R., Zhang, X.: Nonconservative wave equations with unobserved Neumann B.C.: Global uniqueness and observability in one shot. Contemp. Math. **268**, 227–325 (2000)

36. Lavrentev, M.M., Romanov, V.G., Shishataskii, S.P.: Ill-Posed Problems of Mathematical Physics and Analysis, vol. 64. The American Mathematical Society, Providence (1986)

37. Lions, J.L.: Controlabilite Exacte, Perturbations et Stabilisation de Systemes Distribues, vol. 1. Masson, Paris (1988)

38. Liu, S.: Inverse problem for a structural acoustic interaction. Nonlinear Anal. **74**, 2647–2662 (2011)

39. Liu, S., Triggiani, R.: Global uniqueness and stability in determining the damping and potential coefficients of an inverse hyperbolic problem. Nonlinear Anal. Real World Appl. **12**, 1562–1590 (2011)

40. Liu, S., Triggiani, R.: Global uniqueness and stability in determining the damping coefficient of an inverse hyperbolic problem with non-homogeneous Neumann B.C. through an additional Dirichlet boundary trace. SIAM J. Math. Anal. **43**, 1631–1666 (2011)
41. Liu, S., Triggiani, R.: Global uniqueness in determining electric potentials for a system of strongly coupled Schrödinger equations with magnetic potential terms. J. Inverse Ill-Posed Probl. **19**, 223–254 (2011)
42. Liu, S., Triggiani, R.: Recovering the damping coefficients for a system of coupled wave equations with Neumann BC: uniqueness and stability. Chin. Ann. Math. Ser. B **32**, 669–698 (2011)
43. Liu, S., Triggiani, R.: Determining damping and potential coefficients of an inverse problem for a system of two coupled hyperbolic equations. Part I: global uniqueness. DCDS Supplement, 1001–1014 (2011)
44. Liu, S., Triggiani, R.: Global uniqueness and stability in determining the damping coefficient of an inverse hyperbolic problem with non-homogeneous Dirichlet B.C. through an additional localized Neumann boundary trace. Appl. Anal.**91**(8), 1551–1581 (2012)
45. Liu, S., Triggiani, R.: Recovering damping and potential coefficients for an inverse non-homogeneous second-order hyperbolic problem via a localized Neumann boundary trace. Discrete Contin. Dyn. Syst. Ser. A **33**(11–12), 5217–5252 (2013)
46. Liu, S., Triggiani, R.: Boundary control and boundary inverse theory for non-homogeneous second-order hyperbolic equations: a common Carleman estimates approach. HCDTE Lecture notes, AIMS Book Series on Applied Mathematics, vol. 6, pp. 227–343 (2013)
47. Liu, S., Triggiani, R.: An inverse problem for a third order PDE arising in high-intensity ultrasound: global uniqueness and stability by one boundary measurement. J. Inverse Ill-Posed Probl. **21**, 825–869 (2013)
48. Marchand, R., McDevitt, T., R. Triggiani, R.: An abstract semigroup approach to the third-order Moore-Gibson-Thompson partial differential equation arising in high-intensity ultrasound: structural decomposition, spectral analysis, exponential stability. Math. Methods Appl. Sci. **35**, 1896–1929 (2012)
49. Mazya, V.G., Shaposhnikova, T.O.: Theory of Multipliers in Spaces of Differentiable Functions, vol. 23. Monographs and Studies in Mathematics, Pitman (1985)
50. Pazy, A.: Semigroups of linear operators and applications to partial differential equations. Applied Mathematical Sciences, vol. 44. Springer, New York (1983)
51. Tataru, D.: A-priori estimates of Carleman's type in domains with boundary. J. Math. Pures. et Appl. **73**, 355–387 (1994)
52. Tataru, D.: Boundary controllability for conservative PDE's. Appl. Math. & Optimiz. **31**, 257–295 (1995); Based on a Ph.D. dissertation, University of Virginia (1992)
53. Tataru, D.: Carleman estimates and unique continuation for solutions to boundary value problems. J. Math. Pures Appl. **75**, 367–408 (1996)
54. Tataru, D.: On the regularity of boundary traces for the wave equation. Ann. Scuola Norm. Sup. Pisa Cl. Sci. (4) **26**, 185–206 (1998)
55. Taylor, M.: Pseudodifferential Operators. Princeton University Press, Princeton (1981)
56. Triggiani, R.: Exact boundary controllability of $L_2(\Omega) \times H^{-1}(\Omega)$ of the wave equation with Dirichlet boundary control acting on a portion of the boundary and related problems. Appl. Math. Optim. **18**(3), 241–277 (1988)
57. Triggiani, R.: Wave equation on a bounded domain with boundary dissipation: an operator approach. J. Math. Anal. Appl. **137**, 438–461 (1989)
58. Triggiani, R., Yao, P.F.: Carleman estimates with no lower order terms for general Riemannian wave equations: global uniqueness and observability in one shot. Appl. Math. Optim. **46**, 331–375 (2002)
59. Yamamoto, M.: Uniqueness and stability in multidimensional hyperbolic inverse problems. J. Math. Pures Appl. **78**, 65–98 (1999)

Chapter 16
Solutions of Stochastic Systems Generalized Over Temporal and Spatial Variables

Irina V. Melnikova, Uliana A. Alekseeva, and Vadim A. Bovkun

The paper is dedicated to the blessed memory of Alfredo Lorenzi, brilliant mathematician and bright personality

Abstract The Cauchy problem for systems of differential equations with white noise type random perturbations is considered as a particular case of the first order abstract Cauchy problem with generators of R-semigroups in Hilbert spaces and with Hilbert space valued random processes. A generalized Q-white noise and cylindrical white noise are introduced as generalized derivatives of Q-Wiener and cylindrical Wiener processes in special spaces of distributions; R-semigroups generated by differential operators of the systems are defined; solutions generalized over the temporal and spacial variables are constructed in spaces of type $\mathscr{D}'(\Psi')$, where topological spaces Ψ' are chosen in dependence on singularities of solution operators to the corresponding homogeneous systems.

16.1 Introduction and Setting of the Problem

Models of various evolution processes considered with regard to random perturbations lead to problems for equations with inhomogeneities of white noise type in infinite dimensional spaces. Among them, important for applications are the first-order Cauchy problems with operators A being generators of different type semigroups in a Hilbert space H and a white noise $\mathscr{W}$ with values in another Hilbert space $\mathbb{H}$:

$$X'(t) = AX(t) + B\mathscr{W}(t), \quad t \geq 0, \quad X(0) = \xi, \tag{16.1}$$

especially with differential operators $A = A\left(i\frac{\partial}{\partial x}\right)$.

I.V. Melnikova (✉) • U.A. Alekseeva • V.A. Bovkun
Ural Federal University, Lenin av. 51, Ekaterinburg 620083, Russia
e-mail: Irina.Melnikova@urfu.ru; Uliana.Alekseeva@urfu.ru; 123456m@inbox.ru

© Springer International Publishing Switzerland 2014
A. Favini et al. (eds.), *New Prospects in Direct, Inverse and Control Problems for Evolution Equations*, Springer INdAM Series 10,
DOI 10.1007/978-3-319-11406-4_16

Let $(\Omega, \mathscr{F}, P)$ be a random space. We consider the Cauchy problem for the stochastic partial differential equation

$$\frac{\partial X(t, x, \omega)}{\partial t} = A\left(i\frac{\partial}{\partial x}\right) X(t, x, \omega) + B\mathscr{W}(t, x, \omega), \quad t \geq 0, \ x \in \mathbb{R}^n, \ \omega \in \Omega,$$

$X(0, x, \omega) = \xi(x, \omega)$. The equations are understood here almost surely with respect to P ($P_{a.s.}$). Further we usually omit ω and write the problem as follows

$$\frac{\partial X(t, x)}{\partial t} = A\left(i\frac{\partial}{\partial x}\right) X(t, x) + B\mathscr{W}(t, x), \quad t \geq 0, \quad X(0, x) = \xi(x), \ x \in \mathbb{R}^n,$$

$$(16.2)$$

still implying the equalities hold $P_{a.s.}$. White noise $\mathscr{W}$, being informally defined in the finite-dimensional case, is a process identically distributed with independent at different moments t_1, t_2 random values, with zero expectation and infinite variation. Because of the definition, a white noise has irregular properties and in the "classical theory" stochastic equations are considered in an integral form with the stochastic integral with respect to a Brownian motion being "a primitive" of $\mathscr{W}$. The infinite-dimensional Brownian motion is used to call a Wiener process. Then a white nose can be defined as a generalized derivative of a Wiener process.

The operator of the equation is a matrix-operator $A\left(i\frac{\partial}{\partial x}\right) = \{A_{jk}\left(i\frac{\partial}{\partial x}\right)\}_{j,k=1}^m$ generating different type systems in the Gelfand–Shilov classification [6]. $A_{jk}\left(i\frac{\partial}{\partial x}\right)$ are linear finite orders differential operators with constant coefficients. The corresponding homogeneous (deterministic) Cauchy problem:

$$\frac{\partial u(t, x)}{\partial t} = A\left(i\frac{\partial}{\partial x}\right) u(t, x), \quad t \in [0, T], \ x \in \mathbb{R}^n, \qquad u(0, x) = f(x),$$

$$(16.3)$$

is not well-posed in the classical sense. It is well-posed in a generalized sense, precisely in spaces Ψ' of generalized over x functions, determined by properties of operator $A\left(i\frac{\partial}{\partial x}\right)$, that is by properties of the system. Considered as an abstract Cauchy problem, this problem has unbounded solution operators and A generates an R-semigroup of operators in H.

By the reason of unboundedness of solution operators to (16.3) and white noise singularities, we consider (16.2) as a generalized stochastic Cauchy problem with $\mathscr{W}$ defined as the derivative of a Wiener process W in the sense of distributions over time variable t and with values in appropriate spaces Ψ'. So, for each test function $\varphi \in \mathscr{D}$, we have a function of two variables $\langle\varphi(\cdot), \mathscr{W}(\cdot, x, \omega)\rangle$: for each $x \in \mathbb{R}^n$, it is a random value with respect to $\omega \in \Omega$ square integrable on Ω with respect to P and a $P_{a.s.}$ function of $x \in \mathbb{R}^n$ with properties which we will discuss later.

Stochastic problems (16.1) and (16.2) are of great interest from point of view both theoretical study and applications. As we have mentioned above, the main known approach to the investigation of the problems is studying them in the integral form, where the inhomogeneity is considered as the stochastic Ito integral with respect to a Wiener process (see e.g. [3,5,10]). In this framework weak R-solutions of problems with A generating R-semigroups were obtained [9]. Generalized over t solutions of the Cauchy problem (16.1) with white noise $\mathscr{W}$ and A generating integrated and convoluted semigroups were constructed [1]. Relations between weak (weak regularized) solutions and generalized ones are shown in [7].

To obtain a solution to (16.2) in the case under consideration we need to construct solutions generalized over several variables. In the present paper we propose to consider the problem (16.2) in a sense of generalized functions over two variables t and x: in spaces of type $\mathscr{D}'(\Psi')$ for a topological space Ψ'; there are a lot of questions here, even in finite dimensional case (see, e.g. [12]). On the way of studying and solving the problem we construct a Q-white noise and a cylindrical white noise, R-semigroups generated by different differential operators, spaces of the type $\mathscr{D}'(\Psi')$, and finally obtain generalized over t and x solutions in $\mathscr{D}'(\Psi')$, where topological spaces Ψ' are chosen in dependence on singularities of solution operators to (16.3). Operator B is supposed to be in $\mathscr{L}(\mathbb{H}, H)$ for the case of a Q-white noise and $B \in \mathscr{L}_{\mathrm{HS}}(\mathbb{H}, H)$ (Hilbert–Schmidt operators) for the case of a cylindrical white noise.

R-semigroup technique for ill-posed deterministic Cauchy problems with the Gelfand–Shilov operators $A\left(i\frac{\partial}{\partial x}\right)$ firstly was used in [11] for construction of regularizing algorithms with help of the generalized Fourier transform and a regularizing function $K = K_\varepsilon(\sigma), \sigma \in \mathbb{R}$, defining an R-semigroup and depending on a regularizing parameter $\varepsilon > 0$. This technique turns out to be also useful for solving the stochastic Cauchy problem (16.2): it makes possible to construct an operator that defines generalized solutions of this problem via weak regularized solutions, which in turn is determined by $K = K(\sigma), \sigma \in \mathbb{R}$, corresponding to different types of systems in the Gelfand–Shilov classification and defining corresponding R-semigroup.

16.2 Wiener Processes and White Noises in Hilbert Spaces

Let $\mathbb{H}$ be a separable Hilbert space and Q be a linear symmetric nonnegative trace class operator in $\mathbb{H}$.

Definition 16.1 An $\mathbb{H}$-valued stochastic process $\{W_Q(t), t \geq 0\}$ is called a Q-Wiener process if

(QW1) $W_Q(0) = 0 \; P_{\mathrm{a.s.}}$;
(QW2) the process has independent increments $W_Q(t) - W_Q(s), 0 \leq s \leq t$, with the normal distribution $\mathscr{N}(0, (t-s)Q)$;
(QW3) $W_Q(t)$ has continuous trajectories $P_{\mathrm{a.s.}}$

Since Q is a nonnegative trace class operator, its eigenvectors $\{e_j\}$ form an orthonormal basis in $\mathbb{H}$ and $\mathrm{Tr}\, Q = \sum_{j=1}^{\infty} \sigma_j^2 < \infty$, where $Qe_j = \sigma_j^2 e_j$. Then the Q-Wiener process $\{W_Q(t), t \geq 0\}$ is represented by the convergent in $L_2(\Omega; \mathbb{H})$ series:

$$W_Q(t) = \sum_{j=1}^{\infty} \sigma_j \beta_j(t) e_j, \quad t \geq 0,$$

where $\beta_j(t)$ are independent Brownian motions. Thus, an $\mathbb{H}$-valued Q-white noise $\mathscr{W}_Q$ can be defined as follows:

$$\langle \varphi(\cdot), \mathscr{W}_Q(\cdot) \rangle := -\langle \varphi'(\cdot), W_Q(\cdot) \rangle, \quad \varphi \in \mathscr{D}, \ P_{\text{a.s.}}. \tag{16.4}$$

Here $W_Q(\cdot)$ means Q-Wiener process $\{W_Q(t), \ t \geq 0\}$ continued by zero for $t < 0$.

As for a cylindrical Wiener process $\{W(t), t \geq 0\}$ in $\mathbb{H}$, where $Q = I$ and $\text{Tr}\,I = \infty$, that corresponds to $\sigma_j^2 = 1$, it is formally defined by a divergent in $L_2(\Omega; \mathbb{H})$ series $\sum_{j=1}^\infty \beta_j(t)e_j =: W(t), t \geq 0, P_{\text{a.s.}}$ [3, 10]. However, $\{W(t), t \geq 0\}$ can be treated as a Q_1-Wiener process in a wider Hilbert space $\mathbb{H}_1 \supset \mathbb{H}$ with a trace class operator $Q_1 : \mathbb{H}_1 \to \mathbb{H}_1$ and can be represented by a series, convergent $L_2(\Omega; \mathbb{H}_1)$. Let us show it beginning with construction of $\mathbb{H}_1$ in such a way that $\mathbb{H} = J(\mathbb{H}_1)$, where J is a positive self-adjoint Hilbert–Schmidt embedding operator. Define a norm on $\mathbb{H}_1 := J^{-1}(\mathbb{H})$ by $\|f\|_{\mathbb{H}_1} := \|Jf\|_{\mathbb{H}}$. Then $\{g_j = J^{-1}e_j\}$ is an orthonormal basis in $\mathbb{H}_1$. Among all such operators J we choose one defined on the basis as $Jg_j := \eta_j g_j$ where, by properties of Hilbert–Schmidt operators, $\sum_{j=1}^\infty \eta_j^2 < \infty$, that is we construct $\mathbb{H}_1$ as closed linear span of the vectors $\{g_j = \frac{e_j}{\eta_j}\}$. As a result, we obtain that $\{W(t), t \geq 0\}$ defined in $\mathbb{H}$ by the formal series $\sum_{j=1}^\infty \beta_j(t)e_j$ is a Q_1-Wiener process in $\mathbb{H}_1$:

$$W(t) = \sum_{j=1}^\infty \eta_j \beta_j(t) g_j, \quad t \geq 0, \tag{16.5}$$

with zero expectation and covariation operator $Q_1 := J^2$ being a trace class operator as the product of Hilbert–Schmidt operators. Thus, an $\mathbb{H}_1$-valued white noise $\mathscr{W}(\cdot)$ in (16.1) can be defined similarly (16.5):

$$\langle \varphi(\cdot), \mathscr{W}(\cdot) \rangle := -\langle \varphi'(\cdot), W(\cdot) \rangle, \quad \varphi \in \mathscr{D}, \ P_{\text{a.s.}}, \tag{16.6}$$

where $W(\cdot)$ means Wiener process (16.5) continued by zero for $t < 0$. In this connection we can expand $B \in \mathscr{L}_{\text{HS}}(\mathbb{H}, H)$ to $\mathbb{H}_1$ as an operator from $\mathscr{L}(\mathbb{H}_1, H)$.

Note, that the well known condition for existence of the stochastic integral $\int_0^t \Phi(s)dW(s)$ with respect to a cylindrical Wiener process (see, e.g. [3, 10]) is formally defined under the same conditions as with respect to a Q-Wiener process.

16.3 R-semigroups Generated by Differential Operators

Construction of R-semigroups is related to the homogeneous Cauchy problem (16.3). We consider it as a particular case of the deterministic abstract Cauchy problem

$$u'(t) = Au(t), \quad t \in [0, \tau), \qquad u(0) = f, \tag{16.7}$$

in a Banach space H.

Definition 16.2 Let A be a closed linear operator and R be a bounded linear operator in a Banach space H with densely defined R^{-1}. A strongly continuous by t family $\{S(t), \ t \in [0, \tau), \ \tau \le \infty\}$ of bounded linear operators in H is called an R-*semigroup* generated by A if

$$S(t)Af = AS(t)f, \quad t \in [0, \tau), \ f \in \text{dom } A, \tag{16.8}$$

$$S(t)f = A \int_0^t S(\tau)f \, ds + Rf, \quad t \in [0, \tau), \ f \in H. \tag{16.9}$$

The semigroup is called *local* if $\tau < \infty$.

If A generates an R-semigroup, then it commutes with R on dom A and if in addition $\overline{A|_{R(\text{dom}A)}} = A$, then

$$u(t) = R^{-1}S(t)f, \quad f \in R(\text{dom}A), \tag{16.10}$$

is a unique solution of the Cauchy problem (16.7) and the problem is R-well-posed, that is $\|u(t)\| \le C \|R^{-1}f\|, t \in [0, \tau)$ (see, e.g. [2, 8, 13]).

Note that solution operators $R^{-1}S(t)$ of the Cauchy problem (16.7) with the generator of an R-semigroup are not defined on the whole H, as it takes place for generators of C_0 class semigroups; operators $R^{-1}S(t)$ are defined only on $R(H)$ and are not bounded on this set. Thus $u(t), t \in [0, T]$ is not stable with respect to small changes of initial data in H.

Let us return to (16.3). We will show that $A\left(i\frac{\partial}{\partial x}\right)$, in dependence on its properties, generates different R-semigroups in the space $H = L_2^m(\mathbb{R}^n) :=$ $L_2(\mathbb{R}^n) \times \cdots \times L_2(\mathbb{R}^n)$. Following [6], we apply the generalized Fourier transform to the system (16.3) and consider the dual one:

$$\frac{\partial \tilde{u}(t, s)}{\partial t} = A(s)\tilde{u}(t, s), \quad t \in [0, T], \ s \in \mathbb{C}^n. \tag{16.11}$$

Here the matrix-function $A(s), s \in \mathbb{C}^n$ is the Fourier transform of $A\left(i\frac{\partial}{\partial x}\right)$ and vector-function $\tilde{u}(t) = \tilde{u}(t, s), s \in \mathbb{C}^n$ is the Fourier transform of $u(t) = u(t, x)$, $x \in \mathbb{R}^n$.

Let the functions $\lambda_1(\cdot), \ldots, \lambda_m(\cdot)$ be characteristic roots of the matrix-operator $A(\cdot)$ in (16.11) and $\Lambda(s) := \max_{1 \le k \le m} Re \, \lambda_k(s), s \in \mathbb{C}^n$. If p is the maximal order of the differential operators $A_{jk}\left(i\frac{\partial}{\partial x}\right)$, then solution operators of (16.11) satisfy the estimate [6] :

$$e^{t\Lambda(s)} \leq \left\|e^{tA(s)}\right\|_{m\times m} \leq C(1+|s|)^{p(m-1)}e^{t\Lambda(s)}, \quad t \geq 0, \; s \in \mathbb{C}^n, \qquad (16.12)$$

where $\|\cdot\|_{m\times m}$ is the norm of the $m \times m$ matrix-function $e^{tA(s)}$ constructed as the corresponding series for the matrix $A(s)$: $e^{tA(s)} = I + tA(s) + \frac{t^2A^2(s)}{2!} + \dots$

Definition 16.3 Let $\sigma = Re\, s$. The system (16.3) is called

1) *correct by Petrovsky* if there exists such a $C > 0$ that $\Lambda(\sigma) \leq C$, $\sigma \in \mathbb{R}^n$;
2) *conditionally-correct* if there exist such constants $C > 0$, $0 < h < 1$, $C_1 > 0$ that $\Lambda(\sigma) \leq C|\sigma|^h + C_1$, $\sigma \in \mathbb{R}^n$;
3) *incorrect* if the function $\Lambda(\cdot)$ grows for real $s = \sigma$ in the same way as for complex ones: $\Lambda(\sigma) \leq C|\sigma|^{p_0} + C_1$, $\sigma \in \mathbb{R}^n$, where the number $p_0 := \inf\{\rho : |\Lambda(s)| \leq C_\rho(1+|s|)^\rho, \; s \in \mathbb{C}^n\}$ is called a *reduced order of system (16.3)*.

Theorem 16.1 *Let $\tau > 0$. Let $K(\sigma)$, $\sigma \in \mathbb{R}^n$, satisfy the conditions:*

1. $K(\cdot)e^{TA(\cdot)} \in L_2^m(\mathbb{R}^n) \times L_2^m(\mathbb{R}^n)$;
2. $K(\sigma)e^{tA(\sigma)}$ *is a bounded matrix-function of $(n+1)$ variables on $[0,T] \times \mathbb{R}^n$ for arbitrary $T \in (0,\tau)$, that is*

$$\exists C > 0: \quad \|K(\sigma)e^{tA(\sigma)}\|_{m\times m} \leq C, \quad t \in [0,T], \sigma \in \mathbb{R}^n.$$

Then the family of convolution operators

$$[S(t)f](x) := G_R(t,x) * f(x), \quad t \in [0,\tau), \; x \in \mathbb{R}^n, \qquad (16.13)$$

where

$$G_R(t,x) := \frac{1}{(2\pi)^n}\int_{\mathbb{R}^n} e^{-i\sigma x}K(\sigma)e^{tA(\sigma)}\,d\sigma, \quad t \in [0,\tau), \; x \in \mathbb{R}^n, \qquad (16.14)$$

forms a local R-semigroup in $L_2^m(\mathbb{R}^n)$ with generator $A\left(i\frac{\partial}{\partial x}\right)$ and

$$Rf(x) = \frac{1}{(2\pi)^n}\int_{\mathbb{R}^n} e^{-i\sigma x}K(\sigma)\tilde{f}(\sigma)\,d\sigma, \quad x \in \mathbb{R}^n. \qquad (16.15)$$

Proof For the sake of simplicity we give the proof for $n = 1$.

First note that to prove operators $S(t), t \in [0,\tau)$ form an R-semigroup, it is sufficient to prove all the properties of Definition 16.2 on arbitrary $[0,T] \subset [0,\tau)$. So, take $T \in (0,\tau)$. Since $K(\cdot)e^{TA(\cdot)}$ satisfies condition (1) of the theorem, the integral in (16.14)

$$\frac{1}{2\pi}\int_{-\infty}^{\infty} e^{-i\sigma x}K(\sigma)e^{tA(\sigma)}\,d\sigma, \quad t \in [0,T],$$

is convergent in $L_2^m(\mathbb{R}) \times L_2^m(\mathbb{R})$. Moreover, this convergence is uniform with respect to $t \in [0, T]$ and the obtained matrix-function $G_R(t, x)$, called regularized Green function, is defined. Since, in addition, the matrix-function $K(\sigma)e^{tA(\sigma)}$ is bounded, the integral

$$\int_{-\infty}^{\infty} e^{-i\sigma x} K(\sigma) e^{tA(\sigma)} \tilde{f}(\sigma) \, d\sigma, \quad t \in [0, T], \tag{16.16}$$

is an element of $L_2^m(\mathbb{R})$ for each $\tilde{f} \in L_2^m(\mathbb{R})$.

Now we are ready to check that the family (16.13) forms a local R-semigroup in $L_2^m(\mathbb{R})$. First, we verify the strong continuity property of the family $\{S(t), t \in [0, T]\}$: for each $f \in L_2^m(\mathbb{R})$ and $T < \tau$ we show that $\|S(t)f - S(t_0)f\| \to 0$ as $t \to t_0$, $t_0 \in [0, T]$.[1]

$$\|S(t)f - S(t_0)f\|^2 = \int_{\mathbb{R}} \left\| \frac{1}{2\pi} \int_{-\infty}^{\infty} e^{-i\sigma x} K(\sigma) \left[e^{tA(\sigma)} \tilde{f}(\sigma) \right. \right.$$
$$\left. \left. - e^{t_0 A(\sigma)} \tilde{f}(\sigma) \right] d\sigma \right\|_m^2 dx.$$

Let us split the inner integral into the three ones:

$$\int_{|\sigma| \geq N} e^{-i\sigma x} K(\sigma) e^{tA(\sigma)} \tilde{f}(\sigma) \, d\sigma - \int_{|\sigma| \geq N} e^{-i\sigma x} K(\sigma) e^{t_0 A(\sigma)} \tilde{f}(\sigma) \, d\sigma +$$
$$+ \int_{|\sigma| \leq N} e^{-i\sigma x} K(\sigma) \left[e^{tA(\sigma)} - e^{t_0 A(\sigma)} \right] \tilde{f}(\sigma) \, d\sigma. \tag{16.17}$$

Here the functions $h_N(x, t) := \displaystyle\int_{|\sigma| \geq N} e^{-i\sigma x} K(\sigma) e^{tA(\sigma)} \tilde{f}(\sigma) \, d\sigma$ and

$$g_N(x, t) := \int_{|\sigma| \leq N} e^{-i\sigma x} K(\sigma) \left[e^{tA(\sigma)} - e^{t_0 A(\sigma)} \right] \tilde{f}(\sigma) \, d\sigma$$

are elements of $L_2^m(\mathbb{R})$ for each $t \in [0, T]$ as the inverse Fourier transform of the functions from $L_2^m(\mathbb{R})$

$$\tilde{h}_N(\sigma, t) = \begin{cases} 0, & |\sigma| \leq N, \\ K(\sigma) e^{tA(\sigma)} \tilde{f}(\sigma), & |\sigma| > N, \end{cases}$$

---

[1] Throughout this proof the norm $\| \cdot \|$ denotes the norm in $L_2^m(\mathbb{R})$. This means that $\|f\|^2 = \sum_{j=1}^m \int_{\mathbb{R}} |f_j(x)|^2 \, dx = \int_{\mathbb{R}} \left(\sum_{j=1}^m |f_j(x)|^2 \right) dx = \int_{\mathbb{R}} \|f(x)\|_m^2 \, dx$, where $\| \cdot \|_m$ denotes the norm of vector in $\mathbb{C}^m$.

and $\tilde{g}_N(\sigma,t) = K(\sigma)e^{tA(\sigma)}\tilde{f}(\sigma) - \tilde{h}_N(\sigma,t) - K(\sigma)e^{t_0 A(\sigma)}\tilde{f}(\sigma) + \tilde{h}_N(\sigma,t_0)$,
respectively. The integral (16.16) is convergent uniformly with respect to both $x \in \mathbb{R}$
and $t \in [0,T]$, that provided by $K(\cdot)e^{tA(\cdot)} \in L_2^m(\mathbb{R}) \times L_2^m(\mathbb{R})$ and $\tilde{f}(\cdot) \in L_2^m(\mathbb{R})$.
Hence for any $\varepsilon > 0$

$$\|h_N(x,t)\|_m < \varepsilon/4, \quad x \in \mathbb{R},\ t \in [0,T],$$

by the choice of N and the sum of absolute values of the first two integrals in (16.17)
is less than $\varepsilon/2$. Now fix N. Since $\left(e^{(t-t_0)A(\sigma)} - 1\right) \to 0$ as $t \to t_0$ uniformly with
respect to $\sigma \in [-N,N]$, we can take $\|g_N(x,t)\|_m < \varepsilon/2$, $x \in \mathbb{R}$, $t \in O(t_0)$. To
obtain the estimate for

$$\|S(t)f - S(t_0)f\|^2 = \frac{1}{4\pi^2}\int_{\mathbb{R}} \|h_N(x,t) - h_N(x,t_0) + g_N(x,t)\|_m^2 dx$$

we consider the difference $h_N(x,t) - h_N(x,t_0) =: \Delta_N(x,t,t_0)$, $t,t_0 \in [0,T]$, as
a single function, then $\Delta_N(\cdot,t,t_0) \in L_2^m(\mathbb{R})$ and for a fixed N by the choice of t_0,
$\|\Delta_N(x,t,t_0)\|_m < \varepsilon/2$, $x \in \mathbb{R}$. In these notations we have:

$$4\pi^2\|S(t)f - S(t_0)f\|^2 \le \int_{\mathbb{R}} \|\Delta_N^2(x,t,t_0)\|_m\, dx+$$

$$+ 2\int_{\mathbb{R}} \|\Delta_N(x,t)g_N(x,t,t_0)\|_m\, dx + \int_{\mathbb{R}} \|g_N^2(x,t)\|_m\, dx.$$

On the way described above one can show that each of these three integrals is an
infinitesimal value. Really, the integrals over the infinite intervals $|x| > M$ are small
by the choice of M because of their uniform convergence with respect to $t \in [0,T]$.
Integrals on compacts $[-M,M]$ are small because the integrands are small, that is
provided by the sequential choice of M and $t \in O(t_0)$. This completes the proof
that operators of the family (16.13) are strongly continuous.

Next, we show that the obtained operators commute with $A\left(i\frac{\partial}{\partial x}\right)$ on $f \in$
$dom\, A\left(i\frac{\partial}{\partial x}\right)$. By properties of convolution, a differential operator may be applied
to any components of convolution, so we apply $A\left(i\frac{\partial}{\partial x}\right)$ to $f \in dom\, A\left(i\frac{\partial}{\partial x}\right)$:

$$A\left(i\frac{\partial}{\partial x}\right)[S(t)f](x) = G_R(t,x) * A\left(i\frac{\partial}{\partial x}\right)f(x) = S(t)A\left(i\frac{\partial}{\partial x}\right)f(x),$$

and obtain the equality (16.8) hold.

In conclusion, we show the R-semigroup equation (16.9). For an arbitrary $f \in$
$dom\, A\left(i\frac{\partial}{\partial x}\right)$ consider the equality:

$$\frac{\partial}{\partial t}[S(t)f](x) = \frac{\partial}{\partial t}[G_R(t,x) * f(x)] = \frac{1}{2\pi}\frac{\partial}{\partial t}\int_{-\infty}^{\infty} e^{-i\sigma x}K(\sigma)e^{tA(\sigma)}\tilde{f}(\sigma)\, d\sigma.$$

To differentiate under the integral sign we apply the dominated convergence theorem. Due to conditions on $K(\cdot)$, the difference quotient is uniformly bounded with respect to $t \in [0, T]$:

$$\left\| e^{-i\sigma x} K(\sigma) \left(\frac{e^{tA(\sigma)} - e^{t_0 A(\sigma)}}{t - t_0} \right) \tilde{f}(\sigma) \right\|_m =$$

$$= \left\| e^{-i\sigma x} K(\sigma) e^{(t_0 + \theta(t - t_0))A(\sigma)} A(\sigma) \tilde{f}(\sigma) \right\|_m \leq C \left\| A(\sigma) \tilde{f}(\sigma) \right\|_m$$

and the condition $f \in dom\, A\left(i \frac{\partial}{\partial x}\right)$ provides $A(\cdot) \tilde{f}(\cdot) \in L_2^m(\mathbb{R})$. Hence the conditions of the dominated convergence theorem hold and

$$\frac{\partial}{\partial t} [S(t) f](x) = \frac{1}{2\pi} \int_{-\infty}^{\infty} e^{-i\sigma x} K(\sigma) e^{tA(\sigma)} A(\sigma) \tilde{f}(\sigma)\, d\sigma.$$

Taking into account that the inverse Fourier transform of $A(\sigma) \tilde{f}(\sigma)$ is $A\left(i \frac{\partial}{\partial x}\right) f(x)$ we obtain

$$\frac{\partial}{\partial t} [S(t) f](x) = G_R(t, x) * A\left(i \frac{\partial}{\partial x}\right) f(x) = A\left(i \frac{\partial}{\partial x}\right) [G_R(t, x) * f(x)] =$$

$$= A\left(i \frac{\partial}{\partial x}\right) [S(t) f](x).$$

Integration with respect to t gives the equality

$$[S(t) f](x) - [S(0) f](x) = \int_0^t A\left(i \frac{\partial}{\partial x}\right) [S(\tau) f](x)\, d\tau.$$

Since $A\left(i \frac{\partial}{\partial x}\right)$ is closed in $L_2^m(\mathbb{R})$ and differentiable functions are dense there, the equality holds for any $f \in L_2^m(\mathbb{R})$:

$$[S(t) f](x) - [S(0) f](x) = A\left(i \frac{\partial}{\partial x}\right) \int_0^t [S(\tau) f](x)\, d\tau, \quad t \in [0, T].$$

Let operator R in $L_2^m(\mathbb{R})$ equal to $S(0)$, then by the strong continuity property,

$$Rf(x) = \frac{1}{2\pi} \int_{-\infty}^{\infty} e^{-i\sigma x} K(\sigma) \tilde{f}(\sigma)\, d\sigma.$$

So, we have proved that operators (16.13) form in $L_2^m(\mathbb{R})$ the R-semigroup generated by $A\left(i \frac{\partial}{\partial x}\right)$ with R defined by (16.15). $\square$

Note that in [11], where as we mentioned in introduction, the R-semigroup technique was related to Gelfand–Shilov systems, a proof of the R-semigroup property was not given.

Corollary 16.1 *If the system (16.3) is correct by Petrovsky, the function $K(\sigma) = \frac{1}{(1+\sigma^2)^{d/2+1}}$ satisfies the conditions of the theorem with $d = p(m-1)$.*

If the system (16.3) is conditionally-correct, we can choose $K(\sigma) = e^{-a|\sigma|^h}$ with $a > a_0 T$, where $a_0 = a_0(p, m, C_1, C_2)$ is defined by the corresponding parameters of Definition 16.3 and the estimate (16.12).

If the system (16.3) is incorrect, we can choose $K(\sigma) = e^{-a|\sigma|^{p_0}}$ with $a > a_0 T$, where $a_0 = a_0(p, m, C_1, C_2)$ is defined by the corresponding parameters of Definition 16.3 and (16.12).

Remark 16.1 Let us note, that the inverse operator to R is determined by solving the equation $g(x) = Rf(x) = F^{-1}[K(\sigma)\tilde{f}(\sigma)]$, that implies

$$R^{-1}g(x) = F^{-1}\left[\frac{\tilde{g}(\sigma)}{K(\sigma)}\right], \quad x \in \mathbb{R}^n.$$

16.4   The Space $\mathscr{D}'(\Psi')$

Define the space $\mathscr{D}'(\Psi')$ as the space $\mathscr{L}(\mathscr{D}, \Psi')$ of linear continuous operators from $\mathscr{D}$ to Ψ'. Here $\mathscr{D}$ is the space of infinitely differentiable functions with compact supports in $\mathbb{R}$. Ψ is a locally convex space and Ψ' is its adjoint with the weak topology, i.e. the topology corresponding to the convergence of a sequence on each element of a test space. We assume $\mathscr{D}'(\Psi')$ equipped with the strong topology, i.e. that corresponding to the uniform convergence on bounded sets of $\mathscr{D}$.

For the case of L.Schwartz space $\mathscr{D}' = \mathscr{L}(\mathscr{D}, \mathbb{R})$ the well known fact is that for any compact $K \subset \mathbb{R}$ and any $f \in \mathscr{D}'$ there exist such $p \in \mathbb{N}_0$ and $C > 0$ that $|f(\varphi)| \leq C \|\varphi\|_p$, where $\|\varphi\|_p = \sup_{K \subset \mathbb{R}} \|\varphi\|_{K,p}$, $\varphi \in \mathscr{D}$. This is a reflection of the structure of $\mathscr{D}' = \bigcap_K \bigcup_p \mathscr{D}'_{K,p}$, where the space $\mathscr{D}_{K,p}$ consisting of all p times differentiable functions with supports in the compact $K \subset \mathbb{R}$ is complete with the norm

$$\|\varphi\|_{K,p} = \sup_{q \leq p} \sup_{t \in K} |\varphi^{(q)}(t)|,$$

and $\mathscr{D}'_{K,p} = \mathscr{L}(\mathscr{D}_{K,p}, \mathbb{R})$.

We prove a similar statement for $\mathscr{D}'(\Psi')$.

Proposition 16.1 *Let $K \subset \mathbb{R}$ be a compact set. For any $f \in \mathscr{D}'(\Psi')$ there exists such $p \in \mathbb{N}_0$ that for any bounded set $\mathscr{B} \subset \mathscr{D}_{K,p}$ the set $f(\mathscr{B})$ is bounded in Ψ'.*

Proof Suppose the opposite: let for each $p \in \mathbb{N}_0$ there exists such a set $\mathscr{B}_p$ bounded in $\mathscr{D}_{K,p}$ that $f(\mathscr{B}_p)$ is not bounded in Ψ', that is there exists a weak neighborhood

V in Ψ' such that $\lambda f(\mathscr{B}_p) \not\subset V$ for any $\lambda > 0$. Thus, we obtain

$$\forall p \in \mathbb{N}_0 \; \exists \mathscr{B}_p \subset \mathscr{D}_{K,p}, \; \exists V \subset \Psi' \; : \; \left(\forall \lambda > 0 \; \exists \varphi_\lambda \in \mathscr{B}_p \; : \; \lambda f(\varphi_\lambda) \notin V \right).$$

Take $\lambda = \frac{1}{n} \to 0$, then $\frac{1}{n}\varphi_n \to 0$ in $\mathscr{D}_{K,p}$ because φ_n are from a bounded set. Therefore $\frac{1}{n} f(\varphi_n) = f(\frac{\varphi_n}{n}) \to 0$ in Ψ' weakly, that is on each element of Ψ. But we have a weak neighborhood V in which there no elements of this sequence. This contradiction ends the proof. $\qquad \square$

Due to the proposition, we obtain the structure theorem in the introduced space similarly considered in [4].

Theorem 16.2 *Let* $U \in \mathscr{D}'(\Psi')$ *and* $\mathscr{G} \subset \mathbb{R}$ *ba an open bounded set. Then there exist a continuous function* $f : \mathbb{R} \to \Psi'$ *and an integer* $m \in \mathbb{N}_0$ *such that for any* $\varphi \in \mathscr{D}$ *with* $supp\,\varphi \subset \mathscr{G}$

$$U(\varphi) = f^{(m)}(\varphi).$$

Proof Let $K = \overline{\mathscr{G}}$ and $\varepsilon > 0$. Denote $K_\varepsilon = \{t \in \mathbb{R} : \rho(t, K) \leq \varepsilon\}$ and let p be that provided by Proposition 16.1 for K_ε and U. We extend U to $\mathscr{D}_{K_\varepsilon,p}$ by continuity and save after it the same notation. Then for any bounded set $\mathscr{B} \subset \mathscr{D}_{K_\varepsilon,p}$ the set $U(\mathscr{B})$ is bounded in Ψ'.

Let

$$\eta(t) = \begin{cases} \frac{t^{p+1}}{(p+1)!}, & t \geq 0, \\ 0, & t < 0. \end{cases}$$

The function is obviously p times continuously differentiable. Take $\chi \in \mathscr{D}$ with support in $K_\varepsilon, t \in \mathbb{R}$ and consider $\lambda_t(s) := \chi(s)\eta(t-s)$, $s \in \mathbb{R}$. The function $\lambda_t(\cdot)$ is p times continuously differentiable and supp $\lambda_t \subseteq K_\varepsilon$, hence $\lambda_t \in \mathscr{D}_{K_\varepsilon,p}$ for each fixed $t \in \mathbb{R}$. In addition the function is continuous with respect to parameter $t \in \mathbb{R}$, therefore $f(t) := U(\lambda_t)$, $t \in \mathbb{R}$, is a well defined continuous function with values in Ψ'. This function f defines a regular functional with values in Ψ':

$$\int \varphi(t) f(t)\, dt = \int \varphi(t) U(\lambda_t)\, dt, \qquad \varphi \in \mathscr{D}.$$

Considering this integral as the limit of Riemann sums, we arrive at the equality

$$\int \varphi(t) U(\lambda_t)\, dt = U\left(\int \varphi(t)\lambda_t\, dt\right),$$

where $\int \varphi(t)\lambda_t\, dt = \int \varphi(t)\chi(s)\eta(t-s)\, dt = \chi(s)\int \varphi(t)\eta(t-s)\, dt$ is in $\mathscr{D}_{K_\varepsilon,p}$ as a function of s.

To complete the proof we choose $\chi(s) = 1$ for $s \in K$, take $\varphi \in \mathcal{D}$, $supp\varphi \in \mathcal{G}$ and find the generalized derivative of order $p + 2$ from f:

$$f^{(p+2)}(\varphi) := (-1)^{p+2} f(\varphi^{(p+2)}) = (-1)^{p+2} \int \varphi^{(p+2)}(t) f(t) \, dt =$$

$$= (-1)^{p+2} U \left(\chi(s) \int \varphi^{(p+2)}(t) \eta(t - s) \, dt \right).$$

Integrating by parts $p + 1$ times, we obtain

$$\int \varphi^{(p+2)}(t) \eta(t - s) \, dt = (-1)^{p+1} \int \varphi'(t) \eta^{(p+1)}(t - s) \, dt =$$

$$= (-1)^{p+1} \int_{t \geq s} \varphi'(t) \, dt = (-1)^{p+2} \varphi(s).$$

Then

$$f^{(p+2)}(\varphi) = (-1)^{p+2} U \left(\chi(s)(-1)^{p+2} \varphi(s) \right) = U(\varphi),$$

since $\chi \equiv 1$ in $\mathcal{G}$. $\square$

16.5   Solving the Problem in Spaces $\mathcal{D}'(\Psi')$

Now return to the stochastic Cauchy problem (16.2). We defined a Hilbert space valued process $\mathcal{W}$, being a Q-white noise or a cylindrical white noise by (16.4) or (16.6) as the generalized derivative of the corresponding Wiener process $W(t)$. Therefore, the problem (16.2) is understood in the generalized sense over t, this means that for each $\varphi \in \mathcal{D}$, $\xi \in \operatorname{dom} A \left(i \frac{\partial}{\partial x} \right) \subset L_2^m(\mathbb{R}^n)$ and $x \in \mathbb{R}^n$

$$\langle \varphi(t), X_t'(t, x) \rangle = \langle \varphi(t), A \left(i \frac{\partial}{\partial x} \right) X(t, x) \rangle + \varphi(0) \xi(x) + B \langle \varphi(t), W_t'(t, x) \rangle, \ P_{a.s.}.$$

$$(16.18)$$

Since $\mathcal{W}$ is not generally a regular distribution (over t), X is not a regular distribution too. Since the differential operator $A \left(i \frac{\partial}{\partial x} \right)$ generates the R-semigroup $\{S(t, \cdot), \ t \in [0, \tau)\}$ in $L_2^m(\mathbb{R})$ defined by (16.13), the unique solution of the homogeneous Cauchy problem corresponding to (16.2), exists for any $\xi \in R \left(\operatorname{dom} A \left(i \frac{\partial}{\partial x} \right) \right)$ and can be found as follows:

$$R^{-1} S(t, x) \xi(x) = R^{-1} \left[G_R(t, x) * \xi(x) \right] = F^{-1} \left[\frac{K(\sigma) e^{tA(\sigma)} \tilde{\xi}(\sigma)}{K(\sigma)} \right] =$$

$$= F^{-1} \left[e^{tA(\sigma)} \tilde{\xi}(\sigma) \right] =: G(t, x) * \xi(x).$$

Here, since A generally generates an R-semigroup, not a C_0-class semigroup, the Green function $G(t, \cdot)$ is a generalized function with respect to $x \in \mathbb{R}^n$ in an appropriate space Ψ' depending on properties of the system (that is on properties of R-semigroup), the convolution of G with initial data ξ is well-defined for $\xi \in R \left(\text{dom} A \left(i \frac{\partial}{\partial x} \right) \right)$.

Nevertheless, we need to define the convolution with the stochastic inhomogeneity to obtain a solution of (16.2), this inhomogeneity generally is not in the indicated set of well-posedness and we need to construct the convolution on the whole $L_2^m(\mathbb{R}^n)$. This forces to consider the generalized over t problem (16.18) in a generalized sense with respect to $x \in \mathbb{R}^n$, i.e. in a space Ψ'.

Thus, we arrive at the problem: for each $\varphi \in \mathscr{D}$, $\psi \in \Psi$

$$\langle \psi(x), \langle \varphi(t), X_t'(t,x) \rangle \rangle = \langle \psi(x), A \left(i \frac{\partial}{\partial x} \right) \langle \varphi(t), X(t,x) \rangle \rangle +$$

$$+ \varphi(0) \langle \psi(x), \xi(x) \rangle + \langle \psi(x), B \langle \varphi(t), W_t'(t,x) \rangle \rangle, \quad P_{a.s.}. \quad (16.19)$$

Here the notation $X(t,x)$ means that distribution $X(\cdot, \cdot)$ acts on $\varphi(t)$ by the first argument and on $\psi(x)$ by the second one.

Theorem 16.3 *Let $\mathscr{W}$ be a Q-white noise or a cylindrical white noise. Let $A \left(i \frac{\partial}{\partial x} \right)$ generate a local R-semigroup $\{S(t, \cdot), t \in [0, \tau)\}$ in $L_2^m(\mathbb{R}^n)$ and $AR^{-1}S(t, \cdot) : L_2^m(\mathbb{R}^n) \to \Psi'$ is a bounded operator for each $t \in [0, \tau)$. Then*

$$X(t,x) = R^{-1}S(t,x)\xi(x) + R^{-1}S(t,x) * B\mathscr{W}(t,x), \quad \xi \in L_2^m(\mathbb{R}^n), \quad (16.20)$$

solves (16.19) in $\mathscr{D}'(\Psi')$.

Proof Begin with definition the first term of the prospective solution (16.20). Property (16.10) of R-semigroups and boundness of $AR^{-1}S(t, \cdot)$ from $L_2^m(\mathbb{R}^n)$ into Ψ' imply that

$$\langle \psi(x), R^{-1}S(t,x)\xi(x) \rangle = \langle \psi(x), G(t,x) * \xi(x) \rangle, \quad \psi \in \Psi,$$

is a continuous and differentiable function of t, hence we can consider a regular generalized function over $\mathscr{D}$:

$$\langle \varphi(t), \langle \psi(x), R^{-1}S(t,x)\xi(x) \rangle \rangle = \int \varphi(t) \langle \psi(x), G(t,x) * \xi(x) \rangle \, dt =$$

$$= \int \varphi(t) \langle \psi(x), R^{-1}S(t,x)\xi(x) \rangle \, dt, \quad \varphi \in \mathscr{D}.$$

Replacing the last integral by the integral sums, we obtain

$$\sum \varphi(t_i)\langle \psi(x), R^{-1}S(t_i, x)\xi(x)\rangle \Delta t_i = \langle \psi(x), \sum \varphi(t_i)R^{-1}S(t_i, x)\xi(x)\Delta t_i\rangle.$$

Since the left side of the equality has the limit, the right one has too. Passing to the limit, we obtain representation of $R^{-1}S(t, x)\xi(x)$ in $\mathscr{D}'(\Psi')$:

$$\langle \psi(x), \langle \varphi(t), R^{-1}S(t, x)\xi(x)\rangle\rangle = \langle \varphi(t), \langle \psi(x), R^{-1}S(t, x)\xi(x)\rangle\rangle =$$
$$= \int \varphi(t)\langle \psi(x), G(t, x) * \xi(x)\rangle \, dt, \quad \varphi \in \mathscr{D}, \ \psi \in \Psi. \quad (16.21)$$

Now define the second term in (16.20). The convolution with respect to t of generalized functions on $\mathscr{D}$ is defined via convolution of their primitives. Hence we obtain

$$\langle \varphi(t), \langle \psi(x), R^{-1}S(t, x) * BW'_t(t, x)\rangle\rangle = -\langle \varphi'(t), \int_0^t \langle \psi(x), R^{-1}S(t - h, x)$$
$$\times BW(h, x)\rangle \, dh\rangle.$$

As above, $\langle \psi(x), R^{-1}S(t - h, x)BW(h, x)\rangle$ is a continuous and differentiable function with respect to t, hence it defines a regular functional on $\mathscr{D}$. Using the definition of generalized derivative, replacing integrals by Riemann sums due to linearity property of the considered functionals, we obtain

$$\langle \varphi(t), \langle \psi(x), R^{-1}S(t, x) * BW'_t(t, x)\rangle\rangle =$$
$$= -\int \varphi'(t)\int_0^t \langle \psi(x), R^{-1}S(t - h, x)BW(h, x)\rangle \, dh \, dt =$$
$$= \langle \psi(x), \langle -\varphi'(t), \int_0^t R^{-1}S(t - h, x)BW(h, x) \, dh\rangle\rangle =$$
$$= \langle \psi(x), \langle -\varphi'(t), R^{-1}S(t, x) * BW(t, x)\rangle\rangle$$
$$= \langle \psi(x), \langle \varphi(t), R^{-1}S(t, x) * BW'_t(t, x)\rangle\rangle,$$

and get the representation of $R^{-1}S(t, x) * BW'_t(t, x)$ in $\mathscr{D}'(\Psi')$:

$$\langle \psi(x), \langle \varphi(t), R^{-1}S(t, x) * BW'_t(t, x)\rangle\rangle$$
$$= \langle \varphi(t), \langle \psi(x), R^{-1}S(t, x) * BW'_t(t, x)\rangle\rangle$$
$$= -\langle \varphi'(t), \int_0^t \langle \psi(x), G(t - h, x) * BW(h, x)\rangle \, dh\rangle.$$

$$(16.22)$$

Now verify that the defined generalized function (16.20) satisfies (16.19):

$$
\begin{aligned}
\langle\psi(x),\langle\varphi(t),X'_t(t,x)\rangle\rangle &= -\langle\psi(x),\langle\varphi'(t),X(t,x)\rangle\rangle = \\
&= -\langle\psi(x),\langle\varphi'(t),R^{-1}S(t,x)\xi(x) \\
&\quad + R^{-1}S(t,x)*BW'_t(t,x)\rangle\rangle = \\
&= -\langle\psi(x),\langle\varphi'(t),R^{-1}S(t,x)\xi(x)\rangle\rangle \\
&\quad -\langle\psi(x),\langle\varphi'(t),R^{-1}S(t,x)*BW'_t(t,x)\rangle\rangle. \quad (16.23)
\end{aligned}
$$

Due to (16.21) and properties of a solution of the homogeneous Cauchy problem, for the first term in the right-hand side we have

$$
\begin{aligned}
-\langle\psi(x),\langle\varphi'(t),R^{-1}S(t,x)\xi(x)\rangle\rangle &= -\langle\varphi'(t),\langle\psi(x),G(t,x)*\xi(x)\rangle\rangle = \\
&= \langle\varphi(t),\frac{d}{dt}\langle\psi(x),G(t,x)*\xi(x)\rangle\rangle \\
&\quad +\varphi(0)\langle\psi(x),G(0,x)*\xi(x)\rangle = \\
&= \langle\psi(x),A\left(i\frac{\partial}{\partial x}\right)\langle\varphi(t),G(t,x)*\xi(x)\rangle\rangle \\
&\quad +\varphi(0)\langle\psi(x),\xi(x)\rangle = \\
&= \langle\psi(x),A\left(i\frac{\partial}{\partial x}\right)\langle\varphi(t),R^{-1}S(t,x)\xi(x)\rangle\rangle \\
&\quad +\varphi(0)\langle\psi(x),\xi(x)\rangle.
\end{aligned}
$$

Using (16.22), the definition of generalized derivative, properties of solutions of the homogeneous Cauchy problem, properties of R-semigroups and convolutions, and replacing integral by Riemann sums due to linearity of the functional for the second term of (16.23) we get

$$
\begin{aligned}
-\langle\psi(x),\langle\varphi'(t),R^{-1}S(t,x)*BW'_t(t,x)\rangle\rangle &= \\
= \langle\varphi''(t),\int_0^t\langle\psi(x),R^{-1}S(t-h,x)BW(h,x)\rangle\,dh\rangle &= \\
= -\langle\varphi'(t),\langle\psi(x),R^{-1}S(0,x)BW(t,x)\rangle\rangle &- \\
-\langle\varphi'(t),\int_0^t\frac{d}{dt}\langle\psi(x),G(t-h,x)*BW(h,x)\rangle\,dh\rangle &= \\
= \langle\psi(x),\langle\varphi(t),BW'_t(t,x)\rangle\rangle &+ \\
+\langle\psi(x),A\left(i\frac{\partial}{\partial x}\right)\langle\varphi(t),R^{-1}S(t,x)*BW'_t(t,x)\rangle\rangle. &
\end{aligned}
$$

Joining both parts of (16.23) we obtain

$$\langle \psi(x), \langle \varphi(t), X'_t(t, x) \rangle \rangle = \langle \psi(x), A \left(i \frac{\partial}{\partial x} \right) \langle \varphi(t), R^{-1} S(t, x) \xi(x) \rangle \rangle +$$

$$+ \varphi(0) \langle \psi(x), \xi(x) \rangle$$

$$+ \langle \psi(x), \langle \varphi(t), BW'_t(t, x) \rangle \rangle + \langle \psi(x),$$

$$A \left(i \frac{\partial}{\partial x} \right) \langle \varphi(t), R^{-1} S(t, x) * BW'_t(t, x) \rangle \rangle =$$

$$= \langle \psi(x), A \left(i \frac{\partial}{\partial x} \right) \langle \varphi(t), X(t, x) \rangle \rangle + \varphi(0) \langle \psi(x), \xi(x) \rangle +$$

$$+ \langle \psi(x), \langle \varphi(t), BW'_t(t, x) \rangle \rangle,$$

that completes the proof.   □

Remark 16.2 Note that the proof of uniqueness for the linear stochastic equation is not so easy as for the deterministic one, even in the basic case of C_0 class semigroups [3]. In the paper [9] we proved that a weak regularized solution is unique. In [7] is proved the relationship between weak (weak regularized) and generalized solutions. Based on these results the uniqueness of the obtained generalized solutions can be proved.

In conclusion we define spaces Ψ' for each class of systems under consideration in accordance with Gel'fund–Shilov classification given in Definition 16.3.

These spaces provide operators $R^{-1} S(t, \cdot) = G(t, \cdot)*$ to be bounded ones from $L_2^m(\mathbb{R}^n)$ into Ψ' for $t \in [0, T]$. This holds if dual operators $e^{A(\cdot)t}$ (as well as $A(\cdot) e^{A(\cdot)t}$ and $A^2(\cdot) e^{A(\cdot)t}$) are bounded operators of multiplication from $L_2^m(\mathbb{R}^n)$ into $\widetilde{\Psi}'$ defined as Fourier transforms of spaces Ψ': $\widetilde{\Psi}' = \mathscr{F}[\Psi']$.

On the base of results from [6] we obtain the following spaces $\widetilde{\Psi}'$ and Ψ':

1) *for systems correct by Petrovsky* $\widetilde{\Psi}' = S'$, the well-known space of tempered distributions. In this case $\Psi' = S'$;
2) *for conditionally-correct systems* $\widetilde{\Psi}' = S'_\alpha$ with $\alpha = \alpha(h)$, where S'_α is adjoint to the space S_α. The space S_α ($\alpha \geq 0$) consists of all infinitely differentiable functions $\varphi(\cdot)$ of argument $x \in \mathbb{R}$ satisfying inequalities:

$$|x^k \varphi^{(q)}(x)| \leq C_q a^k k^{k\alpha}, \qquad k, q \in \mathbb{N}_0, \quad x \in \mathbb{R},$$

with constants $a = a(\varphi)$ and $C_q = C_q(\varphi)$ ($k^{k\alpha} = 1$ as $k = 0$). In particular, $S_0 = \mathscr{D}$. In this case $\Psi' = (S^\alpha)'$, where $(S^\alpha)'$ is adjoint to the space S^α. The space S^α ($\alpha \geq 0$) consists of all infinitely differentiable functions $\varphi(\cdot)$ of argument $x \in \mathbb{R}$ satisfying to the inequalities

$$|x^k \varphi^{(q)}(x)| \leq C_k b^q q^{q\beta}, \qquad k, q \in \mathbb{N}_0, \quad x \in \mathbb{R}, \tag{16.24}$$

with some constants $b = b(\varphi)$, $C_k = C_k(\varphi)$. (For $q = 0$ we suppose $q^{q\beta} = 1$.);

3) *for incorrect systems* $\widetilde{\Psi}' = W'_{M,a}$ with function $M(\cdot)$ chosen with respect to p_0, where $W'_{M,a}$ is adjoint to the space $W_{M,a}$. The space $W_{M,a}$ is defined as follows.

Let $\mu(\cdot)$ be an increasing continuous function on $[0;\infty)$ and let $\mu(0) = 0$, $\lim_{\xi \to \infty} \mu(\xi) = \infty$. Set

$$M(x) := \int_0^x \mu(\xi)\,d\xi, \quad x \geq 0, \qquad M(x) := M(-x), \quad x < 0.$$

Then $M(\cdot)$ increases at infinity faster than any linear function. The space $W_{M,a}$ consists of all infinitely differentiable functions $\varphi(\cdot)$ of argument $x \in \mathbb{R}$ satisfying for any $\delta > 0$ the inequalities

$$\left| \varphi^{(q)}(x) \right| \leq C_{q,\delta}\, e^{-M((a-\delta)x)}, \qquad x \in \mathbb{R}, \quad q \in \mathbb{N}_0,$$

with some constants $C_{q,\delta} = C_{q,\delta}(\varphi)$. The functions from W_M decrease at infinity faster than any exponent of type $e^{-a|x|}$. In this case $\Psi' = (W^{\Omega,1/a})'$, where $\Omega(\cdot)$ is dual by Young to $M(\cdot)$ [6].

Acknowledgements This work is partially supported by RFBR, project 13-01-00090, and by the Program of state support of RF leading universities (agreement no. 02.A03.21.0006 from 27.08.2013).

References

1. Alshanskiy, M.A., Melnikova, I.V.: Regularized and generalized solutions of infinite-dimensional stochastic problems. Matematicheskii Sbornik **202**(11), 3–30 (2011). doi:10.1070/SM2011v202n11ABEH004199
2. Davis, E.B., Pang, M.M.: The Cauchy problem and a generalization of the Hille–Yosida theorem. Proc. London Math. Soc. **55**, 181–208 (1987)
3. Da Prato, G., Zabczyk, J.: Stochastic Equations in Infinite Dimensions. Encyclopedia: Applied and Mathematics, vol. 45. Cambridge University Press, Cambridge (1992)
4. Fattorini, H.O.: The Cauchy Problem. Encyclopedia: Applied and Mathematics, vol. 18. Addison-Wesley, Reading (1983)
5. Gawarecki, L., Mandrekar, V.: Stochastic Differential Equations in Infnite Dimensions with Applications to Stochastic Partial Differential Equations. Springer, Berlin/Heidelberg (2011)
6. Gel'fand, I.M., Shilov, G.E.: Generalezed Functions. Some questions of the theory of differential equations (English translation), vol. 3. Academic Press, New York (1968)
7. Melnikova, I.V., Starkova, O.S.: Connection between weak and generalized solutions to infinite-dimensional stochastic problems. Russian Math. **58**(5), 8–20 (2014)
8. Melnikova, I.V., Filinkov, A.I.: The Cauchy Problem: Three Approaches. Monographs and Surveys in Pure and Applied Mathematics, vol. 120. Chapman & Hall, London/New York (2001)
9. Melnikova, I.V., Filinkov, A.I.: Abstract stochastic problems with generators of regularized semigroups. Commun. Appl. Anal. **13**(2), 195–212 (2009)
10. Melnikova, I.V., Filinkov, A.I., Anufrieva, U.A.: Abstract stochastic equations I. Classical and generalized solutions. J. Math. Sci. **111**(2), 3430–3475 (2002)

11. Melnikova, I.V., Zheng, Q., Zhang, J.: Regularization of weakly ill–posed Cauchy problems. J. Inverse and Ill–Posed Probl. **10** (5), 503–511 (2002)
12. Pilipović, S., Seleši, D.: Structure theorems for generalized random processes. Acta Math. Hungar. **117**(3), 251–274 (2007)
13. Tanaka, N., Okazawa, N.: Local C-semigroups and local integrated semigroups. Proc. Lond. Math. Soc. **61**(3), 63–90 (1990)

Chapter 17
Recovering the Reaction Coefficient in a Linear Parabolic Equation

Gianluca Mola

The results are devoted in memory of Alfredo Lorenzi, my
mentor and friend

Abstract We study the inverse problem consisting in the identification of the reaction constant in a linear parabolic equation under quadratic overdetermining conditions, which can be local and nonlocal with respect to the time variable. The results we provide concern the well-posedness and the stability of the solutions under singular limit of the parameters. The argument is constructive, and relies on an approximation scheme that computes the solutions. This extends and deepens our previous investigations on the same problem.

17.1 Introduction

As it is well-known, the linear parabolic reaction-diffusion equation

$$\partial_t u - \Delta u = \mu\, u \tag{17.1}$$

prescribed in the bounded and smooth domain $\Omega \subset \mathbf{R}^d$ and endowed with homogeneous boundary conditions, admits global L^2-well-posedness for all the real values of the *reaction* coefficient μ, supposed to be a known constant parameter. More precisely, it is possible to show that, if $u(\cdot, 0) \in L^2(\Omega)$, then for any arbitrary $T > 0$ there exists a unique (weak) solution

$$u \in C([0, T]; L^2(\Omega)) \cap L^2(0, T; H^1(\Omega)) \cap H^1(0, T; H^{-1}(\Omega)),$$

G. Mola (✉)
Dipartimento di Matematica "F. Brioschi", Politecnico di Milano, Via Bonardi 9,
20133 Milano, Italy
e-mail: gianluca.mola@polimi.it

© Springer International Publishing Switzerland 2014
A. Favini et al. (eds.), *New Prospects in Direct, Inverse and Control Problems for Evolution Equations*, Springer INdAM Series 10,
DOI 10.1007/978-3-319-11406-4_17

to the above problem, for all *given* μ. This will be called the *standard direct problem*. We stress that the corresponding proof (cf., for instance [7]) is constructive, and relies on a suitable finite-dimensional approximation of the problem—the Faedo-Galerkin scheme—which can also be applied for computer-based numerical simulations.

Nevertheless, assuming a datum as given, roughly speaking, means that it can be measured. On the other hand, in practical cases, measurements can be extremely difficult, or too expensive, to be performed. This is the motivation for considering as unknown one parameter, or more, appearing in a mathematical model (like, in our present case, constant coefficient μ), along with the "solution" u. Thus, aiming for recovering well-posedness results, it is necessary to feed the entry data of the problem with suitable overdeterminating conditions. This is the *inverse problem*.

In the recent paper [10] we successfully study the inverse problem consisting in the identification of parameter μ in equation (17.1) (together with u), under the additional measurement of the norm of u in the "final" instant $t = T$; that is, considering the optimal regularity for the solution of the associated direct problem, we require

$$\alpha \int_\Omega |u(x,T)|^2 dx + \beta \int_0^T d\tau \int_\Omega |\nabla u(x,\tau)|^2 dx = \rho, \qquad (17.2)$$

where $\alpha, \beta \geq 0$, $\alpha + \beta > 0$ and $\rho > 0$ are given constants. The main idea we develop consists in introducing the functional $\mathcal{M}$ of the form

$$\mathcal{M}(u) = \frac{\dfrac{1}{2}\int_\Omega |u(x,T)|^2 dx - \dfrac{1}{2}\int_\Omega |u(x,0)|^2 dx + \displaystyle\int_0^T d\tau \int_\Omega |\nabla u(x,\tau)|^2 dx}{\displaystyle\int_0^T d\tau \int_\Omega |u(x,\tau)|^2 dx}$$

that is readily seen, by the standard energy identity, to compute the constant μ in terms of $u \neq 0$. Then, the inverse problem of recovering u and μ is modified in the following way: *find a function* $u : \Omega \times [0,T] \to \mathbf{R}$ *fulfilling the (nonlinear) equation*

$$\partial_t u - \Delta u = \mathcal{M}(u) u \quad \text{in} \quad \Omega \times (0,T), \qquad (17.3)$$

endowed with the same initial-boundary conditions prescribed in the standard direct problem, and fulfilling the additional constraint (17.2).

Once again, we stress that equation (17.3) is *not* a differential (forward or backward) equation, due to the nonlocal and noncausal nature of the functional $\mathcal{M}(u)$, which requires the knowledge of the global dynamic on the whole interval $[0,T]$, and—to the best of our knowledge—seems to have no general results to lay on. Nevertheless, in [10] we proved that the same approximation scheme as the standard direct problem can be adapted to the new one, thus leading to existence and uniqueness of the couple (u, μ) under the further regularity assumption $u(\cdot, 0) \in H^s(\Omega)$, for some $s > 0$. Those techniques will be fully explained and extended in the course of the paper (cf. Sect. 17.3).

Also, it is important to notice that the above techniques have been introduced on the abstract equation

$$u'(t) + Au(t) = \mathcal{M}(u)A^\sigma u(t), \quad t \in (0, T),$$

where u is a vector-valued function in an (abstract) Hilbert space H, $A : \mathcal{D}(A) \to H$ is an unbounded closed linear operator as well as its fractional power A^σ, $\sigma < 1$, and $\mathcal{M}$ is a nonlocal functional defined on the domain $C([0, T]; H) \cap L^2(0, T; \mathcal{D}(A^{1/2}))$, with $u \neq 0$ (see Sect. 17.2). In this setting, the additional datum reads

$$\alpha \|u(T)\|^2 + \beta \int_0^T \|A^{1/2}u(\tau)\|^2 d\tau = \rho.$$

This formulation of the problem lets to prove the same statements in a wider class of applications such as, for instance, the fourth-order parabolic equations.

The aims of the present paper are twofold. The first one, which is preparatory for the second, consists into generalizing the well-posedness results in [10], in the same abstract setting, under the extended additional condition

$$\alpha \|A^{q/2}u(T_0)\|^2 + \beta \int_{T_1}^{T_2} \|A^{r/2}u(\tau)\|^2 d\tau = \rho, \tag{17.4}$$

that allows measurements of any energy level ($q, r \in \mathbf{R}$) in arbitrary time-instant $T_0 \in (0, T]$ and time-interval $[T_1, T_2] \subseteq [0, T]$. The second aim, that is the main goal of the paper, is the one to deduce informations about the qualitative behavior of u and μ as the given data approaches values that can be related to some meaningful, possibly singular, state of our problem. To this purpose, we point out that the additional measurement (17.4) is allowed to degenerate to one summand only in the limiting cases

- $\beta = 0$: $\quad \alpha \|A^{q/2}u(T_0)\|^2 = \rho \quad$ *instantaneous measurement* at T_0;
- $\alpha = 0$: $\quad \beta \int_{T_1}^{T_2} \|A^{r/2}u(\tau)\|^2 d\tau = \rho \quad$ *nonlocal measurement* on $[T_1, T_2]$.

Focusing on the second case, we now observe that, if we choose $\beta = (T_2 - T_1)^{-1}$ and $q = r = s$, then the formal limiting relation holds

$$\frac{1}{T_2 - T_1} \int_{T_1}^{T_2} \|A^{s/2}u(\tau)\|^2 d\tau \to \|A^{s/2}u(T_0)\|^2 \quad \text{as} \quad T_1 - T_2 \to 0$$

when $T_1 \to T_0$ and $T_2 \to T_0$. That is, the interval $[T_1, T_2]$ is shrinking to the singleton $\{T_0\}$, as we shall assume from now on. This means, roughly speaking, that the instantaneous measurement can be approximated by an averaged integral one in the same energy norm $\|A^{s/2} \cdot \|$ ($s \in \mathbf{R}$), provided that the related solution—if it exists and is unique—is regular enough.

Thus, if we denote by (w, ω) the solution to our inverse problem with instantaneous additional data, and by (v, ν) the one related to the integral data, it is natural to expect that, if $T_2 - T_1 \to 0$ and the respective data are close, then v and w (as well as ν and ω) will be close, in the proper function spaces. This can be considered as an extension of the usual continuous dependence on the data, in the more involved case when the model is modifying its shape by transforming a nonlocal datum into a localized one, with respect to the time variable. We shall call such a process as *the singular limit*, and its rigorous statement is the main task of this paper. For our purposes, the datum we are interested in transforming under a singular limit process is the overdetermining condition, which is the main feature of the inverse problem. We also point out that considering additional data which are localized with respect to the space variable requires a different approach and will be analyzed in forthcoming papers.

It is useful to point out that the problem of recovering unknown function-parameters entering a parabolic problem has been widely studied. In particular, we can quote the contribution of Cannon and DuChateau in [5], where a time-dependent coefficient is identified in the space of Hölder-continuous functions, under a linear (integral) overdetermining condition, by means of fixed-point techniques. Also, in papers [2, 6, 19] the existence of numerical solutions for problem of this kind have been investigated, and convergence results have been established for time-approximation schemes. It is important to mention that, in all of these cases, the underlying domain Ω is either a segment, either a rectangle. The main advantage of using space-approximations consists in the fact that both analytical and numerical results can be proven even for more complicate geometries for the domain Ω. Moreover, the approach we propose concerns the existence of *weak* solutions, other than classical ones, which as well as in direct problems—allow to deal with more general sets of initial data.

Besides, it is also useful to recall our recent contribution [11], which is concerned with the identification of the unknown constant *diffusion* coefficient $\lambda > 0$, along with u and μ as above, in the same abstract Cauchy problem

$$u'(t) + \lambda Au(t) = \mu u(t), \quad t \in (0, T), \quad u(0) = u_0,$$

when u fulfills two "final time" overdetermining conditions

$$\|A^{r/2}u(T)\|^2 = \varphi \quad \text{and} \quad \|A^{s/2}u(T)\|^2 = \psi \qquad (r \neq s)$$

provided that the data u_0, and $\varphi, \psi > 0$ satisfy proper *a priori* limitations. Furthermore, in [15] the identification of the diffusion coefficient solely is studied in the case of vanishing reaction coefficient.

We conclude this introduction by mentioning that, to the best of our knowledge, there are only few papers (see [1, 4, 9, 12, 14, 18]) where a constant rather than a function is identified. Nevertheless, the statement of singular limits in the field of inverse problems appears to be new in literature.

17.1.1 Plan of the Paper

In Sect. 17.2 we introduce the basic notations on the abstract setting and we state rigorously the problem we shall investigate as well as the main results of the paper (i.e. the well-posedness and the singular limit). In Sect. 17.3 we prove the well-posedness results by constructing the suitable finite-dimensional approximation scheme. Section 17.4 is devoted to study the behavior of the singular limit. Finally, in Sect. 17.5 we provide two concrete applications, which generalize the one devised in the introduction. The appendix is devoted to exploit some remark on sequence of implicit defined functions that we use in the previous sections.

17.2 Preliminaries

Let H be a (real) separable Hilbert space endowed with the scalar product $\langle \cdot, \cdot \rangle$ and the related norm $\| \cdot \|$. Let also $A : \mathscr{D}(A) \to H$ be a self-adjoint (unbounded) operator. We shall also require A to be strictly positive, i.e. $\langle Au, u \rangle \geq \lambda_1 \|u\|^2$ for some $\lambda_1 > 0$ and for all $u \in \mathscr{D}(A)$. Under the further assumption that $\mathscr{D}(A)$ is compactly embedded in H, we recall that by means of the spectral decomposition theorem (see [17] for details), there exist two sequences

$$\{h_n\}_{n=1}^{\infty} \subset \mathscr{D}(A) \quad \text{(eigenfunctions)} \quad \text{and} \quad \{\lambda_n\}_{n=1}^{\infty} \subset (0, \infty) \quad \text{(eigenvalues)},$$

where the set of h_n forms an orthonormal complete system in H, i.e.

$$H_{\infty} = \bigcup_{n=1}^{\infty} H_n, \quad H_n = \text{span}\{h_1, \ldots, h_n\}$$

is dense in H with $\langle h_i, h_j \rangle = \delta_{i,j}$, and $0 < \lambda_1 \leq \lambda_2 \leq \ldots \leq \lambda_{n-1} \leq \lambda_n \uparrow \infty$ as $n \to \infty$, such that

$$Ah_n = \lambda_n h_n, \quad n \geq 1.$$

Next we define the fractional positive powers A^{σ} by the formula

$$A^{\sigma} u = \sum_{n=1}^{\infty} \lambda_n^{\sigma} \langle u, h_n \rangle h_n, \quad \sigma \geq 0$$

on the domain

$$\mathscr{D}(A^{\sigma}) = \left\{ u \in H : \sum_{n=1}^{\infty} \lambda_n^{2\sigma} |\langle u, h_n \rangle|^2 < \infty \right\}.$$

In order to state the proper formulation of our identification problem, we introduce the family of Hilbert spaces $V_\sigma = \mathscr{D}(A^{\sigma/2})$, $\sigma \geq 0$, endowed with the scalar product $\langle A^{\sigma/2}\cdot, A^{\sigma/2}\cdot\rangle$ and the related norm $\| \cdot \|_\sigma = \langle A^{\sigma/2}\cdot, A^{\sigma/2}\cdot\rangle^{1/2}$. Finally, we define the spaces $V_{-\sigma}$ with a negative index as $V_{-\sigma} = (V_\sigma)^*$, $\sigma \geq 0$ as well as the fractional negative powers $A^{-\sigma}$ by the formula

$$A^{-\sigma}u = \sum_{n=1}^{\infty} \lambda_n^{-\sigma} \langle u, h_n\rangle h_n, \quad \sigma \geq 0,$$

on the domain $V_{-2\sigma}$ (cf., in particular, Sect. 17.3.3). Notice also that, as A^r is an isomorphism of V_s onto V_{s-r} for all $s, r \in \mathbf{R}$, then for all $r < s$, V_s is compactly embedded in V_r. Moreover, the following Poincaré-type inequality holds

$$\lambda_1^{s-r}\|u\|_r^2 \leq \|u\|_s^2, \quad u \in V_s. \tag{17.5}$$

In the course of our investigation, we shall also use the following interpolation inequality (see [11, Lemma 2.3])

Lemma 17.1 *Let $a < b < c$ and $u \in V_c$. Then there holds*

$$\|u\|_b \leq \|u\|_a^{\frac{c-b}{c-a}} \|u\|_c^{\frac{b-a}{c-a}}.$$

On account to the functional setting above introduced, we can state the rigorous abstract formulation of the general inverse problem we aim at investigating. From now on we set $T_0 > 0$, $T_1 < T_2$ and $q, r \in \mathbf{R}$. Then we have

Inverse problem *Find a vector-valued function $u : [0, T] \to H$ and a real number μ fulfilling the abstract Chaucy problem*

$$u'(t) + Au(t) = \mu A^\sigma u(t), \quad t \in (0, T), \quad u(0) = u_0,$$

and the additional constraint

$$\alpha\|u(T_0)\|_q^2 + \beta \int_{T_1}^{T_2} \|u(\tau)\|_r^2 d\tau = \rho \tag{17.6}$$

where u_0, $\alpha, \beta \geq 0$ $(\alpha + \beta > 0)$ and $\rho > 0$ are given.

17.2.1 Weak Formulation and Well-Posedness

We can now define the real functional $\mathscr{M}$, which, as discussed in the introduction, computes the constant μ. We define the continuous functional

$$\mathscr{M}(u) = \frac{\frac{1}{2}\|u(T)\|_\xi^2 - \frac{1}{2}\|u(0)\|_\xi^2 + \int_0^T \|u(\tau)\|_{1+\xi}^2 d\tau}{\int_0^T \|u(\tau)\|_{\sigma+\xi}^2 d\tau} \tag{17.7}$$

on the domain $u \in C([0, T]; V_\xi) \cap L^2(0, T; V_{1+\xi})$, $u \neq 0$. It is important to notice that, although functional $\mathscr{M}$ depends on the particular choice of ξ, in the following we shall prove that its value computed in a solution of Problem **P** do not depend on ξ (see Remarks 17.1 and 17.2).

We are now in a position to state the weak formulation of our problem as

Problem P *Find a function* $u : [0, T] \to V_s$ *(s $\in$ R) that fulfills the equation*

$$\langle u', h \rangle + \langle A^{1/2}u, A^{1/2}h \rangle = \mathscr{M}(u)\langle A^{\sigma/2}u, A^{\sigma/2}h \rangle, \quad \forall h \in V_{1+s}, \quad a.e.t \in (0, T) \tag{17.8}$$

the initial datum $u(0) = u_0$ *and the additional condition (17.6).*

Then, we state the existence result in the next

Theorem 17.1 *Let* $\sigma < 1$ *and* $s \geq q$, $s > r - 1$. *Then, for all* $u_0 \in V_s$, $u_0 \neq 0$ *and all* $\rho > 0$ *there exists*

$$u \in C([0, T]; V_s) \cap L^2(0, T; V_{1+s}) \cap H^1(0, T; V_{-1+s})$$

solving Problem **P**.

Moreover, the solution u depends continuously on the initial data, as stated in the next theorem, which, as a byproduct, proves the uniqueness of the solution.

Theorem 17.2 *For* $u_{0,j} \in V_s$, $u_{0,j} \neq 0$ *and* $\rho_j > 0$ $(j = 1, 2)$, *define* u_j *be the corresponding solution to Problem* **P** *devised in Theorem 17.1. Then*

$$\|u_1(t) - u_2(t)\|_s^2 + \int_0^t \left[\|u_1(\tau) - u_2(\tau)\|_{1+s}^2 + \|u_1'(\tau) - u_2'(\tau)\|_{-1+s}^2 \right] d\tau$$

$$\leq \mathscr{C} \cdot \left(\|u_{0,1} - u_{0,2}\|_s^2 + |\rho_1 - \rho_2| \right)$$

for all $t \in [0, T]$, *where* $\mathscr{C}$ *is a positive constant depending on the parameters* s, σ, T_0, T_1, T_2 *and on the data* $u_{0,j}, \rho_j$.

Notice that the above well-posedness theorems generalize the ones in [10], where the additional condition is the same as (17.6) under the restrictions on the measurement times $T_0 = T_2 = T$ and $T_1 = 0$, and on the regularity parameters $q = s$ and $r = s + 1$, with $s > 0$. Furthermore, it is important to mention that the results provided by Theorems 17.1 and 17.2 hold also in the limiting case $s = q$, which requires a nontrivial improvement of the techniques devised in [10].

17.2.2 Main Results: The Singular Limit

We now state the main result in the paper; that is, the regular behaviour of the solutions under the singular limit. To this purpose, we denote by v and w the solutions to Problem P, originating by the nonvanishing initial data

$$v(0) = v_0 \quad \text{and} \quad w(0) = w_0$$

and fulfilling, respectively, the additional conditions

$$\frac{1}{T_2 - T_1} \int_{T_1}^{T_2} \|v(\tau)\|_s^2 d\tau = \rho_v \quad \text{and} \quad \|w(T_0)\|_s^2 = \rho_w.$$

We recall that for every $v_0, w_0 \in V_s$ and $\rho_v, \rho_w > 0$ existence and uniqueness of v and w, along with the regularity therein stated, is ensured by Theorems 17.1 and 17.2, under the choices

$$\text{for } v: \begin{cases} \alpha = 0, \\ \beta = (T_2 - T_1)^{-1}, \\ r = s, \end{cases} \quad \text{for } w: \begin{cases} \alpha = 1, \\ \beta = 0, \\ q = s. \end{cases} \qquad (17.9)$$

Then, we are in a position to state our theorems.

Theorem 17.3 *Let v and w as above. Then*

$$v \to w \quad \text{strongly} \quad \text{in} \quad C([0, T]; V_s) \cap L^2(0, T; V_{1+s}) \cap H^1(0, T; V_{-1+s})$$

as

$$v_0 \to w_0 \quad \text{in} \quad V_s, \quad \rho_v \to \rho_w \quad \text{and} \quad T_1, T_2 \to T_0.$$

The next step consists into deepening the convergence result above stated by displaying a continuous dependence estimate. To this purpose, we define

$$\omega_0 = \frac{\|w(T_0)\|_{1+s}^2}{\|w(T_0)\|_{\sigma+s}^2},$$

and we show that, whenever $\omega = \mathscr{M}(w)$ is different from the singular value ω_0, then the convergence in Theorem 17.3 is of Lipschitz-type with respect to the data. More precisely

Theorem 17.4 *Under the assumption*

$$\omega = \mathscr{M}(w) \neq \omega_0$$

there exists a positive constant $\mathscr{K}$ (depending on the parameters s, σ, T_0, T_1, T_2 and on the data v_0, w_0, ρ_v, ρ_w, but independent of $T_2 - T_1$) such that the estimate

$$\|v(t) - w(t)\|_s^2 + \int_0^t \left[\|v(\tau) - w(\tau)\|_{1+s}^2 + \|v'(\tau) - w'(\tau)\|_{-1+s}^2 \right] d\tau$$

$$\leq \mathscr{K} \cdot \left(\|v_0 - w_0\|_s^2 + |\rho_v - \rho_w| + T_2 - T_1 \right)$$

is fulfilled for all $t \in [0, T]$.

Notice that, due to the parabolic nature of our problem, both numbers $\|w(T_0)\|_{1+s}$ and $\|w(T_0)\|_{\sigma+s}$ are finite. Therefore, it is not possible to exclude *a priori* that the equality $\omega = \omega_0$ can be verified. We also stress that, in this case, establishing some weaker regularity result for the convergence $v \to w$ (possibly, of Hölder-type) is still an open problem that will be analyzed in the future.

17.3 Well-Posedness

We now provide a constructive proof of Theorems 17.1 and 17.2. In order to achieve the assertion, following the main idea in [10], we shall adapt a Faedo-Galerkin (finite-dimensional) approximation scheme to our problem.

17.3.1 The Finite-Dimensional Problem

We set

$$u_n(t) = \sum_{i=1}^{n} v_i(t) v_i,$$

where $\mathbf{u}_n = (v_1, \ldots, v_n) : [0, T] \to \mathbb{R}^n$ is an unknown vector-valued function. We also define $\mathbf{x}_n = (x_1, \ldots, x_n)$, having set the compact notation $x_i = \langle u_0, h_i \rangle$ (i.e. $\mathbf{x}_n$ is the vector whose coordinates are the coefficients of the projection of u_0 on the finite-dimensional subspace $H_n = \text{span}\{h_1, \ldots, h_n\}$). Notice that condition $u_0 \neq 0$

implies that, eventually, $\mathbf{x}_n \neq \mathbf{0}$. Accordingly, we define $N = N(u_0) = \min\{n \geq 1 : x_n \neq 0\}$.

Then, the approximate problem reads

Problem P_n *Find a function $u_n : [0, T] \to H_n$ that fulfills the equation*

$$\langle u_n', h \rangle + \langle A^{1/2}u_n, A^{1/2}h \rangle = \mathcal{M}(u_n)\langle A^{\sigma/2}u_n, A^{\sigma/2}h \rangle, \quad \forall h \in H_n, \qquad (17.10)$$

for almost every $t \in (0, T)$, the initial datum

$$u_n(0) = \sum_{i=N}^{n} x_i h_i,$$

and the additional condition

$$\alpha \|u_n(T_0)\|_q^2 + \beta \int_{T_1}^{T_2} \|u_n(\tau)\|_r^2 d\tau = \rho.$$

The first step towards our main results consists in proving the next

Theorem 17.5 *Let $\sigma < 1$ and let $n \geq N$. Then, for all initial data $\mathbf{x}_n \neq \mathbf{0}$ and all $\rho > 0$ there exists a unique*

$$u_n \in C^{\infty}([0, T]; H_n)$$

solving to problem P_n.

Choosing $h = h_i, i = 1, \ldots, n$ in (17.10), problem P_n can be equivalently stated as: *find a function $\mathbf{u}_n = (v_1, \ldots, v_n) \in C^1([0, T]; \mathbb{R}^n)$ fulfilling the vector equation*

$$\mathbf{u}_n'(t) + A_n\mathbf{u}_n(t) = \mu_n A_n^{\sigma}\mathbf{u}_n(t), \quad \forall t \in (0, T),$$

where we set $A_n = \mathrm{diag}(\lambda_1, \ldots, \lambda_n)$ (diagonal matrix) and $\mu_n = \mathcal{M}(u_n)$, the initial datum $\mathbf{u}_n(0) = \mathbf{x}_n$ and the additional condition

$$\alpha \sum_{i=N}^{n} \lambda_i^q |v_i(T_0)|^2 + \beta \sum_{i=N}^{n} \lambda_i^r \int_{T_1}^{T_2} |v_i(\tau)|^2 d\tau = \rho. \qquad (17.11)$$

In order to prove Theorem 17.5, we use the well-known exponential representation formula on $[0, T]$

$$\mathbf{u}_n(t) = \mathbf{e}^{(-A_n + \mu_n A_n^{\sigma})t}\mathbf{x}_n = \left(x_1 e^{(-\lambda_1 + \mu_n \lambda_1^{\sigma})t}, \ldots, x_n e^{(-\lambda_n + \mu_n \lambda_n^{\sigma})t} \right).$$

Then, by replacing the expressions for $u_i(t)$ in (17.11), we have

$$\alpha \sum_{i=N}^{n} \lambda_i^q \, x_i^2 \, e^{2(-\lambda_i + \mu_n \lambda_i^\sigma)T_0} + \beta \sum_{i=N}^{n} \lambda_i^r \, x_i^2 \int_{T_1}^{T_2} e^{2(-\lambda_i + \mu_n \lambda_i^\sigma)\tau} d\tau = \rho.$$

Therefore, we can prove the existence of a solution of the finite-dimensional problem if we can find (at least) one real solution μ_n of the equation

$$\alpha \, \phi_n(\mu) + \beta \, \psi_n(\mu) = \rho, \quad \mu \in \mathbb{R}, \tag{17.12}$$

for all fixed $\rho > 0$, having set

$$\phi_n(\mu) = \sum_{i=N}^{n} \lambda_i^q \, x_i^2 \, e^{2(-\lambda_i + \mu \lambda_i^\sigma)T_0} \quad \text{and} \quad \psi_n(\mu) = \sum_{i=N}^{n} \lambda_i^r \, x_i^2 \int_{T_1}^{T_2} e^{2(-\lambda_i + \mu \lambda_i^\sigma)\tau} d\tau.$$

Notice that the functions ϕ_n and ψ_n above introduced, as it is apparent, depend also on the given data x_n and T_k $(k = 0, 1, 2)$, without stressing in the notation. Such dependences will play a fundamental role in the following (cf. Sects. 17.3.4 and 17.4).

The proof of next lemma is straightforward

Lemma 17.2 *Let $x_n \neq 0$, $T_0 > 0$ and $T_1 < T_2$ to be fixed. Then the real functions ϕ_n and ψ_n are positive, smooth and strictly increasing on $\mathbb{R}$. Moreover*

$$\phi_n(\mu) \to +\infty \quad \text{and} \quad \psi_n(\mu) \to +\infty \quad \text{as} \quad \mu \to +\infty$$

and

$$\phi_n(\mu) \to 0^+ \quad \text{and} \quad \psi_n(\mu) \to 0^+ \quad \text{as} \quad \mu \to -\infty.$$

As an immediate consequence of Lemma 17.2, we deduce that for any $\rho > 0$ there exists of a unique $\mu_n = \mu_n(q, r, \sigma, T_0, T_1, T_2, x_n, \rho)$ solution to equation (17.12). Thus, by setting μ_n as the constant in the exponential representation formula above, we have constructed a solution to problem P_n. Thus, Lemma 17.5 is proved.

Remark 17.1 As, by definition, $\mu_n = \mathcal{M}(u_n)$ in problem P_n, then uniqueness result of the solution to equation (17.12) implies that functional $\mathcal{M}$ is *independent* on the choice of ξ, if calculated in u_n.

17.3.2 A Priori Estimates

We first provide an upper bound for the sequence $\{\mu_n\}$, that can be proved as in (cf. [10, Lemma 3.4]). We have

Lemma 17.3 *Let μ_n, for $n \geq N = \min\{n \geq 1 : x_n \neq 0\}$ be the solution to equation (17.12). Then μ_n is monotone nonincreasing. In particular, μ_n satisfies*

$$\mu_n = \mathcal{M}(u_n) \leq \mu_N, \quad \forall n \geq N.$$

Thanks to the upper limitation provided by Lemma 17.3, the proof of the next result and can be easily recovered by standard arguments.

Lemma 17.4 *The function u_n exploited in Theorem 17.5 satisfies the estimate*

$$\|u_n(t)\|_s^2 + \int_0^t \left[\|u_n(\tau)\|_{1+s}^2 + \|u_n'(\tau)\|_{-1+s}^2 \right] d\tau \leq c_s, \quad \forall t \in [0, T],$$

where the positive constant c_s depends on $s, \sigma, T_0, T_1, T_2, u_0$ and ρ, but is independent of n.

17.3.3 Passage to the Limit

The last step in order to prove Theorem 17.1 consists in showing that u_n tends in a proper sense to a suitable function u, which is a solution to problem **P**. First, starting from the estimate in Lemma 17.4, we invoke the weak compactness theorem, to deduce (up to subsequences) the limit relations

$$u_n \xrightarrow{w*} u \text{ in } L^\infty(0, T; V_s), \quad u_n \xrightarrow{w} u \text{ in } L^2(0, T; V_{1+s}),$$

$$u_n' \xrightarrow{w} u' \text{ in } L^2(0, T; V_{-1+s})$$

as $n \to \infty$, for some $u \in L^\infty(0, T; V_s) \cap L^2(0, T; V_{1+s}) \cap H^1(0, T; V_{-1+s})$. Now, notice that the embedding $V_{1+s} \hookrightarrow V_s$ is compact, then we can apply the classic compactness result due to J. Simon (see [16, Corollary 4]) to get $u_n \to u$ strongly in $C([0, T]; V_s)$ which, as $s \geq q$, implies $\|u_n(T_0)\|_q \to \|u(T_0)\|_q$. Also, under limitation $s > r - 1$, the same argument yields $u_n \to u$ strongly in $L^2(0, T; V_r)$, and therefore we have $\int_{T_1}^{T_2} \|u_n(\tau)\|_r^2 d\tau \to \int_{T_1}^{T_2} \|u(\tau)\|_r^2 d\tau$. This proves that u fulfills (17.6). Moreover, by setting $\xi < s$ in the definition (17.7) of $\mathcal{M}(u_n)$ (which, as pointed out in Remark 17.2, is independent of ξ), it is immediate to deduce $\mathcal{M}(u_n) \to \mathcal{M}(u)$. Thus, we are left to prove that u fulfills equation (17.8) and the initial datum $u(0) = u_0$. This can be done by integrating the equation over $(0, T)$ and using the convergence stated above.

This concludes the proof of Theorem 17.1.   □

Remark 17.2 Since the sequence $\mu_n = \mathcal{M}(u_n)$ is *independent* on the choice of ξ then its limit $\mu = \mathcal{M}(u)$ enjoys the same property.

17.3.4 Uniqueness

We aim at proving the continuous dependence estimate provided in Theorem 17.2 and once again our proof relies on the finite-dimensional approximation of the solutions. Back to equation (17.12), we need to exploit the dependence of its solution μ_n on $\mathbf{x}_n$ (as the one on ρ is trivial). For this purpose, we invoke the implicit function theorem, to deduce the relation

$$\nabla_{\mathbf{x}_n} \mu_n = -\frac{\nabla_{\mathbf{x}_n}(\alpha\phi_n + \beta\psi_n)}{\partial_\mu(\alpha\phi_n + \beta\psi_n)},$$

having set the notation $\nabla_{\mathbf{x}_n} = (\partial_{x_1}, \ldots, \partial_{x_n})$. First notice that

$$\partial_\mu \phi_n(\mu) = 2T_0 \sum_{i=N}^{n} \lambda_i^{\sigma+q} x_i^2 e^{2(-\lambda_i + \mu\lambda_i^\sigma)T_0}$$

and

$$\partial_\mu \psi_n(\mu) = 2 \sum_{i=N}^{n} \lambda_i^{\sigma+r} x_i^2 \int_{T_1}^{T_2} \tau e^{2(-\lambda_i + \mu\lambda_i^\sigma)\tau} d\tau.$$

Consequently, limiting the above sums to the first summand only, we deduce the lower bound

$$\partial_\mu [\alpha\phi_n(\mu) + \beta\psi_n(\mu)] \geq \gamma(\mu)|x_N|^2, \tag{17.13}$$

having set

$$\gamma(\mu) = 2\alpha T_0 \lambda_N^{\sigma+q} e^{2(-\lambda_N + \mu\lambda_N^\sigma)T_0} + 2\beta \int_{T_1}^{T_2} \tau e^{2(-\lambda_N + \mu\lambda_N^\sigma)\tau} d\tau > 0.$$

Analogously, we see that for all $i = N, \ldots, n$ there holds

$$\partial_{x_i} \phi_n(\mu) = 2\lambda_i^q x_i e^{2(-\lambda_i + \mu\lambda_i^\sigma)T_1} \quad \text{and} \quad \partial_{x_i} \psi_n(\mu) = 2\lambda_i^r x_i \int_{T_1}^{T_2} \tau e^{2(-\lambda_i + \mu\lambda_i^\sigma)\tau} d\tau.$$

As the sequence $-\lambda_n + \mu\lambda_n^\sigma$ is eventually negative, then the (continuous and positive) function

$$\delta(\mu) = 2\left(\max_{i \geq N}\left\{\alpha^2\lambda_i^q e^{4(-\lambda_i + \mu\lambda_i^\sigma)T_0} + \beta^2\lambda_i^r \int_{T_1}^{T_2} \tau e^{2(-\lambda_i + \mu\lambda_i^\sigma)\tau}d\tau\right\}\right)^{1/2}$$

is well-defined and, consequently, we deduce

$$|\nabla_{\mathbf{x}_n}[\alpha\phi_n(\mu) + \beta\psi_n(\mu)]| \leq \delta(\mu)|\mathbf{x}_n|. \tag{17.14}$$

Finally, collecting (17.13) and (17.14) with $\mu = \mu_n$, we deduce the upper bound

$$|\nabla_{\mathbf{x}_n}\mu_n| \leq c(\mu_n)|x_N|^{-2}|\mathbf{x}_n|,$$

for some continuous and positive function c which is independent of n. Then, by setting $\mathbf{x}_{j,n} = (x_{j,1},\ldots,x_{j,n}) \neq \mathbf{0}$ and $\rho_j > 0$ $(j = 1,2)$, the corresponding solution $\mu_{j,n}$ to equation (17.12) fulfills the estimate

$$|\mu_{1,n} - \mu_{2,n}| \leq C(|\mathbf{x}_{1,n} - \mathbf{x}_{2,n}| + |\rho_1 - \rho_2|),$$

for positive constant C independent of n. Moreover, as C depends continuously on $\mu_{j,n}$ and $\mathbf{x}_{j,n}$, the above estimate con passes to the limit as $n \to \infty$, to yield

$$|\mathcal{M}(u_1) - \mathcal{M}(u_2)|^2 \leq C^2\left(\|u_{0,1} - u_{0,2}\|^2 + |\rho_1 - \rho_2|\right).$$

Then, Theorem 17.2 can be proven by standard methods, following [10, Sect. 4].

Remark 17.3 Notice that the Lipschitz constant $\mathcal{C}$ appearing in the continuous dependence estimate in Theorem, 17.2 which have been constructed above can be computed as

$$\mathcal{C} = \max\left\{\frac{\delta(\mu_1)}{\gamma(\mu_1)}\frac{\|u_{0,1}\|}{|x_{N_1,1}|^2}, \frac{\delta(\mu_2)}{\gamma(\mu_2)}\frac{\|u_{0,2}\|}{|x_{N_2,2}|^2}\right\}^2$$

where $N_j = N_j(\mathbf{x}_j) = \min\{i = 1,\ldots,n : x_{i,j} \neq 0\}$ $(j = 1,2)$. In particular, we stress that the behavior of such a quantity may be singular as $u_{0,j} \to 0$ in V_s. This is not surprising, as the problem P makes no sense for a vanishing initial datum, due to the definition of functional $\mathcal{M}$. Nevertheless, if we choose $M > 0$ large enough such that

$$M^{-1} \leq |x_{j,N_j}| \leq |\mathbf{x}_j| \leq M,$$

then $\mathcal{C}$ turns out to be uniformly bounded by M^6.

17.3.5 The Special Case $\beta = \sigma = 0$

It is worth to mention that, in the case when $\beta = 0$ and $\sigma = 0$, then the unique solution μ_n to equation (17.12) con be explicitly computed by the expression

$$\mu_n = \frac{1}{2T_0} \ln \left(\frac{\alpha^{-1}\rho}{\sum_{i=1}^{n} \lambda_i^s \, x_i^2 \, e^{-2\lambda_i T_0}} \right).$$

Consequently, as $n \to \infty$, we recover the representation formula

$$\mu = \frac{1}{2T_0} \ln \left(\frac{\alpha^{-1}\rho}{\sum_{n=1}^{\infty} \lambda_n^s \, x_n^2 \, e^{-2\lambda_n T_0}} \right).$$

17.4 The Singular Limit

In this section we prove the convergence results provided by Theorems 17.3 and 17.4. This will be done by deducing further properties of the solutions of equation (17.12) that will let, using the approximation scheme constructed in the previous section, to deduce the assertions. This will require a few steps.

From now and till the end of this section, we shall denote by c a generic positive constant, independent of $T_2 - T_1$, which may vary line to line.

Moreover, whenever necessary, we shall always assume to work within the regularization scheme provided by the constructive Faedo-Galerkin method, devised in Sect. 17.3. Such a remark is crucial, in order to make rigorous the formal derivation of most of the estimates in what follows (see, in particular, the multiplication argument in the proof of Lemma 17.7).

Step 1. Setting the measurement instant $T_0 \in (0, T)$, for the sake of simplicity, we choose

$$T_1 = T_0 - \varepsilon \quad \text{and} \quad T_2 = T_0 + \varepsilon$$

for any $\varepsilon > 0$ small enough, as we shall assume from now on. Consequently, we introduce the auxiliary function w_ε as the solutions to Problem P, originating by the initial datum w_0 and fulfilling the additional condition

$$\frac{1}{2\varepsilon} \int_{T_0-\varepsilon}^{T_0+\varepsilon} \|w_\varepsilon(\tau)\|_s^2 d\tau = \rho_w,$$

where $w_0 \neq 0$ and $\rho_w > 0$ are the same as in Sect. 17.2.2. Coherently, we set the short notation $\omega_\varepsilon = \mathcal{M}(w_\varepsilon)$. Next, we need to deduce some information on the behavior of ω_ε as $\varepsilon \to 0^+$ that will play an important role in the sequel. Recalling that $\omega = \mathcal{M}(w)$, then we have

Lemma 17.5 *Let ω and ω_ε be as above. Then $\omega_\varepsilon \to \omega$ as $\varepsilon \to 0^+$.*

More is true. In fact, if ω_0 is the same constant as in the statement of Theorem 17.4, then

Lemma 17.6 *Let $\omega \neq \omega_0$. Then the estimate*

$$|\omega - \omega_\varepsilon| \leq c \cdot \varepsilon$$

is fulfilled for every small enough ε.

We mention that the same results as in Lemmas 17.5 and 17.6 (with minor modifications of the proofs) hold also in the cases when the integral condition consists in the measurement of one among the quantities

$$\frac{1}{\varepsilon}\int_{T_0}^{T_0+\varepsilon} \|w_\varepsilon(\tau)\|_s^2 d\tau, \quad \frac{1}{\varepsilon}\int_{T_0+\varepsilon}^{T_0+2\varepsilon} \|w_\varepsilon(\tau)\|_s^2 d\tau,$$

and

$$\frac{1}{\varepsilon}\int_{T_0-2\varepsilon}^{T_0-\varepsilon} \|w_\varepsilon(\tau)\|_s^2 d\tau, \quad \frac{1}{\varepsilon}\int_{T_0-\varepsilon}^{T_0} \|w_\varepsilon(\tau)\|_s^2 d\tau$$

thus leading to the general case. Notice also that, in the two latter expressions, T_0 is also allowed to assume the final value T.

We now shall prove Lemma 17.5. In the curse of the proof, a crucial role is played by the convergence results that are stated in the appendix.

Proof (of Lemma 17.5) In order to prove the claim, we exploit the regularization scheme constructed in Sect. 17.3. To this purpose, we define w_n and $w_{\varepsilon,n}$ to be, respectively, the approximations of w and w_ε in the subspace H_n and, accordingly, we set

$$\omega_n = \mathcal{M}(w_n) \quad \text{and} \quad \omega_{\varepsilon,n} = \mathcal{M}(w_{\varepsilon,n}).$$

Then, following Sect. 17.3.1, we recall that ω_n and $\omega_{\varepsilon,n}$ are the unique solutions to the nonlinear equations (cf. equation (17.12))

$$\phi_n(\mu) = \rho_w \quad \text{and} \quad \psi_n(\mu) = \rho_w, \quad \mu \in \mathbb{R},$$

where, accordingly to the choices our data in (17.9), we have

$$\phi_n(\mu) = \sum_{i=N}^{n} \lambda_i^s x_i^2\, e^{2(-\lambda_i + \mu\lambda_i^\sigma)T_0} \text{ and } \psi_n(\mu) = \frac{1}{2\varepsilon}\sum_{i=N}^{n} \lambda_i^s x_i^2 \int_{T_0-\varepsilon}^{T_0+\varepsilon} e^{2(-\lambda_i+\mu\lambda_i^\sigma)\tau} d\tau,$$

having posed, by convenience, $x_i = \langle w_0, h_i \rangle$ and $N = N(w_0) = \min\{n \geq 1 : x_n \neq 0\}$. Notice immediately that, as the summand functions in the definition of ψ_n are continuous, then, by the mean value theorem, we deduce the identity $\psi_n(\mu) = \phi_{\varepsilon,n}(\mu)$ for all μ, where we define

$$\phi_{\varepsilon,n}(\mu) = \sum_{i=N}^{n} \lambda_i^s \, x_i^2 \, e^{2(-\lambda_i + \mu\lambda_i^{\sigma})T_{\varepsilon}}$$

for some $T_0 - \varepsilon < T_{\varepsilon} < T_0 + \varepsilon$. We now introduce the sequence of functions Φ_n and its uniform limit Φ on the compact set $[0, T] \times [\omega - 1, \omega + 1]$, defined by

$$\Phi_n(\tau, \mu) = \sum_{i=N}^{n} \lambda_i^s \, x_i^2 \, e^{2(-\lambda_i + \mu\lambda_i^{\sigma})\tau} \quad \text{and} \quad \Phi(\tau, \mu) = \sum_{n=N}^{\infty} \lambda_n^s \, x_n^2 \, e^{2(-\lambda_n + \mu\lambda_n^{\sigma})\tau}.$$

Since $\Phi(T_0, \omega) = \rho_w$ and the maps $\mu \mapsto \Phi_n(\tau, \mu)$ and $\mu \mapsto \Phi(\tau, \mu)$ are continuous and monotone increasing for any fixed τ, then Φ_n and Φ are readily seen to fulfill conditions (i) and (ii) of Theorem A in the appendix. Therefore, there exist continuous functions

$$f_n, f : [0, T] \to [\omega - 1, \omega + 1]$$

uniquely solving, respectively, the implicit equations

$$\Phi_n(\tau, f_n(\tau)) = \rho_w \quad \text{and} \quad \Phi(\tau, f(\tau)) = \rho_w \quad \text{on} \quad [0, T],$$

such that f_n converges pointwise to f on $[0, T]$, as $n \to \infty$. Moreover, as Φ_n is monotone increasing, by Theorem B in the appendix we deduce that the convergence is uniform. Since, by construction

$$\Phi_n(T_{\varepsilon}, \mu) = \phi_{\varepsilon,n}(\mu) \quad \text{and} \quad \Phi_n(T_0, \mu) = \phi_n(\mu),$$

recalling that (see Lemma 17.2) $\omega_n = \phi_n^{-1}(\rho_w)$ and $\omega_{\varepsilon,n} = \phi_{\varepsilon,n}^{-1}(\rho_w)$, we have

$$\omega_{\varepsilon,n} = f_n(T_{\varepsilon}) \quad \text{and} \quad \omega_n = f_n(T_0).$$

Thus, by uniformity

$$\lim_{\varepsilon \to 0+} \omega_{\varepsilon} = \lim_{\varepsilon \to 0+} \left(\lim_{n \to \infty} \omega_{\varepsilon,n} \right) = \lim_{n \to \infty} \left(\lim_{\varepsilon \to 0+} \omega_{\varepsilon,n} \right) = \lim_{n \to \infty} \omega_n = \omega.$$

This proves the assertion. $\square$

Proof (of Lemma 17.6)

Relying on the argument introduced in the previous proof, we now need to exploit an estimate from below of

$$\partial_\tau \Phi_n(\tau, \mu) = 2\mu \sum_{i=N}^{n} \lambda_i^{\sigma+s} x_i^2 \, e^{2(-\lambda_i + \mu\lambda_i^\sigma)\tau} - 2 \sum_{i=N}^{n} \lambda_i^{1+s} x_i^2 \, e^{2(-\lambda_i + \mu\lambda_i^\sigma)\tau},$$

which holds uniformly in n and in a neighborhood of (T_0, ω). To this purpose, notice that

$$\partial_\tau \Phi_n(\tau, \mu) = 0 \quad \text{if and only if} \quad \mu = \frac{\displaystyle\sum_{i=N}^{n} \lambda_i^{1+s} x_i^2 \, e^{2(-\lambda_i + \mu\lambda_i^\sigma)\tau}}{\displaystyle\sum_{i=N}^{n} \lambda_i^{\sigma+s} x_i^2 \, e^{2(-\lambda_i + \mu\lambda_i^\sigma)\tau}}.$$

Moreover, by definition of w, we learn that for any $\tau \in (0, T]$ there holds

$$\sum_{i=N}^{n} \lambda_i^{1+s} x_i^2 \, e^{2(-\lambda_i + \mu\lambda_i^\sigma)\tau} \to \|w(\tau)\|_{1+s}^2 \quad \text{and} \quad \sum_{i=N}^{n} \lambda_i^{\sigma+s} x_i^2 \, e^{2(-\lambda_i + \mu\lambda_i^\sigma)\tau} \to \|w(\tau)\|_{\sigma+s}^2$$

as $n \to \infty$ and $\mu \to \omega$. We now recall that, applying the standard parabolic smoothness results (see [13, Chap. 4]), it is possible to prove the further regularity $w \in C((0, T], V_{1+s})$, thus implying

$$\|w(\tau)\|_{1+s} \to \|w(T_0)\|_{1+s} \quad \text{and} \quad \|w(\tau)\|_{\sigma+s} \to \|w(T_0)\|_{\sigma+s} \quad \text{as} \quad \tau \to T_0.$$

Consequently, letting $n \to \infty$, $\mu \to \omega$ and $\tau \to T_0$, and by means of the main assumption of Theorem 17.4 (i.e. $\omega \neq \omega_0 = \|w(T_0)\|_{1+s}^2 / \|w(T_0)\|_{\sigma+s}^2$), we deduce

$$\partial_\tau \Phi_n(\tau, \mu) \to 2\omega \|w(T_0)\|_{\sigma+s}^2 - 2\|w(T_0)\|_{\sigma+s}^2 \neq 0.$$

Then, there exist $N_0 \geq N$ and a neighborhood $N(T_0, \omega)$ such that

$$\partial_\tau \Phi_n(\tau, \mu) \neq 0 \quad \forall n \geq N_0 \quad \text{and} \quad \forall \, (\tau, \mu) \in N(T_0, \omega).$$

Therefore, we are finally led to the lower bound

$$c_0 = \frac{1}{2} \cdot \inf_{(\tau,\mu)\in N(T_0,\omega)} \left(\inf_{n\geq N_0} \left| \partial_\tau \left(\sum_{i=N}^{n} \lambda_i^s x_i^2 \, e^{2(-\lambda_i + \mu\lambda_i^\sigma)\tau} \right) \right| \right) > 0,$$

so that, setting $\tau = T_\varepsilon$ we get

$$|\phi_n(\mu) - \phi_{\varepsilon,n}(\mu)| \geq c_0\varepsilon,$$

which, choosing $n \geq N_0$ and ε small enough, implies

$$|\phi_n^{-1}(\rho_w) - \phi_{\varepsilon,n}^{-1}(\rho_w)| = |\omega_n - \omega_{\varepsilon,n}| \leq c_0^{-1}\varepsilon.$$

This, passing to the limit as $n \to \infty$, yields the limitation

$$|\omega - \omega_\varepsilon| \leq c_0^{-1}\varepsilon$$

that concludes the proof. $\square$

Step 2. We now have to study, in the proper function spaces, the behavior of function w_ε, as $\varepsilon \to 0^+$. This is done in the next

Lemma 17.7 *The next estimate*

$$\|w(t)-w_\varepsilon(t)\|_s^2 + \int_0^t \left[\|w(\tau) - w_\varepsilon(\tau)\|_{1+s}^2 + \|w'(\tau) - w_\varepsilon'(\tau)\|_{-1+s}^2 \right] d\tau \leq c \cdot |\omega - \omega_\varepsilon|$$

is fulfilled for all $t \in [0, T]$ and for every small enough ε.

Proof First, notice that function $w - w_\varepsilon$ fulfills the variational equation

$$\langle (w - w_\varepsilon)', h \rangle + \langle A^{1/2}(w - w_\varepsilon), A^{1/2}h \rangle = \qquad (17.15)$$

$$(\omega - \omega_\varepsilon) \langle A^{\sigma/2}w, A^{\sigma/2}h \rangle + \omega_\varepsilon \langle A^{\sigma/2}(w - w_\varepsilon), A^{\sigma/2}h \rangle,$$

for all $h \in V_s$ and a.e. $t \in (0, T)$. Then, setting $h = A^s(w - w_\varepsilon)$ in (17.15), we deduce

$$\tfrac{1}{2}\tfrac{d}{dt}\|w(t) - w_\varepsilon(t)\|_s^2 + \|w(t) - w_\varepsilon(t)\|_{1+s}^2 = \qquad (17.16)$$

$$(\omega - \omega_\varepsilon) \langle A^{(\sigma+s)/2}w(t), A^{(\sigma+s)/2}[w(t) - w_\varepsilon(t)] \rangle + \omega_\varepsilon \|w(t) - w_\varepsilon(t)\|_{\sigma+s}^2.$$

We now proceed in estimating the right-hand side summands in the above equality. Using inequality (17.5) and bounds provided in Lemma 17.4, we easily get

$$(\omega - \omega_\varepsilon) \langle A^{(\sigma+s)/2}w(t), A^{(\sigma+s)/2}[w(t) - w_\varepsilon(t)] \rangle$$

$$\leq |\omega - \omega_\varepsilon| \, \|w(t)\|_{\sigma+s} \|w(t) - w_\varepsilon(t)\|_{\sigma+s}$$

$$\leq \lambda_1^{1-\sigma} \, |\omega - \omega_\varepsilon| \, \|w(t)\|_{1+s} \|w(t) - w_\varepsilon(t)\|_{1+s}$$

$$\leq c \, |\omega - \omega_\varepsilon|^2 \, \|w(t)\|_{1+s}^2 + \frac{1}{4}\|w(t) - w_\varepsilon(t)\|_{1+s}^2.$$

Concerning the second summand, as from Lemma 17.5 we learn that $|\omega_\varepsilon| \le c$, then using interpolation Lemma 17.1, we learn

$$\omega_\varepsilon \|w(t) - w_\varepsilon(t)\|_{\sigma+s}^2 \le |\omega_\varepsilon| \|w(t) - w_\varepsilon(t)\|_s^{2(1-\sigma)} \|w(t) - w_\varepsilon(t)\|_{1+s}^{2\sigma}, \quad \text{if} \quad 0 < \sigma < 1$$

and

$$\omega_\varepsilon \|w(t) - w_\varepsilon(t)\|_{\sigma+s}^2 \le \lambda_1^{-2\sigma} |\omega_\varepsilon| \|w(t) - w_\varepsilon(t)\|_s^2, \text{if} \quad \sigma \le 0,$$

so that, by the Young's inequality, we get

$$\omega_\varepsilon \|w(t) - w_\varepsilon(t)\|_{\sigma+s}^2 \le c \|w(t) - w_\varepsilon(t)\|_s^2 + \frac{1}{4} \|w(t) - w_\varepsilon(t)\|_{1+s}^2.$$

Summing up, back to (17.16), we get the differential inequality

$$\frac{d}{dt} \|w(t) - w_\varepsilon(t)\|_s^2 + \|w(t) - w_\varepsilon(t)\|_{1+s}^2 \le c \|w(t) - w_\varepsilon(t)\|_s^2 + c \, |\omega - \omega_\varepsilon|^2 \, \|w(t)\|_{1+s}^2,$$
$$(17.17)$$

that, by means of the Gronwall lemma, yields

$$\|w(t) - w_\varepsilon(t)\|_s^2 \le c \left(\int_0^t \|w(\tau)\|_{1+s}^2 d\tau \right) |\omega - \omega_\varepsilon|^2 \le c |\omega - \omega_\varepsilon|, \qquad (17.18)$$

having used the integral bound in Lemma 17.4 and the relation $w(0) = w_\varepsilon(0) = w_0$. Moreover, integrating both members of inequality (17.17) on $(0,t)$ and using again (17.18), we easily obtain

$$\int_0^t \|w(\tau) - w_\varepsilon(\tau)\|_{1+s}^2 d\tau \le c |\omega - \omega_\varepsilon|.$$

Finally, setting $h = A^{-1+s}(w' - w'_\varepsilon)$ in (17.15) and integrating on $(0,t)$, the above inequalities allow, similarly, to recover the integral estimate

$$\int_0^t \|w'(\tau) - w'_\varepsilon(\tau)\|_{-1+s}^2 d\tau \le c |\omega - \omega_\varepsilon|,$$

which completes the proof. $\square$

Step 3.   Finally, we are in a position to prove Theorems 17.3 and 17.4. To this purpose, we use bounds provided by Theorem 17.2 and Lemma 17.7, to deduce

$$\|v(t) - w(t)\|_s^2 \le \|v(t) - w_\varepsilon(t)\|_s^2 + \|w(t) - w_\varepsilon(t)\|_s^2$$

$$\le c \left(\|v_0 - w_0\|_s^2 + |\rho_v - \rho_w| + |\omega - \omega_\varepsilon| \right),$$

whereas the same control holds for the term

$$\int_0^t \left[\|v(\tau) - w(\tau)\|_{1+s}^2 + \|v'(\tau) - w'(\tau)\|_{-1+s}^2 \right] d\tau.$$

Therefore, recalling that in our notations $2\varepsilon = T_2 - T_1$, then the assertions of Theorems 17.3 and 17.4 immediately follows from Lemmas 17.5 and 17.6. This completes the proof. □

Remark 17.4 Concerning the Lipschitz Constant $\mathcal{K}$, as it is apparent from *Step 3* in the previous proof, it can be chosen by the sum of $\mathcal{C}$ (constant appearing in Theorem 17.2) and constant c of Lemma 17.7. As a consequence, it displays the same singular behavior with respect to v_0 and w_0 devised in Remark 17.3.

17.5 Applications

We now display concrete applications of our abstract results by applying Theorems 17.1, 17.2, 17.3 and 17.4 to the parabolic initial-boundary value problems devised in the introduction, once properly stated. To this purpose, in the following, we denote by $\Omega \subseteq \mathbb{R}^d$ $(d \geq 1)$ a bounded domain with regular boundary $\partial\Omega$, with normal direction $\mathbf{n} = (n_1, \ldots, n_d)$ and outward normal derivative $\partial_{\mathbf{n}}$ at $x \in \partial\Omega$. Then, we define the basic space as $L^2(\Omega)$, endowed with the usual scalar product $\langle \cdot, \cdot \rangle$ and norm $\| \cdot \|$. As usual, we shall denote by $H^s(\Omega)$ the Hilbertian Sobolev space with real exponent s.

17.5.1 Second-Order Parabolic Equations

Denote by A_γ, for $\gamma \in [0, 1]$, the realization of Laplace operator $-\Delta$ in $L^2(\Omega)$ on the domains, respectively

$$\mathscr{D}(A_\gamma) = \begin{cases} H^2(\Omega) \cap H_0^1(\Omega), & \gamma = 0 \text{ (Dirichlet)}, \\ \{u \in H^2(\Omega) : \gamma \, \partial_{\mathbf{n}} u \\ \quad +(1 - \gamma) c \, u = 0 \text{ on } \partial\Omega\}, & \gamma \in (0, 1) \text{ (Robin)}, \\ \{u \in H^2(\Omega) : \partial_{\mathbf{n}} u = 0 \text{ on } \partial\Omega \text{ and} \\ \quad \int_\Omega u(x)dx = 0\}, & \gamma = 1 \text{ (Neumann)} \end{cases}$$

where c is a continuous and strictly positive function on $\partial\Omega$. Thanks to the above definitions, it is standard matter to check that A_γ fulfills requirements of Sect. 17.2, and denote by $\lambda_n(\gamma)$ its n-th eigenvalue. Also, it is worth to mention that, when s is positive, the subspaces $V_s^\gamma = \mathscr{D}(A_\gamma^s)$ (previously defined in the abstract setting) admit the following representations (cf. [8, Theorem 8.1])

$$
V_s^0 = \begin{cases} H^s(\Omega), & s \in (0, 1/2), \\ H^s(\Omega) \cap H_0^1(\Omega), & s > 1/2, \end{cases}
$$

$$
V_s^\gamma = \begin{cases} H^s(\Omega), & s \in (0, 3/2), \\ \{u \in H^s(\Omega) : \gamma \partial_n u + (1 - \gamma)u = 0 \ \text{ on } \ \partial\Omega\}, & s > 3/2, \end{cases}
$$

and

$$
V_s^1 = \begin{cases} \{u \in H^s(\Omega) : \int_\Omega u(x)dx = 0\}, & s \in (0, 3/2), \\ \{u \in H^s(\Omega) : \partial_n u = 0 \ \text{ on } \ \partial\Omega \ \text{ and } \ \int_\Omega u(x)dx = 0\}, & s > 3/2. \end{cases}
$$

In either cases, notice that the norm in V_s^γ is the one in $H^s(\Omega)$.

Thus, assuming $\sigma < 1$, v_0, $w_0 : \Omega \to \mathbb{R}$ and ρ_v, $\rho_w > 0$ to be given, we state the following second-order inverse problems, which generalize the ones in introduction.

Problem 1 *Find the function* $v : \Omega \times [0, T] \to \mathbb{R}$ *and the real constant* v *such that the following parabolic initial-boundary value problem is satisfied*

$$
\begin{cases} \partial_t v(x,t) + A_\gamma v(x,t) = v\, A_\gamma^\sigma v(x,t), & (x,t) \in \Omega \times (0, T), \\ v(x,0) = v_0(x), & x \in \Omega, \\ \gamma\, \partial_n v(x,t) + (1 - \gamma)\, c(x)\, v(x,t) = 0, & (x,t) \in \partial\Omega \times (0, T), \end{cases}
$$

along with the global measurement

$$
\frac{1}{T_2 - T_1} \int_{T_1}^{T_2} \|v(\cdot, \tau)\|_{H^s(\Omega)}^2 d\tau = \rho_v.
$$

And, taking an instantaneous additional measurement

Problem 1' *Find the function* $w : \Omega \times [0, T] \to \mathbb{R}$ *and the real constant* ω *such that the following parabolic initial-boundary value problem is satisfied*

$$
\begin{cases} \partial_t w(x,t) + A_\gamma w(x,t) = \omega\, A_\gamma^\sigma w(x,t), & (x,t) \in \Omega \times (0, T), \\ w(x,0) = w_0(x), & x \in \Omega, \\ \gamma\, \partial_n w(x,t) + (1 - \gamma)\, c(x)\, w(x,t) = 0, & (x,t) \in \partial\Omega \times (0, T), \end{cases}
$$

along with the instantaneous measurement

$$
\|w(\cdot, T_0)\|_{H^s(\Omega)}^2 = \rho_w.
$$

Then, abstract Theorems 17.1–17.4 yield

Theorem 17.6 *Let v_0, $w_0 \in V_s^\gamma \setminus \{0\}$ and ρ_v, $\rho_w > 0$. Then there exist the couples*

$$(v, v), (w, \omega) \in \left(C([0, T]; V_s^\gamma) \cap L^2(0, T; V_{1+s}^\gamma) \cap H^1(0, T; V_{-1+s}^\gamma)\right) \times \mathbb{R},$$

unique (weak) solutions to Problems 1 and 1', respectively, depending continuously on the data (v_0, ρ_v) and (w_0, ρ_w). Moreover, as

$$v_0 \to w_0 \quad \text{in} \quad H^s(\Omega), \quad \rho_v \to \rho_w \quad \text{and} \quad T_2 - T_1 \to 0$$

there holds

$$v \to w \text{ strongly in } C([0, T]; V_s^\gamma) \cap L^2(0, T; V_{1+s}^\gamma) \cap H^1(0, T; V_{-1+s}^\gamma) \quad \text{and} \quad v \to \omega.$$

Furthermore, under the assumption

$$\omega \neq \frac{\|w(\cdot, T_0)\|_{H^{1+s}(\Omega)}^2}{\|w(\cdot, T_0)\|_{H^{\sigma+s}(\Omega)}^2},$$

the above convergence is of Lipschitz-type.

For the sake of computations, we point out the following concrete representations of the fractional power operator A_γ^σ, when

- $\sigma = 0$, then $A_\gamma^0 u = u$ on the domain $L^2(\Omega)$. Recalling Sect. 17.3.5, in this case ω in Problem 1' can be explicitly computed by the formula therein provided. For example, in the one-dimensional case $\Omega = [0, L]$, the Dirichlet problem $(\gamma = 0)$, as it is well-known, yields

$$\lambda_n = \pi^2 L^{-2} n \quad \text{and} \quad h_n(x) = \sqrt{2L^{-1}} \sin\left(\pi L^{-1} nx\right), \, x \in [0, L].$$

Thus, we have

$$\omega = \frac{1}{2T_0} \ln\left(\frac{L^{2s} \rho_w}{\pi^{2s} \sum_{n=1}^{\infty} n^{2s} x_n^2 \, e^{-2\pi^2 L^{-2} n^2 T_0}}\right)$$

and, as a consequence

$$w(x, t) = \sqrt{2L^{-1}} \, e^{\omega t} \sum_{n=1}^{\infty} x_n \, e^{-\pi^2 L^{-2} n^2 t} \, \sin\left(\pi L^{-1} nx\right), \, (x, t) \in [0, L] \times [0, T],$$

where, following the previous notations, for any $n \geq 1$ we define

$$x_n = \sqrt{2L^{-1}} \int_0^L w_0(x) \sin\left(\pi L^{-1} nx\right) dx;$$

- $\sigma = -1$, then $A_\gamma^{-1} u = \int_\Omega \mathscr{G}(\cdot, y) u(y) dy$ on the domain $H^{-1}(\Omega)$, $\mathscr{G}$ being the Green function associated to the homogeneous Poisson problem $-\Delta h = u$ in Ω, $h \in V_2^\gamma$.

17.5.2   Fourth-Order Parabolic Equations

Relying on the notations introduced in the previous Subsection, for any $\gamma, \delta \in [0, 1]$ we define the operator $B_{\gamma,\delta} = A_\gamma \circ A_\delta$ (composition of operators), on the domain

$$\mathscr{D}(B_{\gamma,\delta}) = \left\{ u \in \mathscr{D}(A_\gamma) : \Delta u \in \mathscr{D}(A_\delta) \right\},$$

which can be explicitly represented, following the representations of $\mathscr{D}(A_\gamma)$ and $\mathscr{D}(A_\delta)$. Even in this case, $B_{\gamma,\delta}$ is positive and self-adjoint, and its eigenvalues are $\lambda_n(\gamma) \cdot \lambda_n(\delta)$. Furthermore, the intermediate spaces $V_s^{\gamma,\delta} = \mathscr{D}(B_{\gamma,\delta}^s)$ can be represented starting from V_s^γ and V_s^δ. Here, we point out that the norm in $V_s^{\gamma,\delta}$ is the same as in $H^{2s}(\Omega)$.

Consequently, the fourth-order inverse problems in introduction can be generalized to the following

Problem 2 *Find the function* $v : \Omega \times [0, T] \to \mathbb{R}$ *and the real constant* v *such that the following parabolic initial-boundary value problem is satisfied*

$$\begin{cases} \partial_t v(x, t) + B_{\gamma,\delta} v(x, t) = v \, B_{\gamma,\delta}^\sigma v(x, t), & (x, t) \in \Omega \times (0, T), \\ v(x, 0) = v_0(x), & x \in \Omega, \\ \gamma \, \partial_n v(x, t) + (1 - \gamma) c(x) v(x, t) = 0, & (x, t) \in \partial\Omega \times (0, T), \\ \delta \, \partial_n \Delta v(x, t) + (1 - \delta) c(x) \Delta v(x, t) = 0, & (x, t) \in \partial\Omega \times (0, T), \end{cases}$$

along with the global measurement

$$\frac{1}{T_2 - T_1} \int_{T_1}^{T_2} \|v(\cdot, \tau)\|_{H^{2s}(\Omega)}^2 d\tau = \rho_v.$$

And

Problem 2′ *Find the function* $w : \Omega \times [0, T] \to \mathbb{R}$ *and the real constant* ω *such that the following parabolic initial-boundary value problem is satisfied*

$$\begin{cases} \partial_t w(x, t) + B_{\gamma,\delta} w(x, t) = \omega \, B_{\gamma,\delta}^\sigma w(x, t), & (x, t) \in \Omega \times (0, T), \\ w(x, 0) = w_0(x), & x \in \Omega, \\ \gamma \, \partial_n w(x, t) + (1 - \gamma) c(x) w(x, t) = 0, & (x, t) \in \partial\Omega \times (0, T), \\ \delta \, \partial_n \Delta w(x, t) + (1 - \delta) c(x) \Delta w(x, t) = 0, & (x, t) \in \partial\Omega \times (0, T), \end{cases}$$

along with the instantaneous measurement

$$\|w(\cdot, T_0)\|^2_{H^{2s}(\Omega)} = \rho_w.$$

Once again, Theorems 17.1–17.4 yield

Theorem 17.7 *Let $v_0, w_0 \in V_s^{\gamma,\delta} \setminus \{0\}$ and $\rho_v, \rho_w > 0$. Then there exist the couples*

$$(v, v), (w, \omega) \in \left(C([0, T]; V_s^{\gamma,\delta}) \cap L^2(0, T; V_{1+s}^{\gamma,\delta}) \cap H^1(0, T; V_{-1+s}^{\gamma,\delta})\right) \times \mathbb{R},$$

unique (weak) solutions to Problems 2 and 2', respectively, depending continuously on the data (v_0, ρ_v) and (w_0, ρ_w). Moreover, as

$$v_0 \to w_0 \quad in \quad H^s(\Omega), \quad \rho_v \to \rho_w \quad and \quad T_2 - T_1 \to 0$$

there holds

$$v \to w \; strongly \; in \; C([0, T]; V_s^{\gamma,\delta}) \cap L^2(0, T; V_{1+s}^{\gamma,\delta}) \cap H^1(0, T; V_{-1+s}^{\gamma,\delta}) \quad and \quad v \to \omega.$$

Furthermore, under the assumption

$$\omega \neq \frac{\|w(\cdot, T_0)\|^2_{H^{2(1+s)}(\Omega)}}{\|w(\cdot, T_0)\|^2_{H^{2(\sigma+s)}(\Omega)}},$$

the above convergence is of Lipschitz-type.

Even for the fourth-order problems, we distinguish the immediate case

- $\sigma = 1/2$, then $B_{\gamma,\delta}^{1/2} u = A_\gamma u = -\Delta u$ on the domain V_2^γ, that corresponds to a parabolic modification of the plate equation.

17.5.3 Further Remarks

We conclude this section by recalling that, following [10, Sect. 5] all the applications so far devised can be generalized to the wider family of differential operators on Ω of the second-order

$$\mathcal{A}(x, \partial_x) = -\sum_{i,j=1}^{d} \partial_{x_i} \left(a_{i,j}(x) \partial_{x_j}\right) + a_0(x)$$

and fourth-order

$$\mathcal{B}(x, \partial_x) = \sum_{i,j,k,l=1}^{d} \partial_{x_i} \partial_{x_j} \left[b_{i,j,k,l}(x) \partial_{x_k} \partial_{x_l}\right] - \sum_{i,j=1}^{d} \partial_{x_i} \left[b_{i,j}(x) \partial_{x_j}\right] + b_0(x),$$

provided that coefficients fulfill suitable assumptions, which ensure that their realizations in $L^2(\Omega)$ are positive and self-adjoint.

Acknowledgements This paper is devoted to my mentor and friend, Alfredo Lorenzi. Also, I would like to acknowledge the anonymous referee for his/her valuable comments and remarks.

Appendix: Sequences of Implicit Defined Functions

Here we need to exploit some fact regarding implicit functions that is of use in the paper (cf. Sect. 17.4). In particular, we shall investigate the pointwise and the uniform convergence of sequences of implicit defined functions, under the assumption that the generator maps are converging, in some sense.

Let $X \subseteq \mathbb{R}^d$ and α, β be continuous functions on X such that $\alpha(x) < \beta(x)$. Define

$$\mathscr{X} = \left\{(x, y) \in \mathbb{R}^{d+1} : x \in X \quad \text{and} \quad \alpha(x) \leq y \leq \beta(x)\right\}.$$

We shall prove the next

Theorem A *Let F_n, $F : \mathscr{X} \to \mathbb{R}$ be continuous functions such that*

(i) F_n converges pointwise to F on $\mathscr{X}$, as $n \to \infty$;
(ii) the maps $y \mapsto F_n(x, y)$ and $y \mapsto F(x, y)$ are (strictly) increasing for any $x \in X$.

Then, for every $c \in (\operatorname{Im} F)^\circ$ and for every large enough n, there exist continuous functions f_n, $f : X \to \mathbb{R}$ uniquely solving, respectively, the implicit equations

$$F_n(x, f_n(x)) = c \quad \text{and} \quad F(x, f(x)) = c \quad \text{on} \quad X$$

such that f_n converges pointwise to f on X, as $n \to \infty$.

Furthermore, by requiring that F_n is monotone, we can obtain more informations on f_n and its mode of convergence to f, as we state in

Theorem B *Under the assumptions of Theorem A, if the sequence $\{F_n(x, y)\}$ is increasing for any $(x, y) \in \mathscr{X}$, then $\{f_n(x)\}$ is decreasing for any $x \in X$. In this case, the convergence to f is uniform on every compact subset of X.*

Preliminary Fact

Here we need to point out one property of the uniform convergence that will be useful in the following.

Proposition 1 *Let f_n a sequence of continuous functions uniformly convergent to f on a compact interval K, as $n \to \infty$. If $\bigcap_n \operatorname{Im} f_n = \{c\}$, then $f \equiv c$ identically on K.*

Proof Let a_n and b_n, respectively, minimum and maximum points of f_n in K, and suppose that, up to subsequences $a_n \to a$ and $b_n \to b$ for some $a, b \in K$. By uniformity, we see that, as $n \to \infty$

$$f_n(a_n) \le f_n(x) \le f_n(b_n) \quad \text{implies} \quad f(a) \le f(x) \le f(b) \quad \forall x \in K.$$

Moreover, assumption

$$\bigcap_n \operatorname{Im} f_n = \bigcap_n [f_n(a_n), f_n(b_n)] = \{c\}$$

yields $\lim_{n\to\infty} f_n(a_n) = f(a) = \lim_{n\to\infty} f_n(b_n) = f(b) = c$. This concludes the proof.

$\square$

Previous Results

The next results will play a fundamental role in the course of our investigation. As the former is well-known, the latter can be found in [3, Theorem 1].

Lemma 1 *Let f_n be a sequence of functions on a compact interval K, pointwise converging to a continuous function f on K, as $n \to \infty$. If f is increasing, then the convergence is uniform on K.*

Lemma 2 *Let f_n be a sequence of invertible functions on an interval I, uniformly converging to a continuous function f on I, as $n \to \infty$. If f is invertible, then the sequence f_n^{-1} converges uniformly to f^{-1} on any sub-interval of $\bigcap_n \operatorname{Im} f_n$.*

Proof of Theorem A

We are now in a position to prove Theorem A.

Step 1. We show that the requirement $c \in (\operatorname{Im} F)^\circ$ implies that, eventually, $c \in (\operatorname{Im} F_n)^\circ$. In fact, choose a positive δ such that $(c - \delta, c + \delta) \subset \operatorname{Im} F$, and let $(x_-, y_-) \in F^{-1}((c - \delta, c))$ and $(x_+, y_+) \in F^{-1}((c, c + \delta))$. Then, by (i), for any large enough n there holds

$$c - \delta < F_n(x_-, y_-) < c < F_n(x_+, y_+) < c + \delta,$$

which, by continuity of F_n, implies $c \in (\operatorname{Im} F_n)^\circ$.

Step 2. By the implicit function theorem, existence, uniqueness and continuity of f_n and f for n large enough are consequence of assumption *(ii)*.

Step 3. Using (ii) again, by Lemma 1, we deduce that $F_n(x, \cdot)$ converges uniformly to $F(x, \cdot)$ on $[\alpha(x), \beta(x)]$. Also, by monotonicity, $c \in (\operatorname{Im} F_n(x, \cdot))^\circ$ and $c \in (\operatorname{Im} F(x, \cdot))^\circ$. Now, suppose $\{c\} = \bigcap_n \operatorname{Im} F_n(x, \cdot)$, then Proposition 1 implies that $F(x, \cdot)$ is constant on $[\alpha(x), \beta(x)]$, leading to a contradiction. Consequently, it is possible to choose a positive δ_x (depending on x but independent of n) small enough so that

$$(c - \delta_x, c + \delta_x) \subset \bigcap_n \operatorname{Im} F_n(x, \cdot).$$

Step 4. We need to prove that f_n converges to f pointwise on X. To this purpose, let $x \in X$ to be fixed. As the maps $F_n^{-1}(x, \cdot)$ and $F^{-1}(x, \cdot)$ are well-defined and continuous on the domains $\operatorname{Im} F_n(x, \cdot)$ and $\operatorname{Im} F(x, \cdot)$, respectively, we can apply Lemma 2 to deduce that $F_n^{-1}(x, \cdot)$ converges uniformly to $F^{-1}(x, \cdot)$ on $(c - \delta_x, c + \delta_x)$. This, yields, in particular

$$f_n(x) = F_n^{-1}(x, \cdot)(c) \to F^{-1}(x, \cdot)(c) = f(x),$$

as $n \to \infty$ for any fixed x, which concludes the proof. $\square$

Proof of Theorem B

In order to prove that f_n is monotone decreasing in n, consider the identity

$$0 = F_{n+1}(x, f_{n+1}(x)) - F_n(x, f_n(x)) =$$

$$= \Big[F_{n+1}(x, f_{n+1}(x)) - F_{n+1}(x, f_n(x)) \Big] + \Big[F_{n+1}(x, f_n(x)) - F_n(x, f_n(x)) \Big].$$

As, by the monotonicity of F_n, the second summand in the above expression is nonnegative, there holds

$$F_{n+1}(x, f_{n+1}(x)) \le F_{n+1}(x, f_n(x)).$$

This, again by (ii), yields $f_{n+1}(x) \le f_n(x)$ for any $x \in X$.

Finally, if K is a compact subset of X, then the convergence of f_n to f is uniform on K as a straightforward consequence of Dini's Lemma. $\square$

References

1. Artyukhin, E.A., Okhapkin, A.S.: Determination of the parameters in the generalized heat-conduction equation from transient experimental data. J. Eng. Phys. Thermophys. **42**, 693–698 (1982)
2. Azari, H., Zhang, S.: Global superconvergence of finite element methods for parabolic inverse problems. Appl. Math. **3**, 285–294 (2009)
3. Barvínek, E., Daler, I., Franku, J.: Convergence of sequences of inverse functions. Archiv. Math. **27**(3–4), 201–204 (1991)
4. Cannon, J.R.: Determination of certain parameters in heat conduction problems. J. Math. Anal. Appl. **8**, 188–201 (1964)
5. Cannon, J.R., DuChateau, P.: Structural identification of an unknown source term in a heat equation. Inverse Probl. **14**, 535–551 (1998)
6. Dehghan, M.: Identifying a control function in two-dimensional parabolic inverse problems. Appl. Math. Comput. **143**, 375–391 (2003)
7. Evans, L.C.: Partial Differential Equations. American Mathematical Society, Providence (1998)
8. Grisvard, P.: Caractérisation de quelques espaces d'interpolation (French). Arch. Rational Mech. Anal. **25**, 40–63 (1967)
9. Lorenzi, A.: Recovering two constants in a parabolic linear equation. J. Phys. **73**, (2007)
10. Lorenzi, A., Mola, G.: Identification of a real constant in linear evolution equations in Hilbert spaces. Inverse Probl. Imaging **5**(3), 695–714 (2011)
11. Lorenzi, A., Mola, G.: Recovering the reaction and the diffusion coefficients in a linear parabolic equation. Inverse Probl. **28**(7), (2012)
12. Lorenzi, L.: An identification problem for the Ornstein–Uhlenbeck operator. J. Inverse Ill-posed Probl. **19**(2), 293–326 (2011)
13. Lunardi, A.: Analytic Semigroups and Optimal Regularity in Parabolic Problems. Birkäuser, Basel (1995)
14. Lyubanova, A. Sh.: Identification of a constant coefficient in an elliptic equation. Appl. Anal. **87**, 1121–1128 (2008)
15. Mola, G.: Identification of the diffusion coefficient in linear evolution equations in Hilbert spaces. J. Abstr. Differ. Equ. Appl. **2**, 18–28 (2011)
16. Simon, J.: Compact sets in the space $L^p(0, T; B)$. Ann. Mat. Pura Appl. (IV) **146**, 65–96 (1987)
17. Temam, R.: Infinite-Dimensional Dynamical Systems in Mechanics and Physics. Springer, New York 1988
18. Yamamoto, M.: Determination of constant parameters in some semilinear parabolic equations. In: Ill-posed problems in natural sciences (Moscow, 1991), VSP, Utrecht, 439–445 (1992)
19. Ye, C., Sun, Z.: Global superconvergence of finite element methods for parabolic inverse problems. Appl. Math. Comput. **188**, 214–225 (2007)

Chapter 18
L^p-Theory for Schrödinger Operators Perturbed by Singular Drift Terms

Noboru Okazawa and Motohiro Sobajima

Dedicated to the memory of late Professor Alfredo Lorenzi

Abstract Our concern is the essential m-accretivity in $L^p(\mathbf{R}^N)$ ($1 < p < \infty, N \in$ **N**) of the minimal realization of second-order elliptic operator with strongly singular drift term $A_{p,\min}u = -\Delta u + b|x|^{-2}(x \cdot \nabla)u + Vu$, where $b \in \mathbb{R}$ is a constant and $V \in L^p_{\mathrm{loc}}(\mathbb{R}^N \setminus \{0\})$ is bounded below by the inverse-square potential: $V(x) \geq c_0|x|^{-2}$ with a new critical constant $c_0 = c_0(b, p, N) \in \mathbb{R}$. Namely, we shall generalize the result on the Schrödinger operators (that is, $A_{p,\min}$ with $b = 0$) due to Kalf, Walter, Schmincke and Simon in Edmunds and Evans (Spectral Theory and Differential Operators, The Clarendon Press, Oxford, 1987) ($p = 2$) and Okazawa (Jpn. J. Math. **22**, 199–239, 1996) ($p \in (1, \infty)$) to that on the general case of $A_{p,\min}$ with $b \neq 0$. The proof is based on those techniques in singular perturbation of m-accretive operators and resolvent-positivity together with Kato's inequality. These ideas in operator theory are summarized in Okazawa (Jpn. J. Math. **22**, 199–239, 1996).

18.1 Introduction and Result

In this paper we consider the following second-order elliptic operator in $L^p = L^p(\mathbb{R}^N)$ ($1 < p < \infty$ and $N \in$ **N**) with a strongly singular drift term:

$$
\begin{cases}
(A_{p,\min}u)(x) := -\Delta u(x) + \dfrac{b}{|x|^2}(x \cdot \nabla)u(x) + V(x)u(x), \\
D(A_{p,\min}) := C_0^\infty(\mathbb{R}^N \setminus \{0\}),
\end{cases}
\tag{18.1}
$$

N. Okazawa (✉)
Tokyo University of Science, Tokyo 162-8601, Japan
e-mail: okazawa@ma.kagu.tus.ac.jp

M. Sobajima
Università del Salento, Lecce 73100, Italy
e-mail: msobajima1984@gmail.com

© Springer International Publishing Switzerland 2014
A. Favini et al. (eds.), *New Prospects in Direct, Inverse and Control Problems for Evolution Equations*, Springer INdAM Series 10,
DOI 10.1007/978-3-319-11406-4_18

where $b \in \mathbb{R}$ is a constant and $V \in L_{loc}^p(\mathbb{R}^N \setminus \{0\})$ is a real-valued function. Our concern is the essential m-accretivity of $A_{p,min}$ in L^p. Here we discuss it under the following lower bound on V:

$$V(x) \geq \frac{c_0(b, p, N)}{|x|^2} \quad \text{a.a. } x \in \mathbb{R}^N \setminus \{0\} \tag{18.2}$$

for a new critical value $c_0 = c_0(b, p, N) \in \mathbb{R}$.

In particular, if $b = 0$, then $A_{p,min}$ is nothing but a Schrödinger operator. In this case it is well-known that if $p = 2$ and V satisfies (18.2) with

$$c_0(0, 2, N) := -\frac{N(N-4)}{4},$$

then $A_{2,min}$ is nonnegative and essentially selfadjoint. This is known as the Kalf-Walter-Schmincke Simon theorem (see Edmunds and Evans [1, Theorem VII.4.2] or Reed and Simon [11, Theorem X.11]). The result depends essentially on both of Hardy's and Rellich's inequalities. In fact, on the one hand, $-c_0(0, 2, N) = N(N-4)/4$ is known as the best constant of the classical Rellich inequality:

$$\frac{N(N-4)}{4} \left\| \frac{u}{|x|^2} \right\|_{L^2} \leq \|\Delta u\|_{L^2}, \quad u \in H^2(\mathbb{R}^N), \ N \geq 5;$$

in this connection note that the modified Rellich inequality holds for *every* $N \in \mathbb{N}$:

$$\frac{N(N-4)}{4} \int_{\mathbb{R}^N} \frac{|u|^2}{(|x|^2 + \varepsilon)^2} \, dx \leq \text{Re} \int_{\mathbb{R}^N} \frac{(-\Delta u)\bar{u}}{|x|^2 + \varepsilon} \, dx, \quad u \in H^2(\mathbb{R}^N), \ \varepsilon > 0$$

(see Okazawa et al. [10, Lemma 3.3] in which the original proof in Okazawa [7] is simplified). On the other hand, the nonnegativity of $A_{2,min}$ is verified by the classical Hardy inequality

$$\frac{(N-2)^2}{4} \int_{\mathbb{R}^N} \frac{|u|^2}{|x|^2} \, dx \leq \int_{\mathbb{R}^N} (-\Delta u)\bar{u} \, dx, \quad u \in H^2(\mathbb{R}^N), \ N \geq 3.$$

It is worth noticing that the positive difference $(N-2)^2 - N(N-4) = 4$ implies that

$$c_0(0, 2, N) = \max \left\{ -\frac{(N-2)^2}{4}, -\frac{N(N-4)}{4} \right\} = -\frac{N(N-4)}{4}.$$

Later, Okazawa [8] generalized the essential selfadjointness of $A_{2,min}$ to the essential m-accretivity of $A_{p,min}$ $(1 < p < \infty)$. More precisely, it is stated as follows:

Theorem 18.0 ([8, Theorem 4.1]) *Let p' be the Hölder conjugate of $p \in (1, \infty)$. For $N \in \mathbf{N}$ define $\alpha_N(p)$ and $\beta_N(p)$ respectively as follows:*

$$\alpha_N(p) := -\frac{(p-1)(N-2)^2}{p^2} = -\frac{N-2}{p} \cdot \frac{N-2}{p'},$$

$$\beta_N(p) := -\frac{(p-1)N(N-2p)}{p^2} = -\left(\frac{N}{p} - 2\right)\frac{N}{p'}.$$

Assume that $V \in L^p_{\mathrm{loc}}(\mathbb{R}^N \setminus \{0\})$ satisfies

$$V(x) \geq \frac{c_0(0, p, N)}{|x|^2} \quad \text{a.a. } x \in \mathbb{R}^N \setminus \{0\}$$

with $c_0(0, p, N) := \max\{\alpha_N(p), \beta_N(p)\}$. Then $-\Delta + V$ with domain $C_0^\infty(\mathbb{R}^N)$ ($A_{p,\min}$ with $b = 0$) is essentially m-accretive in L^p.

It seems that this theorem determines the critical value $c_0(0, p, N)$ completely. In a way similar to the case of $p = 2$ the result depends again on L^p-versions of modified Hardy's and Rellich's inequalities both. In terms of the family $\{(|x|^2 + \varepsilon)^{-1}; \varepsilon > 0\}$ of Yosida approximations to the inverse-square potential $|x|^{-2}$ they are respectively stated as follows: for $u \in W^{2,p}(\mathbb{R}^N)$ and $\varepsilon > 0$,

$$-\alpha_N(p) \int_{\mathbb{R}^N} \frac{|u|^p}{|x|^2 + \varepsilon}\, dx \leq \mathrm{Re} \int_{\mathbb{R}^N} (-\Delta u)\,\bar{u}\,|u|^{p-2}\, dx, \qquad (18.3)$$

$$-\beta_N(p) \int_{\mathbb{R}^N} \frac{|u|^p}{(|x|^2 + \varepsilon)^p}\, dx \leq \mathrm{Re} \int_{\mathbb{R}^N} \frac{(-\Delta u)\,\bar{u}\,|u|^{p-2}}{(|x|^2 + \varepsilon)^{p-1}}\, dx. \qquad (18.4)$$

The original proof of (18.4) in [8] is simplified in Maeda and Okazawa [5]. The inequality (18.3) represents the *accretivity* of $A_{p,\min}$ with $b = 0$ (cf. (18.19) below), while (18.4) implies the separation property of the approximate operator $-\Delta + c(|x|^2 + \varepsilon)^{-1}$:

$$\|\Delta u\|_{L^p} + \left\|\frac{u}{|x|^2 + \varepsilon}\right\|_{L^p} \leq C(c)\left\|-\Delta u + \frac{c}{|x|^2 + \varepsilon}u\right\|_{L^p}, \quad u \in W^{2,p}(\mathbb{R}^N),$$

where $C(c) := 1 + (1+c)(c - \beta_N(p))^{-1}$ for $c > \beta_N(p)$. This ensures the closedness and *maximality* of $-\Delta + c|x|^{-2}$ in L^p; note that $D(|x|^{-2}) := \{u \in L^p; |x|^{-2}u \in L^p\}$.

By virtue of the scaling argument the essential m-accretivity of $A_{p,\min}$ with $b \neq 0$ can be also dealt with under condition (18.2), possibly, with a new critical value. However, there is no previous work on this problem with $b \neq 0$.

In this context the purpose of this paper is to present a sharp condition for the essential m-accretivity of $A_{p,\min}$. More precisely, we give the new critical value $c_0 = c_0(b, p, N)$ for the essential m-accretivity of $A_{p,\min}$ with $b \neq 0$.

Theorem 18.1 *For* $p \in (1, \infty)$, $N \in \mathbf{N}$ *and* $b \in \mathbb{R}$ *define two new constants* $\alpha_{N,b}(p)$ *and* $\beta_{N,b}(p)$ *as*

$$\alpha_{N,b}(p) := -\frac{N-2}{p}\left(\frac{N-2}{p'} - b\right),$$

$$\beta_{N,b}(p) := -\left(\frac{N}{p} - 2\right)\left(\frac{N}{p'} - b\right),$$

respectively. Assume that $V \in L^p_{\text{loc}}(\mathbb{R}^N \setminus \{0\})$ *satisfies*

$$V(x) \geq \frac{c_0(b, p, N)}{|x|^2} \quad \text{a.a.} \ x \in \mathbb{R}^N \setminus \{0\},$$

with $c_0(b, p, N) := \max\{\alpha_{N,b}(p), \beta_{N,b}(p)\}$. *Then* $A_{p,\min}$ *is essentially* m-accretive *in* L^p.

The formal adjoint of the differential expression

$$A := -\Delta + b|x|^{-2}(x \cdot \nabla) + V(x)$$

is given by

$$B := -\Delta - b|x|^{-2}(x \cdot \nabla) + V(x) - b(N-2)|x|^{-2}.$$

Therefore Theorem 1.1 implies that $B_{p',\min} := B$ with $D(B_{p',\min}) := C_0^\infty(\mathbb{R}^N \setminus \{0\})$ is essentially m-accretive in $L^{p'}$ if $V \in L^{p'}_{\text{loc}}(\mathbf{R}^N \setminus \{0\})$ satisfies

$$V(x) - \frac{b(N-2)}{|x|^2} \geq \frac{\max\{\alpha_{N,-b}(p'), \beta_{N,-b}(p')\}}{|x|^2}.$$

Noting that $b(N-2) + \alpha_{N,-b}(p') = \alpha_{N,b}(p)$, we obtain the following

Corollary 18.1 *Let* $\alpha_{N,b}(p)$ *and* $\beta_{N,b}(p)$ *be as in Theorem 18.1. Assume that*

$$V \in L^p_{\text{loc}}(\mathbb{R}^N \setminus \{0\}) \cap L^{p'}_{\text{loc}}(\mathbb{R}^N \setminus \{0\}) \tag{18.5}$$

satisfies

$$V(x) \geq \frac{c_1(b, p, N)}{|x|^2} \quad \text{a.a.} \ x \in \mathbb{R}^N \setminus \{0\} \tag{18.6}$$

with

$$c_1(b, p, N) := \max\{\alpha_{N,b}(p), \beta_{N,b}(p), b(N-2) + \beta_{N,-b}(p')\}.$$

Then $B_{p',\min}$ is essentially m-accretive in $L^{p'}$ and the adjoint of $B_{p',\min}$ is equal to $(A_{p,\min})\tilde{\,}$, the closure of $A_{p,\min} : (B_{p',\min})^* = (A_{p,\min})\tilde{\,}$, where $A_{p,\min}$ is defined by (18.1) in which V satisfies (18.5) and (18.6).

Remark 18.1 The equality $\beta_{N,b}(p) = b(N-2) + \beta_{N,-b}(p')$ holds if $b = N(p'^{-1} - p^{-1})$.

18.2   Preliminaries

First we prepare several inequalities related to the relative boundedness of $|x|^{-1}\nabla$ with respect to $-\Delta + c|x|^{-2}$ in L^p ($1 < p < \infty$).

Lemma 18.1 *Let $\varepsilon > 0$:*

(a) if $1 < p \leq 2$, then for every $u \in W^{2,p}(\mathbb{R}^N)$,

$$\left\|\frac{\nabla u}{(|x|^2 + \varepsilon)^{1/2}}\right\|_{L^p}^2 \leq p'p \left\|\frac{u}{|x|^2 + \varepsilon}\right\|_{L^p}^2 + p'\|\Delta u\|_{L^p}\left\|\frac{u}{|x|^2 + \varepsilon}\right\|_{L^p}; \tag{18.7}$$

(b) if $2 < p < \infty$, then for every $u \in W^{2,p}(\mathbb{R}^N)$,

$$\left\|\frac{\nabla u}{(|x|^2 + \varepsilon)^{1/2}}\right\|_{L^p}^2 \leq p \left\|\frac{\nabla u}{(|x|^2 + \varepsilon)^{1/2}}\right\|_{L^p}\left\|\frac{u}{|x|^2 + \varepsilon}\right\|_{L^p}$$
$$+ (1 + (p-2)N^2 H_p)\|\Delta u\|_{L^p}\left\|\frac{u}{|x|^2 + \varepsilon}\right\|_{L^p}, \tag{18.8}$$

where H_p is the constant in the Calderón-Zygmund estimate:

$$\left\|\frac{\partial^2 u}{\partial x_j \partial x_k}\right\|_{L^p} \leq H_p\|\Delta u\|_{L^p}, \quad H_p := \begin{cases} \tan(\pi/2p) & \text{when } 1 < p \leq 2, \\ \cot(\pi/2p) & \text{when } 2 \leq p < \infty. \end{cases} \tag{18.9}$$

Proof (a) First we prove (18.7) when $1 < p \leq 2$. Applying the same argument as in the proof of [12, Lemma 3] in which we set

$$a_{jk} = \delta_{jk} \text{ (the Kronecker delta)}, \quad F \equiv 0 = \lambda, \quad V(x) = \Psi_p(x) = p(|x|^2 + \varepsilon)^{-1},$$

we can verify that

$$\left\|\left(\frac{p}{|x|^2+\varepsilon}\right)^{1/2}\nabla u\right\|_{L^p}^2 \le \frac{p}{p-1}\left\|(-\Delta)u+\frac{p}{|x|^2+\varepsilon}u\right\|_{L^p}\left\|\frac{p}{|x|^2+\varepsilon}u\right\|_{L^p}$$

$$\le \frac{p}{p-1}\left(p^2\left\|\frac{u}{|x|^2+\varepsilon}\right\|_{L^p}^2+p\,\|\Delta u\|_{L^p}\left\|\frac{u}{|x|^2+\varepsilon}\right\|_{L^p}\right).$$

Hence, dividing both sides by p, we obtain (18.7).

(b) Next we assume $2<p<\infty$. Then integration by parts gives

$$\left\|\frac{\nabla u}{(|x|^2+\varepsilon)^{1/2}}\right\|_{L^p}^p$$

$$=\int_{\mathbb{R}^N}\nabla\bar u\cdot\frac{|\nabla u|^{p-2}\nabla u}{(|x|^2+\varepsilon)^{p/2}}\,dx$$

$$=-\int_{\mathbb{R}^N}\bar u\,\mathrm{div}\left(\frac{|\nabla u|^{p-2}\nabla u}{(|x|^2+\varepsilon)^{p/2}}\right)dx$$

$$=-\int_{\mathbb{R}^N}\frac{|\nabla u|^{p-2}\bar u\,(\Delta u)}{(|x|^2+\varepsilon)^{p/2}}\,dx-(p-2)\,\mathrm{Re}\int_{\mathbb{R}^N}\frac{|\nabla u|^{p-4}\bar u\,\langle D^2u\nabla u,\nabla\bar u\rangle}{(|x|^2+\varepsilon)^{p/2}}\,dx$$

$$-p\,\mathrm{Re}\int_{\mathbb{R}^N}\frac{|\nabla u|^{p-2}\bar u\,(x\cdot\nabla)u}{(|x|^2+\varepsilon)^{p/2+1}}\,dx,$$

where $\langle\cdot,\cdot\rangle$ denotes the usual hermitian product over $\mathbf{C}^N$ and D^2u is the $N\times N$ matrix given by

$$D^2u:=\left(\frac{\partial^2 u}{\partial x_j\,\partial x_k}\right)_{jk},$$

$$\|D^2u\|_{(L^p)^{N\times N}}\le N^2\max_{1\le j,k\le N}\left\|\frac{\partial^2 u}{\partial x_j\,\partial x_k}\right\|_{L^p}. \tag{18.10}$$

Using Hölder's inequality, we see that

$$\left\|\frac{\nabla u}{(|x|^2+\varepsilon)^{1/2}}\right\|_{L^p}^p\le\|\Delta u\|_{L^p}\left\|\frac{\nabla u}{(|x|^2+\varepsilon)^{1/2}}\right\|_{L^p}^{p-2}\left\|\frac{u}{|x|^2+\varepsilon}\right\|_{L^p}$$

$$+(p-2)\|D^2u\|_{(L^p)^{N\times N}}\left\|\frac{\nabla u}{(|x|^2+\varepsilon)^{1/2}}\right\|_{L^p}^{p-2}\left\|\frac{u}{|x|^2+\varepsilon}\right\|_{L^p}$$

$$+p\left\|\frac{\nabla u}{(|x|^2+\varepsilon)^{1/2}}\right\|_{L^p}^{p-1}\left\|\frac{u}{|x|^2+\varepsilon}\right\|_{L^p}.$$

Hence we have

$$\left\|\frac{\nabla u}{(|x|^2+\varepsilon)^{1/2}}\right\|_{L^p}^2 \le p\left\|\frac{\nabla u}{(|x|^2+\varepsilon)^{1/2}}\right\|_{L^p}\left\|\frac{u}{|x|^2+\varepsilon}\right\|_{L^p}$$

$$+\left(\|\Delta u\|_{L^p} + (p-2)\|D^2 u\|_{(L^p)^{N\times N}}\right)\left\|\frac{u}{|x|^2+\varepsilon}\right\|_{L^p}.$$

Finally, in view of (18.10), the well-known Calderón-Zygmund estimate (18.9) applies to give (18.8) (for the constant H_p see Iwaniec and Martin [2] or [3]). □

Remark 18.2 Let $2 < p < \infty$. Then we see from (18.8) that for $u \in W^{2,p}(\mathbb{R}^N)$,

$$\left\|\frac{\nabla u}{(|x|^2+\varepsilon)^{1/2}}\right\|_{L^p} \le p\left\|\frac{u}{|x|^2+\varepsilon}\right\|_{L^p}$$

$$+\left(1+(p-2)N^2 H_p\right)^{1/2}\|\Delta u\|_{L^p}^{1/2}\left\|\frac{u}{|x|^2+\varepsilon}\right\|_{L^p}^{1/2}.$$

$$(18.11)$$

The computation is done by completing the square (as in (18.24) below).

Corollary 18.2 *Let $p \in (1,\infty)$. Then for a fixed $\varepsilon > 0$, $(|x|^2+\varepsilon)^{-1}(x\cdot\nabla)$ is $(-\Delta)$-bounded with $(-\Delta)$-bound 0, that is, for any $\eta > 0$ there exists $C = C(\eta,\varepsilon) > 0$ such that*

$$\left\|\frac{(x\cdot\nabla)u}{|x|^2+\varepsilon}\right\|_{L^p} \le \eta\|(-\Delta)u\|_{L^p} + C(\eta,\varepsilon)\|u\|_{L^p}, \quad u \in W^{2,p}(\mathbb{R}^N). \quad (18.12)$$

Proof Let $u \in W^{2,p}(\mathbb{R}^N)$. Then we note that (18.7) and (18.11) are roughly unified as

$$\left\|\frac{\nabla u}{(|x|^2+\varepsilon)^{1/2}}\right\|_{L^p} \le k_1\left\|\frac{u}{|x|^2+\varepsilon}\right\|_{L^p} + 2k_2\|\Delta u\|_{L^p}^{1/2}\left\|\frac{u}{|x|^2+\varepsilon}\right\|_{L^p}^{1/2}. \quad (18.13)$$

Since $|(x\cdot\nabla)u| \le (|x|^2+\varepsilon)^{1/2}|\nabla u|$, we see from (18.13) that

$$\left\|\frac{(x\cdot\nabla)u}{|x|^2+\varepsilon}\right\|_{L^p} \le \left\|\frac{\nabla u}{(|x|^2+\varepsilon)^{1/2}}\right\|_{L^p} \le \frac{k_1}{\varepsilon}\|u\|_{L^p} + \frac{2k_2}{\sqrt{\varepsilon}}\|\Delta u\|_{L^p}^{1/2}\|u\|_{L^p}^{1/2}$$

$$\le \eta\|(-\Delta)u\|_{L^p} + \varepsilon^{-1}(k_1 + k_2^2\eta^{-1})\|u\|_{L^p}.$$

Setting $C(\eta,\varepsilon) := \varepsilon^{-1}(k_1 + k_2^2\eta^{-1})$, we obtain (18.12). □

18.3   The Family of Operators with Special Structure

In this section we consider second-order elliptic operators in L^p with special form, that is, operators of scale-invariant structure:

$$\begin{cases} T_p(c)u(x) := -\Delta u(x) + \dfrac{b}{|x|^2}(x \cdot \nabla)u(x) + \dfrac{c}{|x|^2}u(x), \\[2mm] D(T_p(c)) = D(T_p) := W^{2,p}(\mathbb{R}^N) \cap D(|x|^{-2}). \end{cases}$$

The proof of the main theorem is based on

Proposition 18.1 *Let $c \geq c_0 = c_0(b, p, N)$. Here $c_0(b, p, N)$ is the same constant as in Theorem 18.1. Then $T_p(c)$ is m-accretive in L^p when $c > c_0$, while $T_p(c_0)$ is essentially m-accretive on $D(T_p)$. Let $\tilde{T}_p(c)$ denote the closure of $T_p(c)$. Then $C_0^\infty(\mathbb{R}^N \setminus \{0\})$ is a core for $\tilde{T}_p(c)$ in both cases and $\tilde{T}_p(c)$ has a positive resolvent, that is,*

$$\overline{(1 + T_p(c))[C_0^\infty(\mathbb{R}^N \setminus \{0\})_+]} = L_+^p, \quad c \geq c_0,$$

where $C_0^\infty(\mathbb{R}^N \setminus \{0\})_+$ and L_+^p are the positive cones in the respective spaces.

To prove Proposition 18.1 we introduce a family $\{T_{p,\varepsilon}\}_{\varepsilon>0} = \{-\Delta + R_{p,\varepsilon}\}_{\varepsilon>0}$ of operators approximate to $T_p(c)$ in L^p, where

$$\begin{cases} R_{p,\varepsilon}u(x) := \dfrac{b}{|x|^2 + \varepsilon}(x \cdot \nabla)u(x) + \dfrac{c}{|x|^2 + \varepsilon}u(x) + \varepsilon \dfrac{(1 + b/2)^2}{(|x|^2 + \varepsilon)^2}u(x), \\[2mm] \hspace{8.5cm} (18.14) \\[1mm] D(R_{p,\varepsilon}) := W^{1,p}(\mathbb{R}^N). \end{cases}$$

Since $D(T_{p,\varepsilon}) = D(-\Delta) = W^{2,p}(\mathbb{R}^N)$, it follows from (18.12) that the approximate operator $T_{p,\varepsilon} = -\Delta + R_{p,\varepsilon}$ is closed in L^p (see Kato[4, Theorem IV.1.1]). When $p \geq 2$, it is not so difficult to verify the accretivity of $T_{p,\varepsilon}$. In fact, let $u \in C_0^\infty(\mathbb{R}^N)$. Then we have

$$\text{Re} \int_{\mathbb{R}^N} (-\Delta u)\, \bar{u}\, |u|^{p-2}\, dx$$

$$\geq (p-1) \int_{\mathbb{R}^N} |\nabla|u(x)||^2 |u(x)|^{(p-2)/2}\, dx$$

$$\geq \frac{(N-2)^2}{p'p} \int_{\mathbb{R}^N} \frac{|u(x)|^p}{|x|^2 + \varepsilon}\, dx + \frac{N^2 - 4}{p'p}\varepsilon \int_{\mathbb{R}^N} \frac{|u(x)|^p}{(|x|^2 + \varepsilon)^2}\, dx, \quad (18.15)$$

$$\operatorname{Re} \int_{\mathbb{R}^N} \frac{(x \cdot \nabla)u}{|x|^2 + \varepsilon} \, \bar{u} \, |u|^{p-2} \, dx$$

$$= -\frac{N-2}{p} \int_{\mathbb{R}^N} \frac{|u(x)|^p}{|x|^2 + \varepsilon} \, dx - \frac{2}{p} \varepsilon \int_{\mathbb{R}^N} \frac{|u(x)|^p}{(|x|^2 + \varepsilon)^2} \, dx. \tag{18.16}$$

These are the modified versions of [8, Lemma 2.2] and [10, Lemma 3.1]; the second terms (multiplied by ε) on the right-hand sides are new. When $1 < p < 2$, $|u(x)|^{p-2}$ may be approximated by $(|u(x)|^2 + \delta e^{-|x|^2})^{-(2-p)/2}$, $\delta > 0$:

$$\operatorname{Re} \int_{\mathbb{R}^N} \frac{\bar{u}(-\Delta u)}{(|u|^2 + \delta e^{-|x|^2})^{(2-p)/2}} \, dx \geq I(\delta) + (p-1) \int_{\mathbb{R}^N} \frac{|\nabla u(x)|^2}{(|u(x)|^2 + \delta e^{-|x|^2})^{(2-p)/2}} \, dx,$$

where

$$I(\delta) := 2(2-p) \operatorname{Re} \int_{\mathbb{R}^N} \frac{\delta e^{-|x|^2} \overline{u(x)} (x \cdot \nabla)u(x)}{(|u(x)|^2 + \delta e^{-|x|^2})^{(4-p)/2}} \, dx = O(\delta^{(p-1)/2}) \quad (\delta \downarrow 0).$$

Therefore we need to compute a little bit more carefully.

Lemma 18.2 *Let $1 < p < 2$. Then for $u \in C_0^\infty(\mathbb{R}^N)$ and $\varepsilon > 0$ one has*

(a) *(Accretivity of $-\Delta$ with error term).*

$$\operatorname{Re} \int_{\mathbb{R}^N} (-\Delta u) \, \bar{u} \, |u|^{p-2} \, dx = \lim_{\delta \downarrow 0} \operatorname{Re} \int_{\mathbb{R}^N} \frac{\bar{u}(-\Delta u)}{(|u|^2 + \delta e^{-|x|^2})^{(2-p)/2}} \, dx$$

$$\geq \frac{(N-2)^2}{p'p} \int_{\mathbb{R}^N} \frac{|u(x)|^p}{|x|^2 + \varepsilon} \, dx + \frac{N^2 - 4}{p'p} \varepsilon \int_{\mathbb{R}^N} \frac{|u(x)|^p}{(|x|^2 + \varepsilon)^2} \, dx - \lim_{\delta \downarrow 0} J(\delta),$$

$$\tag{18.17}$$

where $J(\delta)$ is given by

$$J(\delta) := \frac{2(N-2)}{p'} J_1(\delta) + \frac{(N-2)^2}{p'p} J_2(\delta),$$

$$J_k(\delta) := \int_{\mathbb{R}^N} \frac{\delta e^{-|x|^2} |x|^2 (|x|^2 + \varepsilon)^{-k}}{(|u(x)|^2 + \delta e^{-|x|^2})^{(2-p)/2}} \, dx = O(\delta^{p/2}) \quad (\delta \downarrow 0) \quad (k = 1, 2);$$

(b) *(Accretivity of $(|x|^2 + \varepsilon)^{-1}(x \cdot \nabla)$ with error term). Let $J_1(\delta)$ be as in (a). Then*

$$\operatorname{Re} \int_{\mathbb{R}^N} \frac{\overline{u(x)}(x \cdot \nabla)u(x)}{(|x|^2 + \varepsilon)(|u(x)|^2 + \delta e^{-|x|^2})^{(2-p)/2}} \, dx$$

$$= -\frac{N-2}{p} \int_{\mathbb{R}^N} \frac{(|u|^2 + \delta e^{-|x|^2})^{p/2}}{|x|^2 + \varepsilon} \, dx$$

$$-\frac{2}{p}\varepsilon \int_{\mathbb{R}^N} \frac{(|u|^2 + \delta e^{-|x|^2})^{p/2}}{(|x|^2 + \varepsilon)^2} \, dx + J_1(\delta). \tag{18.18}$$

Here we can understand the meaning of extra term multiplied by ε in the definition of $R_{p,\varepsilon}u$. In fact, the above-mentioned computation shows that for $u \in C_0^\infty(\mathbb{R}^N)$,

$$\mathrm{Re} \int_{\mathbb{R}^N} (T_{p,\varepsilon}u)\bar{u}|u|^{p-2} \, dx \geq M_1 \int_{\mathbb{R}^N} \frac{|u|^p}{|x|^2 + \varepsilon} \, dx + M_2 \, \varepsilon \int_{\mathbb{R}^N} \frac{|u|^p}{(|x|^2 + \varepsilon)^2} \, dx, \tag{18.19}$$

where M_1 and M_2 are computed by (18.17), (18.18) and the remaining inequalities (18.15), (18.16):

$$M_1 := c - \alpha_{N,b}(p) = \frac{(N-2)^2}{p'p} - \frac{N-2}{p}b + c,$$

$$M_2 := \frac{N^2}{p'p} + \left(1 + \frac{b}{2} - \frac{2}{p}\right)^2 = \frac{N^2 - 4}{p'p} - \frac{2}{p}b + \left(1 + \frac{b}{2}\right)^2 \geq 0.$$

Since $C_0^\infty(\mathbb{R}^N)$ is a core for $-\Delta$, it follows from (18.19) that $T_{p,\varepsilon} = -\Delta + R_{p,\varepsilon}$ is accretive in L^p if $M_1 \geq 0$. For an fixed $\varepsilon > 0$ the basic properties of the operator $T_{p,\varepsilon}$ are summarized as follows:

Lemma 18.3 *Let $\varepsilon > 0$ be fixed and $1 < p < \infty$. If $c \geq \alpha_{N,b}(p)$, then $T_{p,\varepsilon} = -\Delta + R_{p,\varepsilon}$ is m-accretive in L^p. Moreover, the resolvent of $T_{p,\varepsilon}$ is positive.*

Proof To prove the maximality of closed and accretive operator $T_{p,\varepsilon}$ put

$$Au := -\Delta u \quad \text{for } u \in D(A) := W^{2,p}(\mathbb{R}^N),$$

$$Bu := R_{p,\varepsilon}u \quad \text{for } u \in D(B) := W^{1,p}(\mathbb{R}^N).$$

Then the pair of A and B with $D(A) \subset D(B)$ satisfies the assumption of [7, Lemma 3.1]. Namely, for all $t \in [0,1]$, $A + tB = (1-t)(-\Delta) + tT_{p,\varepsilon}$ is closed and accretive in L^p. Therefore the maximality of $A + B = T_{p,\varepsilon}$ ($t = 1$) is reduced to that of $A = -\Delta$ ($t = 0$).

Since $T_{p,\varepsilon}$ is m-accretive in L^p, the resolvent-positivity is equivalent to the dispersivity: $\langle T_{p,\varepsilon}u, (u_+)^{p-1}\rangle_{L^p, L^{p'}} \geq 0$ for all $u \in W^{2,p}(\mathbb{R}^N)$. The dispersivity of $-\Delta$ is nothing but the positivity of $e^{t\Delta}$. Therefore it suffices to show that $\pm(|x|^2 + \varepsilon)^{-1}(x \cdot \nabla)$ is dispersive in real L^p-spaces. But the computation is quite similar to that for accretivity as was done in [8, Lemma 2.6] (see also Wong-Dzung [13]; note that the boundedness of the potential in [13] is later removed by Miyajima and Okazawa [6, Theorem 4.1]). $\qquad\qquad\square$

Lemma 18.4 *Let $\varepsilon > 0$ be fixed and $1 < p < \infty$. If $c > \beta_{N,b}(p)$, then one has*

$$\left\| (|x|^2 + \varepsilon)^{-1} u \right\|_{L^p} \le \left(c - \beta_{N,b}(p) \right)^{-1} \left\| T_{p,\varepsilon} u \right\|_{L^p} \quad \forall\, u \in W^{2,p}(\mathbb{R}^N) \qquad (18.20)$$

and hence

$$\left\| (|x|^2 + \varepsilon)^{-1/2} \nabla u \right\|_{L^p} \le K_1 \left\| T_{p,\varepsilon} u \right\|_{L^p} \quad \forall\, u \in W^{2,p}(\mathbb{R}^N), \qquad (18.21)$$

where K_1 is written as

$$K_1 := \begin{cases} \dfrac{p'|b| + \sqrt{p'[p + |c| + (1 + b/2)^2]}}{c - \beta_{N,b}(p)} + \sqrt{\dfrac{p'}{c - \beta_{N,b}(p)}} & (1 < p \le 2), \\[3ex] \dfrac{p + K_2|b| + \sqrt{K_2[|c| + (1 + b/2)^2]}}{c - \beta_{N,b}(p)} + \sqrt{\dfrac{K_2}{c - \beta_{N,b}(p)}} & (2 < p < \infty). \end{cases}$$

$$\qquad (18.22)$$

Here $K_2 := 1 + (p - 2)N^2 H_p$ is the constant in Lemma 2.1(b).

Proof It suffices to prove (18.20) and (18.21) both on $C_0^\infty(\mathbb{R}^N)$. Let $u \in C_0^\infty(\mathbb{R}^N)$. Then in a previous paper Okazawa et al. [9] we have computed the accretivity of $T_{p,\varepsilon}$ in weighted L^p-spaces. In particular, employing [9, Lemma 2.3 with (α, β, c) replaced with $(2(p - 1), b, -c)$] and then applying Hölder's inequality, we have

$$(c - \beta_{N,b}(p)) \left\| \frac{u}{|x|^2 + \varepsilon} \right\|_{L^p}^p \le \operatorname{Re} \int_{\mathbb{R}^N} \frac{(T_{p,\varepsilon} u)\, \overline{u}\, |u|^{p-2}}{(|x|^2 + \varepsilon)^{p-1}}\, dx$$

$$\le \left\| T_{p,\varepsilon} u \right\|_{L^p} \left\| \frac{u}{|x|^2 + \varepsilon} \right\|_{L^p}^{p-1}.$$

This yields (18.20). Next we prove (18.21). To this end we replace Δu in (18.7) and (18.8) with $-T_{p,\varepsilon} u$ together with the remainder term $R_{p,\varepsilon} u$ defined by (18.14):

$$\Delta u = -T_{p,\varepsilon} u + R_{p,\varepsilon} u. \qquad (18.23)$$

First we consider the case with $p \in (2, \infty)$. Put $K_2 = 1 + (p - 2)N^2 H_p$ as in the statement of this lemma. Then we see from (18.8) that

$$\left\| \frac{\nabla u}{(|x|^2 + \varepsilon)^{1/2}} \right\|_{L^p}^2 \le p \left\| \frac{\nabla u}{(|x|^2 + \varepsilon)^{1/2}} \right\|_{L^p} \left\| \frac{u}{|x|^2 + \varepsilon} \right\|_{L^p} + K_2 \left\| \Delta u \right\|_{L^p} \left\| \frac{u}{|x|^2 + \varepsilon} \right\|_{L^p}$$

$$\le (p + K_2|b|) \left\| \frac{\nabla u}{(|x|^2 + \varepsilon)^{1/2}} \right\|_{L^p} \left\| \frac{u}{|x|^2 + \varepsilon} \right\|_{L^p}$$

$$+ K_2 \left\| T_{p,\varepsilon} u \right\|_{L^p} \left\| \frac{u}{|x|^2 + \varepsilon} \right\|_{L^p} + K_3 \left\| \frac{u}{|x|^2 + \varepsilon} \right\|_{L^p}^2,$$

where $K_3 := K_2[|c| + (1 + b/2)^2]$. By completing the square we have

$$\left(\left\| \frac{\nabla u}{(|x|^2 + \varepsilon)^{1/2}} \right\|_{L^p} - \frac{p + K_2|b|}{2} \left\| \frac{u}{|x|^2 + \varepsilon} \right\|_{L^p} \right)^2$$

$$\leq K_2 \|T_{p,\varepsilon}u\|_{L^p} \left\| \frac{u}{|x|^2 + \varepsilon} \right\|_{L^p} + K_3 \left\| \frac{u}{|x|^2 + \varepsilon} \right\|_{L^p}^2$$

$$+ \frac{(p + K_2|b|)^2}{4} \left\| \frac{u}{|x|^2 + \varepsilon} \right\|_{L^p}^2 \tag{18.24}$$

which implies that

$$\left\| \frac{\nabla u}{(|x|^2 + \varepsilon)^{1/2}} \right\|_{L^p} \leq \sqrt{K_2} \|T_{p,\varepsilon}u\|_{L^p}^{1/2} \left\| \frac{u}{|x|^2 + \varepsilon} \right\|_{L^p}^{1/2}$$

$$+ \left(p + K_2|b| + \sqrt{K_3} \right) \left\| \frac{u}{|x|^2 + \varepsilon} \right\|_{L^p}.$$

Finally, applying (18.20) to the right-hand side, we obtain (18.21) with the constant $K_1 > 0$ given as in (18.22).

Next we consider the case with $p \in (1, 2]$. In the same way as above we see from (18.7) and (18.23) that

$$\left(\left\| \frac{\nabla u}{(|x|^2 + \varepsilon)^{1/2}} \right\|_{L^p} - \frac{p'|b|}{2} \left\| \frac{u}{|x|^2 + \varepsilon} \right\|_{L^p} \right)^2$$

$$\leq p' \|T_{p,\varepsilon}u\|_{L^p} \left\| \frac{u}{|x|^2 + \varepsilon} \right\|_{L^p} + K_4 \left\| \frac{u}{|x|^2 + \varepsilon} \right\|_{L^p}^2 + \frac{(p'|b|)^2}{4} \left\| \frac{u}{|x|^2 + \varepsilon} \right\|_{L^p}^2,$$

where $K_4 := p'[p + |c| + (1 + b/2)^2]$. This leads us to (18.21) with $K_1 > 0$ given as in (18.22).

This completes the proof of (18.21) for all $p \in (1, \infty)$.   □

Now we are in a position to give a proof of Proposition 18.1.

Proof (of Proposition 18.1) The proof is divided into two cases.

The first case (i): $c \geq \alpha_{N,b}(p)$ and $c > \beta_{N,b}(p)$. First we show that $T_{p,\varepsilon}$ is m-accretive in L^p and has a positive resolvent. Note that the first-order (drift) term is well-defined on $D(T_p)$, that is, $D(T_p) \subset D(|x|^{-2}(x \cdot \nabla))$ as a consequence of Lemma 18.1. Let $u \in D(T_p)$. Then it is easily seen that

$$T_{p,\varepsilon}u \to T_p(c)u \quad (\varepsilon \downarrow 0).$$

This implies by (18.19) that

$$\mathrm{Re} \int_{\mathbb{R}^N} (T_p(c)u)\bar{u}|u|^{p-2}\, dx \geq (c - \alpha_{N,b}(p)) \int_{\mathbb{R}^N} |x|^{-2}|u|^p\, dx \quad \forall\, u \in D(T_p).$$

Thus $T_p(c)$ is accretive in L^p if $c \geq \alpha_{N,b}(p)$. To prove the maximality of $T_p(c)$ let $f \in L^p$. Then we see by Lemma 18.3 that there exists a family $\{u_\varepsilon;\ \varepsilon > 0\} \subset W^{2,p}(\mathbb{R}^N)$ such that

$$u_\varepsilon + T_{p,\varepsilon}u_\varepsilon = f, \quad \|u_\varepsilon\|_{L^p} \leq \|f\|_{L^p}. \tag{18.25}$$

Moreover, $u_\varepsilon \geq 0$ (for every $\varepsilon > 0$) if $f \geq 0$. By virtue of (18.9) and Lemma 18.4 we have

$$\|u_\varepsilon\|_{W^{2,p}(\mathbb{R}^N)} + \left\| \frac{u_\varepsilon}{|x|^2 + \varepsilon} \right\|_{L^p}$$

$$\leq C_1(\|u_\varepsilon\|_{L^p} + \|\Delta u_\varepsilon\|_{L^p}) + \left\| \frac{\nabla u_\varepsilon}{(|x|^2 + \varepsilon)^{1/2}} \right\|_{L^p} + \left\| \frac{u_\varepsilon}{|x|^2 + \varepsilon} \right\|_{L^p}$$

$$\leq C_2 \|f\|_{L^p}. \tag{18.26}$$

Thus there exist a subsequence $\{u_{\varepsilon_n}\} \subset \{u_\varepsilon\}$ and $u \in W^{2,p}(\mathbb{R}^N)$ such that

$$u_{\varepsilon_n} \to u \text{ weakly } (n \to \infty) \text{ in } W^{2,p}(\mathbb{R}^N).$$

Furthermore, (18.26) implies that $u \in D(S_p) (\subset D(T_p))$ and the equation

$$u + T_p(c)u = u + (-\Delta)u + b\,|x|^{-2}(x \cdot \nabla)u + c\,S_p u = f$$

holds as the limit of (18.25), where $S_p := |x|^{-2}$. This proves the maximality of $T_p(c)$ with $c > \beta_{N,b}(p)$. Letting $\varepsilon \downarrow 0$ in (18.20), we have a necessary estimate in the second case (**ii**):

$$\|S_p u\|_{L^p} \leq (c - \beta_{N,b}(p))^{-1}\|T_p(c)u\|_{L^p}, \quad u \in D(T_p),\ c > \beta_{N,b}(p). \tag{18.27}$$

Additionally, if $f \geq 0$ in (18.25) then the locally weak compactness of L^p_+ in L^p implies that $u = \text{w-lim}\, u_{\varepsilon_n} \geq 0$. This finishes the proof of m-accretivity and resolvent-positivity of $T_p(c)$ when $c \geq \alpha_{N,b}(p)$ and $c > \beta_{N,b}(p)$.

Next we prove that $C_0^\infty(\mathbb{R}^N \setminus \{0\})$ is a core for $T_p(c)$. Define a new norm on $D(T_p)$ as

$$\|w\| := \|w\|_{L^p} + \|\Delta w\|_{L^p} + \left\| \frac{w}{|x|^2} \right\|_{L^p}, \quad w \in D(T_p) = W^{2,p}(\mathbb{R}^N) \cap D(|x|^{-2});$$

this is equivalent to the graph norm of $T_p(c)$. Then it is proved in [8, Lemma 4.8] that $C_0^\infty(\mathbb{R}^N \setminus \{0\})$ is dense in $W^{2,p}(\mathbb{R}^N) \cap D(|x|^{-2})$ with respect to the norm $\|\|\cdot\|\|$. Hence it suffices to show that for every $u \in C_0^\infty(\mathbb{R}^N \setminus \{0\})$,

$$m_1(\|u\|_{L^p} + \|T_{p,\min}(c)u\|_{L^p}) \le \|\|u\|\| \le m_2(\|u\|_{L^p} + \|T_{p,\min}(c)u\|_{L^p}), \quad (18.28)$$

where $T_{p,\min}(c)$ is the restriction of $T_p(c)$ to $C_0^\infty(\mathbb{R}^N \setminus \{0\})$:

$$T_{p,\min}(c) := T_p(c)|C_0^\infty(\mathbb{R}^N \setminus \{0\}), \quad c \ge c_0.$$

Let $u \in C_0^\infty(\mathbb{R}^N \setminus \{0\})$. Then proceeding in the same way as in the proof of (18.26), we can prove the second inequality in (18.28) with $m_2 = 2C_2$. Conversely, letting $\varepsilon \downarrow 0$ in (18.13), we have

$$\left\||x|^{-1}\nabla u\right\|_{L^p} \le k_1 \left\||x|^{-2}u\right\|_{L^p} + 2k_2 \|\Delta u\|_{L^p}^{1/2} \left\||x|^{-2}u\right\|_{L^p}^{1/2}$$

$$\le (k_1 + k_2)\left(\|\Delta u\|_{L^p} + \left\||x|^{-2}u\right\|_{L^p}\right). \quad (18.29)$$

Thus we obtain

$$\|u\|_{L^p} + \|T_{p,\min}(c)u\|_{L^p}$$

$$\le \|u\|_{L^p} + \|\Delta u\|_{L^p} + |b|\left\||x|^{-1}\nabla u\right\|_{L^p} + |c|\left\||x|^{-2}u\right\|_{L^p}$$

$$\le \|u\|_{L^p} + \left(1 + |b|(k_1 + k_2) + |c|\right)\left(\|\Delta u\|_{L^p} + \left\||x|^{-2}u\right\|_{L^p}\right)$$

$$\le \left(2 + |b|(k_1 + k_2) + |c|\right)\|\|u\|\|.$$

This is nothing but the first inequality in (18.28). Consequently, we can conclude that $T_p(c)$ is the closure of $T_{p,\min}(c)$ when $c \ge \alpha_{N,b}(p)$ and $c > \beta_{N,b}(p)$.

The second case (ii): $c = c_0 = \beta_{N,b}(p) \ge \alpha_{N,b}(p)$. We apply [8, Theorem 1.2] to $T_p(c_0)$ and $S_p = |x|^{-2}$. Namely, we consider the sequence of perturbed operators

$$\begin{cases} (T_p(c_0) + n^{-1}S_p)u(x) = T_p(c_0 + n^{-1})u(x) \\ \qquad = -\Delta u(x) + \dfrac{b}{|x|^2}(x \cdot \nabla)u(x) + \dfrac{c_0 + n^{-1}}{|x|^2}u(x), \\ D(T_p(c_0) + n^{-1}S_p) = D(T_p) = W^{2,p}(\mathbb{R}^N) \cap D(|x|^{-2}). \end{cases}$$

Then the assertion in the first case (i) implies that $T_p(c_0) + n^{-1}S_p$ $(n \in \mathbf{N})$ is m-accretive in L^p and has a positive resolvent (because $c_0 + n^{-1} > \beta_{N,b}(p)$). Moreover, it follows from (18.27) with $c = c_0 + n^{-1}$ that

$$\|T_p(c_0)u\|_{L^p}$$

$$\le \|(T_p(c_0) + n^{-1}S_p)u\|_{L^p} + n^{-1}\|S_p u\|_{L^p}$$

$$\le \|(T_p(c_0) + n^{-1}S_p)u\|_{L^p} + \frac{n^{-1}}{(c_0 + n^{-1}) - \beta_{N,b}(p)}\|(T_p(c_0) + n^{-1}S_p)u\|_{L^p}$$

$$\le 2\|(T_p(c_0) + n^{-1}S_p)u\|_{L^p}, \quad u \in D(T_p).$$

Thus applying [8, Theorem 1.2] to the pair of $T_p(c_0)$ and S_p, we see that $T_p(c_0)$ is essentially m-accretive on $D(T_p)$, and the resolvent of $\tilde{T}_p(c_0)$ has the following expression:

$$(\xi + \tilde{T}_p(c_0))^{-1}f = \lim_{n\to\infty} (\xi + T_p(c_0) + n^{-1}S_p)^{-1}f, \quad \xi > 0, \quad f \in L^p. \tag{18.30}$$

Since $(\xi + T_p(c_0) + n^{-1}S_p)^{-1}$ is positive, we see from (18.30) that $(\xi + \tilde{T}_p(c_0))^{-1}$ is also positive on L^p.

Finally, since $D(T_p)$ is a core for $\tilde{T}_p(c_0)$, it follows from (18.29) and [8, Lemma 4.8] that $C_0^\infty(\mathbb{R}^N \setminus \{0\})$ is also a core for $\tilde{T}_p(c_0)$. This completes the proof of Proposition 18.1.   □

Remark 18.3 It follows from [8, Theorem 1.2] that for every $f \in L^p$,

$$n^{-1}S_p(\xi + T_p(c_0) + n^{-1}S_p)^{-1}f \to 0 \quad \text{weakly} \ (n \to \infty). \tag{18.31}$$

Now let $g \in L^p_+$. Then $u := (1 + \tilde{T}_p(c_0))^{-1}g \in L^p_+$ as shown above. Let $\{g_n\}$ be a sequence in $C_0^\infty(\mathbb{R}^N \setminus \{0\})_+$ such that $g_n \to g \ (n \to \infty)$. Then it follows that $\varphi_n := (1 + T_p(c_0) + n^{-1}S_p)^{-1}g_n \in C_0^\infty(\mathbb{R}^N \setminus \{0\})_+$, with

$$g = (1 + \tilde{T}_p(c_0))u = \text{w-}\lim_{n\to\infty} (1 + T_{p,\min}(c_0))\varphi_n. \tag{18.32}$$

In fact, since $(1 + T_{p,\min}(c_0))\varphi_n = g_n - n^{-1}S_p(1 + T_p(c_0) + n^{-1}S_p)^{-1}g_n$, (18.32) is a consequence of (18.31). (18.32) seems to be the essence of Proposition 18.1 because (18.32) is exactly what we need in the proof of the main theorem in the next section.

18.4 Proof of Theorem 18.1

Proof (of Theorem 18.1) Let $c_0 = \max\{\alpha_{N,b}(p), \beta_{N,b}(p)\}$. Then it follows from condition (18.2) that the accretivity of $A_{p,\min}$ is reduced to that of $T_{p,\min}(c_0) = T_p(c_0)|C_0^\infty(\mathbb{R}^N \setminus \{0\})$:

$$\text{Re}\int_{\mathbb{R}^N} (A_{p,\min}u)\bar{u}|u|^{p-2}\,dx \ge \text{Re}\int_{\mathbb{R}^N} (T_{p,\min}(c_0)u)\bar{u}|u|^{p-2}\,dx \ge 0.$$

We shall show the maximality of the closure of $A_{p,\min}$. Suppose that $v \in L^{p'}$ annihilates the range $R(1 + A_{p,\min})$ of $1 + A_{p,\min}$. Setting

$$E_{p,\min}\varphi(x) := -\Delta\varphi(x) + b|x|^{-2}(x \cdot \nabla)\varphi(x), \quad \varphi \in C_0^\infty(\mathbb{R}^N \setminus \{0\}),$$

we have that for $\varphi \in C_0^\infty(\mathbb{R}^N \setminus \{0\})$,

$$0 = \int_{\mathbb{R}^N} \overline{v(x)}(1 + A_{p,\min})\varphi(x)\, dx$$

$$= \int_{\mathbb{R}^N} \overline{v(x)}\big((1 + V(x))\varphi(x) + E_{p,\min}\varphi(x)\big)\, dx. \tag{18.33}$$

We have to show that $v = 0$. The proof depends on Kato's inequality (see, e.g., [1] or [11]). In fact, (18.33) and (18.2) imply by Kato's inequality that

$$\int_{\mathbb{R}^N} |v(x)|(1 + T_{p,\min}(c_0))\varphi(x)\, dx$$

$$= \int_{\mathbb{R}^N} |v(x)|\big((1 + \frac{c_0}{|x|^2}) + E_{p,\min}\big)\varphi(x)\, dx \le 0 \tag{18.34}$$

for all $\varphi \in C_0^\infty(\mathbb{R}^N \setminus \{0\})_+$. By virtue of (18.32) we have

$$\int_{\mathbb{R}^N} |v(x)||g(x)\, dx \le 0 \quad \forall\, g \in L_+^p.$$

This completes the proof because $\|v\|_{L^{p'}} = 0$ when we choose $g := |v|^{p'-1} \in L_+^p$.

Therefore it remains to prove (18.34). Noting that $|x|^{-b/2}\varphi(x) \in C_0^\infty(\mathbb{R}^N \setminus \{0\})$, we have

$$|x|^{b/2}\Delta(|x|^{-b/2}\varphi(x)) = |x|^{b/2}\mathrm{div}\,[\nabla(|x|^{-b/2}\varphi(x))]$$

$$= -E_{p,\min}\varphi(x) - b_N|x|^{-2}\varphi(x),$$

where $b_N := (b/2)(N - 2 - b/2)$. Thus we obtain

$$\overline{v(x)}E_{p,\min}\varphi(x) = -|x|^{b/2}\overline{v(x)}\Delta(|x|^{-b/2}\varphi(x)) - b_N|x|^{-2}\overline{v(x)}\varphi(x)$$

$$= -\overline{v_b(x)}\Delta\varphi_b(x) - b_N|x|^{-2}\overline{v(x)}\varphi(x) \tag{18.35}$$

when we define

$$v_b(x) := |x|^{b/2}v(x) \in L_{loc}^{p'}(\mathbb{R}^N \setminus \{0\}) \subset L_{loc}^1(\mathbb{R}^N \setminus \{0\}),$$

$$\varphi_b(x) := |x|^{-b/2}\varphi(x) \in C_0^\infty(\mathbb{R}^N \setminus \{0\}).$$

It follows from (18.33) and (18.35) that

$$\int_{\mathbb{R}^N} \overline{v(x)}(1 + V(x) - b_N|x|^{-2})\varphi(x)\, dx = \int_{\mathbb{R}^N} \overline{v_b(x)}\Delta\varphi_b(x)\, dx. \qquad (18.36)$$

Noting that $\overline{v(x)}\varphi(x) = \overline{v_b(x)}\varphi_b(x)$, we see that (18.36) can be written as

$$\int_{\mathbb{R}^N} \overline{v_b(x)}\Delta\varphi_b(x)\, dx = \int_{\mathbb{R}^N} \overline{v_b(x)}(1 + V(x) - b_N|x|^{-2})\varphi_b(x)\, dx.$$

Since $V \in L^p_{\mathrm{loc}}(\mathbb{R}^N \setminus \{0\})$, this implies that

$$\Delta v_b = (1 + V(x) - b_N|x|^{-2})v_b \in L^1_{\mathrm{loc}}(\mathbb{R}^N \setminus \{0\}). \qquad (18.37)$$

Applying Kato's inequality to v_b, we have that for every $\varphi_b = |x|^{-b/2}\varphi$ with $\varphi \in C_0^\infty(\mathbb{R}^N \setminus \{0\})_+$,

$$\int_{\mathbb{R}^N} |v_b(x)|\Delta\varphi_b(x)\, dx \geq \mathrm{Re}\int_{\mathbb{R}^N} [\mathrm{sign}\,\overline{v_b(x)}]\Delta v_b(x) \cdot \varphi_b(x)\, dx. \qquad (18.38)$$

Here (18.37) yields by (18.2) that

$$[\mathrm{sign}\,\overline{v_b(x)}]\Delta v_b(x) = |v_b(x)|(1 + V(x) - b_N|x|^{-2})$$
$$\geq |v_b(x)|((1 + c_0|x|^{-2}) - b_N|x|^{-2}). \qquad (18.39)$$

Since $|v_b(x)|\varphi_b(x) = |v(x)|\varphi(x)$, it follows from (18.38) and (18.39) that

$$\int_{\mathbb{R}^N} |v_b(x)|\left(\Delta\varphi_b(x) + \frac{b_N}{|x|^2}\right)\varphi_b(x)\, dx \geq \int_{\mathbb{R}^N} |v_b(x)|\left(1 + \frac{c_0}{|x|^2}\right)\varphi_b(x)\, dx$$
$$= \int_{\mathbb{R}^N} |v(x)|\left(1 + \frac{c_0}{|x|^2}\right)\varphi(x)\, dx.$$

Finally, we see from (18.35) with $\overline{v(x)}$ replaced with $|v(x)|$ that

$$\int_{\mathbb{R}^N} |v(x)|\left(1 + \frac{c_0}{|x|^2}\right)\varphi(x)\, dx \leq -\int_{\mathbb{R}^N} |v(x)|E_{p,\min}\varphi(x)\, dx \quad \forall\, \varphi \in C_0^\infty(\mathbb{R}^N \setminus \{0\})_+.$$

This is nothing but (18.34). Thus the proof of Theorem 18.1 has been finished. □

Acknowledgements N.O. was partially supported by Grant-in-Aid for Scientific Research (C), No.25400182, Japan Society for the Promotion of Science.

References

1. Edmunds, D.E., Evans, W.D.: Spectral Theory and Differential Operators. The Clarendon Press, Oxford (1987)
2. Iwaniec, T., Martin, G.: Riesz transforms and related singular integrals. J. Reine Angew. Math. **473**, 25–57 (1996)
3. Iwaniec, T., Martin, G.: Geometric Function Theory and Non-linear Analysis. The Clarendon Press/Oxford University Press, New York (2001)
4. Kato, T.: Perturbation Theory for Linear Operators. Grundlehren Math. Wissenschaften, vol. 132. Springer, Berlin (1966); Classics in Mathematics (1995)
5. Maeda, Y., Okazawa, N.: Holomorphic families of Schrödinger operators in L^p. SUT J. Math. **47**, 185–216 (2011)
6. Miyajima, S., Okazawa, N.: Generators of positive C_0-semigroups. Pac. J. Math. **125**, 161–176 (1986)
7. Okazawa, N.: On the perturbation of linear operators in Banach and Hilbert spaces. J. Math. Soc. Jpn. **34**, 677–701 (1982)
8. Okazawa, N.: L^p-theory of Schrödinger operators with strongly singular potentials. Jpn. J. Math. **22**, 199–239 (1996)
9. Okazawa, N., Sobajima, M., Yokota, T.: Existence of solutions to heat equations with singular lower order terms. J. Differ. Equ. **256**, 3568–3593 (2014)
10. Okazawa, N., Tamura, H., Yokota, T.: Square Laplacian perturbed by inverse fourth-power potential, I. self-adjointness (real case). Proc. Roy. Soc. Edinburgh **141A**, 409–416 (2011)
11. Reed, M., Simon, B.: Methods of Modern Mathematical Physics. Fourier Analysis, Self-adjointness, vol. II. Academic Press, New York/London (1975)
12. Sobajima, M., Okazawa, N., Yokota, T.: Domain characterization for a class of second-order elliptic operators with unbounded coefficients in L^p, Nonlinear phenomena with energy dissipation. GAKUTO Int. Ser. Math. Sci. Appl. **36** pp. 203–214. Gakkōtosho, Tokyo (2013)
13. Wong-Dzung, B.: L^p-theory of degenerate-elliptic and parabolic operators of second order. Proc. Roy. Soc. Edinburgh **95A**, 95–113 (1983)

Chapter 19
Semilinear Delay Evolution Equations with Nonlocal Initial Conditions

Ioan I. Vrabie

In memory of Professor Alfredo Lorenzi

Abstract An existence and asymptotic behaviour result for a class of semilinear delay evolution equations subjected to nonlocal initial conditions is established. An application to a semilinear wave equation is also discussed.

19.1 Introduction

In this paper we prove an existence and uniform asymptotic stability result for mild solutions to a semilinear delay differential evolution equation with nonlocal initial data in a Banach space X, i.e.

$$\begin{cases} u'(t) = Au(t) + f(t, u_t), & t \in [0, +\infty), \\ u(t) = g(u)(t), & t \in [-\tau, 0]. \end{cases} \tag{19.1}$$

Here $A : D(A) \subseteq X \to X$ is the infinitesimal generator of a C_0-semigroup of contractions, $\{S(t) : X \to X; t \geq 0\}$, $\tau \geq 0$, $f : [0, +\infty) \times C([-\tau, 0]; X) \to X$ is a compact function which is jointly continuous and Lipschitz with respect to its second argument and $g : C_b([-\tau, +\infty); X) \to C([-\tau, 0]; X)$ is continuous and has affine growth. In the limiting case $\tau = 0$, i.e. when the delay is absent, $C([-\tau, 0]; X) = X$ and so, in this case, $f : [0, +\infty) \times X \to X$ and $g : C_b([0, +\infty); X) \to X$. If I is an unbounded interval, $C_b(I; X)$ denotes the space of all bounded and continuous functions from I to X, equipped with the sup-norm $\| \cdot \|_{C_b(I;X)}$, while $\tilde{C}_b(I; X)$ stands for the space of all bounded

---

I.I. Vrabie (✉)
Faculty of Mathematics, Al. Cuza University, Iași, Romania and Octav Mayer Institute of Mathematics (Romanian Academy), Iași, Romania
e-mail: ivrabie@uaic.ro

© Springer International Publishing Switzerland 2014
A. Favini et al. (eds.), *New Prospects in Direct, Inverse and Control Problems for Evolution Equations*, Springer INdAM Series 10,
DOI 10.1007/978-3-319-11406-4_19

and continuous functions from I to X, endowed with the uniform convergence on compacta topology. Further, $C([a,b]; X)$ denotes the space of all continuous functions from $[a,b]$ to X endowed with the sup-norm $\| \cdot \|_{C([a,b];X)}$. As usual, if $u \in C([-\tau,+\infty); X)$ and $t \in [0,+\infty)$, $u_t \in C([-\tau,0]; X)$ is defined by $u_t(s) := u(t+s)$ for each $s \in [-\tau,0]$.

It should be noticed that, as far as the nondelayed case, i.e. $\tau = 0$, is concerned, in many situations, the nonlocal problem (19.1) has proved more reliable than its classical initial-value counterpart. This is the case, for instance, of long-term weather forecasting in meteorology. See Rabier et al. [23]. In addition, (19.1), with $\tau = 0$, is nothing but the abstract form of various mathematical models for: wave propagation—see Avalishvili and Avalishvili [3], diffusion processes—see Deng [10], Gordeziani [13] and Olmstead and Roberts [21], fluid dynamics—see Gordeziani et al. [14] and Shelukhin [25, 26], or pharmacokinetics—see McKibben [18, Model II.6, p. 395].

Abstract nondelayed evolution equations subjected to nonlocal initial conditions were investigated by Aizicovici and Lee [1], Aizicovici and McKibben [2], Bryszewski [8], Garcia-Falset and Reich [12] and Paicu and Vrabie [22], to cite only a few.

Since the presence of a delay in the source term of an evolution equation is more realistic than an instantaneous feedback, there is an increasing literature on such kind of problems, i.e. functional evolution equations with delay. This explains why, in recent years, abstract delayed evolution equations or inclusions subjected to nonlocal initial conditions were considered by many authors from which we mention Burlică and Roşu [5], Burlică et al. [6, 7] and Vrabie [28–33] and the references therein. For previous results on initial-value problems for delay evolution equations, see Mitidieri and Vrabie [19, 20]. Some existence, uniqueness and continuity with respect to the data theorems concerning source identification for semilinear delay evolution equations were recently obtained, among others, by Di Blasio and Lorenzi [11] and Lorenzi and Vrabie [16, 17].

19.2 Preliminaries

Definition 19.1 Let X, Y be Banach spaces and Z a subset of Y. We say that a mapping $Q : Z \to X$ is *compact* if it carries bounded subsets in Z into relatively compact in X.

Definition 19.2 We say that the Banach space X is C_0-*compact* if there exists a family of linear, compact operators $\{I_\varepsilon; \ \varepsilon \in (0,1)\} \subseteq \mathcal{L}(X)$ with $\|I_\varepsilon\|_{\mathcal{L}(X)} \leq 1$ for each $\varepsilon \in (0,1)$ and $\lim_{\varepsilon \downarrow 0} I_\varepsilon x = x$ for each $x \in X$.

Remark 19.1 Let $p \in [1,+\infty)$ and let Ω be a nonempty and bounded domain in $\mathbb{R}^d$, $d \geq 1$. Then $L^p(\Omega)$ is C_0-compact. This follows from the fact that, for each $p \in [1,+\infty)$, the Laplace operator subjected to Dirichlet boundary conditions generates a compact C_0-semigroup on $L^p(\Omega)$. See [27, Theorem 4.1.3, p. 81]. Furthermore, if X is a separable Hilbert space then it is C_0-compact. Indeed, let $\{e_k; \ k \in \mathbb{N}\}$ be an orthonormal system, let $\varepsilon \in (0, 1)$ and let us define $I_\varepsilon : X \to X$ by:

$$I_\varepsilon(x) := \sum_{k=0}^{\infty} e^{-\varepsilon k} \langle x, e_k \rangle e_k$$

for each $x \in X$. Since for each $m \in \mathbb{N}$, the operator $I_\varepsilon^m : X \to X$, defined by

$$I_\varepsilon^m(x) := \sum_{k=0}^{m} e^{-\varepsilon k} \langle x, e_k \rangle e_k$$

for each $x \in X$, has finite dimensional range and $\lim_{m \to \infty} I_\varepsilon^m = I_\varepsilon$ uniformly on bounded subsets, it readily follows that I_ε is a compact operator for each $\varepsilon \in (0, 1)$. Finally, observing that $\lim_{\varepsilon \downarrow 0} I_\varepsilon(x) = x$ for each $x \in X$, we conclude that X is C_0-compact.

We recall for easy reference the following result due to Schaefer [24].

Theorem 19.1 *Let Y be a real Banach space and let $Q : Y \to Y$ be a continuous, compact operator and let*

$$\mathcal{E}(Q) = \{x \in Y; \exists \lambda \in [0, 1], \ such \ that \ x = \lambda Q(x)\}.$$

If $\mathcal{E}(Q)$ is bounded, then Q has at least one fixed-point.

Let us denote by $\mathcal{M}(\xi, h)$ the unique mild solution u of the Cauchy problem

$$\begin{cases} u'(t) = Au(t) + h(t), & t \in [0, T] \\ u(a) = \xi, \end{cases}$$

corresponding to $\xi \in X$ and $h \in L^1(0, T; X)$, i.e.

$$u(t) = S(t)\xi + \int_0^t S(t - s)h(s)\, ds$$

for $t \in [0, T]$.

The next slight extension of a compactness result due to Becker [4] is a direct consequence of Vrabie [27, Theorem 2.8.4, p. 194] and Cârjă et al. [9, Lemma 1.5.1, p. 14].

Theorem 19.2 *Let $A : D(A) \subseteq X \to X$ be the infinitesimal generator of a C_0-semigroup $\{S(t) : X \to X; \ t \geq 0\}$, let $\mathcal{D}$ be a bounded subset in X, and $\mathcal{F}$ a subset in $L^1(0, T; X)$ for which there exists a compact set $K \subseteq X$ such that $f(t) \in K$ for each $f \in \mathcal{F}$ and a.e. for $t \in [0, T]$. Then, for each $\theta \in (0, T)$, $\mathcal{M}(\mathcal{D}, \mathcal{F})$ is* *relatively compact in* $C([\theta, T]; X)$. *If, in addition, $\mathcal{D}$ is relatively compact, then* $\mathcal{M}(\mathcal{D}, \mathcal{F})$ *is relatively compact even in* $C([0, T]; X)$.

19.3   The General Framework

We assume familiarity with the basic concepts and results of the linear semigroup theory, as well as with the theory of delay evolution equations. However, we recall for easy reference some concepts we will frequently use in the sequel and we refer to Vrabie [27] for details concerning linear semigroups and to Hale [15] for details on delay evolution equations.

We recall that a function $u \in C_b([-\tau, +\infty); X)$ is called a *mild solution* of the problem (19.1) if it is given by the *variation of constants formula*, i.e.

$$u(t) = \begin{cases} S(t)g(u)(0) + \displaystyle\int_0^t S(t-s)f(s, u_s)\, ds, \ t \in [0, +\infty) \\ u(t) = g(u)(t), \hspace{4.3cm} t \in [-\tau, 0]. \end{cases} \tag{19.2}$$

Definition 19.3 We say that the function $g : C_b([-\tau, +\infty); X) \to C([-\tau, 0]; X)$ *has affine growth* if there exists $m_0 \geq 0$ such that, for each $u \in C_b([-\tau, +\infty); X)$, we have

$$\|g(u)\|_{C([-\tau,0];X)} \leq \|u\|_{C_b([-\tau,+\infty);X)} + m_0.$$

We begin with the general assumptions we need in the sequel.

(H_X) the Banach space X is C_0-compact (cf. Definition 19.2);

(H_A) the operator $A : D(A) \subseteq X \to X$ generates a C_0-semigroup, $\{S(t) : X \to X; \ t \geq 0\}$, and there exists $\omega > 0$ such

$$\|S(t)\| \leq e^{-\omega t},$$

for each $t \in [0, +\infty)$;

(H_f) the function $f : [0, +\infty) \times C([-\tau, 0]; X) \to X$ is jointly continuous on its domain, compact (cf. Definition 19.1) and:

(f_1) there exists $\ell > 0$ such that

$$\|f(t, v) - f(t, \tilde{v})\| \leq \ell \|v - \tilde{v}\|_{C([-\tau,0];X)}$$

for each $t \in [0, +\infty)$ and $v, \tilde{v} \in C([-\tau, 0]; X)$;

(f_2)   there exists $m > 0$ such that

$$\| f(t, 0) \| \leq m$$

for each $t \in [0, +\infty)$;

(H_c)   the constants ℓ and ω satisfy the nonresonance condition :

$$\ell < \omega;$$

(H_g)   the function $g : C_b([-\tau, +\infty); X) \to C([-\tau, 0]; X)$ satisfies:

(g_1) g has affine growth (cf. Definition 19.3);
(g_2)   there exists $a > 0$ such that, for each $u, v \in C_b([-\tau, +\infty); X)$, from $u(t) = v(t)$ for each $t \in [a, +\infty)$, it follows that $g(u) = g(v)$;
(g_3) g is continuous from $\tilde{C}_b([-\tau, +\infty); X)$ to $C([-\tau, 0]; X)$.

Remark 19.2 From (g_1) and (g_2), we get that for each $u \in C_b([-\tau, +\infty); X)$, we have

$$\| g(u) \|_{C([-\tau,0];X)} \leq \| u \|_{C_b([a,+\infty);X)} + m_0, \qquad (19.3)$$

where a is given by (g_2). Indeed, if we assume by contradiction that there exists $u \in C_b([-\tau, +\infty); X)$ such that $\| u \|_{C_b([a,+\infty);X)} + m_0 < \| g(u) \|_{C([-\tau,0];X)}$, then the function $\tilde{u} : [-\tau, +\infty) \to X$, defined by

$$\tilde{u}(t) = \begin{cases} u(t), \ t \in [a, +\infty), \\ u(a), \ t \in [-\tau, a), \end{cases}$$

satisfies $u(t) = \tilde{u}(t)$ for each $t \in [a, +\infty)$ and thus $g(u) = g(\tilde{u})$. So,

$$\| u \|_{C_b([a,+\infty);X)} + m_0 < \| g(u) \|_{C([-\tau,0];X)} = \| g(\tilde{u}) \|_{C([-\tau,0];X)}$$

$$\leq \| \tilde{u} \|_{C_b([-\tau,+\infty);X)} + m_0 = \| \tilde{u} \|_{C_b([a,+\infty);X)} + m_0 = \| u \|_{C_b([a,+\infty);X)} + m_0.$$

This contradiction can be eliminated only if (19.3) holds true, as claimed.

Remark 19.3 The class of functions g satisfying (H_g) is very large and includes several important specific cases. More precisely:

• let $\mathcal{N} : X \to X$ be a possibly nonlinear operator having linear growth, i.e.

$$\| \mathcal{N}(x) \| \leq \| x \|$$

for each $x \in X$, let μ is a σ-finite and complete measure on $[0, +\infty)$, satisfying

$$\operatorname{supp} \mu = [b, +\infty),$$

where $b > \tau$ and $\mu([0, +\infty)) = 1$, and let $\psi \in C([-\tau, 0]; X)$. Then, the function g, defined by

$$g(u)(t) = \int_0^{+\infty} \mathcal{N}(u(t + \theta)) \, d\mu(\theta) + \psi(t), \qquad (19.4)$$

for each $u \in C_b([-\tau, +\infty); X)$ and $t \in [-\tau, 0]$, satisfies hypothesis (H_g) with $m_0 = \|\psi\|_{C([-\tau, 0]; X)}$ and $a = b - \tau$;

- let $T > \tau$ and let us consider the T-periodic condition, i.e.

$$g(u)(t) = u(t + T),$$

for $u \in C_b([-\tau, +\infty); X)$ and $t \in [-\tau, 0]$. Clearly g satisfies (H_g), with $m_0 = 0$ and $a = T - \tau$;

- let $T > \tau$ and let us consider the T-anti-periodic condition, i.e.

$$g(u)(t) = -u(t + T),$$

for $u \in C_b([-\tau, +\infty); X)$ and $t \in [-\tau, 0]$. Also in this case g satisfies (H_g), with $m_0 = 0$ and $a = T - \tau$;

- let us consider the multi-point discrete mean condition, i.e.

$$g(u)(t) = \sum_{i=1}^{n} \alpha_i u(t + t_i)$$

for $u \in C_b([-\tau, +\infty); X)$ and $t \in [-\tau, 0]$, where $\alpha_i \in (0, 1)$, for $i = 1, 2, \ldots, n$, $\sum_{i=1}^{n} \alpha_i \leq 1$ and $0 < t_1 < t_2 < \cdots < t_n$ are arbitrary but fixed. In this case, g satisfies (H_g) with $m_0 = 0$ and $a = t_1$.

19.4 The Main Result and Some Auxiliary Lemmas

The main result of this paper is:

Theorem 19.3 *If (H_X), (H_A), (H_f), (H_g) and (H_c) hold true, then the problem (19.1) has at least one mild solution, $u \in C_b([-\tau, +\infty); X)$. Moreover, for each mild solution of (19.1), we have*

$$\|u\|_{C_b([-\tau, +\infty); X)} \leq \frac{m}{\omega - \ell} + \left[\frac{\omega}{\omega - \ell} \cdot \left(\frac{1}{e^{\omega a} - 1} + \frac{\ell}{\omega} \right) + 1 \right] \cdot m_0. \qquad (19.5)$$

If, in addition, instead of (H_c), *the stronger nonresonance condition* $\ell e^{\omega\tau} < \omega$ *is satisfied, then each mild solution of* (19.1) *is globally asymptotically stable.*

The proof of Lemma 19.1 below can be found in Vrabie [33, Lemma 6.2].

Lemma 19.1 *If* (H_c) *is satisfied and* $u \in C_b([-\tau, +\infty); X)$ *is such that*

$$\|u(t)\| \le e^{-\omega t}\|u(0)\| + (1 - e^{-\omega t})\frac{\ell}{\omega}\left[\|u\|_{C_b([-\tau,+\infty);X)} + \frac{m}{\ell}\right]$$

for each $t \in [0, +\infty)$ *and*

$$\|u\|_{C([-\tau,0];X)} \le \|u\|_{C_b([a,+\infty);X)} + m_0,$$

then u *satisfies* (19.5).

The lemma below is a specific form of general result in Vrabie [32, Lemma 4.3], where X is a general Banach space, A is a nonlinear m-dissipative operator and f is globally Lipschitz.

Lemma 19.2 *Let us assume that* (H_A), (H_f) *and* (H_c) *are satisfied. Then, for each* $\varphi \in C([-\tau, 0]; X)$, *the problem*

$$\begin{cases} u'(t) = Au(t) + f(t, u_t), & t \in [0, +\infty), \\ u(t) = \varphi(t), & t \in [-\tau, 0], \end{cases} \tag{19.6}$$

has a unique mild solution $u \in C_b([-\tau, +\infty); X)$.

We conclude this section with a Bellman Lemma for integral inequalities with delay proved in Burlică and Roşu [5].

Lemma 19.3 *Let* $y : [-\tau, +\infty) \to \mathbb{R}_+$ *and* $\alpha_0, \beta : [0, +\infty) \to \mathbb{R}_+$ *be continuous functions with* α_0 *nondecreasing. If*

$$y(t) \le \alpha_0(t) + \int_0^t \beta(s)\|y_s\|_{C([-\tau,0];\mathbb{R})}\,ds \tag{19.7}$$

for each $t \in [0, +\infty)$, *then*

$$y(t) \le \alpha(t) + \int_0^t \alpha(s)\beta(s)e^{\int_s^t \beta(\sigma)\,d\sigma}\,ds, \tag{19.8}$$

for each $t \in [0, +\infty)$, *where* $\alpha(t) := \|y_0\|_{C([-\tau,0];\mathbb{R})} + \alpha_0(t)$ *for each* $t \in [0, +\infty)$.

19.5   Proof of the Main Result

19.5.1   The Approximate Problem

We shall use an interplay of two fixed point arguments and an approximation procedure. Let $\varepsilon > 0$, let $g_\varepsilon = I_\varepsilon g$, where I_ε is given by (H_X), and let us consider the ε-approximate problem

$$\begin{cases} u'(t) = Au(t) + f(t, u_t), & t \in [0, +\infty), \\ u(t) = g_\varepsilon(u)(t), & t \in [-\tau, 0]. \end{cases} \tag{19.9}$$

Remark 19.4 In view of (g_2), for each $v \in C_b([-\tau, +\infty); X)$, $g(v)$ depends only on the values of v on $[a, +\infty)$. In fact we have

$$g(v) = g(\tilde{v}_{|[0,+\infty)}) = g(\tilde{w}_{|[a,+\infty)}), \tag{19.10}$$

where $a > 0$ is given by (g_2), $\tilde{v}$ is any function in $C_b([-\tau, +\infty); X)$ which coincides with v on $[0, +\infty)$ and $\tilde{w}$ is any function in $C_b([-\tau, +\infty); X)$ which coincides with v on $[a, +\infty)$. This explains why, in that follows, we will assume with no loss of generality that g is defined merely on $C_b([0, +\infty); X)$, or even on $C_b([a, +\infty); X)$. From (g_3) and (19.10), we deduce that g is continuous from both $\tilde{C}_b([0, +\infty); X)$ and $\tilde{C}_b([a, +\infty); X)$ to $C([-\tau, 0]; X)$.

Lemma 19.4 *Let us assume that* (H_X), (H_A), (H_f) *and* (H_g) *are satisfied, let* $\varepsilon > 0$ *be arbitrary and let* g_ε *be defined as above. Then the approximate problem* (19.9) *has at least one mild solution* $u_\varepsilon \in C_b([-\tau, +\infty); X)$.

Proof Let us first consider the problem

$$\begin{cases} u'(t) = Au(t) + f(t, u_t), & t \in [0, +\infty), \\ u(t) = g_\varepsilon(v)(t), & t \in [-\tau, 0]. \end{cases} \tag{19.11}$$

By Lemma 19.2 and Remark 19.4, it follows that, for each $v \in C_b([0, +\infty); X)$, (19.11) has a unique mild solution $u \in C_b([-\tau, +\infty); X)$. Thus, we can define the operator

$$\mathcal{S}_\varepsilon : C_b([0, +\infty); X) \to C_b([0, +\infty); X)$$

by

$$\mathcal{S}_\varepsilon(v) := u_{|[0,+\infty)},$$

where u is the unique mild solution of (19.11) corresponding to v. We will complete the proof, by showing that the operator $\mathcal{S}_\varepsilon$ defined as above satisfies the hypotheses of Schaefer Fixed Point Theorem 19.1 and thus (19.9) has at least one mild solution. So, we have to check out that $\mathcal{S}_\varepsilon$ is continuous with respect to norm topology of $C_b([0, +\infty); X)$, is compact and

$$\mathcal{E}(\mathcal{S}_\varepsilon) = \{u \in C_b([0, +\infty); X); \exists \lambda \in [0, 1], \text{ such that } u = \lambda\mathcal{S}_\varepsilon(u)\}$$

is bounded.

To prove the continuity of $\mathcal{S}_\varepsilon$ let $v, \tilde{v} \in C_b([0, +\infty); X)$, set $u(t) = \mathcal{S}_\varepsilon(v)(t)$ and $\tilde{u}(t) = \mathcal{S}_\varepsilon(\tilde{v})(t)$ for $t \in [0, +\infty)$ and let us observe that

$$\|u(t) - \tilde{u}(t)\| \leq e^{-\omega t}\|g_\varepsilon(v)(0) - g_\varepsilon(\tilde{v})(0)\| + \int_0^t e^{-\omega(t-s)}\|f(s, u_s) - f(s, \tilde{u}_s)\|\, ds$$

$$\leq e^{-\omega t}\|g(v)(0) - g(\tilde{v})(0)\| + \ell \int_0^t e^{-\omega(t-s)}\|u_s - \tilde{u}_s\|_{C([-\tau,0];X)}\, ds$$

for each $t \in [0, +\infty)$. So, we have

$$\|u(t) - \tilde{u}(t)\| \leq e^{-\omega t}\|g(v)(0) - g(\tilde{v})(0)\| + (1 - e^{-\omega t})\frac{\ell}{\omega}\|u - \tilde{u}\|_{C_b([-\tau,+\infty);X)}$$

for each $t \in [0, +\infty)$.

But

$$\|u - \tilde{u}\|_{C_b([-\tau,+\infty);X)} \leq \|u - \tilde{u}\|_{C([-\tau,0];X)} + \|u - \tilde{u}\|_{C_b([0,+\infty);X)}.$$

On the other hand, by the nonlocal initial condition, we have

$$\|u - \tilde{u}\|_{C([-\tau,0];X)} = \|g_\varepsilon(v) - g_\varepsilon(\tilde{v})\|_{C([-\tau,0];X)} \leq \|g(v) - g(\tilde{v})\|_{C([-\tau,0];X)}.$$

Thus

$$\|u - \tilde{u}\|_{C_b([0,+\infty);X)} \leq \frac{\omega + \ell}{\omega - \ell}\|g(v) - g(\tilde{v})\|_{C([-\tau,0];X)}$$

for each $v, \tilde{v} \in C_b([0, +\infty); X)$. So, from (g_3) in (H_g) and Remark 19.4, we conclude that $\mathcal{S}_\varepsilon$ is continuous on $C_b([0, +\infty); X)$ in the norm topology.

The next step is to show that, for each bounded set $\mathcal{K}$ in $C_b([0, +\infty); X)$, $\mathcal{S}_\varepsilon(\mathcal{K})$ is relatively compact in $C_b([0, +\infty); X)$. To this aim, let $\mathcal{K}$ be a bounded set in $C_b([0, +\infty); X)$ and let $(v_k)_k$ be an arbitrary sequence in $\mathcal{K}$. We show first that $u_k = \mathcal{S}_\varepsilon(v_k)$, $k = 1, 2, \ldots$, is bounded. Indeed, from (H_A), (H_f) and (19.2), we get

$$\|u_k(t)\| \leq e^{-\omega t}\|u_k(0)\| + (1 - e^{-\omega t})\frac{\ell}{\omega}\left[\|u_k\|_{C_b([-\tau,+\infty);X)} + \frac{m}{\ell}\right]$$

for each $t \in (0, +\infty)$. For $t \in [-\tau, 0]$, we have

$$\|u_k(t)\| = \|g_\varepsilon(v_k)(t)\| \leq \|v_k\|_{C_b([0,+\infty);X)} + m_0$$

for each $k \in \mathbb{N}$ and, since $(v_k)_k$ is bounded, there exists $m_1 > 0$ such that

$$\|u_k\|_{C([-\tau,0];X)} \leq m_1$$

for each $k \in \mathbb{N}$. On the other hand

$$\|u_k\|_{C_b([-\tau,+\infty);X)} = \max\{\|u_k\|_{C([-\tau,0];X)}, \|u_k\|_{C_b([0,+\infty);X)}\}$$

$$\leq m_1 + \|u_k\|_{C_b([0,+\infty);X)}.$$

Accordingly

$$\|u_k(t)\| \leq e^{-\omega t} m_1 + (1 - e^{-\omega t}) \frac{\ell}{\omega} \left[\|u_k\|_{C_b([0,+\infty);X)} + m_1 + \frac{m}{\ell} \right]$$

$$\leq m_1 + \frac{\ell}{\omega} \left[\|u_k\|_{C_b([0,+\infty);X)} + m_1 + \frac{m}{\ell} \right]$$

for each $k \in \mathbb{N}$ and $t \in (0, +\infty)$. Hence

$$\left(1 - \frac{\ell}{\omega}\right) \|u_k\|_{C_b([0,+\infty);X)} \leq m_1 \left(1 + \frac{\ell}{\omega}\right) + \frac{m}{\omega}$$

for each $k \in \mathbb{N}$. Finally, we deduce that

$$\|u_k\|_{C_b([0,+\infty);X)} \leq \frac{\omega + \ell}{\omega - \ell} \cdot m_1 + \frac{m}{\omega - \ell}$$

for each $k \in \mathbb{N}$. So $(\mathcal{S}_\varepsilon(v_k))_k$ is bounded and consequently, for each $T > 0$, the set

$$\{f(t, \mathcal{S}_\varepsilon(v_k)_t); \ k \in \mathbb{N}, \ t \in [0, T]\}$$

is relatively compact. By virtue of Theorem 19.2, it follows that $\{\mathcal{S}_\varepsilon(v_k); \ k \in \mathbb{N}\}$ is relatively compact in $C([\delta, T]; X)$ for each $T > 0$ and $\delta \in (0, T)$.

By (H_X), it follows that $\{\mathcal{S}_\varepsilon(v_k)(0); \ k \in \mathbb{N}\} = \{g_\varepsilon(v_k)(0); \ k \in \mathbb{N}\}$ is relatively compact in X, simply because $(v_k)_k$ is bounded and I_ε is compact. Using once again Theorem 19.2, we conclude that $\{\mathcal{S}_\varepsilon(v_k)); \ k \in \mathbb{N}\}$ is relatively compact in $C([0, T]; X)$ for each $T > 0$ and thus in $\tilde{C}_b([0, +\infty); X)$.

For the sake of simplicity, let us denote also by $(u_k)_k = (\mathcal{S}_\varepsilon(v_k))_k$ a convergent subsequence of $(\mathcal{S}_\varepsilon(v_k))_k$ in $\tilde{C}_b([0, +\infty); X)$ to some function u. Then

$$\|u_k(t) - u(t)\| \le e^{-\omega(t-\tau)}\|u_k(\tau) - u(\tau)\| + \int_\tau^t \ell e^{-\omega(t-s)}\|u_{ks} - u_s\|_{C([-\tau,0];X)}\,ds$$

for each $t \in [\tau, +\infty)$. Taking $y(t) = e^{\omega(t-\tau)}\|u_k(t) - u(t)\|$, $\alpha_0(t) = \|u_k(\tau) - u(\tau)\|$ and $\beta = \ell$ in Lemma 19.3 applied on the shifted intervals $[0, +\infty)$ and $[\tau, +\infty)$, after some simple calculations and recalling that $\ell < \omega$, we deduce

$$\|u_k(t) - u(t)\| \le e^{(\ell-\omega)(t-\tau)}[\|u_k(\tau) - u(\tau)\| + \|u_k - u\|_{C([0,\tau];X)}]$$

for each $t \in [\tau, +\infty)$. Since $(u_k)_k$ is bounded, this shows that $\lim_k u_k = u$ in $C_b([\tau, +\infty); X)$. Furthermore, from $\lim_k u_k = u$ in $\tilde{C}_b([0, +\infty); X)$, it follows that $(u_k)_k = (\mathcal{S}_\varepsilon(v_k))_k$ is convergent even in $C_b([0, +\infty); X)$. Thus $\mathcal{S}_\varepsilon(\mathcal{K})$ is relatively compact in $C_b([0, +\infty); X)$, as claimed.

It remains merely to prove that $\mathcal{E}(\mathcal{S}_\varepsilon)$ is bounded. To this aim, let $u \in \mathcal{E}(\mathcal{S}_\varepsilon)$, i.e.

$$u = \lambda \mathcal{S}_\varepsilon(u)$$

for some $\lambda \in [0, 1]$. Consequently

$$\|u(t)\| \le \|\mathcal{S}_\varepsilon(u)(t)\|$$

for each $t \in [0, +\infty)$. From (H_A) and (19.2), it follows that

$$\|\mathcal{S}_\varepsilon(u)(t)\| \le e^{-\omega t}\|\mathcal{S}_\varepsilon(u)(0)\| + \int_0^t e^{-\omega(t-s)}\|f(s, u_s)\|\,ds$$

for each $t \in [0, +\infty)$ and thus, by (H_f) and (g_1) in (H_g), we get

$$\|\mathcal{S}_\varepsilon(u)(t)\| \le e^{-\omega t}\|\mathcal{S}_\varepsilon(u)(0)\| + (1 - e^{-\omega t})\frac{\ell}{\omega}\left[\|\mathcal{S}_\varepsilon(u)\|_{C_b([-\tau,+\infty);X)} + \frac{m}{\ell}\right]$$

for each $t \in [0, +\infty)$. Since, by (g_1), (g_2) in (H_g), (H_X) and Remark 19.2, we have

$$\|\mathcal{S}_\varepsilon(u)\|_{C([-\tau,0];X)} \le \|g_\varepsilon(u)\|_{C([-\tau,0];X)}$$

$$\le \|u\|_{C_b([a,+\infty);X)} + m_0 \le \|\mathcal{S}_\varepsilon(u)\|_{C_b([a,+\infty);X)} + m_0,$$

we are in the hypotheses of Lemma 19.1 which shows that $\mathcal{E}(\mathcal{S}_\varepsilon)$ is bounded. Thus Schaefer Fixed Point Theorem 19.1 applies implying that $\mathcal{S}_\varepsilon$ has at least one fixed point which $v = u_{|[0,+\infty)}$, where $u \in C_b([-\tau, +\infty); X)$ is a mild solution of the problem (19.9). This concludes the proof of Lemma 19.4. $\qquad\square$

19.5.2   Proof of the Main Result: Continued

We can now proceed with the final part of the proof of Theorem 19.3.

Proof For each $\varepsilon \in (0, 1)$ let us fix a mild solution u_ε of the problem (19.9) whose existence is ensured by Lemma 19.4. We first prove that the set $\{u_\varepsilon; \ \varepsilon \in (0, 1)\}$ is relatively compact in $\tilde{C}_b([-\tau, +\infty); X)$. As a consequence, there exists a sequence $\varepsilon_n \downarrow 0$ such that the corresponding sequence $(u_{\varepsilon_n})_n$ converges in $\tilde{C}_b([-\tau, +\infty); X)$ to a function $u \in C_b([-\tau, +\infty); X)$ which turns out to be a mild solution of (19.1).

From Lemma 19.1, we know that $\{u_\varepsilon; \ \varepsilon \in (0, 1)\}$ is bounded in $C_b([-\tau, +\infty); X)$. From Theorem 19.2, we conclude that $\{u_\varepsilon; \ \varepsilon \in (0, 1)\}$ is relatively compact in $C([\delta, T]; X)$ for each $T > 0$ and each $\delta \in (0, T)$. Then, it is relatively compact in $\tilde{C}_b([a, +\infty); X)$. In view of (g_2) and (g_3) and Remark 19.4, it follows that $\{g_\varepsilon(u_\varepsilon); \ \varepsilon \in (0, 1)\}$ is relatively compact in $C([-\tau, 0]; X)$. This means that, there exists a subsequence of $(\varepsilon_n)_n$, denoted again by $(\varepsilon_n)_n$, such that $(u_{\varepsilon_n})_n$—denoted for simplicity also by $(u_n)_n$—converges in $\tilde{C}_b([a, +\infty); X)$ to some function $u \in C_b([a, +\infty); X)$ and the restriction of $(u_n)_n$ to $[-\tau, 0]$, i.e. $(g_{\varepsilon_n}(u_n))_n$, converges in $C([-\tau, 0]; X)$ to some element $v \in C([-\tau, 0]; X)$, i.e.

$$
\begin{cases}
\lim_{n \to \infty} u_n = u & \text{in } \tilde{C}_b([a, +\infty); X) \\
\lim_{n \to \infty} u_n = \lim_{n \to \infty} g_{\varepsilon_n}(u_n) = \lim_{n \to \infty} I_{\varepsilon_n} g(u_n) = v \text{ in } C([-\tau, 0]; X).
\end{cases}
$$

We have

$$
\|u_n(t) - u_p(t)\| \le e^{-\omega t} \|u_n(0) - u_p(0)\| + (1 - e^{-\omega t}) \frac{\ell}{\omega} \|u_n - u_p\|_{C_b([-\tau, +\infty); X)}
$$

for each $n, p \in \mathbb{N}$ and each $t \in [0, +\infty)$. Since

$$
\|u_n - u_p\|_{C_b([-\tau, +\infty); X)} \le \|u_n - u_p\|_{C([-\tau, 0]; X)} + \|u_n - u_p\|_{C_b([0, +\infty); X)}
$$

$$
= \|g_{\varepsilon_n}(u_n) - g_{\varepsilon_p}(u_p)\|_{C([-\tau, 0]; X)} + \|u_n - u_p\|_{C_b([0, +\infty); X)}, \tag{19.12}
$$

it follows that

$$
\|u_n(t) - u_p(t)\| \le e^{-\omega t} \|g_{\varepsilon_n}(u_n) - g_{\varepsilon_p}(u_p)\|_{C([-\tau, 0]; X)}
$$

$$
+ (1 - e^{-\omega t}) \frac{\ell}{\omega} \left[\|g_{\varepsilon_n}(u_n) - g_{\varepsilon_p}(u_p)\|_{C([-\tau, 0]; X)} + \|u_n - u_p\|_{C_b([0, +\infty); X)} \right]
$$

for each $n, p \in \mathbb{N}$ and $t \in [0, +\infty)$. Hence

$$
\|u_n - u_p\|_{C_b([0, +\infty); X)} \le \frac{\omega + \ell}{\omega - \ell} \|g_{\varepsilon_n}(u_n) - g_{\varepsilon_p}(u_p)\|_{C([-\tau, 0]; X)}
$$

for each $n, p \in \mathbb{N}$. As $(g_{\varepsilon_n}(u_n))_n$ is fundamental in $C([-\tau, 0]; X)$ being convergent, from the last inequality, if follows that $(u_n)_n$ is fundamental in $C_b([0, +\infty); X)$. By virtue of (19.12), we deduce that $(u_n)_n$ is fundamental even in $C_b([-\tau, +\infty); X)$ and so it is convergent in this space.

Now, from Remark 19.2, it follows that Lemma 19.1 applies and thus we get (19.5). Since the global asymptotic stability, in the case when $\ell e^{\tau \omega} < \omega$, follows very similar arguments as those in the last part of the proof of Burlică and Roşu [5, Theorem 3.1], this completes the proof of Theorem 19.3. □

19.6 The Damped Wave Equation

Let Ω be a nonempty bounded and open subset in $\mathbb{R}^d$, $d \geq 1$, with C^1 boundary Γ, let $Q_+ = [0, +\infty) \times \Omega$, $Q_\tau = [-\tau, 0] \times \Omega$, $\Sigma_+ = [0, +\infty) \times \Gamma$, let $\omega > 0$ and let us consider the following damped wave equation with delay, subjected to nonlocal initial conditions:

$$
\begin{cases}
\dfrac{\partial^2 u}{\partial t^2}(t, x) = \Delta u(t, x) - 2\omega \dfrac{\partial u}{\partial t}(t, x) - \omega^2 u(t, x) + h\left(t, u(t - \tau, \cdot)\right) & \text{in } Q_+, \\[2mm]
u(t, x) = 0 & \text{on } \Sigma_+, \\[2mm]
u(t, x) = \displaystyle\int_0^{+\infty} \alpha(s) u(t + s, x)\, ds + \psi_1(t, x) & \text{in } Q_\tau, \\[2mm]
\dfrac{\partial u}{\partial t}(t, x) = \displaystyle\int_0^{+\infty} \mathcal{N}\left(s, u(t + s, x), \dfrac{\partial u}{\partial t}(t + s, x)\right) ds + \psi_2(t, x) & \text{in } Q_\tau,
\end{cases}
$$
$$(19.13)$$

$h : [0, +\infty) \times H_0^1(\Omega) \to L^2(\Omega)$, $\alpha \in L^2([0, +\infty); \mathbb{R})$, $\mathcal{N} : [0, +\infty) \times \mathbb{R} \times \mathbb{R} \to \mathbb{R}$, while $\psi_1 \in C([-\tau, 0]; H_0^1(\Omega))$ and $\psi_2 \in C([-\tau, 0]; L^2(\Omega))$.

Theorem 19.4 *Let Ω be a nonempty bounded and open subset in $\mathbb{R}^d$, $d \geq 1$, with C^1 boundary Γ, let $\tau \geq 0$, $\psi_1 \in C([-\tau, 0]; H_0^1(\Omega))$, $\psi_2 \in C([-\tau, 0]; L^2(\Omega))$ and let us assume that $h : [0, +\infty) \times \mathbb{R} \to \mathbb{R}$, $\alpha \in L^2([0, +\infty); \mathbb{R})$ and $\mathcal{N} : [0, +\infty) \times \mathbb{R} \times \mathbb{R} \to \mathbb{R}$ are continuous and satisfy:*

(h_1) *there exists $\ell > 0$ such that $\|h(t, w) - h(t, y)\|_{L^2(\Omega)} \leq \ell \|w - y\|_{H_0^1(\Omega)}$ for each $t \in [0, +\infty)$ and $w, y \in H_0^1(\Omega)$;*

(h_2) *there exists $m > 0$ such that $\|h(t, 0)\|_{L^2(\Omega)} \leq m$ for each $t \in [0, +\infty)$;*

(n_1) *there exists a nonnegative continuous function $\eta \in L^2(\mathbb{R}_+; \mathbb{R}_+)$ such that*

$$|\mathcal{N}(t, u, v)| \leq \eta(t)(|u| + |v|),$$

for each $t \in \mathbb{R}_+$ and $u, v \in \mathbb{R}$.

Let λ_1 be the first eigenvalue of $-\Delta$ and let us assume that

(n_2) $\begin{cases} \|\eta\|_{L^2([0,+\infty);\mathbb{R})} \leq 1, \\ (1+\lambda_1^{-1}\omega)\|\alpha\|_{L^2([0,+\infty);\mathbb{R})} + \lambda_1^{-1}(1+\omega)\|\eta\|_{L^2([0,+\infty);\mathbb{R})} \leq 1; \end{cases}$

(n_3) there exists $b > \tau$ such that $\alpha(t) = \eta(t) = 0$ for each $t \in [0,b]$;

(c_1) $\ell < \omega$.

Then the problem (19.13) has at least one mild solution $u \in C_b([-\tau, +\infty);$ $H_0^1(\Omega))$ with $\dfrac{\partial u}{\partial t} \in C_b([-\tau, +\infty); L^2(\Omega))$. In addition, u satisfies

$$\|u\|_{C_b([-\tau,+\infty);H_0^1(\Omega))} + \left\|\frac{\partial u}{\partial t}\right\|_{C_b([-\tau,+\infty);L^2(\Omega))}$$

$$\leq \frac{m}{\omega - \ell} + \left[\frac{\omega}{\omega - \ell} \cdot \left(\frac{1}{e^{\omega a} - 1} + \frac{\ell}{\omega}\right) + 1\right] \cdot m_0,$$

where

$$m_0 = \|\psi_1\|_{C([-\tau,0];H_0^1(\Omega))} + \|\omega\psi_1 + \psi_2\|_{C([-\tau,0];L^2(\Omega))}.$$

If, instead of (c_1), the stronger nonresonance condition $\ell e^{\tau\omega} < \omega$ is satisfied, then each mild solution of (19.13) is globally asymptotically stable.

Proof Let us observe that (19.13) can be equivalently rewritten in the form (19.1) in the Hilbert space $X = \begin{pmatrix} H_0^1(\Omega) \\ \times \\ L^2(\Omega) \end{pmatrix}$, endowed with the usual inner product

$$\left\langle \begin{pmatrix} u \\ v \end{pmatrix}, \begin{pmatrix} \tilde{u} \\ \tilde{v} \end{pmatrix} \right\rangle = \int_\Omega \nabla u(x) \cdot \nabla \tilde{u}(x)\, dx + \int_\Omega v(x)\tilde{v}(x)\, dx$$

for each $\begin{pmatrix} u \\ v \end{pmatrix}, \begin{pmatrix} \tilde{u} \\ \tilde{v} \end{pmatrix} \in X$, where A, f, and g are defined as follows. First, let us define the linear operator $A : D(A) \subseteq X \to X$ by

$$D(A) = \begin{pmatrix} H_0^1(\Omega) \cap H^2(\Omega) \\ \times \\ H_0^1(\Omega) \end{pmatrix},$$

$$A\begin{pmatrix} u \\ v \end{pmatrix} := \begin{pmatrix} -\omega u + v \\ \Delta u - \omega v \end{pmatrix}$$

for each $\begin{pmatrix} u \\ v \end{pmatrix} \in D(A)$. Second, let us define $f : [0, +\infty) \times C([-\tau, 0]; X) \to X$ by

$$f\left(t, \begin{pmatrix} z \\ y \end{pmatrix}\right)(x) = \begin{pmatrix} 0 \\ h(t, z(-\tau)) \end{pmatrix}$$

for each $t \in [0, +\infty)$, $x \in \Omega$ and $\begin{pmatrix} z \\ y \end{pmatrix} \in C([-\tau, 0]; X)$. Third, let the nonlocal constraint $g : C_b([-\tau, +\infty); X) \to C([-\tau, 0]; X)$ be given by

$$\left[g\begin{pmatrix} u \\ v \end{pmatrix}(t)\right](x) = \begin{pmatrix} \displaystyle\int_0^{+\infty} \alpha(s) u(t+s, x)\, ds + \psi_1(t, x) \\ \displaystyle\int_0^{+\infty} \mathcal{M}(s, x, u(t+s, x), w(t+s, x))\, ds + \psi_3(t, x) \end{pmatrix}$$

for each $\begin{pmatrix} u \\ v \end{pmatrix} \in C_b([-\tau, +\infty); X)$, each $t \in [-\tau, 0]$ and a.e. $x \in \Omega$, $w = v - \omega u$, $\mathcal{M}(t, u, w) = \mathcal{N}(t, u, w) + \omega\alpha(t)u$ and $\psi_3 = \psi_2 + \omega\psi_1$.

We begin by observing that, in view of Remark 19.1, X satisfies (H_X). Clearly, the linear operator $B := A + \omega I$ for each $u \in D(B) = D(A)$, where A is defined as above and I is the identity on X, is the infinitesimal generator of a C_0-group of unitary operators $\{G(t) : X \to X; t \in \mathbb{R}\}$ in X. See Vrabie [27, Theorem 4.6.2, p. 93]. Consequently, A generates a C_0-semigroup of contractions $\{S(t) : X \to X; t \geq 0\}$, defined by

$$S(t)\xi = e^{-\omega t} G(t)\xi$$

for each $t \geq 0$ and each $\xi \in X$, i.e. A is m-dissipative. So A satisfies (H_A). Next, let us observe that f is compact. Indeed, if $C = \left\{ \begin{pmatrix} u_\alpha \\ v_\alpha \end{pmatrix}; \alpha \in \Lambda \right\}$ is a bounded subset in $C([-\tau, 0]; X)$, then $\{u_\alpha(-\tau); \alpha \in \Lambda\}$ is bounded in $H_0^1(\Omega)$ which is compactly embedded in $L^2(\Omega)$. So, $\{u_\alpha(-\tau); \alpha \in \Lambda\}$ is relatively compact in $L^2(\Omega)$ and since h is continuous and has linear growth, it follows that, for each $T > 0$, $f([0, T] \times C)$ is relatively compact in X. But this shows that f is compact. Moreover, in view of (h_1), it follows that the function f satisfies (H_f), while from (n_1), (n_2) and (n_3), we deduce that g, which is of the form (19.4) in Remark 19.3 with

$$\psi(t) = \begin{pmatrix} \psi_1(t) \\ \psi_3(t) \end{pmatrix},$$

for $t \in [-\tau, 0]$, and satisfies (H_g) with $m_0 = \|\psi\|_{C([-\tau,0];X)}$ and $a = b - \tau > 0$. Hence, the conclusion of Theorem 19.4 follows from Theorem 19.3. $\qquad\square$

Acknowledgements (a) This work was supported by a grant of the Romanian National Authority for Scientific Research, CNCS–UEFISCDI, project number PN-II-ID-PCE-2011-3-0052.

(b) The author expresses his warmest thanks to the referee for the very careful reading of the paper and for his/her useful suggestions and remarks.

References

1. Aizicovici, S., Lee, H.: Nonlinear nonlocal Cauchy problems in Banach spaces. Appl. Math. Lett. **18**, 401–407 (2005)
2. Aizicovici, S., McKibben, M.: Existence results for a class of abstract nonlocal Cauchy problems. Nonlinear Anal. **39**, 649–668 (2000)
3. Avalishvili, G., Avalishvili, M.: Nonclassical problems with nonlocal initial conditions for abstract second-order evolution equations. Bull. Georgian Natl. Acad. Sci. (N.S.) **5**, 17–24 (2011)
4. Becker, R.I.: Periodic solutions of semilinear equations of evolution of compact type. J. Math. Anal. Appl. **82**, 33–48 (1981)
5. Burlică, M.D., Roşu, D.: A class of nonlinear delay evolution equations with nonlocal initial conditions. Proc. Am. Math. Soc. **142**, 2445–2458 (2014)
6. Burlică, M., Roşu, D., Vrabie, I.I.: Continuity with respect to the data for a delay evolution equation with nonlocal initial conditions. Libertas Math. (New series) **32**, 37–48 (2012)
7. Burlică, M.D., Roşu, D., Vrabie, I.I.: Abstract reaction-diffusion systems with nonlocal initial conditions. Nonlinear Anal. **94**, 107–119 (2014)
8. Byszewski, L.: Theorems about the existence and uniqueness of solutions of semilinear evolution nonlocal Cauchy problems. J. Math. Anal. Appl. **162**, 494–505 (1991)
9. Cârjă, O., Necula, M., Vrabie, I.I.: Viability, Invariance and Applications. Mathematics Studies, vol. 207. Elsevier, North-Holland (2007)
10. Deng, K.: Exponential decay of solutions of semilinear parabolic equations with initial boundary conditions. J. Math. Anal. Appl. **179**, 630–637 (1993)
11. Di Blasio, G., Lorenzi, A.: Identification problems for integro-differential delay equations. Differ. Integral Equ. **16**, 1385–1408 (2003)
12. García-Falset, J., Reich, S.: Integral solutions to a class of nonlocal evolution equations. Commun. Contemp. Math. **12**, 1032–1054 (2010)
13. Gordeziani, D.G.: On some initial conditions for parabolic equations. Rep. Enlarged Sess. Sem. I. Vekua Inst. Appl. Math. **4**, 57–60 (1989)
14. Gordeziani, D.G., Avalishvili, M., Avalishvili, G.: On the investigation of one nonclassical problem for Navier–Stokes equations. AMI **7**, 66–77 (2002)
15. Hale, J.: Functional Differential Equations. Applied Mathematical Sciences, vol. 3. Springer, New York (1971)
16. Lorenzi, A., Vrabie, I.I.: Identification of a source term in a semilinear evolution delay equation. An. Ştiint. Univ. "Al. I. Cuza" din Iaşi (S.N.), Matematică **LXI**, 1–39 (2015)
17. Lorenzi, A., Vrabie, I.I.: An identification problem for a semilinear evolution delay equation. J. Inverse Ill-Posed Probl. **22**, 209–244 (2014)
18. McKibben, M.: Discovering Evolution Equations with Applications, I. Deterministic Models. Appl. Math. Nonlinear Sci. Ser. Chapman & Hall/CRC, Boca Raton (2011)
19. Mitidieri, E., Vrabie, I.I.: Existence for nonlinear functional differential equations. Hiroshima Math. J. **17**, 627–649 (1987)
20. Mitidieri, E., Vrabie, I.I.: A class of strongly nonlinear functional differential equations. Ann. Mat. Pura Appl. **4-CLI**, 125–147 (1988)
21. Olmstead, W.E., Roberts, C.A.: The one-dimensional heat equation with a nonlocal initial condition. Appl. Math. Lett. **10**, 89–94 (1997)

22. Paicu, A., Vrabie, I.I.: A class of nonlinear evolution equations subjected to nonlocal initial conditions. Nonlinear Anal. **72**, 4091–4100 (2010)
23. Rabier, F., Courtier, P., Ehrendorfer, M.: Four-dimensional data assimilation: comparison of variational and sequential algorithms. Q. J. R. Meteorol. Sci. **118**, 673–713 (1992)
24. Schaefer, H.: Über die Methode der a priori-Schranken. Math. Ann. **129**, 415–416 (1955)
25. Shelukhin, V.V.: A nonlocal in time model for radionuclides propagation in stokes fluid, dynamics of fluids with free boundaries. Inst. Hydrodynam. **107**, 180–193 (1993)
26. Shelukhin, V.V.: A problem nonlocal in time for the equations of the dynamics of a barotropic ocean. Siberian Math. J. **36**, 701–724 (1995)
27. Vrabie, I.I.: C_0-semigroups and applications. North-Holland, Amsterdam (2003)
28. Vrabie, I.I.: Existence for nonlinear evolution inclusions with nonlocal retarded initial conditions. Nonlinear Anal. **74**, 7047–7060 (2011)
29. Vrabie, I.I.: Existence in the large for nonlinear delay evolution inclusions with nonlocal initial conditions. J. Funct. Anal. **262**, 1363–1391 (2012)
30. Vrabie, I.I.: Nonlinear retarded evolution equations with nonlocal initial conditions. Dyn. Syst. Appl. **21**, 417–440 (2012)
31. Vrabie, I.I.: Global solutions for nonlinear delay evolution inclusions with nonlocal initial conditions. Set-valued Var. Anal. **20**, 477–497 (2012)
32. Vrabie, I.I.: Almost periodic solutions for nonlinear delay evolutions with nonlocal initial conditions. J. Evol. Equ. **13**, 693–714 (2013)
33. Vrabie, I.I.: Delay evolution equations with mixed nonlocal plus local initial conditions. Commun. Contemp. Math. (2013). doi: 10.1142/S0219199713500351.

Chapter 20
Elliptic Differential-Operator Problems with the Spectral Parameter in Both the Equation and Boundary Conditions and the Corresponding Abstract Parabolic Initial Boundary Value Problems

Yakov Yakubov

Dedicated to the memory of Alfredo Lorenzi

Abstract We consider, in *UMD* Banach spaces, boundary value problems for second order elliptic differential-operator equations with the spectral parameter and boundary conditions containing the parameter in the same order as the equation. An isomorphism and the corresponding estimate of the solution (with respect to the space variable and the parameter) are obtained. Then, an application of the obtained abstract results is given to boundary value problems for second order elliptic differential equations with the parameter in non-smooth domains. Further, the corresponding abstract parabolic initial boundary value problem is treated and an application to initial boundary value problems with time differentiation in boundary conditions is demonstrated.

AMS Subject Classifications: 34G10, 35J25, 47E05, 47D06.

20.1   Introduction and Basic Notations

In the last decade, the study of boundary value problems for elliptic differential-operator equations in a Banach space has been intensively developed. We refer

The author was supported by the Israel Ministry of Absorption.

Y. Yakubov (✉)
Raymond and Beverly Sackler School of Mathematical Sciences, Tel-Aviv University, Tel-Aviv 69978, Israel
e-mail: yakubov@post.tau.ac.il

A. Favini et al. (eds.), *New Prospects in Direct, Inverse and Control Problems for Evolution Equations*, Springer INdAM Series 10,
DOI 10.1007/978-3-319-11406-4_20

the reader to the papers [3–9] and to their introductions with the full reference to the last works on the subject. The corresponding problems with the spectral parameter both in the equation and in boundary conditions, especially with the same power, have been less studied (see [1] and the reference therein). In [1], boundary value problems for second order elliptic differential-operator equations with the linear spectral parameter and operator-boundary conditions containing also the same linear parameter have been considered in a Hilbert space. The main operator of the equation is selfadjoint and positive-definite. It seems that there is no the corresponding study in a Banach space. The main purpose of the present paper is to generalize and improve the corresponding isomorphism result in a Hilbert space from [1] in the framework of *UMD* Banach spaces and non-selfadjoint operators. In particular, we get maximal L_p-regularity for the considered problems. The second aim of the paper, and not less, is to study the corresponding abstract parabolic initial boundary value problems with the time differentiation in abstract boundary conditions. The main abstract results in the paper are illustrated by some suitable applications to PDEs.

We give now some necessary definitions. Let E_1 and E_2 be Banach spaces. The set $E_1 \dotplus E_2$ of all the vectors of the form (u, v), where $u \in E_1, v \in E_2$, with the natural coordinatewise linear operations and the norm $\|(u, v)\|_{E_1 \dotplus E_2} := \|u\|_{E_1} + \|v\|_{E_2}$ is a Banach space and is said to be a *direct sum* of Banach spaces E_1 and E_2.

By $(E_1, E_2)_{\theta, p}, 0 < \theta < 1, 1 \le p \le \infty$, we denote the standard (*real*) *interpolation space* (see, e.g., [14] for definitions and properties).

Let A be a linear closed operator in the Banach space E with domain $D(A)$. The domain $D(A)$ is turned into the Banach space $E(A)$ with respect to the norm $\|u\|_{E(A)} := (\|u\|_E^2 + \|Au\|_E^2)^{\frac{1}{2}}$.

By $B(E_1, E)$ we denote the Banach space of all linear bounded operators from E_1 into E with ordinary operator norm; $B(E) := B(E, E)$.

By $L_p((0, 1); E), 1 < p < \infty$, we denote the Banach space of functions $x \to u(x) : (0, 1) \to E$, strongly measurable and summable in the p-th power, with finite norm $\|u\|_{L_p((0,1);E)} := (\int_0^1 \|u(x)\|_E^p dx)^{\frac{1}{p}}$.

The space $W_p^\ell((0, 1); E), 1 < p < \infty, 0 \le \ell$ is an integer, is the Banach space of functions $u(x)$ on $(0, 1)$ with values in E, which have generalized derivatives up to the ℓ-th order inclusive with finite norm $\|u\|_{W_p^\ell((0,1);E)} := \sum_{k=0}^\ell (\int_0^1 \|u^{(k)}(x)\|_E^p dx)^{\frac{1}{p}}$.

By $W_p^2((0, 1); E(A), E)$ we denote

$$W_p^2((0, 1); E(A), E) := \{u : u \in L_p((0, 1); E(A)), u'' \in L_p((0, 1); E)\}$$

with finite norm $\|u\|_{W_p^2((0,1);E(A),E)} := \|u\|_{L_p((0,1);E(A))} + \|u''\|_{L_p((0,1);E)}$. It is known that this space is a Banach space (see for more general spaces [14, Lemma 1.8.1]; see also [16, Sect. 1.7.7]).

With $0 \le k$ is an integer, $C^k([0, T]; E)$ denotes the Banach space of functions $u(x)$ on $[0, T]$ with values from E which have continuous derivatives up to the k-th order inclusive on $[0, T]$ and the norm $\|u\|_{C^k([0,T];E)} := \sum_{\ell=0}^k \max_{x \in [0,T]} \|u^{(\ell)}(x)\|_E$ is finite. If $k = \infty$ then this is a set of functions $u(x)$ with values from E which have continuous derivatives of any order.

Let an operator A act from a Banach space E into a Banach space F. By F' we denote the dual space to F. The operator A from E into F is *Fredholm*, if

(a) the image $R(A)$ is closed in F;
(b) ker A and coker A are finitely dimensional subspaces in E and F', respectively;
(c) dim ker A = dim coker A.

The following notions are well known but we would like to bring them here for the reader convenience.

A Banach space E is said to be of *class HT*, if the Hilbert transform is bounded on $L_p(\mathbb{R}; E)$ for some (and then all) $p > 1$. Here the Hilbert transform H of a function $f \in S(\mathbb{R}; E)$, the Schwartz space of rapidly decreasing E-valued functions, is defined by $Hf := \frac{1}{\pi} PV(\frac{1}{t}) * f$, i.e., $(Hf)(t) := \frac{1}{\pi} \lim_{\varepsilon \to 0} \int_{|\tau| > \varepsilon} \frac{f(t-\tau)}{\tau} d\tau$. These spaces are often also called *UMD* Banach spaces, where the *UMD* stands for the property of *unconditional martingale differences*.

For a linear operator A in the Banach space E, we will, sometimes, write $\lambda + A$ instead of $\lambda I + A$, where $\lambda \in \mathbb{C}$ and I is the identity operator in E. The operator $R(\lambda, A) := (\lambda I - A)^{-1}$ is the resolvent of the operator A.

Definition 20.1 Let E be a complex Banach space, and A is a closed linear operator in E. The operator A is called *sectorial* if the following conditions are satisfied:

1. $\overline{D(A)} = E$, $\overline{R(A)} = E$, $(-\infty, 0) \subseteq \rho(A)$;
2. $\|\lambda(\lambda + A)^{-1}\| \leq M$ for all $\lambda > 0$, and some $M < \infty$.

Definition 20.2 Let E and F be Banach spaces. A family of operators $\mathcal{T} \subset B(E, F)$ is called $\mathcal{R}$-*bounded* if there is a constant $C > 0$ and $p \geq 1$ such that for each natural number n, $T_j \in \mathcal{T}$, $u_j \in E$ and for all independent, symmetric, $\{-1, 1\}$-valued random variables ε_j on $[0, 1]$ (e.g., the Rademacher functions $\varepsilon_j(t) = \text{sign} \sin(2^j \pi t)$) the inequality

$$\left\| \sum_{j=1}^{n} \varepsilon_j T_j u_j \right\|_{L_p((0,1);F)} \leq C \left\| \sum_{j=1}^{n} \varepsilon_j u_j \right\|_{L_p((0,1);E)}$$

is valid. The smallest such C is called $\mathcal{R}$-*bound* of $\mathcal{T}$ and is denoted by $\mathcal{R}\{\mathcal{T}\}_{E \to F}$. If $E = F$, the $\mathcal{R}$-bound will be denoted by $\mathcal{R}\{\mathcal{T}\}_E$ or simply $\mathcal{R}\{\mathcal{T}\}$.

Remark 20.1 From the definition of $\mathcal{R}$-boundedness it follows that every $\mathcal{R}$-bounded family of operators is (uniformly) bounded (it is enough to take $n = 1$). On the other hand, in a Hilbert space H every bounded set is $\mathcal{R}$-bounded (see, e.g., [12, p. 75]). Therefore, in a Hilbert space, the notion of $\mathcal{R}$-boundedness is equivalent to boundedness of a family of operators (see also [2, p. 26]).

We indicate with $\arg \lambda$ the element of the argument of the complex number λ in $(-\pi, \pi]$ ($\lambda \in \mathbb{C} \setminus \{0\}$; $\arg 0 := 0$).

Definition 20.3 A sectorial operator A in E is called $\mathscr{R}$-*sectorial* if

$$\mathscr{R}_A(0) := \mathscr{R}\{\lambda(\lambda + A)^{-1} : \lambda > 0\} < \infty.$$

The number $\phi_A^{\mathscr{R}} := \inf\{\theta \in (0, \pi) : \mathscr{R}_A(\pi - \theta) < \infty\}$, where $\mathscr{R}_A(\theta) := \mathscr{R}\{\lambda(\lambda + A)^{-1} : |\arg \lambda| \le \theta\}$, is called the $\mathscr{R}$-*angle* of the operator A.

20.2 Auxiliary Results

In this section we set some useful facts.

Lemma 20.1 *Let* λ *and* μ *be elements of* $\mathbb{C} \setminus \{0\}$. *Let* $\epsilon \in \mathbb{R}^+$ *(positive real numbers) and assume that*

$$|| \arg \lambda - \arg \mu| - \pi| \ge \epsilon. \tag{20.1}$$

Then $|\lambda + \mu| \ge C(\epsilon)(|\lambda| + |\mu|)$, *with* $C(\epsilon) \in \mathbb{R}^+$, *depending only on* ϵ *and not on* λ *and* μ.

Proof We set $\Lambda := \arg \lambda$ and $M := \arg \mu$. We have

$$|\lambda + \mu|^2 = |\lambda|^2 + |\mu|^2 + 2|\lambda||\mu| \cos(\Lambda - M).$$

Obviously, $\Lambda - M \in (-2\pi, 2\pi)$ and, if θ is in this interval, $\cos \theta = -1$ if and only if $|\theta| = \pi$. In case $||\theta| - \pi| \ge \epsilon$, the estimate $\cos \theta \ge \delta(\epsilon) - 1$, for some $\delta(\epsilon)$ in $(0, 2)$, holds. We deduce that, in case (20.1),

$$|\lambda + \mu|^2 \ge |\lambda|^2 + |\mu|^2 + 2[\delta(\epsilon) - 1]|\lambda||\mu|.$$

As $\min_{\xi \ge 0} \frac{\xi^2 + 1 + 2[\delta(\epsilon) - 1]\xi}{(\xi + 1)^2} = \frac{\delta(\epsilon)}{2}$, we get

$$|\lambda|^2 + |\mu|^2 + 2[\delta(\epsilon) - 1]|\lambda||\mu| = |\mu|^2\Big(\big(\frac{|\lambda|}{|\mu|}\big)^2 + 1 + 2[\delta(\epsilon) - 1]\frac{|\lambda|}{|\mu|}\Big)$$

$$\ge \frac{\delta(\epsilon)}{2}(|\lambda| + |\mu|)^2.$$

The conclusion follows, with $C(\epsilon) = (\frac{\delta(\epsilon)}{2})^{1/2}$. $\square$

We introduce the following notation:

$$\phi_\lambda := \min\{\phi, \pi - |\arg \lambda|\}. \tag{20.2}$$

Lemma 20.2 *Let A be a linear operator in the complex Banach space E and let $\phi \in (0, \pi)$. Assume that $\{v \in \mathbb{C} \setminus \{0\} : |\arg v| \geq \pi - \phi\} \cup \{0\} \subseteq \rho(A)$ and there exists $M \in \mathbb{R}^+$ such that, for v in this set, $(1 + |v|)\|(v - A)^{-1}\|_{B(E)} \leq M$.*

(i) let $\lambda \in \mathbb{C} \setminus \{0\}$, with $|\arg \lambda| \leq \phi$. Then $-\lambda \in \rho(A)$ and so $(\lambda + A)^{-1} \in B(E)$;

(ii) let $\lambda \in \mathbb{C} \setminus \{0\}$, with $0 \leq \arg \lambda \leq \phi$ (so that $\text{Im}\lambda \geq 0$). If $\mu \in \mathbb{C} \setminus \{0\}$ and $-\phi_\lambda < \arg \mu < \phi$ then $\lambda + \mu \neq 0$ and $|\arg(\lambda + \mu)| \leq \phi$. Moreover, $\forall \epsilon \in (0, \phi_\lambda)$, there exists $C(\epsilon, M)$ in $\mathbb{R}^+$ such that, if $-\phi_\lambda + \epsilon \leq \arg \mu \leq \phi - \epsilon$, we have

$$\|(\mu + \lambda + A)^{-1}\|_{B(E)} \leq C(\epsilon, M)(|\lambda| + |\mu|)^{-1}.$$

The constant $C(\epsilon, M)$ is independent of λ and μ at least $\forall \epsilon \in (0, \min\{\phi, \pi - \phi\})$;

(iii) let λ be as in (ii). Then, the operator $(\lambda + A)^{1/2}$ is well defined. If $z \in \mathbb{C} \setminus \{0\}$ and $-\frac{\pi + \phi_\lambda}{2} < \arg z < \frac{\pi + \phi}{2}$, $-z \in \rho((\lambda + A)^{1/2})$. Moreover, $\forall \epsilon \in (0, \phi_\lambda)$, if $-\frac{\pi + \phi_\lambda}{2} + \epsilon \leq \arg z \leq \frac{\pi + \phi}{2} - \epsilon$,

$$\|(z + (\lambda + A)^{1/2})^{-1}\|_{B(E)} \leq C(\phi, M, \epsilon)|z|^{-1}.$$

The constant $C(\phi, M, \epsilon)$ is independent of λ and z at least $\forall \epsilon \in (0, \min\{\phi, \pi - \phi\})$.

Proof (i) follows from the simple observation that, given $\phi \in [0, \pi]$, $|\arg \lambda| \geq \pi - \phi$ if and only if $|\arg(-\lambda)| \leq \phi$.

Let $\lambda \in \mathbb{C} \setminus \{0\}$, with $0 \leq \arg \lambda \leq \phi < \pi$. It is easily seen that, if $\mu \in \mathbb{C} \setminus \{0\}$ and $-\phi_\lambda < \arg \mu < \phi$, then $\mu + \lambda \neq 0$ and $|\arg(\mu + \lambda)| \leq \phi$, so that $(\mu + \lambda + A)^{-1} \in B(E)$. We observe also that $\forall \epsilon \in (0, \phi_\lambda)$, in case $-\phi_\lambda + \epsilon \leq \arg \mu \leq \phi - \epsilon$, we have $|\arg \lambda - \arg \mu| \leq \pi - \epsilon$, so that (20.1) holds. We deduce that, in such a case,

$$\|(\mu + \lambda + A)^{-1}\|_{B(E)} \leq M|\lambda + \mu|^{-1} \leq C(\epsilon, M)(|\lambda| + |\mu|)^{-1}.$$

So (ii) is proved.

Concerning (iii), we start by remarking that, as a consequence of (ii), $\lambda + A$ is positive and the operator $(\lambda + A)^{1/2}$ is defined. We have also, for $z \in \mathbb{R}^+$,

$$[z + (\lambda + A)^{1/2}]^{-1} = \frac{1}{\pi} \int_0^\infty \frac{t^{1/2}(A + \lambda + t)^{-1}}{t + z^2} dt \tag{20.3}$$

(see [13, formula (2.32)]). We observe that (20.3) is well defined even in the more general case $|\arg z| < \frac{\pi}{2}$. So, by analytic continuation, if $|\arg z| < \pi/2$, $[z + (\lambda + A)^{1/2}]^{-1}$ exists and equals (20.3).

We come back to the case that $z \in \mathbb{R}^+$, and observe that, by analyticity, we can shift the path of integration from $\mathbb{R}^+$ to $\mathbb{R}^+ e^{i\theta}$, if $-\phi_\lambda < \theta < \phi$, to obtain

$$[z + (\lambda + A)^{1/2}]^{-1} = \frac{1}{\pi} \int_0^\infty \frac{t^{1/2} e^{i\theta/2} (A + \lambda + t e^{i\theta})^{-1}}{t e^{i\theta} + z^2} e^{i\theta} dt. \qquad (20.4)$$

Now we observe that the right hand side of (20.4) is well defined if $\frac{\theta - \pi}{2} < \arg z < \frac{\pi + \theta}{2}$, implying that $[z + (\lambda + A)^{1/2}]^{-1} \in B(E)$ whenever $-\frac{\pi + \phi_\lambda}{2} < \arg z < \frac{\pi + \phi}{2}$.

Let $\epsilon \in (0, \phi_\lambda)$ and $z \in \mathbb{C} \setminus \{0\}$, with $-\frac{\pi + \phi_\lambda}{2} + \epsilon \leq \arg z \leq \frac{\pi + \phi}{2} - \epsilon$. Then we can take θ in $[-\phi_\lambda + \epsilon, \phi - \epsilon]$, such that $\frac{\theta - \pi}{2} + \frac{\epsilon}{2} \leq \arg z \leq \frac{\pi + \theta}{2} - \frac{\epsilon}{2}$. We have

$$\arg(z^2) = \begin{cases} 2 \arg z & \text{if } -\frac{\pi}{2} < \arg z \leq \frac{\pi}{2}, \\ 2 \arg z - 2\pi & \text{if } \frac{\pi}{2} < \arg z \leq \pi, \\ 2 \arg z + 2\pi & \text{if } -\pi < \arg z \leq -\frac{\pi}{2}. \end{cases}$$

In the first case, we deduce $-\pi + \epsilon \leq \arg(z^2) - \theta \leq \pi - \epsilon$. In the second case, we have $\arg(z^2) - \theta \leq -\pi - \epsilon$. In the third case, we have $\arg(z^2) - \theta \geq \pi + \epsilon$. So, in each case we obtain $|| \arg(z^2) - \theta| - \pi| \geq \epsilon$. Employing Lemma 20.1, we obtain, from (20.4) and part (ii),

$$\|[z + (\lambda + A)^{1/2}]^{-1}\|_{B(E)} \leq C_1(\phi, M, \epsilon) \int_0^\infty \frac{t^{1/2} (|\lambda| + t)^{-1}}{t + |z|^2} dt$$

$$\leq C_1(\phi, M, \epsilon) \int_0^\infty \frac{t^{-1/2}}{t + |z|^2} dt \leq C_2(\phi, M, \epsilon) |z|^{-1}.$$

The proof of (iii) is complete, too.   □

In a similar way, we can prove the following

Lemma 20.3 *Let A satisfy conditions of Lemma 20.2 and let $\lambda \in \mathbb{C} \setminus \{0\}$, with $0 \geq \arg \lambda \geq -\phi$ (so that $\mathrm{Im}\lambda \leq 0$). Then,*

(i) *if $\mu \in \mathbb{C} \setminus \{0\}$ and $-\phi < \arg \mu < \phi_\lambda$, $\lambda + \mu \neq 0$ and $|\arg(\lambda + \mu)| \leq \phi$. Moreover, $\forall \epsilon \in (0, \phi_\lambda)$, there exists $C(\epsilon, M)$ in $\mathbb{R}^+$, such that, if $-\phi + \epsilon \leq \arg \mu \leq \phi_\lambda - \epsilon$, we have*

$$\|(\mu + \lambda + A)^{-1}\|_{B(E)} \leq C(\epsilon, M)(|\lambda| + |\mu|)^{-1}.$$

The constant $C(\epsilon, M)$ is independent of λ and μ at least $\forall \epsilon \in (0, \min\{\phi, \pi - \phi\})$;

(ii) *the operator $(\lambda + A)^{1/2}$ is well defined. If $z \in \mathbb{C} \setminus \{0\}$ and $-\frac{\pi + \phi}{2} < \arg z < \frac{\pi + \phi_\lambda}{2}$, $-z \in \rho((\lambda + A)^{1/2})$. Moreover, $\forall \epsilon \in (0, \phi_\lambda)$, if $-\frac{\pi + \phi}{2} + \epsilon \leq \arg z \leq \frac{\pi + \phi_\lambda}{2} - \epsilon$,*

$$\|(z + (\lambda + A)^{1/2})^{-1}\|_{B(E)} \leq C(\phi, M, \epsilon) |z|^{-1}.$$

The constant $C(\phi, M, \epsilon)$ is independent of λ and z at least $\forall \epsilon \in (0, \min\{\phi, \pi - \phi\})$.

Theorem 20.1 *Let A satisfy conditions of Lemma 20.2 and let $\alpha \in \mathbb{C} \setminus \{0\}$ and $\lambda \in \mathbb{C} \setminus \{0\}$, with $|\arg \lambda| \leq \phi$, such that*

$$|\arg \lambda - \arg \alpha| \leq \frac{\pi + \phi}{2} - \epsilon, \qquad (20.5)$$

for some ϵ in $(0, \min\{\pi - \arg \alpha, \phi_\lambda\})$. Then, the operator $[\lambda(A + \lambda)^{-1/2} + \alpha]^{-1}$ is defined, it belongs to $B(E)$, and

$$\|[\lambda(A + \lambda)^{-1/2} + \alpha]^{-1}\|_{B(E)} \leq C(\phi, M, \alpha, \epsilon).$$

The constant $C(\phi, M, \alpha, \epsilon)$ is independent of λ at least if ϵ is chosen in $(0, \min\{\pi - \arg \alpha, \phi, \pi - \phi\})$.

Proof Set $z := \frac{\lambda}{\alpha}$. Then $z = |z|e^{i(\arg \lambda - \arg \alpha)}$, so that, by (20.5), $\arg z = \arg \lambda - \arg \alpha$. If $\arg \lambda \geq 0$, we have $-\frac{\pi + \phi_\lambda}{2} = \max\{-\frac{\pi + \phi}{2}, \frac{\arg \lambda}{2} - \pi\}$, $\arg z - (\frac{\arg \lambda}{2} - \pi) = \frac{\arg \lambda}{2} + \pi - \arg \alpha \geq \epsilon$. We deduce that $-\frac{\pi + \phi_\lambda}{2} + \epsilon \leq \arg z \leq \frac{\pi + \phi}{2} - \epsilon$. So, by Lemma 20.2, $[\frac{\lambda}{\alpha} + (\lambda + A)^{1/2}]^{-1}$ is defined and

$$\|[\frac{\lambda}{\alpha} + (\lambda + A)^{1/2}]^{-1}\| \leq C(M, \phi, \epsilon, \alpha)|\lambda|^{-1}. \qquad (20.6)$$

An analogous argument works in case $\arg \lambda \leq 0$, employing Lemma 20.3.

We have $\lambda(A + \lambda)^{-1/2} + \alpha = \alpha[\frac{\lambda}{\alpha} + (A + \lambda)^{1/2}](A + \lambda)^{-1/2}$, so that $[\lambda(A + \lambda)^{-1/2} + \alpha]^{-1}$ is defined and

$$[\lambda(A + \lambda)^{-1/2} + \alpha]^{-1} = \alpha^{-1}(A + \lambda)^{1/2}[\frac{\lambda}{\alpha} + (A + \lambda)^{1/2}]^{-1}$$

$$= \alpha^{-1}\left(1 - \frac{\lambda}{\alpha}[\frac{\lambda}{\alpha} + (A + \lambda)^{1/2}]^{-1}\right).$$

So the conclusion follows from (20.6). □

Remark 20.2 Assume that A is selfadjoint and positive definite in the Hilbert space E. Then $\{v \in \mathbb{C} \setminus \{0\} : |\arg v| \geq \pi - \phi\} \cup \{0\} \subseteq \rho(A)$, for every ϕ in $(0, \pi)$. Assume that $\alpha \in \mathbb{C} \setminus \{0\}$, with $|\arg \alpha| \leq \frac{\pi - \phi}{2}$. If $|\arg \lambda| \leq \phi - \epsilon$, with $0 < \epsilon < \phi$, we have

$$|\arg \lambda - \arg \alpha| \leq |\arg \lambda| + |\arg \alpha| \leq \frac{\pi + \phi}{2} - \epsilon,$$

so that Theorem 20.1 is applicable if $0 < \epsilon < \min\{\phi_\lambda, \pi - \arg \alpha\}$.

20.3 Homogeneous Equations

We start with homogeneous equation and boundary conditions without perturbation terms

$$L_0(\lambda)u := \lambda u(x) - u''(x) + Au(x) = 0, \quad x \in (0, 1), \tag{20.7}$$

$$L_{10}(\lambda)u := \lambda u(0) - \alpha u'(0) = f_1,$$
$$L_{20}(\lambda)u := \lambda u(1) + \beta u'(1) = f_2. \tag{20.8}$$

Theorem 20.2 *Let the following conditions be satisfied:*

1. *an operator A is closed, densely defined and invertible in the complex Banach space E and $\|R(\lambda, A)\|_{B(E)} \leq C(1 + |\lambda|)^{-1}$, for $|\arg \lambda| \geq \pi - \varphi$, for some $\varphi \in (0, \pi)$;*
2. *$\alpha = 0$ or $|\arg \alpha| \leq \frac{\pi - \varphi}{2}$ and $\beta = 0$ or $|\arg \beta| \leq \frac{\pi - \varphi}{2}$.*

Then, the problem (20.7)–(20.8), for $f_k \in (E(A), E)_{\theta_k, p}$, where $\theta_k = \frac{m_k}{2} + \frac{1}{2p}$,

$$p \in (1, \infty), \quad m_1 = \begin{cases} 1, & \alpha \neq 0, \\ 0, & \alpha = 0 \end{cases}, \quad m_2 = \begin{cases} 1, & \beta \neq 0, \\ 0, & \beta = 0 \end{cases}, \quad and \; |\arg \lambda| \leq \varphi - \epsilon,$$

$\epsilon > 0$ is any sufficiently small and $|\lambda|$ is sufficiently large, has the unique solution $u(x)$ that belongs to the space $W_p^2((0, 1); E(A), E)$ and, for these λ, the following estimate holds for the solution:

$$|\lambda| \|u\|_{L_p((0,1);E)} + \|u''\|_{L_p((0,1);E)} + \|Au\|_{L_p((0,1);E)}$$

$$\leq C(\varphi, \epsilon) \sum_{k=1}^{2} |\lambda|^{-1+m_k} \left(\|f_k\|_{(E(A),E)_{\theta_k, p}} + |\lambda|^{1-\theta_k} \|f_k\|_E \right). \tag{20.9}$$

Proof In view of condition (1), for $|\arg \lambda| \leq \varphi < \pi$, there exists the holomorphic, for $x > 0$, and strongly continuous, for $x \geq 0$, semigroup $e^{-x(A+\lambda I)^{\frac{1}{2}}}$ (see, e.g., [16, Lemma 5.4.2/6]). By virtue of [8, Lemma 1], an arbitrary solution of the equation (20.7), for $|\arg \lambda| \leq \varphi$, belonging to $W_p^2((0, 1); E(A), E)$ has the form

$$u(x) = e^{-x(A+\lambda I)^{\frac{1}{2}}} g_1 + e^{-(1-x)(A+\lambda I)^{\frac{1}{2}}} g_2, \tag{20.10}$$

where $g_k \in (E(A), E)_{\frac{1}{2p}, p}$.

Let us now prove the converse, i.e., a function $u(x)$ of the form (20.10), with $g_k \in (E(A), E)_{\frac{1}{2p}, p}$, belongs to the space $W_p^2((0, 1); E(A), E)$. In view of [16,

Theorem 5.4.2/1] (see the appendix), from (20.10), we find, for $|\arg \lambda| \leq \varphi$,

$$\|u\|_{W_p^2((0,1);E(A),E)} \leq \left(\|A(A + \lambda I)^{-1}\|_{B(E)} + 1\right)\left[\left(\int_0^1 \|(A + \lambda I)e^{-x(A+\lambda I)^{\frac{1}{2}}} g_1\|_E^p dx\right)^{\frac{1}{p}}\right.$$

$$\left. + \left(\int_0^1 \|(A + \lambda I)e^{-(1-x)(A+\lambda I)^{\frac{1}{2}}} g_2\|_E^p dx\right)^{\frac{1}{p}}\right]$$

$$\leq C(\varphi) \sum_{k=1}^{2} \left(\|g_k\|_{(E(A),E)_{\frac{1}{2p},p}} + |\lambda|^{1-\frac{1}{2p}}\|g_k\|_E\right).$$

A function $u(x)$ of the form (20.10) satisfies boundary conditions (20.8) if

$$\lambda\left[g_1 + e^{-(A+\lambda I)^{\frac{1}{2}}} g_2\right] - \alpha\left[-(A + \lambda I)^{\frac{1}{2}} g_1 + (A + \lambda I)^{\frac{1}{2}} e^{-(A+\lambda I)^{\frac{1}{2}}} g_2\right] = f_1,$$

$$\lambda\left[e^{-(A+\lambda I)^{\frac{1}{2}}} g_1 + g_2\right] + \beta\left[-(A + \lambda I)^{\frac{1}{2}} e^{-(A+\lambda I)^{\frac{1}{2}}} g_1 + (A + \lambda I)^{\frac{1}{2}} g_2\right] = f_2. \tag{20.11}$$

Rewrite (20.11) in the form

$$\left[\lambda I + \alpha(A + \lambda I)^{\frac{1}{2}}\right]g_1 + \left[\lambda e^{-(A+\lambda I)^{\frac{1}{2}}} - \alpha(A + \lambda I)^{\frac{1}{2}} e^{-(A+\lambda I)^{\frac{1}{2}}}\right]g_2 = f_1,$$

$$\left[\lambda e^{-(A+\lambda I)^{\frac{1}{2}}} - \beta(A + \lambda I)^{\frac{1}{2}} e^{-(A+\lambda I)^{\frac{1}{2}}}\right]g_1 + \left[\lambda I + \beta(A + \lambda I)^{\frac{1}{2}}\right]g_2 = f_2. \tag{20.12}$$

From (20.12) we get

$$\left[\lambda(A + \lambda I)^{-\frac{m_1}{2}} + \alpha\right](A + \lambda I)^{\frac{m_1}{2}} g_1 + \left[\lambda e^{-(A+\lambda I)^{\frac{1}{2}}} (A + \lambda I)^{-\frac{m_1}{2}}\right.$$

$$\left. - \alpha e^{-(A+\lambda I)^{\frac{1}{2}}}\right](A + \lambda I)^{\frac{m_1}{2}} g_2 = f_1,$$

$$\left[\lambda e^{-(A+\lambda I)^{\frac{1}{2}}} (A+\lambda I)^{-\frac{m_2}{2}} - \beta e^{-(A+\lambda I)^{\frac{1}{2}}}\right](A + \lambda I)^{\frac{m_2}{2}} g_1 \tag{20.13}$$

$$+ \left[\lambda(A + \lambda I)^{-\frac{m_2}{2}} + \beta\right](A + \lambda I)^{\frac{m_2}{2}} g_2 = f_2.$$

Denote $v_1 := (A + \lambda I)^{\frac{m}{2}} g_1$, $v_2 := (A + \lambda I)^{\frac{m}{2}} g_2$, where $m = \max\{m_1, m_2\}$. Note that, by Yakubov and Yakubov [16, Lemma 1.7.3/5], v_1 and v_2 are well-defined. Then, (20.13) can be rewritten in the form

$$\left[\lambda(A + \lambda I)^{-\frac{m_1}{2}} + \alpha\right](A + \lambda I)^{\frac{m_1-m}{2}} v_1 + \left[\lambda e^{-(A+\lambda I)^{\frac{1}{2}}} (A + \lambda I)^{-\frac{m_1}{2}}\right.$$

$$\left. - \alpha e^{-(A+\lambda I)^{\frac{1}{2}}}\right](A + \lambda I)^{\frac{m_1-m}{2}} v_2 = f_1,$$

$$\left[\lambda e^{-(A+\lambda I)^{\frac{1}{2}}} (A+\lambda I)^{-\frac{m_2}{2}} - \beta e^{-(A+\lambda I)^{\frac{1}{2}}}\right](A + \lambda I)^{\frac{m_2-m}{2}} v_1 \tag{20.14}$$

$$+ \left[\lambda(A + \lambda I)^{-\frac{m_2}{2}} + \beta\right](A + \lambda I)^{\frac{m_2-m}{2}} v_2 = f_2.$$

System (20.14) has the form

$$\left(A(\lambda) + R(\lambda)\right)\begin{pmatrix} v_1 \\ v_2 \end{pmatrix} = \begin{pmatrix} (A + \lambda I)^{\frac{m-m_1}{2}} f_1 \\ (A + \lambda I)^{\frac{m-m_2}{2}} f_2 \end{pmatrix}, \tag{20.15}$$

where $A(\lambda)$ and $R(\lambda)$ are operator-matrixes 2×2:

$$A(\lambda) = \begin{pmatrix} \lambda(A + \lambda I)^{-\frac{m_1}{2}} + \alpha & 0 \\ 0 & \lambda(A + \lambda I)^{-\frac{m_2}{2}} + \beta \end{pmatrix}$$

and $R(\lambda)$ is equal to

$$\begin{pmatrix} 0 & \lambda e^{-(A+\lambda I)^{\frac{1}{2}}}(A + \lambda I)^{-\frac{m_1}{2}} - \alpha e^{-(A+\lambda I)^{\frac{1}{2}}} \\ \lambda e^{-(A+\lambda I)^{\frac{1}{2}}}(A + \lambda I)^{-\frac{m_2}{2}} - \beta e^{-(A+\lambda I)^{\frac{1}{2}}} & 0 \end{pmatrix}.$$

Note that, by Yakubov and Yakubov [16, Lemma 1.7.3/5], $(A + \lambda I)^{\frac{m-m_k}{2}} f_k$ are well-defined.

By virtue of [16, Lemma 5.4.2/6], for $|\arg \lambda| \le \varphi$ and $|\lambda| \to \infty$,

$$\|R(\lambda)\|_{B(E^2)} \le C(\varphi)e^{-\delta|\lambda|^{\frac{1}{2}}}, \quad \|R(\lambda)\|_{B([E(A)]^2)} \le C(\varphi)e^{-\delta|\lambda|^{\frac{1}{2}}}, \tag{20.16}$$

for some $\delta > 0$. On the other hand, by virtue of the conditions of our theorem, Theorem 20.1 implies, for $|\arg \lambda| \le \varphi - \epsilon$ and any small enough $\epsilon > 0$, that

$$\|(\lambda(A + \lambda I)^{-\frac{m_1}{2}} + \alpha)^{-1}\|_{B(E)} \le C(\varphi, \epsilon)|\lambda|^{-1+m_1},$$
$$\|(\lambda(A + \lambda I)^{-\frac{m_2}{2}} + \beta)^{-1}\|_{B(E)} \le C(\varphi, \epsilon)|\lambda|^{-1+m_2}. \tag{20.17}$$

Since, by (20.17), there exists

$$A(\lambda)^{-1} = \begin{pmatrix} (\lambda(A + \lambda I)^{-\frac{m_1}{2}} + \alpha)^{-1} & 0 \\ 0 & (\lambda(A + \lambda I)^{-\frac{m_2}{2}} + \beta)^{-1} \end{pmatrix}$$

then, by virtue of (20.16) and (20.17), $\|R(\lambda)A(\lambda)^{-1}\| \to 0$ for $|\arg \lambda| \le \varphi - \epsilon$ and $|\lambda| \to \infty$. Hence, by the Neumann identity, for $|\arg \lambda| \le \varphi - \epsilon$ and $|\lambda| \to \infty$,

$$\left(A(\lambda) + R(\lambda)\right)^{-1} = A(\lambda)^{-1}\left(I + R(\lambda)A(\lambda)^{-1}\right)^{-1} = A(\lambda)^{-1}\sum_{k=0}^{\infty}\left(-R(\lambda)A(\lambda)^{-1}\right)^k.$$

Consequently, system (20.15) has a unique solution for $|\arg \lambda| \leq \varphi - \epsilon$ and $|\lambda|$ sufficiently large, and the solution can be expressed in the form, for $k = 1, 2$,

$$v_k = \left[C_{k1}(\lambda) + R_{k1}(\lambda) \right] (A + \lambda I)^{\frac{m-m_1}{2}} f_1 + \left[C_{k2}(\lambda) + R_{k2}(\lambda) \right] (A + \lambda I)^{\frac{m-m_2}{2}} f_2, \tag{20.18}$$

where $C_{11}(\lambda) = (\lambda (A + \lambda I)^{-\frac{m_1}{2}} + \alpha)^{-1}$, $C_{22}(\lambda) = (\lambda (A + \lambda I)^{-\frac{m_2}{2}} + \beta)^{-1}$, $C_{12}(\lambda) = C_{21}(\lambda) = 0$, and $R_{kj}(\lambda)$ are some bounded operators both in E and $E(A)$ (since A is invertible in E). Moreover, by (20.17), for $k = 1, 2$,

$$\| C_{kk}(\lambda) \|_{B(E)} \leq C(\varphi, \epsilon) |\lambda|^{-1+m_k}, \quad \| C_{kk}(\lambda) \|_{B(E(A))} \leq C(\varphi, \epsilon) |\lambda|^{-1+m_k}$$

and, by virtue of (20.16), for $k = 1, 2$ and $j = 1, 2$,

$$\| R_{kj}(\lambda) \|_{B(E)} \leq C(\varphi, \epsilon) e^{-\delta |\lambda|^{\frac{1}{2}}}, \quad \| R_{kj}(\lambda) \|_{B(E(A))} \leq C(\varphi, \epsilon) e^{-\delta |\lambda|^{\frac{1}{2}}}, \tag{20.19}$$

for $|\arg \lambda| \leq \varphi - \epsilon$ and $|\lambda| \to \infty$.

From the form of C_{kj} and R_{kj} it follows that these operators commute with $(A + \lambda I)^{-\frac{m}{2}}$. Then, multiplying (20.18) by $(A + \lambda I)^{-\frac{m}{2}}$, we get (remind that $v_k = (A + \lambda I)^{\frac{m}{2}} g_k$), for $k = 1, 2$,

$$g_k = \left[C_{k1}(\lambda) + R_{k1}(\lambda) \right] (A + \lambda I)^{-\frac{m_1}{2}} f_1 + \left[C_{k2}(\lambda) + R_{k2}(\lambda) \right] (A + \lambda I)^{-\frac{m_2}{2}} f_2. \tag{20.20}$$

Substituting (20.20) into (20.10) we get

$$u(x) = \sum_{k=1}^{2} \left[(A + \lambda I)^{-\frac{m_k}{2}} e^{-x(A+\lambda I)^{\frac{1}{2}}} \left(C_{1k}(\lambda) + R_{1k}(\lambda) \right) \right.$$

$$\left. + (A + \lambda I)^{-\frac{m_k}{2}} e^{-(1-x)(A+\lambda I)^{\frac{1}{2}}} \left(C_{2k}(\lambda) + R_{2k}(\lambda) \right) \right] f_k.$$

Then, for $|\arg \lambda| \leq \varphi - \epsilon$ and $|\lambda|$ sufficiently large, we have

$$|\lambda| \| u \|_{L_p((0,1));E)} + \| u'' \|_{L_p((0,1);E)} + \| Au \|_{L_p((0,1);E)}$$

$$\leq C \sum_{k=1}^{2} \left\{ |\lambda| \left[\left(\int_0^1 \| (A + \lambda I)^{-\frac{m_k}{2}} e^{-x(A+\lambda I)^{\frac{1}{2}}} C_{1k}(\lambda) f_k \|_E^p dx \right)^{\frac{1}{p}} \right. \right.$$

$$+ \left(\int_0^1 \| (A + \lambda I)^{-\frac{m_k}{2}} e^{-x(A+\lambda I)^{\frac{1}{2}}} R_{1k}(\lambda) f_k \|_E^p dx \right)^{\frac{1}{p}}$$

$$+ \left(\int_0^1 \| (A + \lambda I)^{-\frac{m_k}{2}} e^{-(1-x)(A+\lambda I)^{\frac{1}{2}}} C_{2k}(\lambda) f_k \|_E^p dx \right)^{\frac{1}{p}}$$

$$+ \left(\int_0^1 \| (A + \lambda I)^{-\frac{m_k}{2}} e^{-(1-x)(A+\lambda I)^{\frac{1}{2}}} R_{2k}(\lambda) f_k \|_E^p dx \right)^{\frac{1}{p}} \Big]$$

$$+ (1 + \| A(A + \lambda I)^{-1} \|_{B(E)}) \Big[\left(\int_0^1 \| (A + \lambda I)^{1 - \frac{m_k}{2}} e^{-x(A+\lambda I)^{\frac{1}{2}}} C_{1k}(\lambda) f_k \|_E^p dx \right)^{\frac{1}{p}}$$

$$+ \left(\int_0^1 \| (A + \lambda I)^{1 - \frac{m_k}{2}} e^{-x(A+\lambda I)^{\frac{1}{2}}} R_{1k}(\lambda) f_k \|_E^p dx \right)^{\frac{1}{p}}$$

$$+ \left(\int_0^1 \| (A + \lambda I)^{1 - \frac{m_k}{2}} e^{-(1-x)(A+\lambda I)^{\frac{1}{2}}} C_{2k}(\lambda) f_k \|_E^p dx \right)^{\frac{1}{p}}$$

$$+ \left(\int_0^1 \| (A + \lambda I)^{1 - \frac{m_k}{2}} e^{-(1-x)(A+\lambda I)^{\frac{1}{2}}} R_{2k}(\lambda) f_k \|_E^p dx \right)^{\frac{1}{p}} \Big] \Big\}. \tag{20.21}$$

It was mentioned above that the operators $C_{kj}(\lambda)$ both in E and $E(A)$ act bound-edly and $\| C_{kj}(\lambda) \|_{B(E)} \leq C(\varphi, \epsilon) |\lambda|^{-1+m_k}$, $\| C_{kj}(\lambda) \|_{B(E(A))} \leq C(\varphi, \epsilon) |\lambda|^{-1+m_k}$ since $C_{12}(\lambda) = C_{21}(\lambda) = 0$. Then, from [16, Sect. 1.7.9], it follows that the operators $C_{kj}(\lambda)$ act boundedly in the space $(E(A), E)_{\theta, p}$, where $0 < \theta < 1$, too. Moreover, for $|\arg \lambda| \leq \varphi - \epsilon$ and $|\lambda| \to \infty$,

$$\| C_{kj}(\lambda) \|_{B((E(A), E)_{\theta, p})} \leq \| C_{kj}(\lambda) \|_{B(E(A))}^{1-\theta} \| C_{kj}(\lambda) \|_{B(E)}^{\theta} \leq C(\varphi, \epsilon) |\lambda|^{-1+m_k},$$

as well.

By virtue of [16, Theorem 5.4.2/1] (see the appendix) and that $(E_0, E_1)_{\theta, p} = (E_1, E_0)_{1-\theta, p}$ (see, e.g., [16, Lemma 1.7.3/1]), for the first term of the right-hand side of inequality (20.21), we have, for $|\arg \lambda| \leq \varphi - \epsilon$ and $|\lambda| \to \infty$,

$$|\lambda| \left(\int_0^1 \| (A + \lambda I)^{-\frac{m_k}{2}} e^{-x(A+\lambda I)^{\frac{1}{2}}} C_{1k}(\lambda) f_k \|_E^p dx \right)^{\frac{1}{p}}$$

$$\leq C |\lambda| \| (A + \lambda I)^{-1} \|_{B(E)} \left(\int_0^1 \| (A + \lambda I)^{1 - \frac{m_k}{2}} e^{-x(A+\lambda I)^{\frac{1}{2}}} C_{1k}(\lambda) f_k \|_E^p dx \right)^{\frac{1}{p}}$$

$$\leq C(\varphi) \left(\| C_{1k}(\lambda) f_k \|_{(E(A), E)_{\theta_k, p}} + |\lambda|^{1-\theta_k} \| C_{1k}(\lambda) f_k \|_E \right)$$

$$\leq C(\varphi, \epsilon) |\lambda|^{-1+m_k} \left(\| f_k \|_{(E(A), E)_{\theta_k, p}} + |\lambda|^{1-\theta_k} \| f_k \|_E \right).$$

From (20.19) and [16, Sect. 1.7.9], it follows that the operators $R_{kj}(\lambda)$ are bounded in the space $(E(A), E)_{\theta, p}$, $0 < \theta < 1$, and, for $|\arg \lambda| \leq \varphi - \epsilon$, $|\lambda| \to \infty$,

$$\| R_{kj}(\lambda) \|_{B((E(A), E)_{\theta, p})} \leq \| R_{kj}(\lambda) \|_{B(E(A))}^{1-\theta} \| R_{kj}(\lambda) \|_{B(E)}^{\theta} \leq C(\varphi, \epsilon) e^{-\delta |\lambda|^{\frac{1}{2}}}.$$

By the same theorem [16, Theorem 5.4.2/1] (see the appendix), for the second term of the right-hand side of inequality (20.21), we have, for $|\arg \lambda| \le \varphi - \epsilon$, $|\lambda| \to \infty$,

$$|\lambda| \left(\int_0^1 \|(A + \lambda I)^{-\frac{m_k}{2}} e^{-x(A+\lambda I)^{\frac{1}{2}}} R_{1k}(\lambda) f_k\|_E^p dx \right)^{\frac{1}{p}}$$

$$\le C |\lambda| \|(A + \lambda I)^{-1}\|_{B(E)} \left(\int_0^1 \|(A + \lambda I)^{1-\frac{m_k}{2}} e^{-x(A+\lambda I)^{\frac{1}{2}}} R_{1k}(\lambda) f_k\|_E^p dx \right)^{\frac{1}{p}}$$

$$\le C(\varphi) \left(\|R_{1k}(\lambda) f_k\|_{(E(A),E)_{\theta_k,p}} + |\lambda|^{1-\theta_k} \|R_{1k}(\lambda) f_k\|_E \right)$$

$$\le C(\varphi, \epsilon) e^{-\delta |\lambda|^{\frac{1}{2}}} \left(\|f_k\|_{(E(A),E)_{\theta_k,p}} + |\lambda|^{1-\theta_k} \|f_k\|_E \right)$$

$$\le C(\varphi, \epsilon) |\lambda|^{-1+m_k} \left(\|f_k\|_{(E(A),E)_{\theta_k,p}} + |\lambda|^{1-\theta_k} \|f_k\|_E \right).$$

In a similar way, we estimate all the other terms on the right-hand side of inequality (20.21). This proves estimate (20.9). □

20.4 Nonhomogeneous Equations

We start with auxiliary lemma.

Lemma 20.4 *Let an operator A be closed, densely defined, invertible in the complex UMD Banach space E, and $\mathcal{R}\{\lambda R(\lambda, A) : |\arg \lambda| \ge \pi - \varphi\} < \infty$, for some $\varphi \in (0, \pi)$, and $f \in L_p(\mathbb{R}; E)$. Then, if u is a solution of the equation $\lambda u(x) - u''(x) + Au(x) = f(x)$ on $\mathbb{R}$,*

$$\|u\|_{W_p^1(\mathbb{R}; E(A^{\frac{1}{2}}), E)} \le C |\lambda|^{-\frac{1}{2}} \|f\|_{L_p(\mathbb{R}; E)},$$

where the constant C is independent of λ, with $|\arg \lambda| \le \varphi$.

Proof We have

$$\|u\|_{W_p^1(\mathbb{R}; E(A^{\frac{1}{2}}), E)} \le C \left(\|A^{\frac{1}{2}} u\|_{L_p(\mathbb{R}; E)} + \|u'\|_{L_p(\mathbb{R}; E)} \right)$$

$$= C \left(\|F^{-1} A^{\frac{1}{2}} (\lambda + \xi^2 + A)^{-1} F f\|_{L_p(\mathbb{R}; E)} \right.$$

$$\left. + \|F^{-1} \xi (\lambda + \xi^2 + A)^{-1} F f\|_{L_p(\mathbb{R}; E)} \right),$$

where F and F^{-1} denote the Fourier transform and the inverse Fourier transform, respectively. It is enough to show that for nonnegative integers i, j, k, if $i + j + k = 2$,

$$\||\lambda|^{\frac{i}{2}} F^{-1} \xi^j A^{\frac{k}{2}} (\lambda + \xi^2 + A)^{-1} F f\|_{L_p(\mathbb{R}; E)} \le C \|f\|_{L_p(\mathbb{R}; E)},$$

where the constant C is independent of λ, with $|\arg \lambda| \leq \varphi$. This can be done employing [15, Theorem 3.4]. In turn, in order to do that, it is enough to prove that the sets

$$\{m(\lambda, i, j, k, \xi) \,:\, \lambda \in \mathbb{C}\backslash\{0\}, \, |\arg \lambda| \leq \varphi, \, \xi \in \mathbb{R}\}$$

and

$$\{\xi D_\xi m(\lambda, i, j, k, \xi) \,:\, \lambda \in \mathbb{C}\backslash\{0\}, \, |\arg \lambda| \leq \varphi, \, \xi \in \mathbb{R}\}$$

are $\mathscr{R}$-bounded in $B(E)$, where $D_\xi := \frac{\partial}{\partial \xi}$ and $m(\lambda, i, j, k, \xi) := |\lambda|^{\frac{i}{2}} \xi^j A^{\frac{k}{2}} (\lambda + \xi^2 + A)^{-1}$, for $i + j + k = 2$.

We have

$$\xi D_\xi m(\lambda, i, j, k, \xi) = jm(\lambda, i, j, k, \xi) - 2m(\lambda, i, j, k, \xi)[\xi^2 (\lambda + \xi^2 + A)^{-1}].$$

As $\xi^2 \leq C(\varphi)|\lambda + \xi^2|$, for mentioned above λ and ξ, by Kahane's contraction principle, $\{\xi^2(\lambda + \xi^2 + A)^{-1} \,:\, \lambda \in \mathbb{C}\backslash\{0\}, \, |\arg \lambda| \leq \varphi, \, \xi \in \mathbb{R}\}$ is $\mathscr{R}$-bounded in $B(E)$. So, we have only to prove that $\{m(\lambda, i, j, k, \xi) \,:\, \lambda \in \mathbb{C}\backslash\{0\}, \, |\arg \lambda| \leq \varphi, \, \xi \in \mathbb{R}\}$ is $\mathscr{R}$-bounded in $B(E)$. We have already examined the case $i = k = 0$, $j = 2$. In general, the case $k = 0$ follows from the inequality $|\lambda|^{\frac{i}{2}} |\xi|^j \leq C(\varphi)|\lambda + \xi^2|$ and, again, Kahane's contraction principle. The case $i = j = 0, k = 2$ follows from

$$A(\lambda + \xi^2 + A)^{-1} = I - (\lambda + \xi^2)(\lambda + \xi^2 + A)^{-1}.$$

It remains only to consider the situation $k = 1$. By [3, Lemma 1.5(V)], the set

$$\{|\lambda + \xi^2|^{\frac{1}{2}} A^{\frac{1}{2}} (\lambda + \xi^2 + A)^{-1} \,:\, \lambda \in \mathbb{C}\backslash\{0\}, \, |\arg \lambda| \leq \varphi, \, \xi \in \mathbb{R}\}$$

is $\mathscr{R}$-bounded in $B(E)$. So, the conclusion for the last situation follows from Kahane's contraction principle and the fact that, if $i + j = 1$, $|\lambda|^{\frac{i}{2}} |\xi|^j \leq C(\varphi)|\lambda + \xi^2|^{\frac{1}{2}}$.

$\square$

Now consider a boundary value problem for the nonhomogeneous equation with the parameter

$$L_0(\lambda)u := \lambda u(x) - u''(x) + Au(x) = f(x), \quad x \in (0, 1), \tag{20.22}$$

$$L_{10}(\lambda)u := \lambda u(0) - \alpha u'(0) = f_1,$$
$$L_{20}(\lambda)u := \lambda u(1) + \beta u'(1) = f_2. \tag{20.23}$$

Theorem 20.3 *Let the following conditions be satisfied:*

1. *an operator A is closed, densely defined and invertible in the complex UMD Banach space E and*

$$\mathscr{R}\{\lambda R(\lambda, A) : |\arg \lambda| \geq \pi - \varphi\} < \infty,$$

for some $\varphi \in (0, \pi)$[1];
2. *$\alpha = 0$ or $|\arg \alpha| \leq \frac{\pi - \varphi}{2}$ and $\beta = 0$ or $|\arg \beta| \leq \frac{\pi - \varphi}{2}$;*
3. *$m_1 := \begin{cases} 1, & \alpha \neq 0, \\ 0, & \alpha = 0 \end{cases}$, $m_2 := \begin{cases} 1, & \beta \neq 0, \\ 0, & \beta = 0 \end{cases}$; $\theta_k := \frac{m_k}{2} + \frac{1}{2p}$, $p \in (1, \infty)$, $k = 1, 2$.*

Then, the operator $\mathbb{L}_0(\lambda) : u \rightarrow \mathbb{L}_0(\lambda)u := \left(L_0(\lambda)u, L_{10}(\lambda)u, L_{20}(\lambda)u \right)$, for $|\arg \lambda| \leq \varphi - \epsilon$, $\epsilon > 0$ is any sufficiently small and $|\lambda|$ is sufficiently large, is an isomorphism from $W_p^2((0, 1); E(A), E)$ onto $L_p((0, 1); E) \dotplus (E(A), E)_{\theta_1, p} \dotplus (E(A), E)_{\theta_2, p}$ and, for these λ, the following estimate holds for the solution of the problem (20.22)–(20.23):

$$|\lambda| \|u\|_{L_p((0,1);E)} + \|u''\|_{L_p((0,1);E)} + \|Au\|_{L_p((0,1);E)}$$

$$\leq C(\varphi, \epsilon) \Big[|\lambda|^{\frac{\max\{m_k\}}{2}} \|f\|_{L_p((0,1);E)}$$

$$+ \sum_{k=1}^{2} |\lambda|^{-1+m_k} \left(\|f_k\|_{(E(A),E)_{\theta_k,p}} + |\lambda|^{1-\theta_k} \|f_k\|_E \right) \Big]. \tag{20.24}$$

Proof The uniqueness follows from Theorem 20.2. Let us now define $\tilde{f}(x) := f(x)$ if $x \in (0, 1)$ and $\tilde{f}(x) := 0$ if $x \notin (0, 1)$. We now show that a solution of the problem (20.22)–(20.23) belonging to $W_p^2((0, 1); E(A), E)$ can be represented as a sum of the form $u(x) = u_1(x) + u_2(x)$, where $u_1(x)$ is the restriction on $[0, 1]$ of the solution $\tilde{u}_1(x)$ of the equation

$$L_0(\lambda)\tilde{u}_1 = \tilde{f}(x), \quad x \in \mathbb{R}, \tag{20.25}$$

and $u_2(x)$ is a solution of the problem

$$L_0(\lambda)u_2 = 0, \quad L_{k0}(\lambda)u_2 = f_k - L_{k0}(\lambda)u_1, \quad k = 1, 2. \tag{20.26}$$

---

[1]In fact, this condition is equivalent to that A is an invertible $\mathscr{R}$-sectorial operator in E with the $\mathscr{R}$-angle $\phi_A^{\mathscr{R}} < \pi - \varphi$.

It was shown in the proof of [4, Theorem 4] that a solution of the equation (20.25) is given by the formula

$$\tilde{u}_1(x) = \frac{1}{2\pi} \int_{\mathbb{R}} e^{i\mu x} L_0(\lambda, i\mu)^{-1} F \tilde{f}(\mu) d\mu, \qquad (20.27)$$

where $F\tilde{f}$ is the Fourier transform of the function $\tilde{f}(x)$, and $L_0(\lambda, \sigma)$ is the characteristic operator pencil of the equation (20.25), i.e., $L_0(\lambda, \sigma) = -\sigma^2 I + A + \lambda I$. Moreover, the solution belongs to $W_p^2(\mathbb{R}; E(A), E)$ and for the solution it holds the estimate (see [4, formula (4.11)])

$$|\lambda| \|\tilde{u}_1\|_{L_p(\mathbb{R};E)} + \|\tilde{u}_1\|_{W_p^2(\mathbb{R};E(A),E)} \leq C(\varphi) \|\tilde{f}\|_{L_p(\mathbb{R};E)}, \quad |\arg \lambda| \leq \varphi, \quad (20.28)$$

and, therefore, $u_1 \in W_p^2((0, 1); E(A), E)$.

By virtue of [16, Theorem 1.7.7/1] (for $v(x) = u_1(x + x_0)$) and the inequality (20.28), we have $u_1^{(s)}(x_0) \in (E(A), E)_{\frac{s}{2} + \frac{1}{2p}, p}$, $\forall x_0 \in [0, 1]$, $s = 0, 1$. Hence, $L_{k0}(\lambda)u_1 \in (E(A), E)_{\theta_k, p}$ since $(E(A), E)_{\frac{1}{2p}, p} \subset (E(A), E)_{\theta_k, p}$. Thus, by virtue of Theorem 20.2 (together with Remark 20.1), the problem (20.26) has the unique solution $u_2(x)$ that belongs to $W_p^2((0, 1); E(A), E)$ as $|\arg \lambda| \leq \varphi - \epsilon, \epsilon > 0$ is any sufficiently small and $|\lambda|$ is sufficiently large. Moreover, for the solution of the problem (20.26), for $|\arg \lambda| \leq \varphi - \epsilon$, $|\lambda| \to \infty$, we have

$$|\lambda| \|u_2\|_{L_p((0,1);E)} + \|u_2''\|_{L_p((0,1);E)} + \|Au_2\|_{L_p((0,1);E)}$$

$$\leq C(\varphi, \epsilon) \sum_{k=1}^{2} |\lambda|^{-1+m_k} \left(\|f_k - L_{k0}(\lambda)u_1\|_{(E(A),E)_{\theta_k,p}} \right.$$

$$\left. + |\lambda|^{1-\theta_k} \|f_k - L_{k0}(\lambda)u_1\|_E \right)$$

$$\leq C(\varphi, \epsilon) \sum_{k=1}^{2} |\lambda|^{-1+m_k} \left[\|f_k\|_{(E(A),E)_{\theta_k,p}} + |\lambda|^{1-\theta_k} \|f_k\|_E \right.$$

$$+ |\lambda| \|u_1\|_{C([0,1];(E(A),E)_{\theta_k,p})} + |\lambda|^{1-\theta_k} |\lambda| \|u_1\|_{C([0,1];E)}$$

$$+ m_k \|u_1'\|_{C([0,1];(E(A),E)_{\theta_k,p})} + m_k |\lambda|^{1-\theta_k} \|u_1'\|_{C([0,1];E)} \right]. \qquad (20.29)$$

From (20.28), for $|\arg \lambda| \leq \varphi$, it follows that

$$|\lambda| \|u_1\|_{L_p((0,1);E)} + \|u_1\|_{W_p^2((0,1);E(A),E)} \leq C(\varphi) \|f\|_{L_p((0,1);E)}. \qquad (20.30)$$

Hence, from $(E(A), E)_{\frac{1}{2p},p} \subset (E(A), E)_{\theta_k,p}$, [16, Theorem 1.7.7/1] (for $v(x) = u_1(x + x_0)$), and (20.30), for any $x_0 \in [0, 1]$, we have

$$\|u_1^{(s)}(x_0)\|_{(E(A),E)_{\theta_k,p}} \leq C \|u_1\|_{W_p^2((0,1);E(A),E)} \leq C(\varphi)\|f\|_{L_p((0,1);E)}, \quad s = 0, 1.$$
$$(20.31)$$

By virtue of [16, Theorem 1.7.7/2], for $\mu \in \mathbb{C}$, $u_1 \in W_p^2((0, 1); E)$, $m_k = 0, 1$,

$$|\mu|^{2-m_k} \|u_1^{(m_k)}(x_0)\| \leq C\left(|\mu|^{\frac{1}{p}}\|u_1\|_{W_p^2((0,1);E)} + |\mu|^{2+\frac{1}{p}}\|u_1\|_{L_p((0,1);E)}\right). \quad (20.32)$$

Dividing (20.32) by $|\mu|^{\frac{1}{p}}$ and substituting $\lambda = \mu^2$ for $\lambda \in \mathbb{C}$, $|\lambda| \to \infty$, $u_1 \in W_p^2((0, 1); E)$, we have

$$|\lambda|^{1-\theta_k} \|u_1^{(m_k)}(x_0)\| \leq C\left(\|u_1\|_{W_p^2((0,1);E)} + |\lambda|\|u_1\|_{L_p((0,1);E)}\right), \quad m_k = 0, 1.$$
$$(20.33)$$

Then, (20.30) and (20.33) imply, for $|\arg \lambda| \leq \varphi$, $|\lambda| \to \infty$,

$$|\lambda|^{1-\theta_k} \|u_1^{(m_k)}(x_0)\| \leq C\left(\|u_1\|_{W_p^2((0,1);E(A),E)} + |\lambda|\|u_1\|_{L_p((0,1);E)}\right)$$
$$\leq C(\varphi)\|f\|_{L_p((0,1);E)}, \quad m_k = 0, 1. \quad (20.34)$$

From (20.29), (20.31), and (20.34), for $|\arg \lambda| \leq \varphi - \epsilon$, $|\lambda| \to \infty$, we have

$$|\lambda|\|u_2\|_{L_p((0,1);E)} + \|u_2''\|_{L_p((0,1);E)} + \|Au_2\|_{L_p((0,1);E)}$$
$$\leq C(\varphi, \epsilon)\left[|\lambda|^{\max\{m_k\}}\|f\|_{L_p((0,1);E)}\right.$$
$$\left. + \sum_{k=1}^{2}|\lambda|^{-1+m_k}\left(\|f_k\|_{(E(A),E)_{\theta_k,p}} + |\lambda|^{1-\theta_k}\|f_k\|_E\right)\right]. \quad (20.35)$$

Then, (20.30) and (20.35) imply (20.24) for $m_1 = m_2 = 0$.

Let us now at least one $m_k = 1$. Then, the corresponding $\theta_k = \frac{1}{2} + \frac{1}{2p}$. By [14, Theorem 1.15.2], $(E(A), E)_{\theta_k,p} = (E(A^{\frac{1}{2}}), E)_{\frac{1}{p},p}$. In turn, by Favini et al. [3, Lemma 2.1],

$$(E(A^{\frac{1}{2}}), E)_{\frac{1}{p},p} = \{u(0) \, : \, u \in W_p^1(\mathbb{R}^+; E(A^{\frac{1}{2}}), E)\}.$$

Hence, $\|u(0)\|_{(E(A),E)_{\theta_k,p}} \leq C\|u\|_{W_p^1(\mathbb{R};E(A^{\frac{1}{2}}),E)}$. So, by Lemma 20.4,

$$\|u_1\|_{C([0,1];(E(A),E)_{\theta_k,p})} \le C \|\tilde{u}_1\|_{W_p^1(\mathbb{R};E(A^{\frac{1}{2}}),E)} \le C|\lambda|^{-\frac{1}{2}}\|\tilde{f}\|_{L_p(\mathbb{R};E)}$$

$$= C|\lambda|^{-\frac{1}{2}}\|f\|_{L_p((0,1);E)}. \qquad (20.36)$$

Moreover, by (20.32) (take $\mu = \lambda^2$) and (20.30),

$$\|u_1\|_{C([0,1];E)} \le C\left(|\lambda|^{\frac{1}{2p}-1}\|u_1\|_{W_p^2((0,1);E)} + |\lambda|^{\frac{1}{2p}}\|u_1\|_{L_p((0,1);E)}\right)$$

$$\le C|\lambda|^{\frac{1}{2p}-1}\|f\|_{L_p((0,1);E)}. \qquad (20.37)$$

Therefore, from (20.29), (20.31), (20.34), (20.36), and (20.37), for $|\arg \lambda| \le \varphi - \epsilon$, $|\lambda| \to \infty$,

$$|\lambda|\|u_2\|_{L_p((0,1);E)} + \|u_2''\|_{L_p((0,1);E)} + \|Au_2\|_{L_p((0,1);E)}$$

$$\le C(\varphi,\epsilon)\left[|\lambda|^{\frac{1}{2}}\|f\|_{L_p((0,1);E)} + \sum_{k=1}^{2}|\lambda|^{-1+m_k}\left(\|f_k\|_{(E(A),E)_{\theta_k,p}} + |\lambda|^{1-\theta_k}\|f_k\|_E\right)\right]$$

$$= C(\varphi,\epsilon)\left[|\lambda|^{\frac{\max\{m_k\}}{2}}\|f\|_{L_p((0,1);E)}\right.$$

$$\left. + \sum_{k=1}^{2}|\lambda|^{-1+m_k}\left(\|f_k\|_{(E(A),E)_{\theta_k,p}} + |\lambda|^{1-\theta_k}\|f_k\|_E\right)\right]. \qquad (20.38)$$

Then, (20.30) and (20.38) imply (20.24). $\qquad\qquad\qquad\qquad\qquad\qquad\qquad\square$

20.5 Isomorphism of a Problem with the Parameter in Both the Equation and Boundary Conditions and with Perturbation Terms in the Boundary Conditions

First, consider the problem (20.22)–(20.23) with the perturbed both equation and boundary conditions, namely,

$$L(\lambda)u := \lambda u(x) - u''(x) + Au(x) + (A_1u)(x) = f(x), \quad x \in (0,1), \qquad (20.39)$$

$$L_1(\lambda)u := \lambda u(0) - \alpha u'(0) + T_1 u = f_1,$$
$$L_2(\lambda)u := \lambda u(1) + \beta u'(1) + T_2 u = f_2, \qquad (20.40)$$

and prove the following Fredholmness theorem.

Theorem 20.4 *Let the following conditions be satisfied:*

1. *an operator A is closed, densely defined in the complex UMD Banach space E and $\mathscr{R}\{\lambda R(\lambda, A) : |\arg\lambda| \geq \pi - \varphi, |\lambda| \geq M\} < \infty$, for some $\varphi \in (0, \pi)$ and some $M \geq 0$[2];*

2. $\alpha = 0$ *or* $|\arg\alpha| \leq \frac{\pi-\varphi}{2}$ *and* $\beta = 0$ *or* $|\arg\beta| \leq \frac{\pi-\varphi}{2}$;

3. $m_1 := \begin{cases} 1, & \alpha \neq 0, \\ 0, & \alpha = 0 \end{cases}$, $m_2 := \begin{cases} 1, & \beta \neq 0, \\ 0, & \beta = 0 \end{cases}$; $\theta_k := \frac{m_k}{2} + \frac{1}{2p}$, $p \in (1, \infty)$,

   $k = 1, 2$;

4. *the embedding $E(A) \subseteq E$ is compact;*

5. *for any $\delta > 0$ and $u \in W_p^2((0, 1); E(A), E)$,*

$$\|A_1 u\|_{L_p((0,1);E)} \leq \delta \|u\|_{W_p^2((0,1);E(A),E)} + C(\delta)\|u\|_{L_p((0,1);E)};$$

6. *if $m_k = 0$ then $T_k = 0$; if $m_k = 1$ then, for any $\delta > 0$ and $u \in W_p^2((0, 1); E(A), E)$,*

$$\|T_k u\|_{(E(A),E)_{\theta_k,p}} \leq \delta \|u\|_{W_p^2((0,1);E(A),E)} + C(\delta)\|u\|_{L_p((0,1);E)}.$$

Then, the operator $\mathbb{L}(\lambda_0) : u \to \mathbb{L}(\lambda_0)u := \left(L(\lambda_0)u, L_1(\lambda_0)u, L_2(\lambda_0)u\right)$, for any fixed λ_0 with $|\arg\lambda_0| \leq \varphi - \epsilon$, $\epsilon > 0$ is any sufficiently small and $|\lambda_0|$ is sufficiently large, from $W_p^2((0, 1); E(A), E)$ into $L_p((0, 1); E) \dot{+} (E(A), E)_{\theta_1,p} \dot{+} (E(A), E)_{\theta_2,p}$ is bounded and Fredholm.

Proof Without loss of generality, we can assume that condition (1) is satisfied for the whole sector $|\arg\lambda| \geq \pi - \varphi$ and $\lambda = 0$ ($\lambda = 0$ means that A is invertible). Our general case is reduced to the latter if the operator $A + M_1 I$ is considered instead of the operator A, and the operator $A_1 - M_1 I$ is considered instead of the operator A_1 with sufficiently large $M_1 \geq M$.

We can represent the operator $\mathbb{L}(\lambda_0)$ in the form

$$\mathbb{L}(\lambda_0) = \mathbb{L}_0(\lambda_0) + \mathbb{L}_1$$

with $\mathbb{L}_0(\lambda)u := (L_0(\lambda)u, L_{10}(\lambda)u, L_{20}(\lambda)u)$ and $\mathbb{L}_1 u := ((A_1 u)(x), T_1 u, T_2 u)$, where $L_0(\lambda)$, $L_{k0}(\lambda)$, $k = 1, 2$, are defined by (20.22)–(20.23).

By Theorem 20.3, the operator $\mathbb{L}_0(\lambda_0)$ from $W_p^2((0, 1); E(A), E)$ onto $L_p((0, 1); E) \dot{+}(E(A), E)_{\theta_1, p} \dot{+}(E(A), E)_{\theta_2,p}$ is invertible. From condition (4), by Yakubov and Yakubov [16, Theorem 5.2.1/1], it follows that the embedding $W_p^2((0, 1); E(A), E) \subset L_p((0, 1); E)$ is compact. By conditions (5) and (6), for

---

[2]For example, any $\mathscr{R}$-sectorial operator in E with the $\mathscr{R}$-angle $\phi_A^{\mathscr{R}} < \pi - \varphi$ satisfies this condition.

any $\delta_1 > 0$ and $u \in W_p^2((0, 1); E(A), E)$, we have

$$\|L_1 u\|_{L_p((0,1);E)\dotplus(E(A),E)_{\theta_1,p}\dotplus(E(A),E)_{\theta_2,p}} \leq \|A_1 u\|_{L_p((0,1);E)} + \|T_1 u\|_{(E(A),E)_{\theta_1,p}}$$

$$+ \|T_2 u\|_{(E(A),E)_{\theta_2,p}} \leq \delta_1 \|u\|_{W_p^2((0,1);E(A),E)} + C(\delta_1)\|u\|_{L_p((0,1);E)}.$$

Hence, by Yakubov and Yakubov [16, Lemma 1.2.7/2], the operator $\mathbb{L}_1$ from $W_p^2((0, 1); E(A), E)$ into $L_p((0, 1); E)\dotplus(E(A), E)_{\theta_1,p}\dotplus(E(A), E)_{\theta_2,p}$ is compact. Applying [11, Sect. 14, Theorem 14.1] to the operator $\mathbb{L}(\lambda_0)$, we complete the proof.   □

In order to prove our main isomorphism theorem, which, in particular, implies maximal L_p-regularity, we need to define two more Banach spaces.

Let $s > 0$ and let E, E_1 be Banach spaces. Denote, for $0 < s \leq 1$, $p \in (1, \infty)$, the Banach space

$$B_p^s\big((0, 1); (E_1, E)_{1-\frac{s}{2},2}, E\big) := \big(W_p^2((0, 1); E_1, E), L_p((0, 1); E)\big)_{1-\frac{s}{2},2} \tag{20.41}$$

and, for $1 < s < 2$, the Banach space

$$B_p^s\big((0, 1); (E_1, E)_{1-\frac{s}{2},2}, E\big) := \big(W_p^2((0, 1); E_1, E), B_p^1((0, 1); (E_1, E)_{\frac{1}{2},2}, E)\big)_{2-s,2}. \tag{20.42}$$

Consider the problem

$$L_0(\lambda)u := \lambda u(x) - u''(x) + Au(x) = f(x), \quad x \in (0, 1), \tag{20.43}$$

$$L_1(\lambda)u := \lambda u(0) - \alpha u'(0) + T_1 u = f_1,$$
$$L_2(\lambda)u := \lambda u(1) + \beta u'(1) + T_2 u = f_2. \tag{20.44}$$

Theorem 20.5 *Let the following conditions be satisfied:*

1. *an operator A is closed, densely defined and invertible in the complex UMD Banach space E and $\mathscr{R}\{\lambda R(\lambda, A) : |\arg \lambda| \geq \pi - \varphi\} < \infty$, for some $\varphi \in (0, \pi)^3$;*
2. *$\alpha = 0$ or $|\arg \alpha| \leq \frac{\pi - \varphi}{2}$ and $\beta = 0$ or $|\arg \beta| \leq \frac{\pi - \varphi}{2}$;*
3. *$m_1 := \begin{cases} 1, & \alpha \neq 0, \\ 0, & \alpha = 0 \end{cases}$, $m_2 := \begin{cases} 1, & \beta \neq 0, \\ 0, & \beta = 0 \end{cases}$; $\theta_k := \frac{m_k}{2} + \frac{1}{2p}$, $p \in (1, \infty)$, $k = 1, 2$;*
4. *the embedding $E(A) \subseteq E$ is compact;*

---

[3]In fact, this condition is equivalent to that A is an invertible $\mathscr{R}$-sectorial operator in E with the $\mathscr{R}$-angle $\phi_A^{\mathscr{R}} < \pi - \varphi$.

5. *if* $m_k = 0$ *then* $T_k = 0$; *if* $m_k = 1$ *then, for any* $\delta > 0$ *and* $u \in$ $W_p^2((0,1); E(A), E)$,

$$\|T_k u\|_{(E(A),E)_{\theta_k,p}} \leq \delta \|u\|_{W_p^2((0,1);E(A),E)} + C(\delta)\|u\|_{L_p((0,1);E)},$$

$$\|T_k u\|_E \leq \delta \|u\|_{B_p^{1+\frac{1}{p}}((0,1);(E(A),E)_{1-\theta_k,2,E})} + C(\delta)\|u\|_{L_p((0,1);E)}.$$

Then, the operator $\tilde{\mathbb{L}}(\lambda) : u \to \tilde{\mathbb{L}}(\lambda)u := \left(L_0(\lambda)u, L_1(\lambda)u, L_2(\lambda)u\right)$, *for* $|\arg \lambda| \leq \varphi - \epsilon$, $\epsilon > 0$ *is any sufficiently small and* $|\lambda|$ *is sufficiently large, is an isomorphism from* $W_p^2((0,1); E(A), E)$ *onto* $L_p((0,1); E) \dot{+} (E(A), E)_{\theta_1,p} \dot{+}$ $(E(A), E)_{\theta_2,p}$ *and, for these* λ, *the following estimate holds for the solution of the problem* (20.43)–(20.44):

$$|\lambda| \|u\|_{L_p((0,1);E)} + \|u''\|_{L_p((0,1);E)} + \|Au\|_{L_p((0,1);E)}$$

$$\leq C(\varphi, \epsilon) \left[|\lambda|^{\frac{\max\{m_k\}}{2}} \|f\|_{L_p((0,1);E)} \right.$$

$$\left. + \sum_{k=1}^{2} |\lambda|^{-1+m_k} \left(\|f_k\|_{(E(A),E)_{\theta_k,p}} + |\lambda|^{1-\theta_k} \|f_k\|_E \right) \right]. \tag{20.45}$$

Proof Let $u \in W_p^2((0,1); E(A), E)$ be a solution of problem (20.43)–(20.44). Then, $u(x)$ is a solution of the problem $L_0(\lambda)u = f$, $L_{k0}(\lambda)u = f_k - T_k u$, $k = 1, 2$, where $L_{k0}(\lambda)u$, $k = 1, 2$, are determined by the equalities (20.23). Then, by Theorem 20.3, for $|\arg \lambda| \leq \varphi - \epsilon$, $\epsilon > 0$ is any sufficiently small and $|\lambda|$ is sufficiently large, we have

$$|\lambda| \|u\|_{L_p((0,1);E)} + \|u''\|_{L_p((0,1);E)} + \|Au\|_{L_p((0,1);E)}$$

$$\leq C(\varphi, \epsilon) \left[|\lambda|^{\frac{\max\{m_k\}}{2}} \|f\|_{L_p((0,1);E)} \right.$$

$$\left. + \sum_{k=1}^{2} |\lambda|^{-1+m_k} \left(\|f_k - T_k u\|_{(E(A),E)_{\theta_k,p}} + |\lambda|^{1-\theta_k} \|f_k - T_k u\|_E \right) \right].$$

$$\tag{20.46}$$

If $m_k = 0$, $k = 1, 2$, then, by condition (5), $T_k = 0$ and then, from (20.46), we get (20.45). Assume now that at least one $m_k = 1$, i.e., the corresponding $\theta_k = \frac{1}{2} + \frac{1}{2p}$, and show that again (20.46) implies (20.45).

By condition (5), for $\delta > 0$, $u \in W_p^2((0,1); E(A), E)$,

$$\|f_k - T_k u\|_{(E(A),E)_{\theta_k,p}} = \|f_k - T_k u\|_{(E(A),E)_{\frac{1}{2}+\frac{1}{2p},p}} \leq \|f_k\|_{(E(A),E)_{\frac{1}{2}+\frac{1}{2p},p}}$$

$$+ \delta \|u\|_{W_p^2((0,1);E(A),E)} + C(\delta)\|u\|_{L_p((0,1);E)}. \tag{20.47}$$

Obviously,

$$|\lambda|^{1-\theta_k}\|f_k - T_k u\|_E = |\lambda|^{\frac{1}{2}-\frac{1}{2p}}\|f_k - T_k u\|_E \le |\lambda|^{\frac{1}{2}-\frac{1}{2p}}(\|f_k\|_E + \|T_k u\|_E).$$
(20.48)

By virtue of (20.42) and [16, Lemma 1.7.3/8] (take $E_1 = E(A)$ and $s = 1 + \frac{1}{p}$ in (20.42) and $E_0 := W_p^2((0,1); E(A), E)$, $E_1 := B_p^1((0,1); (E(A), E)_{\frac{1}{2},2}, E)$, $\lambda = \mu$, and $\theta = 1 - \frac{1}{p}$ in Yakubov and Yakubov [16, Lemma 1.7.3/8]), we have, for $\mu \in \mathbb{C}$, $u \in W_p^2((0,1); E(A), E)$,

$$|\mu|^{1-\frac{1}{p}}\|u\|_{B_p^{1+\frac{1}{p}}\left((0,1);(E(A),E)_{\frac{1}{2}-\frac{1}{2p},2},E\right)} \le C\Big(|\mu|\|u\|_{B_p^1\left((0,1);(E(A),E)_{\frac{1}{2},2},E\right)}$$
$$+ \|u\|_{W_p^2((0,1);E(A),E)}\Big).$$

Replacing $\lambda = \mu^2$, as a result, we have, for $\lambda \in \mathbb{C}$, $u \in W_p^2((0,1); E(A), E)$,

$$|\lambda|^{\frac{1}{2}-\frac{1}{2p}}\|u\|_{B_p^{1+\frac{1}{p}}\left((0,1);(E(A),E)_{\frac{1}{2}-\frac{1}{2p},2},E\right)} \le C\Big(|\lambda|^{\frac{1}{2}}\|u\|_{B_p^1\left((0,1);(E(A),E)_{\frac{1}{2},2},E\right)}$$
$$+ \|u\|_{W_p^2((0,1);E(A),E)}\Big).$$
(20.49)

Further, in (20.41) we take $E_1 = E(A)$, $s = 1$. Then,

$$B_p^1((0,1); (E(A), E)_{\frac{1}{2},2}, E) := \big(W_p^2((0,1); E(A), E), L_p((0,1); E)\big)_{\frac{1}{2},2}$$

and, by Yakubov and Yakubov [16, Lemma 1.7.3/8] (take there $\theta = \frac{1}{2}$), for $\lambda \in \mathbb{C}$, $u \in W_p^2((0,1); E(A), E)$,

$$|\lambda|^{\frac{1}{2}}\|u\|_{B_p^1\left((0,1);(E(A),E)_{\frac{1}{2},2},E\right)} \le C\big(\|u\|_{W_p^2((0,1);E(A),E)} + |\lambda|\|u\|_{L_p((0,1);E)}\big).$$
(20.50)

Then, taking into account (20.50) into (20.49), we get

$$|\lambda|^{\frac{1}{2}-\frac{1}{2p}}\|u\|_{B_p^{1+\frac{1}{p}}\left((0,1);(E(A),E)_{\frac{1}{2}-\frac{1}{2p},2},E\right)} \le C\big(\|u\|_{W_p^2((0,1);E(A),E)} + |\lambda|\|u\|_{L_p((0,1);E)}\big).$$
(20.51)

From condition (5) and the inequality (20.51), we have

$$|\lambda|^{\frac{1}{2}-\frac{1}{2p}}\|T_k u\|_E \le C\delta\big(\|u\|_{W_p^2((0,1);E(A),E)} + |\lambda|\|u\|_{L_p((0,1);E)}\big)$$
$$+ C(\delta)|\lambda|^{\frac{1}{2}-\frac{1}{2p}}\|u\|_{L_p((0,1);E)}.$$
(20.52)

Then, taking into account (20.52) into (20.48), we have

$$|\lambda|^{1-\theta_k}\|f_k - T_k u\|_E \leq |\lambda|^{\frac{1}{2}-\frac{1}{2p}}\|f_k\|_E + C\delta\big(\|u\|_{W_p^2((0,1);E(A),E)} + |\lambda|\|u\|_{L_p((0,1);E)}\big)$$

$$+ C(\delta)|\lambda|^{\frac{1}{2}-\frac{1}{2p}}\|u\|_{L_p((0,1);E)}. \tag{20.53}$$

Hence, by (20.47) and (20.53), from (20.46), we have (remind that at least one $m_k = 1$, i.e., the corresponding $\theta_k = \frac{1}{2} + \frac{1}{2p}$)

$$\Big(1 - C\delta - C|\lambda|^{-(\frac{1}{2}+\frac{1}{2p})} - C(\delta)|\lambda|^{-(\frac{1}{2}+\frac{1}{2p})}\Big)\Big(|\lambda|\|u\|_{L_p((0,1);E)} + \|u''\|_{L_p((0,1);E)}$$

$$+ \|Au\|_{L_p((0,1);E)}\Big) \leq C\Big[|\lambda|^{\frac{1}{2}}\|f\|_{L_p((0,1);E)}$$

$$+ \sum_{k=1}^{2}|\lambda|^{-1+m_k}\big(\|f_k\|_{(E(A),E)_{\theta_k,p}} + |\lambda|^{1-\theta_k}\|f_k\|_E\big)\Big]. \tag{20.54}$$

Now, first, choose δ_0 such that $C\delta_0 < 1$. Furthermore, choose $|\lambda|$ so that $C\delta_0 + C|\lambda|^{-(\frac{1}{2}+\frac{1}{2p})} + C(\delta_0)|\lambda|^{-(\frac{1}{2}+\frac{1}{2p})} < 1$. Then, from (20.54) we get (20.45).

Consequently, for $|\arg\lambda| \leq \varphi - \epsilon$, $\epsilon > 0$ is any sufficiently small and $|\lambda|$ is sufficiently large, the solution of problem (20.43)–(20.44) in $W_p^2((0,1); E(A), E)$ is unique. By Theorem 20.4, for each such λ, the operator $\tilde{\mathbb{L}}(\lambda)$ from $W_p^2((0,1); E(A), E)$ into $L_p((0,1); E)\dotplus(E(A), E)_{\theta_1,p}\dotplus(E(A), E)_{\theta_2,p}$ is a Fredholm operator. Then, the desired isomorphism follows from the uniqueness and the Fredholm property. $\qquad\square$

20.6 Application of Obtained Abstract Results to Elliptic Boundary Value Problems in Non-smooth Domains

It is well-known that usually boundary value problems for elliptic equations in non-smooth domains do not have the maximal regularity for a solution (see, e.g., [10, Preface]). We have succeeded to find a class of elliptic boundary value problems with the spectral parameter both in the equation and boundary conditions, in cylindrical domains (i.e., in non-smooth domains), which has the maximal regularity for a solution, i.e., the solution belongs to W_p^2-Sobolev spaces for any $1 < p < \infty$. Note, that our considered equations (see below (20.55)) do not contain mixed derivatives on x-variable and y-variable.

Let $\Omega := (0,1) \times G$, where $G \subset \mathbb{R}^r$, $r \geq 2$, be a bounded open domain with an $(r-1)$-dimensional boundary ∂G which locally admits rectification. We will consider, in this section, the Besov space

$$B_{q,p}^s(G) := (W_q^{s_0}(G), W_q^{s_1}(G))_{\theta,p},$$

where $0 \le s_0, s_1$ are integers, $0 < \theta < 1$, $1 < p < \infty$, $1 < q < \infty$ and $s = (1 - \theta)s_0 + \theta s_1$, and the space

$$W_{p,q}^{l,s}(\Omega) := W_p^l\big((0, 1); W_q^s(G), L_q(G)\big),$$

where $0 \le l, s$ are integers, $1 < p < \infty$, $1 < q < \infty$. If $l = s$ then $W_{p,q}^l(\Omega) := W_{p,q}^{l,l}(\Omega)$. Finally, $L_{p,q}(\Omega) := W_{p,q}^{0,0}(\Omega) = L_p\big((0, 1); L_q(G)\big)$.

In the cylindrical domain Ω, let us consider a boundary value problem for second order elliptic equations with the spectral parameter in the equation and in a part of boundary conditions, namely,

$$L_0(\lambda, D_x, D_y)u := \lambda u(x, y) - D_x^2 u(x, y) - \sum_{s,j=1}^r a_{sj}(y) D_s D_j u(x, y) + \lambda_0 u(x, y)$$

$$= f(x, y), \quad (x, y) \in \Omega, \tag{20.55}$$

$$L_1(\lambda, D_x)u := \lambda u(0, y) - \alpha D_x u(0, y) + (T_1 u)(y) = f_1(y), \quad y \in G,$$

$$L_2(\lambda, D_x)u := \lambda u(1, y) + \beta D_x u(1, y) + (T_2 u)(y) = f_2(y), \quad y \in G, \tag{20.56}$$

$$L(D_y)u := \sum_{j=1}^r c_j(y') D_j u(x, y') + c_0(y')u(x, y') = 0, \quad (x, y') \in (0, 1) \times \partial G, \tag{20.57}$$

where $D_x := \frac{\partial}{\partial x}$, $D_j := -i\frac{\partial}{\partial y_j}$, $D_y := (D_1, \dots, D_r)$, α, β are complex numbers, $y := (y_1, \dots, y_r)$, T_k are, generally speaking, unbounded operators from $L_p\big((0, 1); L_q(G)\big)$ into $L_q(G)$, $1 < p, q < \infty$. Let m be the order of the differential expression $L(D_y)$.

We would like to give an exact description of the corresponding interpolation spaces $(E(A), E)_{\theta_k, p}$, so the relation between m and $2 - m_k - \frac{1}{p} - \frac{1}{q}$ becomes important (see the proof of the below Theorem 20.6). Note that in our case, the equality $m = 2 - m_k - \frac{1}{p} - \frac{1}{q}$ is possible only if $m_k = 0, m = 1$ or $m_k = 1, m = 0$. In that case, we will take $p = q = 2$ in order to give the exact description of the corresponding interpolation spaces. Two other possible situations $m_k = m = 0$ or $m_k = m = 1$ do not imply the above equality. Introduce the last notation of

$$B_{q,p,*}^{2-m_k-\frac{1}{p}}(G) := \begin{cases} B_{q,p}^{2-m_k-\frac{1}{p}}(G; L(D_y)u = 0) & \text{if } m < 2 - m_k - \frac{1}{p} - \frac{1}{q}, \\ B_{q,p}^{2-m_k-\frac{1}{p}}(G) & \text{if } m > 2 - m_k - \frac{1}{p} - \frac{1}{q}, \\ B_{p,p}^{2-m_k-\frac{1}{p}}(G; L(D_y)u \in \tilde{B}_{p,p}^{\frac{1}{p}}(G)) & \text{if } m = 2 - m_k - \frac{1}{p} - \frac{1}{q}, \end{cases} \tag{20.58}$$

with $\tilde{B}_{p,p}^{\frac{1}{p}}(G) := \{u \mid u \in B_{p,p}^{\frac{1}{p}}(\mathbb{R}^r), \operatorname{supp}(u) \subseteq \overline{G}\}$. By the above, in the last case we take $p = q = 2$.

Theorem 20.6 *Let the following conditions be satisfied:*

1. *(Smoothness condition)* $|a_{sj}(y) - a_{sj}(z)| \le C|y - z|^{\gamma}$, *for some* $C > 0$ *and* $\gamma \in (0, 1)$, *and* $\forall y, z \in \overline{G}$; $c_j, c_0 \in C^{2-m}(\partial G)$; $\partial G \in C^2$;
2. *(Ellipticity condition for the operator A defined in the below proof) for* $y \in \overline{G}$, $\sigma \in \mathbb{R}^r$, $|\arg \lambda| \ge \pi - \varphi$, *for some* $\varphi \in (0, \pi)$, $|\sigma| + |\lambda| \ne 0$, *one has* $\lambda + \sum_{s,j=1}^{r} a_{sj}(y)\sigma_s\sigma_j \ne 0$;
3. *(Lopatinskii-Shapiro condition for the operator A)* y' *is any point on* ∂G, *the vector* σ' *is tangent and* σ *is a normal vector to* ∂G *at the point* $y' \in \partial G$. *Consider the following ordinary differential problem, for* $|\arg \lambda| \ge \pi - \varphi$,

$$\left[\lambda + \sum_{s,j=1}^{r} a_{sj}(y')\left(\sigma_s' - i\sigma_s \frac{d}{dt}\right)\left(\sigma_j' - i\sigma_j \frac{d}{dt}\right)\right]u(t) = 0, \quad t > 0, \tag{20.59}$$

$$\sum_{j=1}^{r} c_j(y')\left(\sigma_j' - i\sigma_j \frac{d}{dt}\right)u(t)\Big|_{t=0} = h, \quad \text{for } m = 1, \tag{20.60}$$

$$u(0) = h, \quad \text{for } m = 0; \tag{20.61}$$

it is required that, for $m = 1$, *the problem (20.59), (20.60) (for* $m = 0$, *the problem (20.59), (20.61)) has one and only one solution, including all its derivatives, tending to zero as* $t \to \infty$, *for any numbers* $h \in \mathbb{C}$;[4]

4. $\alpha = 0$ *or* $|\arg \alpha| \le \frac{\pi - \varphi}{2}$ *and* $\beta = 0$ *or* $|\arg \beta| \le \frac{\pi - \varphi}{2}$;

5. $m_1 := \begin{cases} 1, & \alpha \ne 0, \\ 0, & \alpha = 0 \end{cases}$, $m_2 := \begin{cases} 1, & \beta \ne 0, \\ 0, & \beta = 0 \end{cases}$; $\theta_k := \frac{m_k}{2} + \frac{1}{2p}$, $p \in (1, \infty)$, $k = 1, 2$;

6. $q \in (1, \infty)$; *if* $\frac{1}{p} + \frac{1}{q} = 1$ *and, for at least one* k, $m_k + m = 1$ *then* $p = q = 2$;

7. *if* $m_k = 0$ *then* $T_k = 0$; *if* $m_k = 1$ *then, for any* $\delta > 0$ *and* $u \in W_{p,q}^2(\Omega; L(D_y)u = 0)$,

$$\|T_k u\|_{B_{q,p}^{1-\frac{1}{p}}(G)} \le \delta \|u\|_{W_{p,q}^2(\Omega)} + C(\delta)\|u\|_{L_{p,q}(\Omega)},$$

$$\|T_k u\|_{L_q(G)} \le \delta \|u\|_{B_p^{1+\frac{1}{p}}\left((0,1); B_{q,2}^{1+\frac{1}{p}}(G), L_q(G)\right)} + C(\delta)\|u\|_{L_{p,q}(\Omega)}, \quad q \in (1, \infty);$$

Then, there exists $\lambda_0 > 0$ *such that the operator*

$$\tilde{\mathbb{L}}(\lambda, D_x, D_y) : u \to \tilde{\mathbb{L}}(\lambda, D_x, D_y)u := \left(L_0(\lambda, D_x, D_y)u, L_1(\lambda, D_x)u, L_2(\lambda, D_x)u\right),$$

---

[4]In the case $m = 0$, the boundary condition (20.57) is transformed into the Dirichlet condition $u(x, y') = 0$.

for $|\arg \lambda| \le \varphi - \epsilon$, $\epsilon > 0$ *is any sufficiently small and* $|\lambda|$ *is sufficiently large, is an isomorphism from* $W_{p,q}^2(\Omega; L(D_y)u = 0)$ *onto* $L_{p,q}(\Omega) \overset{2}{\underset{k=1}{\dotplus}} B_{q,p,*}^{2-m_k-\frac{1}{p}}(G)$ *and, for these* λ, *the following estimate holds for the solution of the problem* (20.55)–(20.57):

$$|\lambda|\|u\|_{L_{p,q}(\Omega)} + \|u\|_{W_{p,q}^2(\Omega)} \le C(\varphi,\epsilon)\Big[|\lambda|^{\frac{\max\{m_k\}}{2}}\|f\|_{L_{p,q}(\Omega)}$$

$$+ \sum_{k=1}^2 |\lambda|^{-1+m_k}\Big(\|f_k\|_{B_{q,p}^{2-m_k-\frac{1}{p}}(G)} + |\lambda|^{1-\frac{m_k}{2}-\frac{1}{2p}}\|f_k\|_{L_q(G)}\Big)\Big].$$

$$(20.62)$$

Proof Let us denote $E := L_q(G)$ and consider an operator A which is defined by the equalities

$$D(A) := W_q^2(G; L(D_y)u = 0), \quad Au := -\sum_{s,j=1}^r a_{sj}(y)D_s D_j u(y) + \lambda_0 u(y).$$

Then, problem (20.55)–(20.57) can be rewritten in the form

$$L_0(\lambda)u := \lambda u(x) - u''(x) + Au(x) = f(x), \quad x \in (0,1),$$
$$L_1(\lambda)u := \lambda u(0) - \alpha u'(0) + T_1 u = f_1, \quad\quad\quad (20.63)$$
$$L_2(\lambda)u := \lambda u(1) + \beta u'(1) + T_2 u = f_2,$$

where $u(x) := u(x,\cdot)$, $f(x) := f(x,\cdot)$ are functions with values in the Banach space $E = L_q(G)$ and $f_k := f_k(\cdot)$. Let us apply Theorem 20.5 to problem (20.63). In view of conditions (1)–(3), by Denk et al. [2, Theorem 8.2], there exists $\lambda_0 > 0$ such that A is an $\mathscr{R}$-sectorial operator in E with the $\mathscr{R}$-angle $\phi_A^{\mathscr{R}} < \pi - \varphi$. So, conditions (1)–(3) of Theorem 20.5 are already true. By virtue of, e.g., [14, Theorem 3.2.5], the embedding $W_q^2(G) \subset L_q(G)$ is compact. Consequently, condition (4) of Theorem 20.5 is fulfilled, too. Let us check the last condition (5) of Theorem 20.5.

First, always, for any $0 < \theta < 1$,

$$(E(A), E)_{\theta,p} = \big(W_q^2(G; L(D_y)u = 0), L_q(G)\big)_{\theta,p}$$

$$\subset \big(W_q^2(G), L_q(G)\big)_{\theta,p} = B_{q,p}^{2(1-\theta)}(G).$$

On the other side, by virtue of [14, Theorem 4.3.3] (see also [16, Theorem 1.7.4/6]), if $m < 2(1 - \theta) - \frac{1}{q}$,

$$(E(A), E)_{\theta,p} = \big(W_q^2(G; L(D_y)u = 0), L_q(G)\big)_{\theta,p} = B_{q,p}^{2(1-\theta)}(G; L(D_y)u = 0)$$

and, if $m > 2(1 - \theta) - \frac{1}{q}$,

$$(E(A), E)_{\theta,p} = \left(W_q^2(G; L(D_y)u = 0), L_q(G)\right)_{\theta,p} = B_{q,p}^{2(1-\theta)}(G).$$

If $m = 2(1 - \theta) - \frac{1}{q}$ and $p = q$ then, by Triebel [14, Theorem 4.3.3],

$$(E(A), E)_{\theta,p} = \left(W_p^2(G; L(D_y)u = 0), L_p(G)\right)_{\theta,p}$$

$$= B_{p,p}^{2(1-\theta)}\left(G; L(D_y)u \in \tilde{B}_{p,p}^{\frac{1}{p}}(G)\right).$$

Therefore, from condition (7), condition (5) of Theorem 20.5 is satisfied. The last calculations explain us also (20.58). □

Let us give some examples of the operators T_k which satisfy condition (7) of Theorem 20.6:

1. $T_k u = \sum\limits_{j=1}^{M_k} \delta_{jk} u(x_{jk}, y)$ with $\delta_{jk} \in \mathbb{C}$, $x_{jk} \in [0, 1]$, $k = 1, 2$ (and $y \in G$);

2. $T_k u = \int_G \int_0^1 \sum_{m+|\ell| \leq 1} T_{km\ell}(x, y, z) \frac{\partial^{m+|\ell|} u(x,z)}{\partial x^m \partial z^\ell} dx dz$ with $k = 1, 2$, $z = (z_1, \ldots, z_r)$, $\ell = (\ell_1, \ldots, \ell_r)$, $|\ell| = \sum\limits_{i=1}^{r} \ell_i$, where all functions $T_{km\ell}(x, y, z) \in L_2((0, 1) \times G \times G)$, $T_{km\ell}(x, y, z)$ are continuously differentiable with respect to y and all the mentioned derivatives also belong to $L_2((0, 1) \times G \times G)$.

It was shown in Aliev and Yakubov [1] that the above operators T_k satisfy condition (7) of Theorem 20.6, for $p = q = 2$.

20.7 Abstract Parabolic Initial Boundary Value Problems with Time Differentiation in Abstract Boundary Conditions

Let X be a Banach space and let A be a linear, closed operator in X. Consider Banach spaces

1. $C_\mu(I; X) := \left\{ f \mid f \in C(I; X), \|f\|_{C_\mu(I;X)} := \sup\limits_{t \in I} \|t^\mu f(t)\|_X < \infty \right\}$, $\mu \geq 0$;

2.

$$C_\mu^\gamma(I; X) := \left\{ f \mid f \in C(I; X), \|f\|_{C_\mu^\gamma(I;X)} := \sup\limits_{t \in I} \|t^\mu f(t)\|_X \right.$$

$$\left. + \sup\limits_{\substack{t < t+h \\ t, t+h \in I}} \|f(t + h) - f(t)\|_X h^{-\gamma} t^\mu < \infty \right\}, \quad \gamma \in (0, 1], \mu \geq 0,$$

where I denotes an interval of the real axis $\mathbb{R}$.

An abstract interpretation of initial boundary value problems for parabolic equations such that some of the boundary value conditions contain the differentiation on the time t is different from those problems which do not contain the differentiation on t in boundary conditions. Let us derive such an abstract interpretation.

Let X and X^k, $k = 1, \ldots, s$, be Banach spaces. Consider the Cauchy problem for a system of differential-operator equations

$$L_0(t, D_t)u := u'(t) + Bu(t) = f(t), \tag{20.64}$$

$$L_k(t, D_t)u := (A_{k0}u(t))' + A_{k1}u(t) = f_k(t), \quad k = 1, \ldots, s, \tag{20.65}$$

$$u(0) = u_0, \tag{20.66}$$

where $t \in [0, T]$; B is an operator in X; A_{k0} and A_{k1} are operators from X into X^k; $f(t)$ from $[0, T]$ into X and $f_k(t)$ from $[0, T]$ into X^k are given functions; $u(t)$ from $[0, T]$ into X is an unknown function. Note that operators B, A_{k0}, and A_{k1} are, generally speaking, unbounded; $D_t := \frac{d}{dt}$. Remind, $X(B)$ denotes the domain $D(B)$ with the graph norm.

Consider a system of characteristic operator pencils corresponding to the system of equations (20.64)–(20.65)

$$L_0(\lambda) := \lambda I + B,$$

$$L_k(\lambda) := \lambda A_{k0} + A_{k1}, \quad k = 1, \ldots, s,$$

where λ is a complex number. The following theorem belongs to Yakubov and Yakubov. The proof of the theorem is the same of that [16, Theorem 7.2.10/1]. One should only to use [16, Theorem 7.2.2/1] instead of [16, Theorem 7.2.6/4].

Theorem 20.7 *Let the following conditions be satisfied:*

1. *the operator B is closed and densely defined in the Banach space X;*
2. *the operators A_{k0} and A_{k1} are bounded from $X(B)$ into X^k, $k = 1, \ldots, s$;*
3. *for some $\eta \in (0, 1]$, $\theta > 0$, all numbers λ from the sector $|\arg \lambda| \leq \frac{\pi}{2} + \theta$ and with sufficiently large moduli are regular points for the operator pencil $\tilde{L}(\lambda) : u \to \tilde{L}(\lambda)u := \left(L_0(\lambda)u, L_1(\lambda)u, \ldots, L_s(\lambda)u\right)$, which acts boundedly from $X(B)$ onto $X \dotplus X^1 \dotplus \cdots \dotplus X^s$, and, for $|\arg \lambda| \leq \frac{\pi}{2} + \theta$, $|\lambda| \to \infty$,*

$$\|\tilde{L}(\lambda)^{-1}\|_{B(X \overset{s}{\underset{k=1}{\dotplus}} X^k, X)} \leq C |\lambda|^{-\eta},$$

$$\|A_{k0}\tilde{L}(\lambda)^{-1}\|_{B(X \overset{s}{\underset{k=1}{\dotplus}} X^k, X^k)} \leq C |\lambda|^{-\eta}, \quad k = 1, \ldots, s;$$

4. *$f \in C_\mu^\gamma((0, T]; X)$, $f_k \in C_\mu^\gamma((0, T]; X^k)$, for some $\gamma \in (1 - \eta, 1]$, $\mu \in [0, \eta)$;*
5. *$u_0 \in X(B)$.*

Then, there exists a unique solution $u(t)$ of problem (20.64)–(20.66) such that the function $t \to (u(t), A_{10}u(t), \ldots, A_{s0}u(t))$ from $(0, T)$ into $X \dotplus X^1 \dotplus \cdots \dotplus X^s$ is continuously differentiable and from $[0, T]$ into $X(B) \dotplus X^1 \dotplus \cdots \dotplus X^s$ is continuous, and, for $t \in (0, T]$, the following uniform on $t \in (0, T]$ estimates hold

$$\|u(t)\|_X + \sum_{k=1}^{s} \|A_{k0}u(t)\|_{X^k} \leq C\Big[\|Au_0\|_X + \|u_0\|_X$$

$$+ \|f\|_{C_\mu((0,t];X)} + \sum_{k=1}^{s} \|f_k\|_{C_\mu((0,t];X^k)}\Big],$$

$$\|u'(t)\|_X + \sum_{k=1}^{s} \|(A_{k0}u(t))'\|_{X^k} + \|Au(t)\|_X \leq C\Big[t^{\eta-1}\big(\|Au_0\|_X + \|u_0\|_X\big)$$

$$+ t^{\eta-\mu-1}\big(\|f\|_{C_\mu^\gamma((0,t];X)} + \sum_{k=1}^{s} \|f_k\|_{C_\mu^\gamma((0,t];X^k)}\big)\Big].$$

Consider now, in a Banach space E, the following abstract initial boundary value problem for a parabolic differential-operator equation

$$\frac{\partial u(t, x)}{\partial t} - \frac{\partial^2 u(t, x)}{\partial x^2} + Au(t, x) = f(t, x), \quad (t, x) \in (0, T) \times (0, 1), \quad (20.67)$$

$$\frac{\partial [u(t, 0)]}{\partial t} - \alpha \frac{\partial u(t, 0)}{\partial x} + T_1 u(t, \cdot) = f_1(t), \quad t \in (0, T),$$

$$\frac{\partial [u(t, 1)]}{\partial t} + \beta \frac{\partial u(t, 1)}{\partial x} + T_2 u(t, \cdot) = f_2(t), \quad t \in (0, T), \quad (20.68)$$

$$u(0, x) = u_0(x), \quad x \in (0, 1). \quad (20.69)$$

Theorem 20.8 *Let the following conditions be satisfied:*

1. *an operator A is closed, densely defined and invertible in the complex UMD Banach space E and $\mathcal{R}\{\lambda R(\lambda, A) : |\arg \lambda| \geq \frac{\pi}{2} - \tau\} < \infty$, for some $\tau \in (0, \frac{\pi}{2})$;*[5]
2. *$|\arg \alpha| \leq \frac{\frac{\pi}{2}-\tau}{2}$ and $|\arg \beta| \leq \frac{\frac{\pi}{2}-\tau}{2}$;*
3. *$m_1 := \begin{cases} 1, & \alpha \neq 0, \\ 0, & \alpha = 0, \end{cases}$, $m_2 := \begin{cases} 1, & \beta \neq 0, \\ 0, & \beta = 0, \end{cases}$; $\theta_k := \frac{m_k}{2} + \frac{1}{2p}$, $p \in (1, \infty)$, $k = 1, 2$;*
4. *the embedding $E(A) \subseteq E$ is compact;*

---

[5]In fact, this condition is equivalent to that A is an invertible $\mathcal{R}$-sectorial operator in E with the $\mathcal{R}$-angle $\phi_A^{\mathcal{R}} < \frac{\pi}{2} - \tau$.

5. if $m_k = 0$ then $T_k = 0$; if $m_k = 1$ then, for any $\delta > 0$ and $u \in W_p^2((0,1); E(A), E)$,

$$\|T_k u\|_{(E(A),E)_{\theta_k,p}} \leq \delta \|u\|_{W_p^2((0,1);E(A),E)} + C(\delta)\|u\|_{L_p((0,1);E)},$$

$$\|T_k u\|_E \leq \delta \|u\|_{B_p^{1+\frac{1}{p}}\left((0,1);(E(A),E)_{1-\theta_k,2},E\right)} + C(\delta)\|u\|_{L_p((0,1);E)};$$

6. $f \in C_\mu^\gamma((0,T]; L_p((0,1); E))$, $f_k \in C_\mu^\gamma((0,T]; (E(A), E)_{\theta_k,p})$, for some

$\gamma \in (\frac{\max\{m_k\}}{2}, 1]$, $\mu \in [0, 1 - \frac{\max\{m_k\}}{2})$;
7. $u_0 \in W_p^2((0,1); E(A), E)$.

Then, there exists a unique solution $u(t,x)$ of problem (20.67)–(20.69) such that the function $t \to (u(t,x), u(t,0), u(t,1))$ from $(0,T)$ into $L_p((0,1); E) \dotplus (E(A), E)_{\theta_1,p} \dotplus (E(A), E)_{\theta_2,p}$ is continuously differentiable and from $[0,T]$ into $W_p^2((0,1); E(A), E) \dotplus (E(A), E)_{\theta_1,p} \dotplus (E(A), E)_{\theta_2,p}$ is continuous, and, for $t \in (0,T]$, the following uniform on $t \in (0,T]$ estimates hold

$$\|u(t,\cdot)\|_{L_p((0,1);E)} + \|u(t,0)\|_{(E(A),E)_{\theta_1,p}} + \|u(t,1)\|_{(E(A),E)_{\theta_2,p}}$$

$$\leq C\Big[\|Au_0(\cdot)\|_{L_p((0,1);E)} + \|u_0(\cdot)\|_{L_p((0,1);E)}$$

$$+ \|f\|_{C_\mu((0,t];L_p((0,1);E))} + \sum_{k=1}^2 \|f_k\|_{C_\mu((0,t];(E(A),E)_{\theta_k,p})}\Big],$$

$$\Big\|\frac{\partial u(t,\cdot)}{\partial t}\Big\|_{L_p((0,1);E)} + \Big\|\frac{\partial}{\partial t}[u(t,0)]\Big\|_{(E(A),E)_{\theta_1,p}} + \Big\|\frac{\partial}{\partial t}[u(t,1)]\Big\|_{(E(A),E)_{\theta_2,p}}$$

$$+ \|Au(t,\cdot)\|_{L_p((0,1);E)} \leq C\Big[t^{-\frac{\max\{m_k\}}{2}}\big(\|Au_0(\cdot)\|_{L_p((0,1);E)}$$

$$+ \|u_0(\cdot)\|_{L_p((0,1);E)}\big) + t^{-\frac{\max\{m_k\}}{2}-\mu}\big(\|f\|_{C_\mu^\gamma((0,t];L_p((0,1);E))}$$

$$+ \sum_{k=1}^2 \|f_k\|_{C_\mu^\gamma((0,t];(E(A),E)_{\theta_k,p})}\big)\Big].$$

Proof Denote the Banach spaces $X := L_p((0,1); E)$, $X(B) := W_p^2((0,1); E(A), E)$, $X^k := (E(A), E)_{\theta_k,p}$, $k = 1, 2$. Consider, in the space X, the operator B defined by the equalities

$$D(B) := W_p^2((0,1); E(A), E),$$

$$Bu := -u''(x) + Au(x),$$

and introduce $A_{10}u := u(0)$, $A_{11}u := -\alpha u'(0) + T_1 u$, $A_{20}u := u(1)$, $A_{21}u := \beta u'(1) + T_2 u$. Then, problem (20.67)–(20.69) in E can be rewritten in the form (20.64)–(20.66) in X with $s = 2$, to which we want to apply Theorem 20.7.

From conditions (1) and (5), and the trace inequality (see, e.g., (20.31)), we get conditions (1) and (2) of Theorem 20.7. Condition (7) implies condition (5) of Theorem 20.7, and condition (6) implies condition (4) of Theorem 20.7 with $\eta = 1 - \frac{\max\{m_k\}}{2}$. The only thing remains is to prove the main condition (3) of Theorem 20.7 with the same $\eta = 1 - \frac{\max\{m_k\}}{2} \in (0, 1]$ and some $\theta > 0$. For this, we are going to use Theorem 20.5.

Choose $0 < \theta < \tau$. Then, by our conditions (1)–(5), from the estimate (20.45), for $|\arg \lambda| \leq \frac{\pi}{2} + \theta$, $|\lambda| \to \infty$, we get

$$|\lambda| \|u\|_{L_p((0,1);E)} \leq C\left(|\lambda|^{\frac{\max\{m_k\}}{2}} \|f\|_{L_p((0,1);E)} + \sum_{k=1}^{2} |\lambda|^{-1+m_k} |\lambda|^{1-\theta_k} \|f_k\|_{(E(A),E)_{\theta_k,p}}\right)$$

$$= C\left(|\lambda|^{\frac{\max\{m_k\}}{2}} \|f\|_{L_p((0,1);E)} + \sum_{k=1}^{2} |\lambda|^{\frac{m_k}{2}-\frac{1}{2p}} \|f_k\|_{(E(A),E)_{\theta_k,p}}\right)$$

$$\leq C|\lambda|^{\frac{\max\{m_k\}}{2}} \left(\|f\|_{L_p((0,1);E)} + \sum_{k=1}^{2} \|f_k\|_{(E(A),E)_{\theta_k,p}}\right),$$

i.e., $\|u\|_X \leq C|\lambda|^{-\left(1-\frac{\max\{m_k\}}{2}\right)} \|(f, f_1, f_2)\|_{X+X^1+X^2}$, which gives us the first inequality in condition (3) of Theorem 20.7 with $\eta = 1 - \frac{\max\{m_k\}}{2} \in (0, 1]$.

On the other hand, from (20.44), we get

$$\|u(0)\|_{(E(A),E)_{\theta_1,p}} \leq C|\lambda|^{-1}\left(\|u'(0)\|_{(E(A),E)_{\theta_1,p}} + \|T_1 u\|_{(E(A),E)_{\theta_1,p}} + \|f_1\|_{(E(A),E)_{\theta_1,p}}\right),$$

$$\|u(1)\|_{(E(A),E)_{\theta_2,p}} \leq C|\lambda|^{-1}\left(\|u'(1)\|_{(E(A),E)_{\theta_2,p}} + \|T_2 u\|_{(E(A),E)_{\theta_2,p}} + \|f_2\|_{(E(A),E)_{\theta_2,p}}\right).$$

Then, using the trace inequality (see, e.g., (20.31)), condition (5), and the estimate (20.45), we conclude

$$\|u(0)\|_{(E(A),E)_{\theta_1,p}} \leq C|\lambda|^{-1}\left(|\lambda|^{\frac{\max\{m_k\}}{2}} \|f\|_{L_p((0,1);E)} + \sum_{k=1}^{2} |\lambda|^{\frac{m_k}{2}-\frac{1}{2p}} \|f_k\|_{(E(A),E)_{\theta_k,p}}\right.$$

$$\left.+ \|f_1\|_{(E(A),E)_{\theta_1,p}}\right) \leq C|\lambda|^{-1}|\lambda|^{\frac{\max\{m_k\}}{2}}\left(\|f\|_{L_p((0,1);E)} + \sum_{k=1}^{2} \|f_k\|_{(E(A),E)_{\theta_k,p}}\right)$$

$$= C|\lambda|^{-\left(1-\frac{\max\{m_k\}}{2}\right)}\left(\|f\|_{L_p((0,1);E)} + \sum_{k=1}^{2} \|f_k\|_{(E(A),E)_{\theta_k,p}}\right),$$

and, similarly,

$$\|u(1)\|_{(E(A),E)_{\theta_2,p}} \le C|\lambda|^{-\left(1-\frac{\max\{m_k\}}{2}\right)}\left(\|f\|_{L_p((0,1);E)} + \sum_{k=1}^{2} \|f_k\|_{(E(A),E)_{\theta_k,p}}\right),$$

i.e., $\|A_{k0}u\|_{X^k} \le C|\lambda|^{-\left(1-\frac{\max\{m_k\}}{2}\right)}\|(f, f_1, f_2)\|_{X+X^1+X^2}, k = 1, 2$, which gives the
second inequality in condition (3) of Theorem 20.7 with $\eta = 1 - \frac{\max\{m_k\}}{2} \in (0, 1]$
and $s = 2$. $\square$

20.8 Application of Obtained Abstract Results to Parabolic Initial Boundary Value Problems with Time Differentiation in Boundary Conditions

Show a relevant application of the previous section. Let $G \subset \mathbb{R}^r$, $r \ge 2$, be a
bounded domain with an $(r-1)$-dimensional boundary ∂G which locally admits
rectification. Consider the following parabolic initial boundary value problem

$$\frac{\partial u(t,x,y)}{\partial t} - \frac{\partial^2 u(t,x,y)}{\partial x^2} - \sum_{s,j=1}^{r} a_{sj}(y)D_s D_j u(t,x,y) + \lambda_0 u(t,x,y)$$

$$= f(t,x,y), \quad (t,x,y) \in (0,T) \times (0,1) \times G, \tag{20.70}$$

$$\frac{\partial[u(t,0,y)]}{\partial t} - \alpha\frac{\partial u(t,0,y)}{\partial x} + (T_1 u(t,\cdot,\cdot))(y) = f_1(t,y), \quad (t,y) \in (0,T) \times G,$$

$$\frac{\partial[u(t,1,y)]}{\partial t} + \beta\frac{\partial u(t,1,y)}{\partial x} + (T_2 u(t,\cdot,\cdot))(y) = f_2(t,y), \quad (t,y) \in (0,T) \times G, \tag{20.71}$$

$$\sum_{j=1}^{r} c_j(y')D_j u(t,x,y') + c_0(y')u(t,x,y') = 0, \quad (t,x,y') \in (0,T) \times (0,1) \times \partial G, \tag{20.72}$$

$$u(0,x,y) = u_0(x,y), \quad (x,y) \in (0,1) \times G, \tag{20.73}$$

where $D_j := -i\frac{\partial}{\partial y_j}$, α, β are complex numbers, $y := (y_1, \ldots, y_r)$, T_k are,
generally speaking, unbounded operators from $L_p((0,1); L_q(G))$ into $L_q(G)$, $1 < p, q < \infty$.

Theorem 20.9 *Let the following assumptions be satisfied:*

1. *conditions of Theorem 20.6 are fulfilled with* $\varphi = \frac{\pi}{2} + \tau$, *for some* $\tau \in (0, \frac{\pi}{2})$;

2. $f \in C_{\mu}^{\gamma}\left((0, T]; L_p\left((0, 1); L_q(G)\right)\right)$, $f_k \in C_{\mu}^{\gamma}\left((0, T]; B_{q,p,*}^{2-m_k-\frac{1}{p}}(G)\right)$, *for some*
   $\gamma \in (\frac{\max\{m_k\}}{2}, 1]$, $\mu \in [0, 1 - \frac{\max\{m_k\}}{2})$, *where* $B_{q,p,*}^{2-m_k-\frac{1}{p}}(G)$ *is defined by (20.58)*;

3. $u_0 \in W_p^2\left((0, 1); W_q^2(G; Lu = 0), L_q(G)\right)$, *where* $Lu := \sum_{j=1}^{r} c_j(y') D_j u(y')$
   $+ c_0(y')u(y')$, $y' \in \partial G$.

Then, there exists $\lambda_0 > 0$ *such that there exists a unique solution* $u(t, x, y)$ *of problem (20.70)–(20.73) such that the function* $t \rightarrow (u(t, x, y), u(t, 0, y), u(t, 1, y))$ *from* $(0, T)$ *into* $L_p\left((0, 1); L_q(G)\right) \dotplus B_{q,p,*}^{2-m_1-\frac{1}{p}}(G) \dotplus B_{q,p,*}^{2-m_2-\frac{1}{p}}(G)$ *is continuously differentiable and from* $[0, T]$ *into* $W_p^2\left((0, 1); W_q^2(G; Lu = 0), L_q(G)\right) \dotplus B_{q,p,*}^{2-m_1-\frac{1}{p}}$
$(G) \dotplus B_{q,p,*}^{2-m_2-\frac{1}{p}}(G)$ *is continuous, and, for* $t \in (0, T]$, *the following uniform on* $t \in (0, T]$ *estimates hold*

$$\|u(t, \cdot, \cdot)\|_{L_p\left((0,1); L_q(G)\right)} + \|u(t, 0, \cdot)\|_{B_{q,p,*}^{2-m_1-\frac{1}{p}}(G)} + \|u(t, 1, \cdot)\|_{B_{q,p,*}^{2-m_2-\frac{1}{p}}(G)}$$

$$\leq C\Big[\|u_0(\cdot, \cdot)\|_{L_p\left((0,1); W_q^2(G)\right)} + \|u_0(\cdot, \cdot)\|_{L_p\left((0,1); L_q(G)\right)}$$

$$+ \|f\|_{C_\mu\left((0,t]; L_p\left((0,1); L_q(G)\right)\right)} + \sum_{k=1}^{2} \|f_k\|_{C_\mu\left((0,t]; B_{q,p,*}^{2-m_k-\frac{1}{p}}(G)\right)}\Big],$$

$$\Big\|\frac{\partial u(t, \cdot, \cdot)}{\partial t}\Big\|_{L_p\left((0,1); L_q(G)\right)} + \Big\|\frac{\partial}{\partial t}[u(t, 0, \cdot)]\Big\|_{B_{q,p,*}^{2-m_1-\frac{1}{p}}(G)} + \Big\|\frac{\partial}{\partial t}[u(t, 1, \cdot)]\Big\|_{B_{q,p,*}^{2-m_2-\frac{1}{p}}(G)}$$

$$+ \|u(t, \cdot, \cdot)\|_{L_p\left((0,1); W_q^2(G)\right)}$$

$$\leq C\Big[t^{-\frac{\max\{m_k\}}{2}}\big(\|u_0(\cdot, \cdot)\|_{L_p\left((0,1); W_q^2(G)\right)} + \|u_0(\cdot, \cdot)\|_{L_p\left((0,1); L_q(G)\right)}\big)$$

$$+ t^{-\frac{\max\{m_k\}}{2}-\mu}\big(\|f\|_{C_\mu^\gamma\left((0,t]; L_p\left((0,1); L_q(G)\right)\right)} + \sum_{k=1}^{2} \|f_k\|_{C_\mu^\gamma\left((0,t]; B_{q,p,*}^{2-m_k-\frac{1}{p}}(G)\right)}\big)\Big].$$

Proof Let us denote $E := L_q(G)$ and consider an operator A which is defined by the equalities

$$D(A) := W_q^2(G; Lu = 0), \quad Au := -\sum_{s,j=1}^{r} a_{sj}(y) D_s D_j u(y) + \lambda_0 u(y).$$

Then, problem (20.70)–(20.73) can be rewritten in the form (20.67)–(20.69), to which Theorem 20.8 is applied. All conditions of Theorem 20.8 have been, actually, checked in the proof of Theorem 20.6. □

The same examples of the operators T_k from the end of Sect. 20.6 can be employed also here.

Acknowledgements I would like to thank Professor Davide Guidetti who kindly allowed me to include into the paper his calculations (Sect. 20.2) which appear in his paper "Abstract elliptic problems depending on a parameter and parabolic problems with dynamic boundary conditions" in this volume.

Appendix

Theorem 20.10 ([16, Theorem 5.4.2/1]) Let an operator A be closed, densely defined and invertible in the complex Banach space E and $\|R(\lambda, A)\|_{B(E)} \leq L(1 + |\lambda|)^{-1}$, for $|\arg \lambda| \geq \pi - \varphi$, for some $\varphi \in (0, \pi)$. Moreover, let m be a positive integer, $p \in (1, \infty)$, and $\alpha \in (\frac{1}{2p}, m + \frac{1}{2p})$.

Then, there exists $C \in \mathbb{R}^+$ (depending only on L, φ, m, α, and p) such that, for every $u \in (E, E(A^m))_{\frac{\alpha}{m} - \frac{1}{2mp}, p}$ and $|\arg \lambda| \leq \varphi$,

$$\int_0^\infty \|(A + \lambda I)^\alpha e^{-x(A+\lambda I)^{\frac{1}{2}}} u\|_E^p dx \leq C \left(\|u\|_{(E, E(A^m))_{\frac{\alpha}{m} - \frac{1}{2mp}, p}}^p + |\lambda|^{p\alpha - \frac{1}{2}} \|u\|_E^p \right).$$

References

1. Aliev, B.A., Yakubov, Y.: Elliptic differential-operator problems with a spectral parameter in both the equation and boundary-operator conditions. Adv. Differ. Equ. **11**(10), 1081–1110 (2006); Erratum in **12**(9), 1079 (2007)
2. Denk, R., Hieber, M., Prüss, J.: R-boundedness, Fourier multipliers and problems of elliptic and parabolic type. Mem. Am. Math. Soc. **166**, 788 (2003)
3. Favini, A., Guidetti, D., Yakubov, Y.: Abstract elliptic and parabolic systems with applications to problems in cylindrical domains. Adv. Differ. Equ. **16**(11–12), 1139–1196 (2011)
4. Favini, A., Shakhmurov, V., Yakubov, Y.: Regular boundary value problems for complete second order elliptic differential-operator equations in UMD Banach spaces. Semigroup Forum **79**, 22–54 (2009)
5. Favini A., Yakubov, Y.: Higher order ordinary differential-operator equations on the whole axis in UMD Banach spaces. Differ. Integral Equ. **21**(5–6), 497–512 (2008)
6. Favini, A., Yakubov, Y.: Regular boundary value problems for elliptic differential-operator equations of the fourth order in UMD Banach spaces. Sci. Math. Japonicae **70**, 183–204 (2009)
7. Favini, A., Yakubov, Y.: Irregular boundary value problems for second order elliptic differential-operator equations in UMD Banach spaces. Math. Ann. **348**, 601–632 (2010)
8. Favini, A., Yakubov, Y.: Regular boundary value problems for ordinary differential-operator equations of higher order in UMD Banach spaces. Discrete Contin. Dyn. Syst. **4**(3), 595–614 (2011)
9. Favini, A., Yakubov, Y.: Isomorphism for regular boundary value problems for elliptic differential-operator equations of the fourth order depending on a parameter. In press in Rivista di Matematica della Università di Parma (2014)
10. Grisvard, P.: Elliptic Problems in Nonsmooth Domains. Pitman, Boston (1985)

11. Krein, S.G.: Linear Equations in Banach Space. Birkhäuser, Boston (1982)
12. Kunstmann, P.C., Weis, L.: Maximal L_p-regularity for Parabolic Equations, Fourier Multiplier Theorems and H^∞-Functional Calculus. In: Iannelli, M., Nagel, R., Piazzera, S. (eds.) Functional Analytic Methods for Evolution Equations. Lecture Notes in Mathematics, vol. 1855, pp. 65–311. Springer, Heidelberg (2004)
13. Tanabe, H.: Equations of Evolution. Pitman, Boston (1979)
14. Triebel, H.: Interpolation Theory. Function Spaces. Differential Operators. North-Holland Mathematical Library, Amsterdam (1978)
15. Weis, L.: Operator-valued Fourier multiplier theorems and maximal L_p-regularity. Math. Ann. **319**, 735–758 (2001)
16. Yakubov, S., Yakubov, Y.: Differential-Operator Equations. Ordinary and Partial Differential Equations. Chapman and Hall/CRC, Boca Raton (2000)